AF477474

Proceedings of
The Rice Meeting

Proceedings of
The Rice Meeting

Volume 1

*1990 Meeting of the Division of Particles and Fields
of the American Physical Society*

Houston, Texas   3 – 6 January 1990

Editors
Billy Bonner and Hannu Miettinen
*Department of Physics
Rice University, Houston*

World Scientific
Singapore • New Jersey • London • Hong Kong

Published by

World Scientific Publishing Co. Pte. Ltd.

P O Box 128, Farrer Road, Singapore 9128

USA office: 687 Hartwell Street, Teaneck, NJ 07666

UK office: 73 Lynton Mead, Totteridge, London N20 8DH

THE RICE MEETING — 1990 MEETING OF THE DIVISION OF PARTICLES AND FIELDS OF THE AMERICAN PHYSICAL SOCIETY

ISBN 981-02-0258-X

Printed in Singapore by Loi Printing Pte. Ltd.

PREFACE

The 1990 General Meeting of the Division of Particles and Fields was held at Rice University in Houston, Texas on January 3–6, 1990. This volume is the written record of the physics that was presented at DPF90.

From the very beginning, we planned DPF90 as a major conference of truly international character which would review recent developments in all areas of particle physics. We chose a conference format similar to that used in the very successful Santa Fe meeting of 1984, the first in the "new series" of DPF meetings. The first and the last day of DPF90 consisted solely of plenary talks, which highlighted exciting recent developments in our field and presented an overview of particle physics. The two middle days consisted of eight parallel mini-conferences, each devoted to a rather broad subject and meeting for the entire two-day period. Each mini-conference was organized by two or three conveners who were responsible for its scientific content.

The overall scientific program of the conference was based on the advice we received from a distinguished 22-member International Program Committee. They recommended appropriate plenary session topics and speakers, as well as suitable subjects and conveners for the mini-conferences. We wish to thank all IPC members for their involvement and thoughtful advice. In our contacts with the IPC we emphasized the crucial role of the conveners: they would determine the detailed scientific program for the conference by their selection of topics and speakers for the mini-conferences. We were extremely fortunate in that we ended up working with a superb group of conveners who were not only broadly knowledgeable in their field but who also worked very hard when the time came to sweat the details. It was their efforts, more than anything else, that made DPF90 an outstanding scientific conference.

Without question, the main physics interest in the meeting was focused on the new Z results, particularly from the SLC and the four LEP experiments. In fact, a special evening session on "Z Highlights", and an extra mini-conference on "Physics at the Z" were added to the program at a late date as it became clear that LEP experiments would be ready to present their first data. We think it is also safe to assume that Z physics had something to do with the large turnout. The conference attracted some 550 participants, 480 of them registered, at a time of year when most people would rather be in Honolulu or Snowbird. The weather did little to change anybody's mind — most attendees will probably always think of Houston as a place where it rains all the time. Oh well, at least it wasn't freezing rain.

In conjuction with DPF90, we held a special symposium on "SSC and Society" intended to showcase the SSC and high energy physics to representatives from education, industry, and the newsmedia. The main attraction was a two-day program of presentations and workshops on "SSC, Partricle Physics and Cosmology" for some 260 high school teachers and students from Texas. Speakers include R. Huson, L. Jones, E.

Kolb, C. Quigg, and R. Schwitters. This event proved to be very successful, with participants and organizers alike enjoying the two days. Our heartfelt thanks to those who really put it all together: R. Clark from Texas A&M, M. Bardeen, W. Schearer and others from Friends of Fermilab, A. Erzberger from Palo Alto High School, the speakers mentioned above, and many others who helped out. The event was sponsored by the DPF and the SSC Directorate.

The full conference schedule left little time for a social program (just as well, said the DPF Executive Committee). The IMAX showing of NASA's "Dream is Alive" attracted over 200 viewers. The conference banquet, in one of Houston's finest hotels, featured superb food and wine, followed by good advice and plenty of jokes, the latter courtesy of Leon Lederman with help from J. D. Jackson. It is hard to top this pair for banquet entertainment!

We would like to thank all the sponsors who helped to underwrite some of the conference expenses. These include the Department of Energy, the National Science Foundation, Mitchell Energy and Development Company, and the SSC Laboratory. Transportation from and to the airports was paid for by Inland Steel of Chicago. Many of the details of the conference organization, including travel, hotel reservations, tours, the banquet and other arrangements were handled most ably by Dolores Amato and her staff at Amato Bourne Travel in Houston.

We also want to acknowledge the invaluable help of William Noblitt and his staff at Rice University Relations who prepared DPF90 posters, letterheads and other information materials for us. We are grateful to Marj Corcoran, Paul Stevenson, and other members of the organizing committee, as well as our graduate students at the T.W. Bonner Laboratory for their crucial help during the hectic weeks preceding the conference and during the conference itself. Lastly, our heartfelt thanks go to our secretarial staff: Rose Berridge, Mary Comerford, Gwen Gann, Arlene Estle, Sandy Mitchell, and especially Helen Viereck who did much towards the success of DPF90.

Billy E. Bonner
Hannu E. Miettinen

Houston, August 1990

TABLE OF CONTENTS

Volume 1

ELECTROWEAK PHYSICS

Volume 2

QCD AND HADRON PHYSICS

NON-ACCELERATOR PHYSICS

HEAVY IONS / LATTICE

ACCELERATORS AND INSTRUMENTATION

SSC PHYSICS AND DETECTORS

PLENARY SESSIONS

INFLATION AND QUANTUM COSMOLOGY

Andrei Linde

CERN, CH-1211 Geneva 23, Switzerland
and
Lebedev Physical Institute, Moscow 117 924, Leninsky prospect 23, USSR

ABSTRACT

We discuss an interplay between elementary particle physics, quantum cosmology and inflation. Both the standard version of the chaotic inflation scenario and the extended chaotic inflation scenario are considered. It is shown that domains of the inflationary universe with sufficiently large energy density permanently produce new inflationary domains. Therefore the evolution of the universe in the inflationary scenario has no end and may have no beginning. After inflation the universe becomes divided into different exponentially large domains inside which properties of elementary particles and dimension of space-time may be different. It is argued that if the present vacuum energy density ρ_v exceeds some extremely small critical value ρ_c ($\rho_c \sim 10^{-10^7} g.cm^{-3}$ for chaotic inflation in the theory $\frac{m^2}{2}\phi^2$), then the lifetime of mankind in the inflationary universe should be finite, even though the universe as a whole will exist without end. A possible way to justify the anthropic principle in the context of the baby universe theory and to apply it to the evaluation of masses of elementary particles, of their coupling constants and of the vacuum energy density is also discussed.

1 Introduction

With the invention of unified theories of strong, weak, electromagnetic and gravitational interactions, elementary particle physics has entered a very interesting and unusual stage of its development. It appears that in the context of these theories we can predict much more than we can actually verify by the standard experimental methods of high energy physics.

The energy scale at which the unified nature of all four fundamental interactions is expected to become manifest is not very different from the Planck mass M_p, about 10^{-5} grams, where quantum gravity effects become important. (The Planck mass is that mass for which the Compton wavelength l_P, about 10^{-33} cm, equals the Schwarzschild radius.) Its rest energy $M_p c^2$, about 10^{19} GeV, corresponds to the kinetic energy of a small airplane. By contrast, the 80-km-circumference Superconducting Super Collider the Americans hope to build in the near future will accelerate particles up to 10^4 GeV. The largest accelerator ring one could build on Earth, with a circumference of 40.000 km, could not accelerate particles beyond about 10^8 GeV, a center-of-mass energy occasionally to be seen in cosmic-ray collisions. But this still leaves us 12 orders of magnitude short of the energy necessary for a direct test of the unified theories. Of course, there are some indirect tests, such as the searches for proton decay and for supersymmetric partners of ordinary particles. But trying to get a correct theory of all fundamental interactions with only such low-energy experiments is like looking for the correct unified electroweak theory by using nothing but radiotelescopes. (Note that $M_p/E_W \sim E_W/E_\gamma$, where $E_W \sim 10^2$ GeV is the unification energy scale in the theory of weak and electromagnetic interactions, $E_\gamma \sim 10^{-5}$eV is the typical energy of photons in a radiowave).

The only accelerator that could ever produce particles energetic enough for a direct testing of the unified theories of all fundamental interactions is our universe itself. The Big Bang scenario as it stood ten years ago, which I will call the hot-universe theory, asserts that the universe was born at some moment $t = 0$ about 15 billion years ago, in a state of infinitely large density ρ. Of course, one cannot really speak of classical space-time for the earliest moments, when kT/c^2 was greater than the Planck mass, and ρ exceeded M_p/l_P^3, making quantum fluctuations of the metric predominant. (We will use a convenient unit system that sets k, c and $\hbar$ all equal to 1, so that $l_P = 1/M_p$ and the Planck density is M_p^4, roughly $10^{94} g/cm^3$.) It is just at such times, when the average particle energy exceeded M_p, that the unity of all four fundamental interactions would have been manifest.

With the rapid expansion of the universe, the average energy of particles, given by the temperature, decreases rapidly, and the universe becomes cold. The temperature falls as the reciprocal of $a(t)$, the scale factor, or "radius", of the universe. This means that particle interactions at extremely large energies can have occurred only at the very early stages of the evolution of the universe. One might think it very difficult to extract useful and reliable information from the unique experiment carried out about 10^{10} years ago. Thus, it came as a great surprise to those who study elementary particles that the investigation of physical processes at the very early stages of the universe can rule out most of the existing unified theories.

For example, all the grand unified theories predict the existence of superheavy stable particles carrying magnetic charge: magnetic monopoles. These objects have a typical mass 10^{16} times that of the proton. According to the standard hot-universe theory, monopoles should appear at the very early stages of the universe, and they should now be as abundant as protons. In that case the mean density of matter in the universe would be about 15 orders of magnitude higher than its present value of about $10^{-29}g/cm^3$ [1].

It was shown that all theories with spontaneous breaking of a discrete symmetry (including the simplest $SU(5)$ model, the models of spontaneous CP violation, most of the theories with axions, etc.) lead to the existence of superheavy domain walls, which would be in a drastic contradiction with observational cosmology [2]. It was shown that in most of the theories based on $N = 1$ supergravity the primordial abundance of gravitinos contradicts cosmological data by about 10 orders of magnitude [3], whereas the energy density stored in the so-called Polonyi fields contradicts cosmological data by 15 orders of magnitude for the simplest models [4], and by 6 orders of magnitude for the no-scale models [5]. Several models based on superstring theory lead to a cosmologically unacceptable type of axions [6]. Most of the Kaluza-Klein theories based on $N = 1 d = 11$ supergravity predict the present vacuum energy to be $O(-M_P^4)$, which would contradict the observational data by 125 orders of magnitude. The situation with superstring theories seemed, at first glance, to be somewhat better. To be fair, however, one must say that no sufficiently good cosmological model based on superstrings has been suggested so far, though some interesting ideas have been already proposed, see e.g. [7].

Despite many efforts, some of the problems listed above still remain unsolved, which probably means that the corresponding theories are actually unrealistic. Fortunately, however, the problems of primordial monopoles, gravitinos, domain walls and some other important cosmological problems such as the flatness problem, the homogeneity and isotropy problem, etc. can be solved simultaneously in the context of a relatively simple scenario of the universe evolution - the inflationary universe scenario [8]-[12]. This scenario makes it possible also to remove some constraints on the parameters of the elementary particle theory, for example, the constraint on the axion mass $m_A \geq 10^{-5}eV$ ($f_A \leq 10^{12}GeV$) [13]. The invention of this scenario has modified considerably the standard cosmological paradigm. One of the most important modifications was performed within the last three years. Therefore, we will start with the discussion of the "standard" inflationary universe scenario. Then we will discuss some recent developments in the inflationary cosmology [14,15] including an extended version of chaotic inflation [16]. We will finish with the discussion of exciting possibilities related to baby-universe theory, anthropic principle and the cosmological constant problem [17]-[29].

2 The "Standard" Inflationary Scenario

2.1 The stage of inflation

Historically, there were several different versions of the inflationary universe scenario. At present, it seems that the simplest and, simultaneously, the most general one is the chaotic inflation scenario [12]. To describe it, let us consider the simplest model based on the theory of a massive non-interacting scalar field ϕ with the Lagrangian

$$L = \frac{M_P^2}{16\pi}R + \frac{1}{2}\partial_\mu\phi\partial^\mu\phi - \frac{m^2}{2}\phi^2 \tag{1}$$

Here $M_P^{-2} = G$ is the gravitational constant, $M_p \sim 10^{19}GeV$ is the Planck mass, R is the curvature scalar, m is the mass of the scalar field $\phi, m \ll M_p$. If the classical field ϕ is sufficiently homogeneous in some domain of the universe (see below), then its behaviour inside this domain is governed by the equations

$$\ddot{\phi} + 3H\dot{\phi} = -dV/d\phi, \tag{2}$$

$$H^2 + \frac{k}{a^2} = \frac{8\pi}{3M_P^2}(\frac{1}{2}\dot{\phi}^2 + V(\phi)). \tag{3}$$

Here $V(\phi)$ is the effective potential of the field ϕ (in our case $V(\phi) = \frac{1}{2}m^2\phi^2$), $H = \dot{a}/a, a(t)$ is the scale factor of the locally Friedmannian universe (inside the domain under consideration), $k = +1, -1$, or 0 for a closed, open or flat universe, respectively. If the field ϕ initially is sufficiently large ($\phi \geq M_p$), then the functions $\phi(t)$ and $a(t)$ rapidly approach the asymptotic regime

$$\phi(t) = \phi_0 - \frac{mM_p}{2(3\pi)^{1/2}}t, \tag{4}$$

$$a(t) = a_o \cdot \exp\left(\frac{2\pi}{M_P^2}(\phi_0^2 - \phi^2(t))\right). \tag{5}$$

According to (2.4) and (2.5), during a time $\tau \sim \phi/mM_p$ the value of the field ϕ remains almost unchanged and the universe expands quasi-exponentially:

$$a(t + \Delta t) \sim a(t)exp(H\Delta t) \tag{6}$$

for $\Delta t \leq \tau = \phi/mM_p$. Here

$$H = \frac{2\pi^{1/2}}{\sqrt{3}}\frac{m\phi}{M_p} \tag{7}$$

Note that $H \gg \tau^{-1}$ for $\phi \gg M_p$.

The regime of quasi-exponential expansion (inflation) occurs for $\phi \geq \frac{1}{5}M_p$. For $\phi \leq \frac{1}{5}M_p$ the field ϕ oscillates rapidly, and if this field interacts with other matter fields (which are not written explicitly in eq. (1)), its potential energy $V(\phi) \sim \frac{m^2\phi^2}{2} \sim m^2 M_P^2$ is transformed into heat. The reheating temperature T_R may be of the order $(mM_p)^{1/2}$ or somewhat smaller, depending on the strength of the interaction of the field ϕ with other fields. It is important that T_R does not depend on the initial value ϕ_0 of the field ϕ. The only parameter which depends on ϕ_0 is the scale factor $a(t)$, which grows $\exp((2\pi/M_P^2)\phi_0^2)$ times during inflation.

If, as is usually assumed, a classical description of the universe becomes possible only when the energy-momentum tensor of matter becomes smaller than M_P^4, then at this moment $\partial_\mu\phi\partial^\mu\phi \lesssim M_P^4$ and $V(\phi) \lesssim M_P^4$.

Therefore, the only constraint on the initial amplitude of the field ϕ is given by $\frac{1}{2}m^2/\phi^2 \leq M_P^4$. This gives a typical initial value of the field ϕ:

$$\phi_0 \sim \frac{M_P^2}{m} \tag{8}$$

Let us consider for definiteness a closed universe of a typical initial size $O(M_p^{-1})$. It can be shown that if initially $\partial_\mu \phi \partial^\mu \phi$ becomes much smaller than $V(\phi)$, the evolution of the universe becomes describable by (2)-(7), and after inflation the total size of the universe becomes larger than

$$l \sim M_p^{-1} exp(\frac{2\pi}{M_P^2}\phi_0^2) \sim M_p^{-1} exp(\frac{2\pi M_P^2}{m^2}) \tag{9}$$

For $m \sim 10^{-6} M_p$ (which is necessary to produce density perturbations $\frac{\delta\rho}{\rho} \sim 10^{-4}$, see below)

$$l \sim M_p^{-1} \exp(2\pi 10^8) \geq 10^{10^8} cm, \tag{10}$$

which is much greater than the size of the observable part of the universe $\sim 10^{28} cm$.

After such a large inflation the term k/a^2 in (3) becomes negligibly small compared with H^2, which means that the universe becomes flat and its geometry locally Euclidean. This implies that the total density of the universe ρ becomes almost exactly equal to the critical density $\rho_c = \frac{3M_P^2}{8\pi}H^2$, i.e.

$$\Omega = \frac{\rho}{\rho_c} \sim 1. \tag{11}$$

For similar reasons the universe becomes locally homogeneous and isotropic. The density of all 'undesirable' objects (monopoles, domain walls, gravitinos) created before or during inflation becomes exponentially small, and they never appear again if the reheating temperature, T_R, is not too large.

We should like to emphasize that for a realisation of this scenario it is sufficient that initially $\partial_\mu\phi\partial^\mu\phi \leq V(\phi) \sim M_P^4$ in a domain of a smallest possible size $l \sim M_P^{-1}$. Since $\partial_\mu\phi\partial^\mu\phi \leq M_P^4, V(\phi) \leq M_P^4$ in any classical spacetime, the above-mentioned initial conditions are quite natural. For a more detailed discussion of initial conditions which are necessary for inflation see [31].

2.2 Scalar field fluctuations, perturbations of density and galaxy formation

According to quantum field theory, empty space is not entirely empty. It is filled with quantum fluctuations of all types of physical fields. These fluctuations can be regarded as waves of physical fields with all possible wave-lengths, moving in all possible directions. If the values of these fields, averaged over some macroscopically large time, vanish, then the space filled with these fields seems to us empty and can be called the vacuum.

In the exponentially expanding universe the vacuum structure is much more complicated. The wave-lengths of all vacuum fluctuations of the scalar field ϕ grow exponentially in the expanding universe. When the wavelength of any particular fluctuation becomes greater than H^{-1}, this fluctuation stops propagating, and its amplitude freezes at some nonzero value $\delta\phi(x)$ because of the large friction term $3H\dot\phi$ in the equation of motion of the field ϕ. The amplitude of this fluctuation then remains almost unchanged for a very long time, whereas its wavelength grows exponentially. Therefore, the appearance of such a frozen fluctuation is equivalent to the appearance of a classical field $\delta\phi(x)$ that does not vanish after averaging over macroscopic intervals of space and time.

Because the vacuum contains fluctuations of all wavelengths, inflation leads to the creation of more and more new perturbations of the classical field with wavelengths greater than H^{-1}. The average amplitude of such perturbations generated during a time interval H^{-1} (in which the universe expands by a factor of e) is given by

$$|\delta\phi(x)| \approx \frac{H}{2\pi}. \tag{12}$$

Perturbations of the field lead to adiabatic perturbations of density $\delta\rho \sim V'(\phi)\delta\phi$, which after inflation grow, and acquire the amplitude [32,33,34]

$$\frac{\delta\rho}{\rho} = \frac{48}{5}\sqrt{\frac{2\pi}{3}}\frac{V_\phi^{3/2}}{M_P^3 V'(\phi)}, \tag{13}$$

where ϕ is the value of the classical field $\phi(t)$ (4), at which the fluctuation we consider has the wavelength $l \sim k^{-1} \sim H^{-1}(\phi)$ and becomes frozen in amplitude. In the theory of the massive scalar field with $V(\phi) = \frac{m^2}{2}\phi^2$

$$\frac{\delta\rho}{\rho} = \frac{24}{5}\sqrt{\frac{\pi}{3}}\frac{m}{M_p}\left(\frac{\phi}{M_p}\right)^2. \tag{14}$$

Taking into account of (4), (5) and also of the expansion of the universe by about 10^{30} times after the end of inflation, one can obtain the following result for the density perturbations with the wavelength $l(cm)$ at the moment when these perturbations begin growing and the process of the galaxy formation starts:

$$\frac{\delta\rho}{\rho} \sim 0.8\frac{m}{M_p}\ln l(cm). \tag{15}$$

At a galaxy scale ($l \sim 10^{21} - 10^{22} cm$)

$$\frac{\delta\rho}{\rho} \sim 40\frac{m}{M_p}, \tag{16}$$

which gives a desirable amplitude $\frac{\delta\rho}{\rho} \sim 10^{-4} - 10^{-5}$ for $m \sim 2.10^{-7} - 2.10^{-6}M_p$. In what follows we will assume that $m \sim 10^{-6}M_p$.

3 Self-Reproducing Universe

In our previous investigation we only considered local properties of the inflationary universe, which is quite sufficient for a description of the observable part of the universe of the present size $l \sim 10^{28}$ cm. Indeed, in accordance to (15) our universe remains relatively homogeneous on a scale

$$l \leq l^* \sim exp(2\pi \frac{M_p}{m})cm \sim 10^{3.10^6} cm. \tag{17}$$

The corresponding density perturbations were formed at the time, when the scalar field $\phi(t)$ was bigger than ϕ^*, where

$$\phi^* \sim M_p \sqrt{\frac{M_p}{m}} \sim 30M_p, \tag{18}$$

see (14). (Note that $V(\phi^*) \sim \frac{m^2}{2}(\phi^*)^2 \sim mM_P^3 << M_p^4$). On a scale $l > l^*$ the universe becomes extremely inhomogenous due to quantum fluctuations produced during inflation. We are coming to a paradoxical conclusion, that the global properties of the inflationary universe are determined not by classical but by quantum effects.

Let us try to understand the origin of such a behaviour of the inflationary universe.

A very unusual feature of the inflationary universe is that processes separated by distances l greater than H^{-1} proceed independently of one another. This is so because during exponential expansion any two objects separated by more than H^{-1} are moving away from each other with a velocity ν exceeding the speed of light. (This does not contradict special relativity because ν is not the speed of any signal; it is just the rate at which the general expansion of the universe separates two distant points.) As a result, any observer in the inflationary universe can see only those processes occurring nearer than H^{-1}.

An important consequence of this general result is that the process of inflation in any spatial domain of radius H^{-1} occurs independently of any events outside it. Any two inflationary domains displaced by more than H^{-1} cannot collide or eat one another, or do each other any damage. Their expansion is due not to the annexation of the territory of their neighbors, but rather to the peaceful (and very rapid) growth in their own volume, as allowed by general relativity. In this sense any inflationary domain of initial size exceeding $2H^{-1}$ can be considered as a separate mini-universe, expanding independently of what occurs outside it.

To investigate the behavior of such a mini-universe, with an account taken of quantum fluctuations, let us consider an inflationary domain of initial size roughly H^{-1} containing a sufficiently homogeneous field whose initial value ϕ greatly exceeds M_p. Eq.(4) tells us that during a typical time interval $\Delta t = H^{-1}$ the field inside this domain will be reduced by

$$\Delta\phi = \frac{M_p^2}{4\pi\phi}. \tag{19}$$

By comparison of (12) and (19) one can easily see that if ϕ is much less than $\phi^* \sim M_p\sqrt{\frac{M_p}{m}}$, the decrease of the field ϕ due to its classical motion is much larger than the amplitude of the quantum fluctuations $\delta\phi$ generated during the same time. But for large ϕ (up to the classical limit of $10^4 M_p$), $\delta\phi(x)$ will exceed $\Delta\phi$, i.e. the Brownian motion of the field ϕ becomes more rapid than its classical motion. Because the typical wavelength of the fluctuation field $\delta\phi(x)$ generated during this time is H^{-1}, the whole domain volume after Δt will effectively have become divided into e^3 separate domains (mini-universes) of diameter H^{-1}. In almost half of these domains the field ϕ grows by $|\delta\phi(x)| - \Delta\phi$, which is not very different from $|\delta\phi(x)|$ or $H/2\pi$, rather than decreases. During the next time interval $\Delta t = H^{-1}$ the field grows again in half of these mini-universes. It can be shown that the total physical volume occupied by a permanently growing field ϕ increases with time like $exp(3 - \ln 2)Ht$, and the total volume occupied by a field that does not decrease grows almost as fast as $\frac{1}{2}e^{3Ht}$.

Because the value of the Hubble constant $H(\phi)$ is proportional to ϕ, the main part of the physical volume of the universe is the result of the expansion of domains with nearly the maximal possible field value, M_p^2/m, for which $V(\phi)$ is close to M_p^4. There are also exponentially many domains with smaller values of ϕ. Those domains in which ϕ eventually becomes smaller than about $30 M_p$ give rise to the mini-universes of *our* type. In such domains, ϕ eventually rolls down to the minimum of $V(\phi)$, and these mini-universes are subsequently describable by the usual Big Bang theory. However a considerable part of the physical volume of the entire universe remains forever in the inflationary phase [13]-[15].

Thus in our scenario the universe, in which there was initially at least one domain of a size on the order of H^{-1} filled with a sufficiently large and homogeneous field ϕ, unceasingly reproduces itself and becomes immortal. One mini-universe produces many others, and this process goes on without end, even if some of the mini-universes eventually collapse.

But this means that it is vanishingly improbable that our mini-universe would have been the first in the sequence of all mini-universes. Moreover, it no longer seems necessary to assume that there actually was some first mini-universe appearing from nothing or from an initial singularity at some moment $t = 0$ before which there was no space-time at all.

From general topological theorems about singularities in cosmology it does not actually follow that our universe was created *as a whole* at some moment before which the universe did not exist. The usual assumption that the whole universe appears from the unique Big Bang singularity at $t = 0$ is based on the implicit assumption that the universe *as a whole* is sufficiently homogeneous. Indeed, the observable part of our universe is very homogeneous. Observed density fluctuations are less than a part in a thousand and there has been no reason to expect that the universe is inhomogeneous on a larger scale beyond the 10^{10}-light-year horizon. In a homogeneous universe one can use the density $\rho(t)$ as a measure of time. In that case it can be shown that the universe appears *as a whole* from a singularity at $t = 0$. The initial density $\rho(0)$ is infinite, and it becomes possible to describe the whole universe in terms of classical space-time after the Planck time M_p^{-1} (about 10^{-43} seconds), when the energy density everywhere simultaneously becomes smaller than the Planck density M_p^4.

With the invention of the inflationary scenario the situation changes drastically. At present only inflation can explain why the observable part of the universe is so homogeneous, but from inflation it also follows that on a much larger scale the universe is extremely *in*homogeneous. In some parts of the universe the energy density ρ is now of the order of M_p^4, 125 orders of magnitude higher than the 10^{-29} or $10^{-30} g/cm^3$ we can see nearby. In such a scenario there is no reason to assume that the universe was initially homogeneous and that all its causally disconnected parts started their expansions simultaneously.

If the universe is infinitely large (like the Friedmann open or flat universe), then it cannot have had a single beginning; a simultaneous creation of infinitely many causally disconnected regions is totally improbable. Therefore the universe cannot be infinite *ab initio*, or it must exist eternally as a huge self-reproducing entity. Some of its parts appear at different times from singularities, or may die in a singular state. New parts are constantly being created from the space-time foam when $V(\phi)$ exceeds the Planck density, or they may revert to the foamlike state again as a result of large fluctuations in ϕ. But the evolution of the universe as a whole has no end, and it may have had no beginning.

4 Extended Chaotic Inflation

The first versions of the inflationary universe scenario were based on the models where the process of inflation was determined by the evolution of just one scalar field ϕ. Even this simplest class of models leads to many unusual and unexpected properties of the universe, such as a non-trivial thermal history of the universe, the existence of a mechanism of formation of a scale-free (flat) spectrum of adiabatic density perturbations, the absence of the global end of the universe evolution etc. (For a review see e.g. [34].)

However, in most of the realistic models of elementary particles there are many different scalar fields. This may lead to existence of two consequent stages of inflation (so-called combined scenario [35], double or multiple inflation [36] - [37]), to generation of non-Gaussian perturbations with a non-flat spectrum [38] - [39], to appearance of exponentially big bubbles or domains with different energy density [37,39], or with almost the same energy density but with a different density of baryons, or with density perturbations or different type inside different domains [34,39,41]. Domains formed at early stages of inflation may contain matter described by different laws of the low-energy physics [34]. In some Kaluza-Klein models even the effective dimensionality of space in different exponentially big domains of the universe may be different [42].

An investigation of such more complicated models is justified if by the study of such models one can learn something qualitatively new about the inflationary cosmology. Many interesting possibilities which may arise in theories with many scalar fields still remain unexplored. One of the most interesting models of that type suggested this year is the extended inflation scenario [43], which is essentially a realization of the old inflationary universe scenario by Guth [44] in the context of the Brans-Dicke theory [45]. In what follows the model of ref. [43] will be called an extended old inflation scenario. In this paper we are going to suggest another scenario, which will be called an extended chaotic inflation scenario, and to discuss some rather unexpected features of the new class of inflationary models.

Let us consider a theory with the action

$$A = \int d^4x \sqrt{-g}\,[\frac{\phi^2 R}{8b} - \frac{1}{2}\partial_\mu \phi \partial^\mu \phi - V(\phi) + L(\sigma)]. \tag{20}$$

Here $\phi^2 = \frac{b}{2\pi}\Phi$, where Φ is the Brans-Dicke field; σ is the inflaton scalar field. The case $V(\phi) = 0$ corresponds to the original version of the BD theory. In the first papers on extended old inflation it was assumed that $V(\phi) = 0$ and that the effective potential of the field σ has a deep minimum at $\sigma = 0$, as in the old inflationary universe scenario [43]. Then it was understood that a successful realization of this scenario is possible only for not very big values of b, which is incompatible with the experimental constraints on this parameter in the BD theory, $b > 500$ [46,47], and to cure this problem it was suggested to add to the theory a nonvanishing effective potential of the field ϕ [46] - [48].

In what follows we will consider this theory with the potentials $V(\sigma)$ of the type used in the chaotic inflationary universe scenario. In such a case the value of the parameter b can be arbitrarily big and there is no need to assume the existence of a nonvanishing potential $V(\phi)$ of the BD field ϕ. We will return to a discussion of theories with a nonvanishing potential $V(\phi)$ in the end of the paper.

From a comparison with the standard Einstein theory it follows that the effective gravitational constant $G(\phi)$ and the effective Planck mass $M_p(\phi)$ in the theory (20) depend on ϕ and are given by

$$G^{-1} = M_p^2 = \frac{2\pi\phi^2}{b}. \tag{21}$$

We will try to find inflationary solutions in the theories with $b \gg 1$. During inflation one can neglect $\frac{\dot{\phi}}{\phi}$ as compared with $H = \frac{\dot{a}}{a}$ (where $a(t)$ is the scale factor of the universe), $\ddot{\phi}$ as compared with $3H\dot{\phi}$ and $\dot{\phi}^2, \dot{\sigma}^2$ as compared with the effective potential $V(\sigma)$. In such a case equations for a, ϕ and σ in the theory (20) with $V(\phi) = 0$ take a very simple form:

$$H = \frac{\dot{a}}{a} = \frac{2}{\phi}\left(\frac{b}{3}V(\sigma)\right)^{1/2}, \tag{22}$$

$$3H\dot{\phi} = \frac{4}{\phi}V(\sigma), \tag{23}$$

$$3H\dot{\sigma} = -V'(\sigma). \tag{24}$$

From eq. (22), (23) it follows that

$$\frac{\dot{a}}{a} = b\frac{\dot{\phi}}{\phi}. \tag{25}$$

This equation is very interesting. It says that independently of a choice of any particular theory of the type of (20)

$$\frac{a(t)}{a_o} = \left(\frac{\phi(t)}{\phi_o}\right)^b \tag{26}$$

during inflation. (The field ϕ practically does not change after inflation.) This result will be very important for us in what follows. From eq. (26) it follows in particular that for large b the value of the effective gravitational constant changes in time slowly as compared with $a(t)$. This implies that all theories $L(\sigma)$ which lead to inflation in the standard Einstein theory, lead to inflation in the theory (20) as well. To understand some novel features of inflation in the theory (20) let us consider e.g. the theory of the scalar field σ with the effective potential

$$V(\sigma) = \lambda\frac{\sigma^{2n}}{2n}. \tag{27}$$

It is known that in such theories the chaotic inflation scenario can be realized in the context of the Einstein theory [12]. This remains true in the theory (20), (27). Equations (23), (24) for this theory can be written in the following form:

$$\dot{\sigma} = -\phi\sigma^{n-1}\left(\frac{n\lambda}{6b}\right)^{1/2} , \qquad (28)$$

$$\dot{\phi} = \sigma^n\left(\frac{2\lambda}{3nb}\right)^{1/2} . \qquad (29)$$

From these equations it follows that

$$\frac{d}{dt}\left(\frac{2}{n}\sigma^2 + \phi^2\right) = 0 . \qquad (30)$$

A solution of this equation can be most conveniently represented after the change of variables, $\varphi = \sigma\sqrt{\frac{2}{n}}$:

$$\varphi^2 + \phi^2 = c^2 , \qquad (31)$$

where c is an arbitrary constant, or, in a parametric form,

$$\phi = c \cdot \sin x(t) , \qquad (32)$$

$$\varphi = c \cdot \cos x(t) , \qquad (33)$$

where $x(t)$ obeys equation

$$\dot{x}(t) = \left(\frac{\lambda}{3b}\right)^{1/2}\left(\frac{nc^2}{2}\right)^{\frac{n-1}{2}} cos^{n-1}x(t) . \qquad (34)$$

Let us consider e.g. the theory of a free massive scalar field σ with the effective potential $V(\sigma) = \frac{m^2\sigma^2}{2}$ (*i.e.* $n = 1, \lambda = m^2$ in eq. (27)). In this case $\varphi = \sigma\sqrt{2}$ and eq. (34) reads

$$\dot{x}(t) = \frac{m}{\sqrt{3b}} , \qquad (35)$$

which yields

$$\phi = c \cdot \sin(x_o + \frac{m}{\sqrt{3b}}t) ,$$

$$\sigma = \frac{\varphi}{\sqrt{2}} = \frac{c}{\sqrt{2}} \cdot \cos(x_o + \frac{m}{\sqrt{3b}}t) ,$$

$$a(t) = a_o \cdot \left(\frac{\sin(x_o + \frac{m}{\sqrt{3b}}t)}{\sin x_o}\right)^b . \qquad (36)$$

For $\frac{m}{\sqrt{3b}}t > x_o$ the last equation describes the power-law inflation, $a(t) \sim t^b$, $b \gg 1$.

To get a complete picture one should know when inflation starts and when it finishes. One can easily check by the use of eqs. (27)-(29) for $V(\sigma) = \lambda\frac{\sigma^{2n}}{2n}$ that $V(\sigma) \gg \dot{\phi}^2, \dot{\sigma}^2$ and inflationary regime (34) is realized only if the field $\varphi = \sigma\sqrt{\frac{2}{n}}$ is bigger than its final value at the inflationary stage φ_f, where

$$\varphi_f \sim \phi_f\sqrt{\frac{n}{3b}} = M_p(\phi_f)\sqrt{\frac{n}{6\pi}} \ . \tag{37}$$

. On the other hand, one can really speak about inflation of a classical space-time only if the energy density is smaller than the Planck density, which implies in particular that $V(\sigma) < M_p^4(\phi)$. Moreover, one can easily show that the gradient energy of inhomogeneities $\delta\phi$ and $\delta\sigma$ produced during inflation is smaller than the potential energy density $V(\sigma)$ (which is necessary for inflation to occur) only if the same condition $V(\sigma) < M_p^4(\phi)$ is satisfied [15,47].

As it is argued in [34], the quantum boundary corresponds to the most natural initial conditions for chaotic inflation. Moreover, as it is obvious from eq. (26), the closer the initial values of the fields to the quantum boundary, the bigger is the volume of the corresponding domains after inflation. Therefore we will assume that typical initial values ϕ_i and φ_i (or σ_i) of the fields ϕ and φ (or σ) are close to the quantum boundary,

$$V(\sigma_i) \sim M_p^4(\phi_i) \sim \frac{4\pi^2\phi_i^4}{b^2} \ . \tag{38}$$

Let us first try to understand possible implications of this condition in a different kind of theories where $V(\sigma)$ has a minimum near $\sigma = 0$, as required e.g. in the old inflationary universe scenario [44] and in the first versions of the new inflationary universe scenario. In such theories eq. (38) yields

$$\phi_i \sim \left(\frac{b^2 V(0)}{4\pi^2}\right)^{1/4} \ . \tag{39}$$

In the extended old scenario [43] the probability of the bubble formation is exponentially suppressed, whereas the field ϕ linearly grows with time. Since the most natural initial value of this field, as we have argued, is fixed by the condition (39), the typical value ϕ_f of the field ϕ at the end of inflation (when tunneling occurs) should be exponentially big and can be expressed through $V(0)$ and those parameters of the theory which give contribution to the Euclidean action corresponding to the probability of the bubble formation. One then could try to use it as a possible explaination of the big value of the Planck mass $M_p \sim \phi_f\sqrt{\frac{2\pi}{b}}$, i.e. of the weakness of the gravitational interactions.

Unfortunately, as we already mentioned, the old extended scenario [43] should be extended even further by introducing an effective potential $V(\phi)$, which is absent in the original BD theory [47,48]. In such a case the final value of the Planck mass is not of a dynamic origin but is just fixed by hand. However, it does seem possible to get an exponential suppression of the effective gravitational constant in more complicated models, e.g. in the theories describing two subsequent stages of inflation: old and chaotic one.

Now let us return to the extended chaotic inflation in the theory (27). In this case the effective Planck mass in different parts of the universe may take all possible values, corresponding to different values of ϕ_f at the boundary (37). Due to inflation the present distribution of the scalar field $\phi(x) \sim \phi_f(x)$ in the observable part of the universe is practically homogeneous, even if the original distribution of the fields $\phi_i(x)$ and $\sigma_i(x)$ was not exactly homogeneous. That is why to a good accuracy we can speak about the gravitational *constant*, even though at extremely large distances the values of this *constant* differ from each other.

However, inhomogeneities of the field ϕ, just as of any other scalar field, are produced due to quantum fluctuations during inflation [49]. This leads to large-scale inhomogeneities of the effective gravitational constant $G(\phi)$ which may serve as an additional source of perturbations necessary for galaxy formation.

On an extremely large length scale the variations of the gravitational constant produced by quantum fluctuations may be very big even if originally the field ϕ was almost homogeneous in the whole universe (which is possible e.g. if the universe originally was closed and small). Indeed, as it was shown in [14,12] in the context of the chaotic inflation scenario, generation of large long-wave perturbations of scalar fields may lead to the process of permanent self-regeneration of inflationary domains. This process makes the distribution of the scalar field inhomogeneous on extremely large length scales (much bigger than the size of the observable part of the universe), so that at each given moment of time the classical scalar fields take *all* their possible values in some parts of the universe.

The same result remains true for the extended chaotic inflation as well. The necessary (and sufficient) condition for the appearance of the self-regeneration regime [14] is that the change of the scalar fields ϕ and σ during the typical time H^{-1} in accordance to the classical equations of motion (28), (29) is smaller that the amplitude of the long-wave inhomogeneities of these fields $|\delta\phi| \sim |\delta\sigma| \sim \frac{H}{2\pi}$ produced at the same time due to freezing of quantum fluctuations with the wavelength bigger than H^{-1} during inflation [49], see Section 3. One can easily check that this condition is satisfied if

$$\frac{3\pi^2 \phi^4}{b^3}\left(1 + \frac{\phi^2}{\varphi^2}\right) < V(\sigma) < \frac{4\pi^2}{b^2}\phi^4 \,, \tag{40}$$

the last inequality corresponding to the quantum (Planck) boundary (38).

From eq. (40) it follows that the inflationary universe with most natural initial conditions (i.e. not far from the quantum boundary (38)) enters regime of self-regeneration. During this regime the universe becomes filled with all possible values of the fields ϕ and σ, independently of their initial values in the region (40). Consequently, the universe after inflation will contain many exponentially large domains in which the Planck mass $M_p(\phi)$ and the gravitational constant $G(\phi)$ take all possible values from 0 to ∞.

But if different parts of the universe are filled by different values of the field ϕ, is there any way to explain why in our part of the universe this field takes a particular value corresponding to the extremely big Planck mass $M_p \sim 10^{19} GeV$?

A simplest idea is that the parts of the universe filled by different fields inflate differently, so we should live in a part of the universe, filled now by the field $\phi \sim \phi_f$, which experienced bigger inflation and consequently occupies bigger volume. According to eq. (26), the total amount of inflation is given by $\left(\frac{\phi_f}{\phi_i}\right)^b$. This result has a simple graphic interpretation: The bigger is the angle between the ray from the origin O and the final pointon (φ_f, ϕ_f) at the plane (φ, ϕ) and the ray showing from the origin O to any particular initial point (φ_i, ϕ_i) at the quantum boundary, the bigger inflation is. The domains of the universe, containing fields (φ_i, ϕ_i) corresponding to the biggest angle, after inflation occupy the biggest fraction of the volume of the universe. Then with the help of eq. (21) and the fact that $\phi_f^2 = \phi_i^2 + \varphi_i^2$ at $b \gg 1$ one can easily calculate the present value of M_p :

$$M_p \sim \sqrt{\frac{2\pi}{b}(\phi_i^2 + \varphi_i^2)} \ . \tag{41}$$

From this point of view the theory with $V(\sigma) = \frac{m^2\sigma^2}{2}$, may seem unfavourable since the maximal inflation corresponds to infinitely large σ_i and, conseqluently, to infinitely large value of M_p. For the theory $\frac{\lambda\sigma^4}{4}$ the answer also does not look particularly good since the rate of inflation does not depend on a particular choice of (φ_i, ϕ_i). The most interesting case is the theory $\frac{m^2\sigma^2}{2} + \frac{\lambda\sigma^4}{4}ln\frac{\sigma}{m}$. Since $ln\frac{\sigma}{m}$ is a slowly changing function, the solutions of the corresponding equations to a good accuracy also can be represented as circles (32), (33), and eq. (41) remains approximately true, with $\varphi \sim \sigma$. In this case it can be shown that for a sufficiently small λ the biggest inflation occurs for $\sigma_i \sim 2m/\sqrt{\lambda}$, which leads to the Planck mass $M_p \sim 5 \cdot 10^7 m/\sqrt{b}$ for $\lambda \sim 10^{-14}$. This means that the small value of the coupling constant $\lambda \sim 10^{-14}$, which is necessary in order to get density perturbations $\frac{\delta\rho}{\rho} \sim 10^{-5}$ on a galaxy scale, and the weakness of gravitational interactions (existence of the very large mass scale $M_p \sim 10^{19}GeV$) in this model are related to each other.

However, this may be not a final answer. Indeed, the volume of the universe after inflation depends on initial conditions only as some power of $\frac{\phi_f}{\phi_i}$. Meanwhile the probability to live and to ask questions about the value of the gravitational constant is most certainly exponentially suppressed for the values of G outside some narrow region. Indeed, the process of formation of stars, galaxies, of light elements and even the existence of baryonic matter in the universe crucially depend on the value of G. For example, with the gravitational constant G much smaller than its present value the speed of the universe expansion at a given temperature becomes more slow, the deviation from thermal equilibrium which is necessary for the baryosynthesis after inflation becomes too small, and the corresponding part of the universe becomes empty and devoid of observers [29]. In this sense, the use of anthropic considerations, which can be justified in the context of the chaotic inflation scenario [34], may save the theories with $V(\sigma) = \frac{m^2\sigma^2}{2}$ or $\frac{\lambda\sigma^4}{4}$ and may be sufficient for the determination of the order of magnitude of the gravitational constant in our part of the universe.

To get a final solution of the problem one needs a more detailed investigation of the stage of self-regeneration of the universe. Indeed, eq. (26) is valid even with an account taken of the process of self-regeneration, but only for a typical degree of inflation near a certain point [15]. If one is interested in a typical history and a typical degree of inflation of a domain of a hypersurface of a given density $\rho \sim 10^{-29}g \cdot cm^{-3}$ (and this is what we are interested in), then the answer may be different. In particular, as it was shown in [15] in the context of the chaotic inflation scenario, the main part of the volume of such a hypersurface is formed during not very probable but volume-producing quantum jumps of the scalar field near the quantum boundary. In the extended chaotic inflation scenario a similar investigation probably will show that in many theories the main part of the physical volume of the universe is formed due to quantum jumps of the fields ϕ and σ near some particular point at the quantum boundary. However, due to a long and complicated process of diffusion of the fields ϕ and σ in the region (40), the distribution of the probability to come from this point to some particular point ϕ_f may be very wide. This again would make anthropic considerations appropriate.

5 Life After Inflation

The results obtained in Section 3 imply that life in the inflationary universe will never disappear. Unfortunately, this conclusion does not automatically mean that one can be very optimistic about the future of mankind. Even though new domains of our type are permanently produced during inflation, later on each particular domain either will become practically empty and unsuitable for life or it will evolve into a huge black hole and collapse [29]. The only possible strategy of survival which we see at the moment is to travel from old domains to the new ones, which are displaced near the self-reproducing inflationary domains and always have large enough density of baryons. In the worst case, if we will be unable to travel to such distant places ourselves, we can try to send some information about us, our life and our knowledge, and maybe even stimulate development of such kinds of life there, which would be able to receive and use this information. In such a case one would have a comforting thought that even though life in our part of the universe will disappear, we will have some inheritors, and in this sense our existence is not entirely meaningless. (At least it would be not worse than what we have here now.)

In order to check whether such a possibility does actually exist, it would be necessary to perform a detailed investigation of the global structure of the universe, and especially of its causal structure, which becomes very nontrivial in the context of inflationary cosmology [15]. One should answer the question whether it is actually possible to travel or to send a signal from our part of the universe to a vicinity of a nearby self-reproducing inflationary domain, and to do it in such a way that after a long time which is necessary for the signal to arrive the density of matter near this domain will be sufficiently large for the existence of life there. At present we do not know any no-go theorems which would imply that it is generally impossible and that one cannot elaborate any successful strategy of survival of life in our part of the universe. We will argue now that the answer to this question may depend crucially on the value of the vacuum energy density, i.e. on the absence or presence of a tiny cosmological constant in the Einstein equations.

Namely, according to [14,15], the self-reproducing domains appear in the inflationary universe independently of the existence or nonexistence of a small energy density ρ_v of the vacuum state in which we live now (if it is not negative and as big as - $10^{-6}M^4{}_p$). However, the causal structure of the universe, the possibility of communication between its different domains and the possibility of the existence of life there depend crucially on the value of ρ_v. For example, if ρ_v is negative and its absolute value is much bigger than the present energy density of matter $\rho_o \sim 10^{-29}g.cm^{-3}$, each domain of the universe after inflation and a sufficiently long subsequent period of expansion behaves as an anti de Sitter space and collapses at a time $t \sim M_p \rho_v^{-1/2} \ll M_p \rho_o^{-1/2} \sim 10^{10}$ years, i.e. much earlier than the time necessary for the development of life of our type. Note, that this process occurs locally, at different times in different places of a self-reproducing universe. Therefore one may escape the local collapse by travelling towards the self-reproducing domains. However, since the typical distance between our place and a nearby self-reproducing domain is extremely large, such a journey is possible only if the lifetime of the anti de Sitter domain $t \sim M_p \rho_v^{-1/2}$ is also extremely large and, consequently, the absolute value of ρ_v is extremely small (see below).

If the vacuum energy density is positive and bigger than $10^{-27} g.cm^{-3}$, formation of galaxies of our type becomes impossible [50]. Smaller values of ρ_v do not affect galaxy formation, but do affect the causal structure of the universe. For example, the univesre with $\rho_v > \rho_0$ at the time $t > 10^{10}$ years behaves as de Sitter space with the event horizon of the size $H^{-1} \sim 10^{28} cm$ (i.e. smaller than the particle horizon $\sim t$ now). In such a universe the distance between the galaxies grows as e^{Ht}, and after a time of the order $H^{-1} < 10^{10}$ years there would remain only one (our own) galaxy inside our event horizon. (The galaxy itself is not stretched by the universe expansion since it is a gravitationally bounded system with the energy density much bigger than ρ_o.) It is well known that in the exponentially expanding universe one can travel or send any information only to those regions which were inside the event horizon at the moment when the information was sent. This means that soon after the beginning of the exponential expansion of the universe with $\rho_v > \rho_0$ there will be no possibility to travel or send information to other galaxies, and it is certainly impossible to send a signal to any self-reproducing inflationary domain, since a typical distance from our place to any such domain is much bigger than H^{-1}.

This result can be considerably strengthened in all realistic models of inflation. To do it one should first make an estimate of a typical distance l^* from our place to a nearby inflationary self-reproducing domain. This can be easily done, since there should be many such domains in each region of initial size $H^{-1}(\phi)$ containing field $\phi > \phi^*$, and there should be no such domains inside the regions of initial size H^{-1} containing field $\phi < \phi^*$. For the theory $m^2 \phi^2 / 2$ this gives, according to eq.(5),

$$l^* \sim H^{-1}(\phi^*) \, exp(\frac{2\pi \phi^{*2}}{M^2_p}) \, (\frac{T_R}{T_o}). \tag{42}$$

Here $H^{-1}(\phi^*) = M_p(\frac{8\pi}{3V(\phi^*)})^{1/2}$, T_R is the reheating temperature of the universe after inflation, T_o is the present temperature of the blackbody radiation in the universe. The last factor appears due to the additional stretching of the universe when the temperature falls down from T_R to T_o. Typically this factor is of the order of 10^{30} and $H^{-1}(\phi^*)$ is of the order of 10^{-30} cm. Therefore from (18), (42) it follows that

$$l^* \sim exp(2\pi M_p/m) \, cm \sim exp((2\pi)10^6) \, cm. \tag{43}$$

for $m \sim 10^{-6} M_p$. The universe becomes exponentially expanding and acquires an event horizon $H^{-1}(\rho_v) = M_n(\frac{8\pi}{3\rho_v})^{-1/2}$ at the late stages of its evolution, when the energy density of matter, which decreases as t^{-2}, becomes smaller than the vacuum energy density ρ_v. It occurs at the time which is bigger than the present age of the universe $t_o \sim 10^{10}$ years by the factor $\left(\frac{\rho_o}{\rho_v}\right)^{1/2}$. At that time the distance l^* from us to the nearby inflationary self-reproducing domain grows by a factor $\left(\frac{\rho_o}{\rho_v}\right)^{1/3}$, since the scale factor of a cold matter dominated universe grows as $t^{2/3}$. The causal connection between us and this region will be possible only if this distance will remain smaller than $H^{-1}(\rho_v)$,

$$M_p \left(\frac{3}{8\pi\rho_v}\right)^{1/2} > M_p \left(\frac{3}{8\pi V(\phi^*)}\right)^{1/2} \exp\left(\frac{2\pi\phi^{*2}}{M_p^2}\right) \left(\frac{T_R}{T_0}\right) \left(\frac{\rho_0}{\rho_v}\right)^{1/3} , \qquad (44)$$

which implies that causal connection is possible only if ρ_v is smaller than some critical value ρ_c, where

$$\rho_c \sim \frac{V^3(\phi^*)}{\rho_0^2} \left(\frac{T_0}{T_R}\right)^6 \exp\left(-\frac{12\pi M_p}{m}\right) \sim 10^{-10^7} g \cdot cm^{-3} . \qquad (45)$$

Here $V(\phi^*) \sim mM^3_p \sim 10^{-6}M^4{}_p$, $\rho_o \sim 10^{-29}g.cm^{-3} \sim 10^{-125}M^4{}_p$, $T_R \sim 10^{10}GeV$. One can easily see that the constraint $\rho_v < \rho_c$ (45) on the value of the vacuum energy density ρ_v is very stringent and practically does not depend on the particular values of $V(\phi^*), \rho_o$ and T_R .

Returning to the case of a negative vacuum energy, it can be shown by methods similar to those used above that in this case as well a causal contact between our part of the universe and a nearby self-reproducing inflationary domain is impossible if $|\rho_v| > \rho_c$, since in this case the time which is necessary for the contact is bigger than the time before the local collapse of our part of the universe.

These results are rather unexpected. It was clear *a priori* that the existence of a nonvanishing vacuum energy may slightly change the conditions of our life , but it was difficult to imagine that the development of life in our part of the universe may depend crucially on the existence or nonexistence of a tiny vacuum energy density $|\rho_v| \geq 10^{-10^7} g.cm^{-3}$.

Of course, the events we are discussing here will occur in a very distant future (at $t \sim 10^{5.10^6}$ years), but it is very difficult not to be curious about our common fate as it can be understood at the present level of the development of science. Another possible application of these results would be to use them in order to obtain a strong anthropic bound $|\rho_v| < \rho_c \sim 10^{-10^7} g.cm^{-3}$, which would help us to solve the cosmological constant problem. At the first glance, such an idea certainly cannot work, since the existence of the vacuum energy $-10^{-29}g.cm^{-3} < \rho_v < 10^{-27}g.cm^{-3}$ does not prevent formation of galaxies and can hardly have any effect on the appearance of life in the universe [50]. Moreover, until very recently the general attitude of physicists to the Anthropic Principle was rather sceptical, to say the least. It was believed that the weak, strong and electromagnetic interactions are the same in all parts of our universe, that the fundamental constants of Nature are universal and it is meaningless to discuss the possibility to live in a universe of a different type. This attitude was somewhat changed first when it was understood that in accordance with the chaotic inflationary universe scenario the universe should contain an exponentially large (or maybe even infinitely large) number of exponentially large domains (mini-universes), in which all kinds of symmetry breaking and all kinds of compactification which are possible in a theory of a given type are actually realized [34]. In other words, the universe becomes divided into many exponentially large domains with different types of low-energy physics and maybe even with different dimensionality [42] inside each of them. This made it possible to get a justification of a kind of weak anthropic principle : If many exponentially large domains with the low-energy physics of our type do exist in the universe described by a given theory, then it is quite natural that we live in one of such domains rather than in a domain where life of our type is impossible [34].

To explain it in a more detailed way let us remember that in realistic theories of elementary particles there exist many different types of scalar fields ϕ_i. The potential energy $V(\phi_i)$ often has many different local minima, in which the universe may live for an extremely long time, much greater than the 10^{10} years of our observable domain. For example, in the supersymmetric $SU(5)$ theory, $V(\phi_i)$ has several different minima of almost equal depth. Because the laws governing the interactions of elementary particles at the low energies at which we do experiments depend on the values of the classical fields ϕ_i, each of these minima corresponds to a different low-energy physics. In one of them the $SU(5)$ symmetry between all types of interactions remains unbroken - that is, the scalar fields ϕ_i remain equal to zero. In other minima various symmetry breaking patterns are realized, and in only one of these minima is the broken symmetry of the weak, strong and electromagnetic interactions that which we in fact observe.

During inflation there are large-scale fluctuations of all the fields ϕ_i. As a result, the inflationary universe becomes divided into an exponentially large number of inflationary mini-universes, with the scalar fields taking all possible values. As inflation ends in some mini-universes, these scalar fields roll down to all possible minima of $V(\phi_i)$. The universe becomes divided into many different exponentially large domains, realizing all possible types of symmetry breaking between the fundamental interactions. In some of these mini-universes the low-energy physics is quite different from our own. We cannot now see them because the size of our own domains is much greater than the size of its 10^{10}-light-year observable portion. We could not live in these domains because our kind of life requires our kind of low-energy physics.

It is very important that in the inflationay universe there is lots of room for all possible types of symmetry breaking and for all possible types of life. There is, therefore, no longer any need to require that in the true theory the minimum of $V(\phi_i)$ corresponding to our type of symmetry breaking be the only one or the deepest one. This new cosmopolitan viewpoint may greatly simplify the task of building realistic models of the elementary particles.

The change of the values of the scalar fields ϕ_i – that is to say, the change of the vacuum state – is the simplest kind of "mutation" that may occur during inflation. Much more interesting possibilities appear if one considers chaotic inflation in the higher-dimensional Kaluza-Klein theories. In those domains in which the energy density of the field ϕ grows to the Planck density, quantum fluctuations of the metric at a length scale of M_p^{-1} become of order unity. In such domains an inflationary d-dimensional universe can squeeze locally into a tube of smaller dimensionality $d-n$ (or vice versa). If this tube is also inflationary (in $d-n$ dimensions) and the initial length of the tube is greater than M_p^{-1} (which is quite probable near the Planck density), then its further expansion proceeds independently of its prehistory and of the fate of its mother universe. In an eternally existing universe such processes should occur even if their probability is very small. In fact the probability of such processes is small only if $V(\phi)$ is far below M_p^4.

Thus the inflationary universe becomes divided into different mini-universes in which all possible types of compactification produce all sorts of dimensionalities [42]. By this argument we find ourselves inside a four-dimensional domain with our kind of low-energy physics not because other kinds of mini-universes are impossible or improbable, but simply because our kind of life cannot exist in other domains.

This may have important implication for the building of realistic Kaluza-Klein and superstring theories. For example, it is extremely complicated, if not impossible, to construct a theory in which only one type of compactification can occur, leading precisely to a four-dimensional inflationary universe with the low-energy particle physics of our experience. But from the point of view discussed here, there is no need to require that the results compactification and inflation have wrought in our realm be the only possible results, or the best. It is enough to find a theory in which such a compactification is possible. This problem is still difficult, but it is much easier than the one we have been trying to solve.

With the invention of the theory of many universes interacting with each other only globally [17], and especially with the development of the baby universe theory [18] -[20] the situation changed even more. As it will be argued below, in the context of the baby universe theory it is possible to justify a more strong version of the anthropic principle as well.

According to the baby universe theory [18] -[20], we may live in different quantum states of the universe $|\alpha_i >$, corresponding to different choices of the "fundamental constants", such as the vacuum energy ρ_v (cosmological constant $\Lambda = 8\pi\rho_v/M^2_p$), gravitational constant $G = M^{-2}_p$, fine structure constant α, etc. By measuring the coupling constants we actually determine in which quantum state $|\alpha_i >$ we live now, and after that we cannot "jump" to another quantum state with other constants. That is why we are used to believe that we live in a classical universe with well-determined and unchanging laws of physics.

It would be very desirable to understand why do we live in a universe in a quantum state $|\alpha_i >$ with an extremely small (or even vanishing) cosmological constant, with $M_p \sim 10^{19}$ GeV, $\alpha = 1/137$, etc. A possible idea is to compute the wave function of the universe and then to prove that it is entirely concentrated near some particular values of the coupling constants [21,25]. However, the validity of the euclidean approach to quantum cosmology used in most of the papers on this problem is far from being clear [34], and it is quite possible also that the wave function of the universe actually does not have any sharp peak at $\Lambda = 0$ [27,28]. Moreover, it is not quite clear whether it is the wave function of the whole universe that is to be computed. It is known that in quantum mechanics only those statements can make sense that can be formulated in a concrete operational way. For example, at the classical level one can speak of the age of the universe t. However, the essence of the Wheeler-DeWitt equation for the wave function of the universe is that this wave function *does not depend on time*, since the total Hamiltonian of the universe, including the Hamiltonian of the gravitational field, vanishes identically [51,52]. Therefore if one would wish to describe the evolution of the universe with the help of its wave function, one would be in trouble. The resolution of this paradox is rather instructive [52]. The notion of evolution is not applicable to the universe as a whole since there is no external observer with respect to the universe, and there is no external clock as well which would not belong to the universe. However, we do not actually ask why the universe *as a whole* is evolving in the way we see it. We are just trying to understand our own experimental data. Thus, a more precisely formulated question is why *we see* the universe evolving in time in a given way. In order to answer this question one should first divide the universe into two main pieces: i) an observer with his clock and other measuring devices and ii) the rest of the universe. Then it can be shown that the wave function of the rest of the universe does depend on the state of the clock of the observer, i.e. on his "time" [52]. This time dependence in some sense is "objective", which means that the results obtained by different (macroscopic) observers living in the same quantum state of the universe and using sufficiently good (macroscopic) measuring apparatus agree with each other [53].

This example teaches us an important lesson. By an investigation of the wave function of the universe *as a whole* one sometimes gets information which has no direct relevance to the observational data, e.g. that the universe does not evolve in time. (Moreover, it is rather difficult to give any probabilistic interpretation to the wave function of the universe as a whole, since typically it is not normalizable.) In order to describe the universe *as we see it* one should divide the universe into several macroscopic pieces and calculate a conditional probability to observe it in a given state under an obvious condition that the observer and his measuring apparatus do exist. This simple condition, however, is very restrictive. In order to understand its meaning one should remember the famous Einstein-Podolsky-Rosen experiment, where two fermions (or photons) are produced by a decay of a spin zero particle. By a measurement of a spin of one of the particles an observer instantaneously determines the spin of another particle even if this particle and the observer are not in a causal contact at the moment of the measurement. This apparent paradox is resolved if one takes into account that prior to the measurement one should describe both particles by the wave function of the pair of particles rather than by separate wave functions of each of them. The knowledge of this wave function (the fact that it corresponds to a state with zero angular momentum) and the knowledge of the spin of one of these particles makes it possible to determine the spin of another particle without making any causal contact with it, just for the reason that there is a strong correlations between spins of particles in a system with a vanishing total angular momentum. For example, the conditional probability to find the second particle in a state with spin s $= -1/2$ is equal to unity under the condition that the first particle has a spin s $= +1/2$.

Similarly, any scientific exploration of the properties of the universe starts from the division of the universe into two subsystems (the observer with his measuring devices and the rest of the universe). Due to a strong correlation between the properties of these two subsystems, an investigation of our own properties, the properties of our environment and our measuring devices makes it possible to say a lot about the properties of the rest of the universe such as the vacuum energy, the values of some coupling constants etc.

It is important also, that in the EPR experiment one can choose different apparatus, e.g. the one which measures a linear polarization of the first photon, or the one which measures its circular polarization. After the measurment the second photon will be either linearly or circularly polarized, depending on our choice of measuring device. Similarly, many kinds of life and many types of measuring devices may exist, but what we are studying is the correlation of *our* kind of life, environment and measuring devices with the properties of the rest of the universe. But this means that in the context of the baby universe theory one can actually make sense of the Strong Anthropic Principle and try to explain why the electromagnetic coupling constant, the gravitational coupling constant and some other parameters of the theory have those particular values which we know: With other values of the coupling constants the existence of life *of our type* and of the measuring devices we use would be impossible. (Note, however, that in this interpretation the Strong Anthropic Principle has nothing in common with designing the universe for the benefit of human beings or with changing the whole universe by our own will, just as the determination of the properties of the second particle in the EPR experiment and the dependence of these properties on our choice of measuring device has nothing to do with the acausal action at a distance.)

If this idea is correct, then there is no need to compute the most probable values of the cosmological constant and of the gravitational constant by the investigation of the wave function of the universe as a whole, and it is not surprising that some of the results of the corresponding investigation look as counterintuitive as the statement that the universe as a whole does not depend on time. Rather one should investigate the probability to measure some particular eigenvalues of the operators corresponding to the "constants" Λ, e, m_e, G in the quantum state of the universe in which we live now under the obvious condition that we can live and perform our measurements here. In some cases (e.g. in the case of the constants e, m_e, G) the corresponding investigation is rather simple and straightforward. For example, according to [54], an increase of m_e by more than 2.5 times would make impossible the existence of atoms. An increase of e by more than 30 % would lead to an instability of proton and nuclei. A change of the strong interaction constant by more than 10 % would lead to the absence of heavy nuclei and of hydrogen in the universe . An increase or decrease of the weak interaction constant by an order of magnitude would lead to the absence of complex elements or of hydrogen respectively. An increase of the gravitational constant G by an order of magnitude would make the lifetime of the Sun so small that no biological molecules would appear on the Earth. An even bigger increase of G would lead to an extremely efficient nuclearsynthesis and to the absence of hydrogen in the universe [54,55]. A smaller value of G would make expansion of the universe more slow. In such a universe the departure from thermal equilibrium during the process of baryosynthesis would be small, this process would be inefficient and the universe now would be practically empty.

This list can be easily continued, but we hope that the results discussed above clearly demonstrate that the correlation between the properties of observers and the properties of the quantum state of the rest of the universe is very strong indeed. Moreover, most of the results discussed above are not so much *anthropic*, they just show that a small deviation from the present values of coupling constants would make impossible the existence of all measuring devices we use. In the context of the baby universe theory this means that the distribution of the conditional probability to make measurements in a quantum state with given values of weak, strong, gravitational and some other "fundamental constants" proves to be sharply peaked near the actual values of these constants we have measured in the quantum state of the universe in which we live now. This peak in the probability distribution should be especially narrow in unified theories of all fundamental interactions where the values of different coupling constants are strongly correlated due to the underlying symmetry of the theory. This indicates that if one wishes to get an order-of-magnitude estimate of the coupling constants, in some cases one can do it without a computation of these constants in the context of the "big fix" paradigm [25], and if one still wishes to make a computation (which, of course, is a good idea), then one should compute the amplitudes corresponding to the actual experimental environment rather than the amplitudes corresponding to the metaphysical "universe in itself" investigated by an imaginary external observer. (The discussion of the time-independence of the wave function of the universe shows that in the context of quantum cosmology the subtle difference between the amplitudes of these two types can be crucially important.)

Now let us return to the discussion of the cosmological constant problem. The possibility to explain the small value of the vacuum energy density $|\rho_v| < 10^{-29} g.cm^{-3}$ with the help of the Anthropic Principle is much less clear, but is not hopeless. From the results of ref. [50] it follows that formation of galaxies and appearance of life of our type is possible only if $-10^{-29} g.cm^{-3} < \rho_v < 10^{-27} g.cm^{-3}$. In the context of the baby universe theory this means that the conditional probability to measure the vacuum energy density ρ_v is concentrated in the interval between $-10^{-29} g.cm^{-3}$ and $10^{-27} g.cm^{-3}$ under the condition that the measurement is made by an observer of our type. This result could be strengthened up to the desirable constraint $|\rho_v| < 10^{-29} g.cm^{-3}$ if it would be possible to prove that life on the Earth has an extragalactic origin. (Any communication between galaxies would be hampered by the exponential expansion of the universe with $\rho_v > 10^{-29} g.cm^{-3}$.) Unfortunately, at the moment we have no evidence in favour of this extravagant hypothesis. (Of course, one can turn this argument upside down and say that the vanishing of the cosmological constant can be considered as an argument in favour of the extragalactic origin of life. However, this will be true only if no other solution of the cosmological constant problem will be found.)

Another possibility is related to the results obtained in the first part of this Section. It may be not unreasonable to assume that the conditional probability to be born in the universe in a quantum state with a given value of ρ_v is proportional to the total volume of those parts of the universe with this value of ρ_v where life of our type can appear. From the investigation performed in the beginning of this Section it follows that in the quantum states with $|\rho_v| < 10^{-29} g.cm^{-3}$ the total volume of the parts of the universe where life can exist is much bigger than in the states with $|\rho_v| > 10^{-29} g.cm^{-3}$, and it may be especially big in the states with $|\rho_v| < 10^{-10^7} g.cm^{-3}$. If there is some relation between the total volume of the parts of the universe where life can exist and the volume where life of our type can appear, then one can argue that the probability that we measure the vacuum energy density ρ_v in our quantum state of the universe $|\alpha_i >$ is peaked in the region $|\rho_v| < 10^{-10^7} g.cm^{-3}$. This would be more than enough to solve the cosmological constant problem.

Unfortunately, this line of reasoning, which may lead to a justification of the so-called final anthropic principle [55] in its strongest form, is still far from being well elaborated. (The original version of the final anthropic principle as suggested in [55] is trivially satisfied in the theory of the self reproducing inflationary universe) There are many obvious and less obvious objections which can be raised against the arguments suggested above. It is quite possible that the cosmological constant problem eventually will be solved in quite a different way, e.g. by some modification of ideas proposed in [17,22,25]. However, the efficiency of the Anthropic Principle for the determination of many other coupling constants in the context of the baby universe theory and the existence of a strong correlation between the value of the cosmological constant and the lifetime of mankind makes it very tempting to continue an investigation of the possibilities discussed above. At the very least, what we have obtained already is a surprising result that the existence of the vacuum energy density as small as $10^{-10^{7}} g.cm^{-3}$ would be fatal for the future evolution of life in our part of the universe. This is the first indication of the potential practical importance of recent attempts to investigate the cosmological constant problem. Unfortunately (or, may be, fortunately), it may take as much as $10^{5.10^{6}}$ years until the significance of the current work on this problem will be fully appreciated.

References

[1] Ya.B. Zeldovich and M.Yu. Khlopov, Phys. Lett. **B 79** (1978) 239.
J.P. Preskill, Phys. Rev. Lett. **43** (1979) 1365.

[2] Ya.B. Zeldovich, M.-Yu. Kobzarev and L.B. Okun, Phys. Lett. **B 50** (1974) 340.
S. Parke and S.-Y. Pi, Phys. Lett. **B 121** (1981) 313.
G. Lazarides, Q. Shafi and T.F. Walsh, Nucl. Phys. **B195**(1982) 157.
P. Sikivie, Phys. Rev. Lett. **48** (1982) 1156.

[3] J. Ellis, A.D. Linde and D.V. Nanopoulos, Phys. Lett. **B 116** (1982) 59.
M.Yu. Khlopov and A.D. Linde, Phys. Lett. **B 138** (1984) 265.

[4] G.D. Coughlan, W. Fischler, E.W. Kolb, S. Raby and G.G. Ross, Phys. Lett. **B 131** (1983) 59.

[5] A.S. Goncharov, A.D. Linde and M.I. Vysotsky, Phys. Lett. **B 147** (1984) 279.

[6] S.M. Barr, K. Choi and J.E. Kim, Nucl. Phys. **B283** (1987) 591.

[7] I. Antoniadis, C. Bachas, J. Ellis and D.V. Nanopoulos, Phys. Lett.**B 211** (1988) 393.

[8] E.B. Gliner, Sov. Phys. JETP **22** (1965) 378; Dokl. Akad. Nauk SSSR **192** (1970) 771.
E.B. Gliner and I.G. Dymnikova, Pis. Astron. Zh. **1** (1975) 7.
I.E. Gurevich, Astrophys. Space Sci. **38** (1975) 67.

[9] A.A. Starobinsky, JETP Lett. **30** (1979) 682; Phys. Lett. **B 91** (1980) 99.

[10] A.H. Guth, Phys. Rev. **D23** (1981) 347.

[11] A.D. Linde, Phys. Lett. **B 108** (1982); **B 114** (1982) 431; **B 116** (1982) 335, 340.
A. Albrecht and P.J. Steinhardt, Phys. Rev. Lett. **48** (1982) 1220.

[12] A.D. Linde, Phys. Lett. **B 129** (1983) 177.

[13] A.D. Linde, Phys. Lett. **B 201** (1988) 437.

[14] A.D. Linde, Phys. Lett. **B 175** (1986) 395; Physica Scripta **T15** (1987) 169; Physics Today **40** (1987) 61.

[15] A.S. Goncharov, A.D. Linde and V.F. Mukhanov, Int. J. Mod. Phys. **A2** (1987) 561.

[16] A.D. Linde, Phys. Lett. **B 238** (1990) 160.

[17] A. D. Linde, Phys. Lett. **B 200** (1988) 272.

[18] S.W. Hawking, Phys. Lett. **B 195** (1987) 277; Phys. Rev. **D37** (1988) 904.

[19] G.V. Lavrelashvili, V.A. Rubakov and P.G. Tinyakov, JETP Lett. **46** (1987) 167; Nucl. Phys. **B299** (1988) 757.

[20] S.B. Giddings and A. Strominger, Nucl. Phys. **B306** (1988) 890.

[21] S. Coleman, Nucl. Phys. **B307** (1988) 867.

[22] T. Banks, Nucl. Phys. **B309** (1988) 493.

[23] S. Giddings and A. Strominger, Nucl. Phys.**B307** (1988) 854.

[24] S.W. Hawking and R. Laflamme, Phys. Lett. **B 209** (1988) 39.

[25] S. Coleman, Nucl. Phys. **B310** (1989) 643.

[26] S. Giddings and A. Strominger, Nucl. Phys. **B321** (1989) 481.

[27] J. Polchinski, Phys. Lett. **B 219** (1989) 251.

[28] W. Fischler, I. Klebanov, J. Polchinski and L. Susskind, Nucl. Phys. **B327** (1989).

[29] A.D. Linde, Phys. Lett. **B 227** (1989) 352.

[30] A.D. Linde, Phys.Lett. **B 211** (1988) 29.

[31] A.D. Linde, Phys. Lett. **B 162** (1985) 281.

[32] V.F. Mukhanov and G.V. Chibisov, JETP Lett. **33** (1981) 523.
S.W. Hawking, Phys. Lett. **B 115** (1982) 339.
A.A. Starobinsky, Phys. Lett. **B 117** (1982) 175.
A.H. Guth and S.-Y. Pi, Phys. Rev. Lett. **49** (1982) 1110.
J. Bardeen, P.J. Steinhardt and M. Turner, Phys. Rev. **D28** (1983) 679.
R. Brandenberger, Rev. Mod. Phys. **57** (1985) 1.

[33] V.F. Mukhanov, JETP Lett. **41** (1985) 493;
V.F. Mukhanov, L.A. Kofman and D.Yu. Pogosyan, Phys. Lett. **B 193** (1987) 427.

[34] A.D. Linde, **Particle Physics and Inflationary Cosmology** (Gordon and Breach, New York, 1990).

[35] L.A. Kofman, A.D. Linde and A.A. Starobinsky, Phys. Lett. **B 157** (1985) 36.

[36] A.A. Starobinsky, JETP Lett. **42** (1985) 152.

[37] J. Silk and M. Turner, Phys. Rev. **D35** (1986) 419.

[38] A.D. Linde, JETP Lett. **40** (1984) 1333; Phys. Lett. **B158** (1985)375; L.A. Kofman, Phys. Lett. **B 174** (1986) 400; T.J. Allen, B. Grinstein and M.B. Wise, Phys. Lett. **B 197** (1987) 66.

[39] L.A. Kofman and A.D. Linde, Nucl. Phys. **B 282** (1987) 555.

[40] L.A. Kofman and D.Y Pogosyan, Phys. Lett. **B 214** (1988) 508; D.S. Salopek, J.R. Bond and J.M. Bardeen, Phys. Rev. **D 40** (1989) 1753; H.M. Hodges, G.R. Blumental, L.A. Kofman and J.R. Primack, Santa Cruz preprint SCIPP 89/31 (1989).

[41] L.A. Kofman, A.D. Linde and J. Einasto, Nature **326** (1987) 48.

[42] A.D. Linde and M.I. Zelnikov, Phys. Lett. **B 215** (1988) 59.

[43] D. La and P.J. Steinhardt, Phys. Rev. Lett. **62** (1989) 376; Phys.Lett. **B 220** (1989) 375.

[44] A.H. Guth, Phys. Rev. **D23** (1981) 347.

[45] C. Brans and C.H. Dicke, Phys. Rev. **124** (1961) 925.

[46] E. Weinberg, Columbia university preprint CU-TP-436 (1989).

[47] D. La, P.J. Steinhardt and E.W. Bertschinger, Phys. Lett. **B 231** (1989) 231.

[48] F.S. Accetta and J.J. Trester, Phys. Rev. **D39** (1989) 2854.

[49] A. Vilenkin and L. Ford, Phys. Rev. **D26** (1982) 1231; A.D. Linde, Phys. Lett. **B 116** (1982) 335; A.A. Starobinsky, Phys. Lett.**B 117** (1982) 175.

[50] S. Weinberg, Phys. Rev. Lett. **59** (1987) 2607; Rev. Mod. Phys. **61** (1989) 1.

[51] J. A. Wheeler, in **Batelleis Recontres**, eds. C. DeWitt and J. A. Wheeler (Benjamin, New York, 1968).

[52] B. S. DeWitt, Phys. Rev. **160** (1967) 113.

[53] H. Everett, Rev. Mod. Phys.,**29** (1957) 454; B. S. DeWitt and N. Graham, The Many-Worlds Interpretation of Quantum Mechanics (Princeton University Press, Princeton, 1973).

[54] I. L. Rosental,**Big Bang, Big Bounce: How Particles and Fields Drive Cosmic Evolution** (Springer, Berlin, 1988).

[55] J. Barrow and F. Tipler, **The Cosmological Anthropic Principle** (Oxford University Press, Oxford, 1986).

High Energy Astrophysics

T.K. Gaisser[*]
Bartol Research Institute
University of Delaware
Newark DE 19716

March 26, 1990

Large non-accelerator experiments now in progress or under construction typically have several objectives that include topics in both particle physics and astrophysics. They are intrinsically interdisciplinary, both as concerns scientific goals and techniques. In this talk I review the background of the astrophysical questions addressed by these experiments.

1   Introduction

There are currently more than 50 large underground and surface detectors investigating problems in high energy particle astrophysics. The primary aims of the surface air shower experiments concern questions such as the origin of very high energy cosmic rays and the related subjct of high energy gamma ray astronomy. The large underground detectors are motivated primarily by problems in high energy physics that require the low background of a deep mine or tunnel, such as the searchs for proton decay, magnetic monopoles and dark matter. These experiments also have an important component of atrophysics. Indeed, the largest rates in the underground detectors are due to cosmic ray muons which may be used as a probe of the chemical composition of the primary cosmic radiation in a high energy region inaccessible to direct experiments.

In table 1 I summarize the main objectives of the large non-accelerator

[*]Invited talk at DPF 90, General meeting of the Division of Particles and Fields of the American Physical Society, Rice University, January 3 1990. Work supported in part by the U.S. Department of Energy.

Table 1: Objectives of large, non-accelerator experiments.

Surface experiments	Underground experients
	nucleon decay search
	monopole search
	dark matter search
	ν-oscillation search
	*high energy neutrino astronomy
*100 TeV γ-astronomy ===	=======================
	stellar collapse neutrinos
	solar neutrinos
*TeV γ-astronomy	
	atmospheric neutrinos
	*cosmic ray muons ($\sim$TeV)
*cosmic ray air showers	

experiments. The double line separates searches from subjects that have been established experimentally. The topics below the line are arranged in order of increasing event rate. The study of photons with $E_\gamma \sim 100$ TeV from point sources is on the borderline, for reasons to be discussed. The subjects marked with an asterisk all address the problem of the origin, acceleration and transport of high energy particles in the galaxy, which is the focus of this talk.

The talk is divided into four sections. I first review the standard theory of cosmic ray origin, which accounts nicely for the majority of galactic cosmic rays, but which fails for particles with energies much above 100 TeV. Next I list some suggestions for the origin of particles with $E > 100$ TeV, with emphasis on the possibility of point sources. This leads to the subject of high energy gamma ray and neutrino astronomy and its experimental status. I conclude with a discussion of the elemental composition of cosmic rays with $E > 100$ TeV, which is the traditional approach to the origin of these high energy particles.

2 Standard theory of cosmic rays.

The first figure shows the energy spectrum of several components of the cosmic radiation[1] below 1 TeV/nucleon (Fig. 1a). After correcting for the

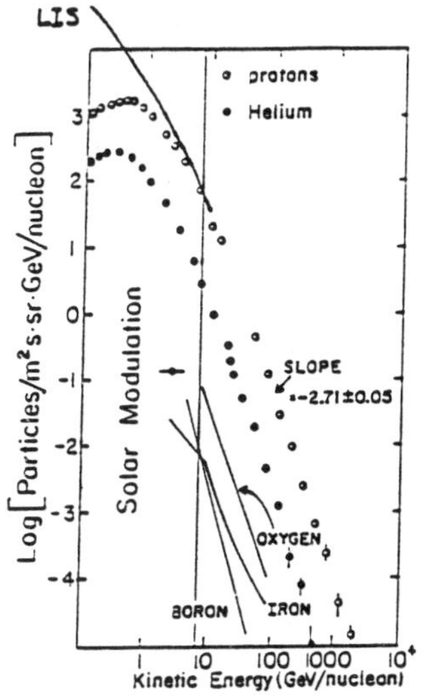

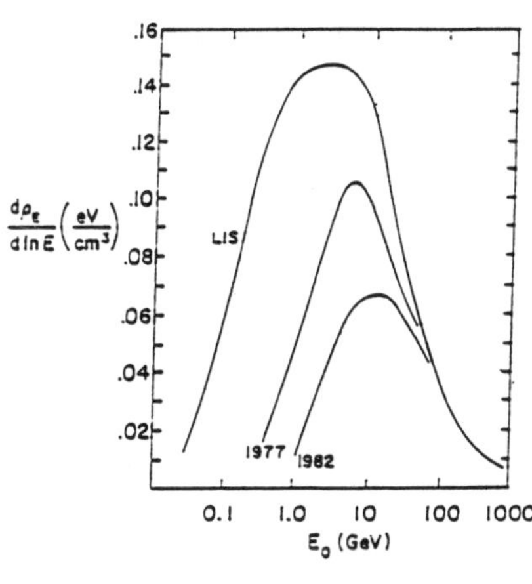

Figure 1: Cosmic ray energy spectrum (a) and (b) energy density carried by cosmic rays as a function of cosmic ray energy.

effect of solar modulation,[2] one obtains an estimate of the local interstellar (LIS) spectrum, which may be integrated to estimate the energy density of cosmic ray particles in interstellar space near the solar system. The energy content is displayed as a function of particle energy in Fig. 1b. The area under the curve (LIS) is $\rho_E \sim 1$ eV cm^{-3}, most of which is carried by particles with energy less than 1 TeV. If the local interstellar medium is typical of the galactic disk, then the total energy of all cosmic rays in the disk ($V \sim 5 \times 10^{66}$ cm^3) is $\sim 8 \times 10^{54}$ erg.

To estimate the power required to maintain this density of relativisitc particles, we need an estimate of the residence time of cosmic rays in the disk. This comes from a measurement of the ratio of secondary to primary cosmic rays.[3] Secondary nuclei, such as Li, Be and B are virtually absent as end products of stellar evolution, yet they are relatively abundant in the cosmic radiation. They are produced by spallation of primary nuclei, such as carbon and oxygen. If this production occurs in collisions with the gas of the interstellar medium (density $\rho \sim 1$ hydrogen atom per cm^3), then the secondary and primary spectra are related by

$$N_S(E) = \rho\, c\, \tau\, \sigma_{P \to S}\, N_P(E). \tag{1}$$

For particles with energies in the 1-10 GeV range, $\tau \sim 5 \times 10^6$ yrs. The light crossing time for the galactic disk is ~ 1000 light years. Since $\tau \gg 1000$ yrs, the interpretation is that cosmic rays undergo diffusion in the turbulent magnetized plasma of the interstellar medium eventually escaping altogher from the galaxy. The total power required to maintain the cosmic ray flux against loss out of the galaxy is

$$\text{power} \sim \frac{\rho_E V}{\tau} \sim 5 \times 10^{40}\ \text{erg/s.} \tag{2}$$

Ginzburg and Syrovatsky[4] noted long ago that this is nicely matched to the power available in supernova explosions. If type II supernovas occur in the galaxy at the rate of one every 30 years with 10^{51} ergs deposited as kinetic energy in the shell, then supernova explosions provide $\sim 10^{42}$ erg/s. The efficiency for particle acceleration needs to be $\sim 5 - 10\%$ to supply all the cosmic rays from this source.

The measured ratio of secondary to primary nuclei is observed to decrease with energy, so from Eq. 1,

$$\tau(E) \propto \frac{N_S(E)}{N_P(E)} \propto E^{-\delta}, \tag{3}$$

with $\delta \sim 0.6$ for $E > 2$ GeV/nucleon.[5] Assuming that this diffusion time applies also to primary cosmic rays, including the dominant proton component, Eq. 3 has important implications for the spectrum of cosmic rays at the source. In equilibrium, the source spectrum, $Q(E)$, and the observed spectrum, $N_P(E)$, are related by

$$Q_P(E) \propto N_P(E)/\tau(E), \tag{4}$$

analogous to Eq. 2 for the source power. The observed differential energy spectrum is approximately an inverse power with $N_P(E) \propto E^{-2.7}$, so from Eqs. 3 and 4, the source spectrum is

$$Q_P(E) \propto E^{-2.1}. \tag{5}$$

Acceleration by diffusion in turbulent magnetic fields[1] at a strong shock (Mach number $M \gg 1$) gives a differential spectrum of accelerated particles $Q(E) \propto E^{-\alpha}$, with $\alpha \approx 2 + 4/M^2$. Since supernova shells expand with highly supersonic velocities, they drive strong blast shocks which can serve as the mechanism for particle acceleration. The combination of sufficient power and a natural mechanism for accelerating particles with the observed spectrum makes the supernova origin of cosmic rays a particularly attractive picture.

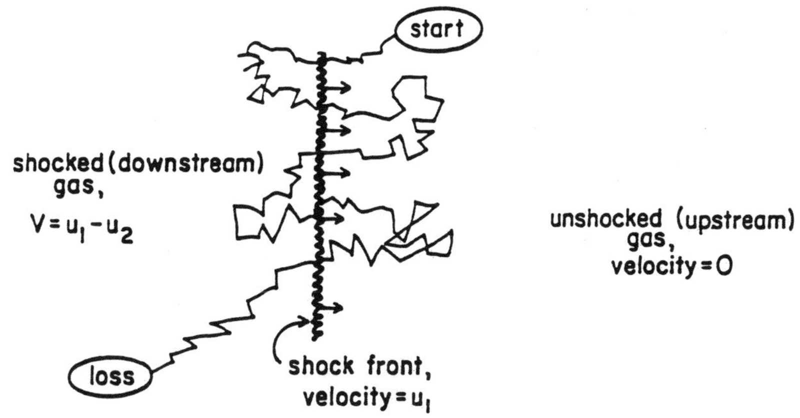

Figure 2: Particle accelerating at a shock.

The idea of shock acceleration[6] is illustrated in Fig. 2. The figure shows a test particle particle diffusing in the turbulent magnetized plasma in the vicinity of the shock and occasionally crossing from one side to the other. The assumption is that there is sufficient diffusion so that the particle distribution is isotropized in the rest frame of the gas before it crosses to the other side. A pair of Lorentz transformations, averaged over all angles

---

[1] Diffusive acceleration at a strong shock is called first order Fermi acceleration. Every back and forth crossing of the shock results in an energy gain for a particle by an amount proportional to its energy before the crossing. The original version of diffusive acceleration due to Fermi is a second order process in which particles sometimes lose and sometimes gain energy in encounters with moving magnetized clouds.

at which a particle can cross the shock, shows that the energy gain per cycle (one crossing from upstream to downstream and back) is proportional to the particle energy,

$$\Delta E = \xi E, \qquad (6)$$

with $\xi \propto V/c$. Continuity at the shock, together with kinetic gas theory, gives

$$V \equiv u_1 - u_2 \sim u_1 \left(1 - 1/M^2\right). \qquad (7)$$

The rate of energy gain is given by

$$\frac{dE}{dt} = \frac{\xi E}{T_{\text{cycle}}}. \qquad (8)$$

The cycle time is estimated as the time it takes the shock to travel one diffusion length, by which time it catches up with the particle diffusing in the upstream region. Since the mechanism for scattering is diffusion in magnetic fields (rather than by collisions with gas, which would destroy the acceleration process) the diffusion length, λ, is bounded by the Larmor radius in the mean field. Thus,

$$T_{\text{cycle}} \sim \frac{\lambda}{u_1} \geq \frac{E}{u_1 \, e \, B}, \qquad (9)$$

and

$$\frac{dE}{dt} > \frac{u_1}{c} \, u_1 \, e \, B. \qquad (10)$$

Because diffusive shock acceleration takes time, there is an upper limit[7] to the energy that can be achieved if the shock has a finite lifetime.[2] A supernova shell begins to slow down after $T \sim 10^3$ years, when $M_{\text{shell}} \sim \rho(U_1 T)^3$. Thus

$$E_{\text{max}} < \frac{u_1}{c} \, e \, B \, u_1 T \sim 100 \text{ TeV} \times \frac{B}{3 \, \mu\text{G}}. \qquad (11)$$

3 Origin of cosmic rays above 100 TeV

Various suggestions have been made to explain the existence of particles with ultra-high energy ($E > 100$ TeV). Note that Eq. 11 is of the form

$$E_{\text{max}} \sim \frac{V}{c} \, e \, B \, R, \qquad (12)$$

---

[2] Another limiting factor arises when a particle's mean gyroradius is comparable to the curvature of a shock.

where the distance scale is $R \sim u_1 T$. Clearly, one needs either to increase the magnetic field strength or the scale on which acceleration occurs. One possibility is to modify the supernova blast mechanism. If the pre-supernova environment into which the supernova explodes is accounted for in more detail (for example,[8] by accounting for regions in which the magnetic field is perpendicular to the propagation of the shock, thus confining the particles more strongly to the shock front), one may obtain a larger E_{max}. It has also been pointed out[9] that a supernova does not expand into the nominal interstellar medium, but into the wind of the progenitor star, which can also increase the estimate of E_{max}. Another possibility, which may be able to accelerate particles close to the highest energies observed (approaching 10^{20} eV), is to postulate[10] acceleration at a shock of galactic scale that occurs in a galactic wind expanding into the intergalactic medium. This wind could ultimately be driven by the cumulative effect of many supernova explosions acting through the pressure of the cosmic rays accelerated at supernova blast waves as described above.

Another possibility is that high energy cosmic rays are accelerated near compact objects in regions of strong magnetic field. Eq. 12 can be thought of as dimensional analysis (eBR) times an efficiency factor (V/c). At the surface of a neutron star with a 10^{12} gauss surface field, $eBR \sim 3 \times 10^{20}$ eV. Although it is not easy to find a mechanism to realize this full potential, it is possible that acceleration to $E > 100$ TeV could be associated with neutron stars in compact systems in some way.

This possibility has received a great deal of attention in recent years because of observations of air showers with $E > 100$ TeV from the direction of x-ray binary stars such as Cygnus X-3, Hercules X-1 and others.[11] These showers cannot be generated by cosmic ray nuclei from the source, because charged particles accelerated at distant sources would diffuse away from their sources along complicated paths in the galactic magnetic field. The signals would have to be carried by neutral particles, presumably photons, produced at the source by some portion of the accelerated beam when it is dumped into nearby concentrations of matter.

Two sources of power have been suggested for acceleration to high energy in X-ray binaries – accretion and rotation of a magnetized neutron star. Fig. 3 shows two different configurations in which acceleration might occur. Not all mechanisms can work in all systems, and it can be difficult to realize the full power that may be present. A detailed discussion of X-ray binaries is beyond the scope of this talk. It is clear, however, that the systematic study of high energy signals from such sources would yield a rich harvest of

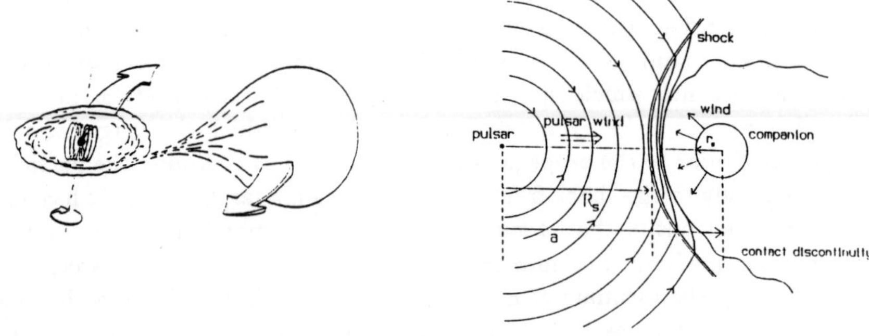

Figure 3: Two configurations of a neutron star in an x-ray binary in which acceleration to high energy might occur: (*a*) a neutron star accreting material from its companion through an accretion disk (sketch courtesy of J. Perrett); (*b*) a pulsar driving a wind that interacts with its companion.[12]

new information about how the systems work. This would complement the extensive knowledge of interacting binary stars already obtained from X-ray astronomy.

4   Astronomy with atmospheric cascades.

Despite a promising beginning, astronomy with > 100 TeV air showers has not yet reached a stage where systematic study of sources has been achieved.[13] One difficulty is that potential sources are likely to be sporadic. The Sun is a helpful analog. Particles are accelerated to GeV energies in solar flares. A fraction of these particles interacts in the solar atmosphere to produce ~ 100 MeV photons, as well as neutrons.[14] It would not be surprising if sources much more energetic than the Sun also accelerated particles in flare-like episodes.

The air shower signals from point sources indeed appear to be sporadic. But this is not the only problem. Fig. 4 shows the evolution of signals[15 – 18] and upper limits[19 – 21] on air showers with energies > 500 TeV from Cygnus X-3. The various experiments have thresholds that differ by more

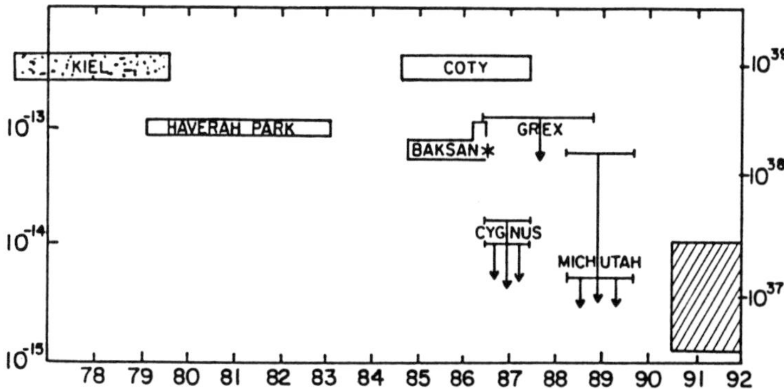

Figure 4: History of signals and limits on air shower signals from Cygnus X-3 (converted to a threshold of 500 TeV assuming an E^{-1} integral energy spectrum).

than an order of magnitude. I have therefore followed Bloomer et al.[19] and converted all results to 500 TeV, assuming an E^{-2} differential energy spectrum. The left ordinate is labelled by flux and the right by the required power in accelerated protons at Cygnus X-3 (distance $\sim 3 \times 10^{22}$ cm) needed to produce the corresponding flux level. The boxes at 1991-92 indicate the level of sensitivity that is expected with detectors now being constructed[22] or expanded.[23] For comparison, the power required to supply all the cosmic rays in the galaxy that have $E > 100$ TeV is $\sim 2 \times 10^{39}$ erg/s.[3]

The more recent air shower experiments have emphasized the capability for muon detection to be able preferentially to select γ-showers, which are expected to be muon-poor relative to background cosmic ray showers.[24] Surprisingly, however, the muon selection criterion does not appear to work for apparent air shower signals from one of the most promising sources, Hercules X-1.[25] A similar anomaly has turned up recently in observations of this source with an air Cherenkov telescope, which is sensitive in a lower energy region, close to 1 TeV. The group of Weekes et al.[26] have developed an imaging method to discriminate between photon- and hadron-induced

---

[3]This is only a crude estimate because of the difficulty of estimating the residence time of the particles in the high energy tail of the cosmic ray energy spectrum.

showers on the basis of differences in their Cherenkov light images. When applied to events from the direction of the Crab Nebula the method appears to improve the signal to background by many standard deviations.[26] Apparent signals from Hercules X-1, however, do not satisfy the photon-like requirement.[27] The image criterion is based on Monte Carlo simulations, which are confirmed experimentally to the extent that the method works for the Crab Nebula.

Much of the experimental activity referred to in the introduction is now aimed at clearing up this confusing situation.[e.g. 22,23] In addition, there are several proposals in the U.S. and Europe for construction of large detectors for the related subject of $\sim$TeV neutrino astronomy.[28] Neutrino astronomy in the TeV energy range is done by detecting upward muons generated outside the detector by interactions of muon neutrinos from extraterrestrial sources. Such neutrinos would originate from decay of $\pi^{\pm}$ produced in the same sources that give $\geq$TeV photons through $p + \text{gas} \rightarrow \pi^0 + \cdots$. Current experiments[29,30] have smaller collection areas than are likely to be necessary to see signals from astrophysical sources, but the technique has been calibrated by detecting neutrinos produced in the atmosphere.

5 Cosmic ray composition at the knee.

Many of the same air shower experiments directed at ultra-high energy gamma ray astronomy also have the potential for improving significantly the traditional cosmic ray approach to the origin of cosmic rays above 100 TeV. This is through detailed studies of the structure and particle content of air showers in order to infer the elemental composition of the primaries in this energy region. Above 100 TeV, the flux is too low to be accessible at present to direct observation with small detectors above the atmosphere.

Figure 5a[31] shows a summary of observations of the "all-particle" spectrum, that is, the total flux of primary particles as a function of total energy per nucleus. The differential energy spectrum (shown here multiplied by $E^{2.5}$) steepens from $\alpha \approx 2.7$ below 10^6 GeV/nucleus to $\alpha \approx 3.0$ above 10^7 GeV/nucleus. This feature is known as the "knee" of the cosmic ray spectrum. If the steepening depends on magnetic rigidity only, with a single population of sources, then the composition should show a characteristic enrichment in heavy nuclei as the all particle spectrum steepens. This follows simply from the relation between magnetic rigidity and total energy

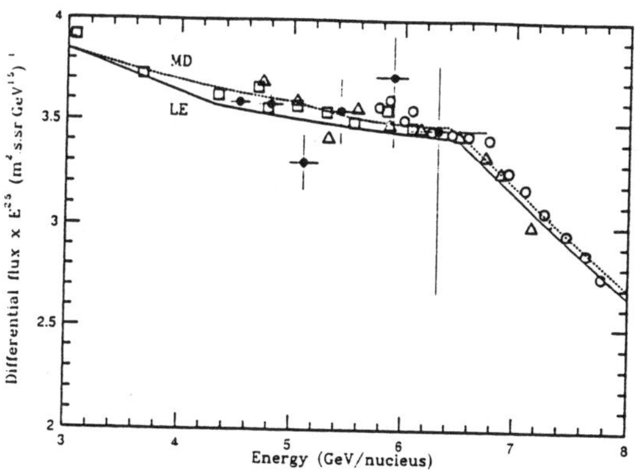

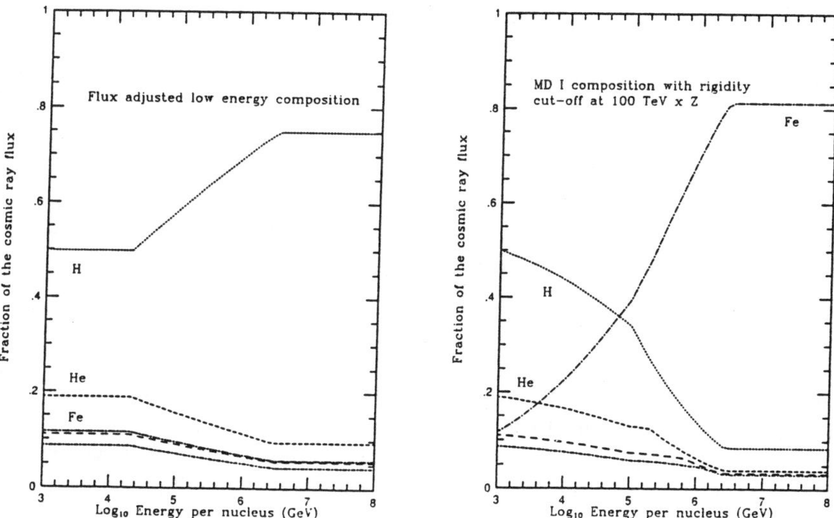

Figure 5: The figure shows (a) a summary of the all-particle spectrum with two fits corresponding to the fractional compositions shown in (b) and (c).

per nucleus,

$$R(\mathrm{GV}) \;=\; \frac{c\,P_{\mathrm{tot}}}{Z\,e} \;\approx\; \frac{2\,E_{\mathrm{tot}}(\mathrm{GeV})}{A}, \qquad (13)$$

where Z and A are the charge and mass of nuclei with $Z > 1$. (For protons $R(\mathrm{GV}) \approx E_{\mathrm{tot}}(\mathrm{GeV})$ at high energy.) If the spectrum steepens at a characteristic rigidity, R_c, then the steepening occurs for each nuclear species at $E_{\mathrm{tot}} \sim \frac{1}{2} A\,R_c$ for all nuclei with $Z > 1$.

A steeper spectrum as a function of rigidity could indicate the upper limit of the acceleration mechanism, or it could reflect an increasingly rapid escape of accelerated particles from the galaxy. In either case, as first pointed out by Peters[32], there would be a characteristic enrichment of the composition in heavy nuclei when the nuclei are classified by total energy per nucleus. This follows from the relation (13) between E_{tot} and R.

Another possibility is that the "knee" of the cosmic ray spectrum indicates the onset of a new source of cosmic rays in the high energy region. In this case, the energy dependence of the composition would in general be different from that characteristic of a rigidity-dependent steepening. At present, the air shower data is very difficult to interpret, and quite different compositions have been proposed in the region of the knee, as illustrated in Fig. 5 b and c. A composition similar to that shown in 5b has been suggested by Fichtel and Linsley[33] on the basis of their analysis of various measures of fluctuations in air showers. The iron rich composition has been suggested by the Maryland group[34] on the basis of observations of slow hadrons in air showers.

There are good prospects for improving the knowledge of composition above 100 TeV with experiments now in progress. Especially promising for the region 10^{14} to 10^{16} eV is the combination of a large deep underground muon detector and a large air shower array at Gran Sasso.[35] The UMC experiment[22] also has the capability for new progress in this energy range as a consequence of its unprecedented ability to measure the ratio of low energy muons to electrons and positrons in the shower front. At higher energies, around 10^{18} eV, the Fly's Eye experiment has sensitivity to composition[36] through its measurement of the distribution of shower maximum.

The best one can expect from these indirect measurements, however, is a determination of the gross structure of the composition. Direct measurements are indispensible for a detailed knowledge of the elemental composition in the region of the knee of the spectrum. One proposal currently under consideration[37] is to send a massive calorimeter aloft with a Soviet "Energia" rocket. Another is to build a large calorimeter on the moon.[38]

6   References

1. The spectra for protons, helium, oxygen, boron and iron are from J. Ormes & P. Freier, *Ap. J.* **222**, 471 (1978).

2. P.A. Evenson, in *Proc. Conf. on Interplanetary Particle Environment* (JPL Pub. 88-28, ed. J. Feynman & S. Gabriel), p. 149 (1988).

3. J.A. Simpson *Ann. Revs. Nucl. Part. Sci* **33**, 323 (1983).

4. V.I. Ginzburg & S.I. Syrovatskii *The Origin of Cosmic Rays* (Pergamon Press, 1964).

5. M. Gupta & W.R. Webber, *Ap. J.* **340**, 1124 (1989).

6. For recent reviews of shock acceleration see R. Blandford & D. Eichler *Phys. Reports* **154**, 1 (1987) and W.I. Axford *Proc. 20th Int. Cosmic Ray Conf.* (Moscow) **8**, 120 (1987).

7. P.O. Lagage & C.J. Cesarsky *Astron. Astrophys.* **125**, 249 (1983).

8. J.R. Jokipii *Proc. 6th Int. Solar Wind Conf.* (NCAR TN-306) vol. 2, p. 481. See also *Ap. J.* **313**, 842 (1987) and **255**, 716 (1982).

9. H.J. Völk & P.L. Biermann *Ap. J.* **333**, L65 (1988).

10. J.R. Jokipii & G. Morfill *Ap. J.* **312**, 170 (1987).

11. For reviews see T.C. Weekes *Phys. Reports* **160**, 1 (1988) and D.E. Nagle, T.K. Gaisser & R.J. Protheroe *Ann. Rev. Nucl. Part. Sci.* **38**, 609 (1988).

12. A.K. Harding & T.K. Gaisser *Ap. J.* to be published (1990).

13. T.K. Gaisser *Science* **247**, 1049 (1990). D. Fegan, Rapporteur talk, *21st Int. Cosmic Ray Conf.* (Adelaide, 1990) to be published.

14. E.L. Chupp *Ann. Rev. Astron. Astrophys.* **22** 359 (1984).

15. M. Samorski & W. Stamm (Kiel) *Ap. J.* **268**, L17 (1983).

16. J. Lloyd-Evans et al. (Haverah Park) *Nature* **305**, 784 (1983).

17. V.V. Alexeenko et al. (Baksan) *Proc. 20th Int. Cosmic Ray Conf.* **1**, 229 (1987).

18. S. Tonwar et al. (Ooty) *Ap. J.* **330**, L107 (1988).

19. S.D. Bloomer et al. (GREX) *Proc. 21st Int. Cosmic Ray Conf.* (Adelaide) **2**, 39 (1990).

20. B. Dingus et al. (Cygnus) *Phys. Rev. Letters* **60**, 1785 (1988).

21. D. Ciampa et al. (Mich.–Utah) *Proc. 21st Int. Cosmic Ray Conf.* (Adelaide) **2**, 35 (1990).

22. K. Gibbs et al. (UMC) *Nuclear Instruments & Methods* **A264**, 67 (1988). I am grateful to Rene Ong (private communication) for estimating the sensitivity of this experiment for me.

23. D. Berley et al. *Proc. 21st Int. Cosmic Ray Conf.* (Adelaide) **10**, 301 (1990).

24. T. Stanev, T.K. Gaisser & F. Halzen, *Phys. Rev.* **D32**, 1244 (1985) and P.G. Edwards et al. *J. Phys. G:Nucl.Phys.* **11**, L101 (1985).

25. B. Dingus et al. *Phys. Rev. Letters* **61**, 1906 (1988).

26. T. Weekes et al. *Ap. J.* **342**, 379 (1989).

27. D. Fegan et al. *Proc. Workshop on Experimental Techniques of High Energy Neutrino and VHE and UHE Gamma-Ray Particle Astrophysics* (ed. D. Wold & G.B. Yodh, *Nuclear Physics B*, in press).

28. The volume *Proc. Workshop on Experimental Techniques of High Energy Neutrino and VHE and UHE Gamma-Ray Particle Astrophysics* (ed. D. Wold & G.B. Yodh, *Nuclear Physics B*, in press) gives an overview of current activity in ground based gamma ray astronomy and neutrino astronomy.

29. R. Svoboda et al. *Ap. J.* **315**, 420 (1987).

30. Y. Oyama et al. *Phys. Rev.* **D39**, 1481 (1989).

31. G. Auriemma, T.K. Gaisser & T. Stanev *Proc. 21st Int. Cosmic Ray Conf.* **9**, 362 (1990).

32. B. Peters *Nuovo Cimento (Suppl.)* **14**, 436 (1959). Also, *Proc. 6th Int. Cosmic Ray Conf.* (Moscow) **3**, 157 (1960).

33. C.E. Fichtel & J. Linsley *Ap. J.* **300**, 474 (1986).

34. J.A. Goodman et al. *Phys. Rev.* **D26**, 1043 (1982) and H.T. Freudenreich et al. UMDCR Preprint 90-024 (1989).

35. EASTOP/MACRO Collaboration *Proc. 21st Int. Cosmic Ray Conf.* (Adelaide) **9** 331 and **9** 335 (1990).

36. G.L. Cassiday et al. *Proc. 21st Int. Cosmic Ray Conf.* (Adelaide) **3**, 154 (1990).

37. "Elisa" a proposal to the European Space Acency for the next medium size project (contact person: Luke O'C. Drury, Dublin Inst. for Advanced Study, unpublished).

38. S.P. Swordy *Proc. Workshop on the NASA Cosmic Ray Program for the 1990s and Beyond* (A.I.P. Conference Proceedings, to be published, 1990).

e^+e^- PHYSICS BELOW THE Z^0

Yorikiyo Nagashima

Physics Department, Osaka University, Toyonaka 560, Japan

ABSTRACT

Two topics, test of electro-weak theory and test of QCD/Jets are covered. Forward-backward asymmetry information is used to test isospin assignment of the quarks and leptons and confirms the existence of the top and the tau neutrino. Basic parameters as determined from e^+e^- reactions are consistent with those derived from other reactions. Possible existence of a new neutral gauge boson is discussed. In QCD, running nature of the coupling constant, derivation of $\Lambda_{\overline{MS}}$, and effect of rescaling are discussed. Self coupling of the gluon has been demonstrated. Finally, some model comparisons are discussed.

1.   INTRODUCTION

At the total center of mass energy between 50 and 60 GeV, TRISTAN is severely handicapped by being at the very bottom of the cross section valley. Coming down like $1/s$ from low energy side, it has yet to make ascent of the Z^0 peak where the counting rate is almost three order of magnitude higher. However, because the lower energy region is purely electromagnetic while the Z^0 region is of weak origin, TRISTAN is ideally suited to look into interference effect of the electro-weak interaction. Since the asymmetry is the direct measure of the axial vector coupling constant of fundamental particles to which Z^0 couples, it provides an excellent test bench of the unified electro-weak theory.

2. TEST OF THE ELECTRO-WEAK THEORY

2.1 Number of Generation

Many people take it for granted that all left handed fermions transform as doublets and all right handed ones as singlets in the SU(2) group. This was true for the first two doublets of the lepton family, (v_e, e), (v_μ, μ) and also for the first quark doublet (u, d'). Evidences exist in μ- and β-decays, leptonic pion and kaon decays. It is not, however, obvious with other fermions [1]. The assumption predicts the existence of top quark and τ neutrino as partners of b-quark and τ-lepton. Then there is a question, "How many generations?". The last question is partially answered by recent measurements of the Z^0 width by LEP groups [2] that the number of light neutrinos is three. That does not, however, exclude the existence of a new generation. Therefore, searches for a fourth charge 1/3 quark b' or a charged lepton L^- with a heavy neutrino L^0 as its partner are valid within the standard framework. No such particles are discovered yet [3, 4]. Figure 1 shows present lower limits of new particles in the (extended) standard model. To test correct assignment of the fermions, we look into forward-backward (abbreviated FB) asymmetry of the fermion pair production cross section.

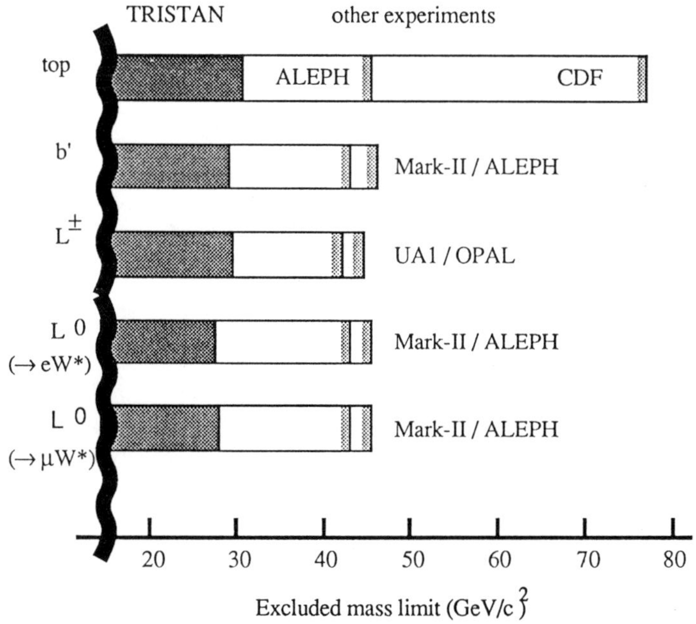

Figure 1

2.2   Forward-Backward Asymmetry

<u>General formula</u>   The lowest order differential cross section of fermion pairs ($f\bar{f}$ $f \neq e$) production in e^+e^--annihilation is written as,

$$\frac{\mathrm{d}\sigma}{\mathrm{d}\Omega} = \frac{\alpha^2}{4s} N_c \left[R_{ff}\left(1+\cos^2\theta\right) + B_{ff}\cos\theta \right],$$

$$N_c = 1 \text{ for leptons} \quad \text{and} \quad N_c = 3\,(1+r_{QCD}) \text{ for quarks},$$

$$R_{ff} = Q_f^2 - 2Q_f v_e v_f \,\mathrm{Re}(\chi) + (v_e^2 + a_e^2)(v_f^2 + a_f^2)|\chi|^2,$$

$$B_{ff} = -4Q_f a_e a_f \,\mathrm{Re}(\chi) - 2v_e v_f a_e a_f |\chi|^2,$$

$$\chi = \frac{1}{16\,\sin^2\theta_W \cos^2\theta_W} \frac{s}{s - M_Z^2 + iM_Z\Gamma_Z},$$

$$v_f = 2\,(t_{3L} + t_{3R} - 2Q_f \sin^2\theta_W),$$

$$a_f = 2\,(t_{3L} - t_{3R}).$$

The FB asymmetry is defined as,

$$A_{ff} = \frac{F-B}{F+B} = \frac{3}{8}\frac{B_{ff}}{R_{ff}}$$

$$\sim \frac{3}{2}\,a_e a_f \,\mathrm{Re}(\chi) \qquad \text{for leptons at TRISTAN.}$$

In the standard model (hereafter SM for short),

$$t_{3L} = \frac{1}{2} \quad \text{and} \quad Q_f = \frac{2}{3} \qquad \text{for } u,c,d,$$

$$t_{3L} = -\frac{1}{2} \quad \text{and} \quad Q_f = -\frac{1}{3} \qquad \text{for } d,s,b,$$

$$t_{3L} = -\frac{1}{2} \quad \text{and} \quad Q_f = -1 \qquad \text{for } \mu \text{ and } \tau,$$

$$t_{3R} = 0.$$

Note, Z^0 effect in R_{ll} (l = lepton) is ~7%, in R_{qq} ~30%, and r_{QCD} ~3–5% at the TRISTAN energy $\sqrt{s} = 61$ GeV.

Since the value of $\sin^2\theta_W$ is almost equal to 1/4, the vector coupling constants of the leptons (v_μ and v_τ) are small. Because of this, A_{ll} is especially sensitive to the axial vector constants and is expected large at TRISTAN energy.

All three experiments at TRISTAN (AMY, TOPAZ, and VENUS) have measured leptonic pair production cross sections and asymmetries based on about 30 pb^{-1} of data between 50 and 60.8 GeV [5]. The combined results together with those of PEP and PETRA [6] are shown in Figure 2 and 3. The data are in general agreement with

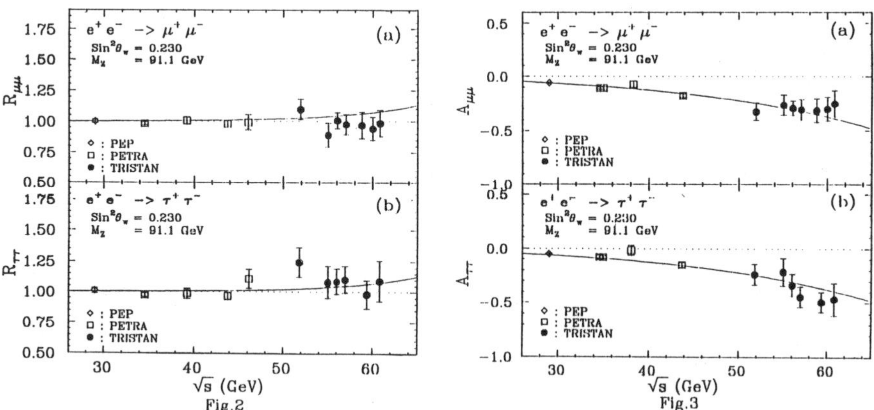

the prediction of SM (solid line). To obtain world average, we cannot combine A_{ll}'s because experimental conditions (cut etc.) are different, but we may combine values of axial coupling constants (a_μ and a_τ) assuming $a_e = -1$ [7],

$$a_\mu = -1.057 \pm 0.051 \quad \text{and} \quad a_\tau = -0.930 \pm 0.07.$$

<u>Dose v_τ exist?</u>　　　The Michel parameter value as determined from energy spectrum of the tau decay ($\tau \to v_\tau e v_e$) is 0.74 ± 0.05 [8]. This means that v_τ is coupled with τ and τ is either left handed or right handed or both. Namely, either t_{3L} or t_{3R} of τ or both of them are $-1/2$. Since a_τ equals $2(t_{3L}-t_{3R})$, it follows that τ is left handed and is in a doublet unless τ_R is paired with a charge $-3/2$ partner. Therefore, v_τ must exist and τ_R is in a singlet.

<u>Is τ a heavy brother of e and μ?</u>　　　To determine the coupling strength of τ, we use the lifetime of τ which is related to that of μ by the relation,

$$\tau_\tau = \left(\frac{G_\mu}{G_\tau}\right)^2 \left(\frac{m_\mu}{m_\tau}\right)^5 \mathrm{BR}(\tau \to v_\tau e v_e)\tau_\mu.$$

Using $\tau_\tau = 303 \pm 8$ fsec, $\mathrm{BR}(\tau \to v_\tau e v_e) = 0.176 \pm 0.004$ [7,9], we obtain $G_\mu/G_\tau = 0.964 \pm 0.018$. From the total cross section, we obtain $R_{\tau\tau} = (\sigma_{\tau\tau}/\sigma_{\mu\mu}) = 1.11 \pm 0.06$ at $\sqrt{s} = 50$–60.8 GeV. We conclude that the τ coupling is the same as μ within 2σ. All things considered we may conclude that properties of τ is identical with its light brothers except the mass.

<u>c- and b-quark asymmetry</u>   The asymmetry of c- and b-quark has not been measured very well because it requires large statistics to obtain an enriched sample of c or b. It is known that D^* is hard. This means that most likely D^* contains the primary c-pair. Since D^* is the main decay channel of the c-quark, the asymmetry of D^* is a good measure of the primary c-quark. D^* can be identified by its characteristic decay mode $D^* \to D^0 + \pi_s$, $D^0 \to K\pi$, $K\pi\pi$, $K\pi\pi\pi$, where π_s essentially follows the motion of D^0, because it is produced with very low Q-value (5.8 MeV). Figure 4 (a) shows an angular distribution of such D^*'s at $\sqrt{s} = 36.2$ GeV [10]. The obtained asymmetry is $A_{cc} = -0.160 \pm 0.072 \pm 0.011$ whereas SM gives -0.157. Similar analysis by JADE [11] gives a value -0.149 ± 0.067 at $\sqrt{s} = 35$ GeV. The fact that the transverse momentum of π_s with respect to D^* direction is very small (max. 49.6 MeV in D^* frame) can also be used to detect the direction of the c-quark. Figure 4 (b) shows an angular distribution of such π_s. Events were chosen from an independent data sample of TASSO at $\sqrt{s} = 35.0$ GeV by making a transverse momentum cut w. r. t. the jet axis ($p_T^2 \leq 0.0075$ (GeV/c)^2). The distribution gives $A_{cc} = -0.168 \pm 0.047 \pm 0.027$ where SM gives -0.145.

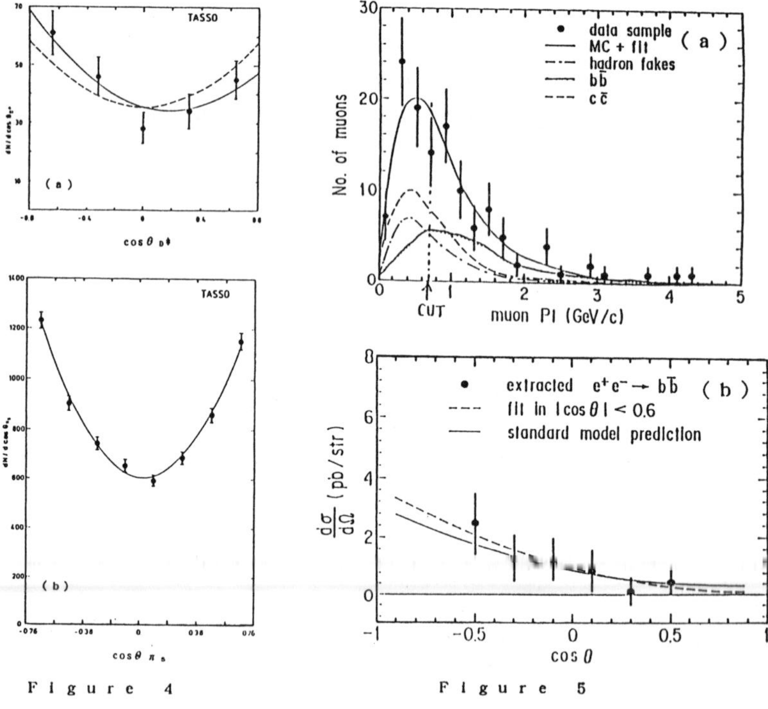

Figure 4   Figure 5

Recently, AMY group measured the asymmetry of b-quark [12]. One hundred and twenty seven hadronic events with an isolated muon ($p_T \geq 0.7$ GeV w. r. t. thrust axis) were chosen from data of integrated luminosity 18.6 pb^{-1} at $\sqrt{s}$ = 52–57 GeV. Charm quark contribution and other backgrounds are estimated by Monte Carlo and subtracted (Figure 5). The result gives $R_{bb} = 0.57 \pm 0.16 \pm 0.10$ and $A_{bb} = -0.72 \pm 0.28$ (stat.) ± 0.13 (syst.) where SM predicts -0.56. The result does not take into account $b\bar{b}$ mixing effect. With the mixing, SM prediction is changed to $-0.2 \sim -0.5$.

The $b\bar{b}$ mixing correction can be applied as follows.

$$A_{bb}(\text{obs}) = (1-2\chi)A_{bb}(\text{nomix}), \quad \chi = \frac{1}{2.3}\chi_d + \frac{0.3}{2.3}\chi_s \quad \text{for } u{:}d{:}s = 1{:}1{:}0.3.$$

With ARGUS and CLEO data (χ_d=0.17±0.05, χ_s=0.25±0.25) [13], the correction factor $1-2\chi$ equals 0.787.

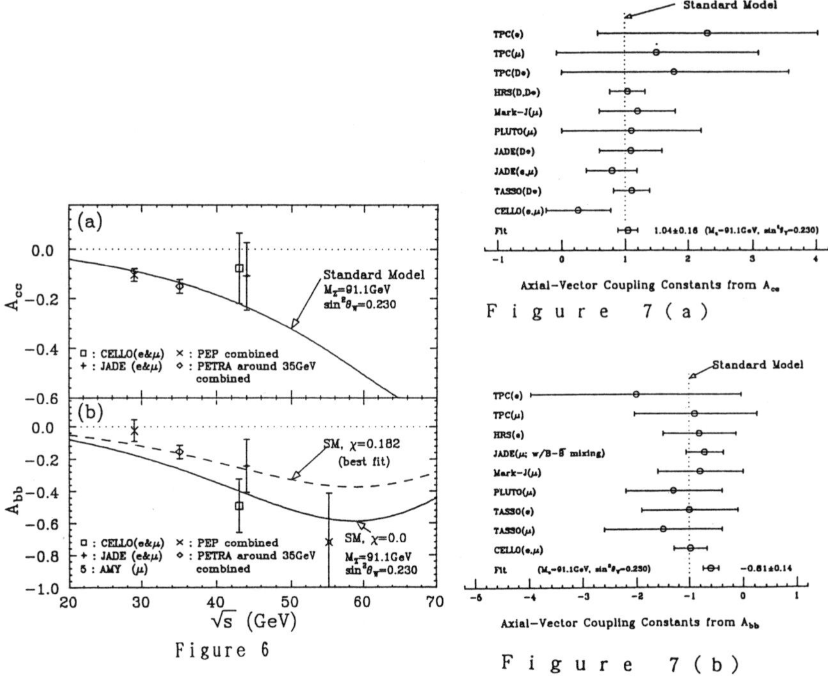

Figure 6

F i g u r e   7 (a)

F i g u r e   7 (b)

<u>Does the top exist?</u>   Figure 6 shows energy dependence of A_{cc} and A_{bb}. The results, in particular, those of A_{bb} are poor, but are in general agreement with SM prediction. Figure 7 shows a compilation of world data on a_c and a_b derived from A_{cc} and A_{bb}

[7,12,14]. The world average is $a_c = 1.04 \pm 0.16$ and $a_b = -0.61 \pm 0.14$ without mixing or -0.78 ± 0.2 with mixing. The conclusion is that both c- and b-quark are in doublets. Therefore the "top quark" must exist.

Without the top, little can be said about the coupling of c and b. From known values of Kobayashi-Maskawa matrix elements [15],

u-coupling: $|V_{ud}|^2 + |V_{us}|^2 + |V_{ub}|^2 = 0.9979 \pm 0.0021$

c-coupling: $|V_{cd}|^2 + |V_{cs}|^2 + |V_{cb}|^2 = 0.91 \pm 0.17$

b-coupling: $|V_{ub}|^2 + |V_{cb}|^2 + |V_{tb}|^2 = ?$

Although there is no proof that the coupling of b is the same as u and c, so far we have seen all of the data (R_{ff}, A_{ff}) are consistent with SM predictions.

2.3 Total Hadronic Cross section, R_{had}

The relative production cross section of all the hadrons is obtained by summing R_{ff} over all quarks (u, d, s, c, and b below $\sqrt{s} = 60$ GeV) and is expressed as,

$$R_{had} = 3 \sum Q_q^2 \left(1 + (Z^0\text{-pole term})\right)\left(1 + r_{QCD}\right)$$
$$= \frac{N_{obs} - N_{bg}}{\varepsilon (1+\delta) L \, \sigma_{\mu\mu},}$$

where N_{obs} (N_{bg}) is a number of observed (background) events, ε, detection efficiency, δ, radiative correction, L, integrated luminosity, and $\sigma_{\mu\mu}$ is μ-pair production cross section via one-photon exchange.

Table 1

	AMY	TOPAZ	VENUS
ε	0.63	0.75	0.73
L [pb^{-1}]	33.7	28.9	29.2
source	systematic error [%]		
L	2.5	< 4.0	2.6
$1+\delta$	1.3	1.9	2.1
ε	1.3	1.0	1.8
bkg, cuts	1.7	2.3	1.7
total	3.3	3.3	4.2

From May 1987 until July 1989, the energy of TRISTAN gradually went up from $\sqrt{s} = 50$ to 61.4 GeV, during which period each experimental group accumulated total of ~ 30 pb^{-1}. Table 1 summarizes efficiency, integrated luminosity, and systematic errors of each group. Figure 8 shows combined data [16, 17] together with those of PEP and PETRA.

Determination of basic parameters of the standard model   At TRISTAN energy the electro-weak effect amounts to 30 % and QCD correction to about 5 %. In extracting

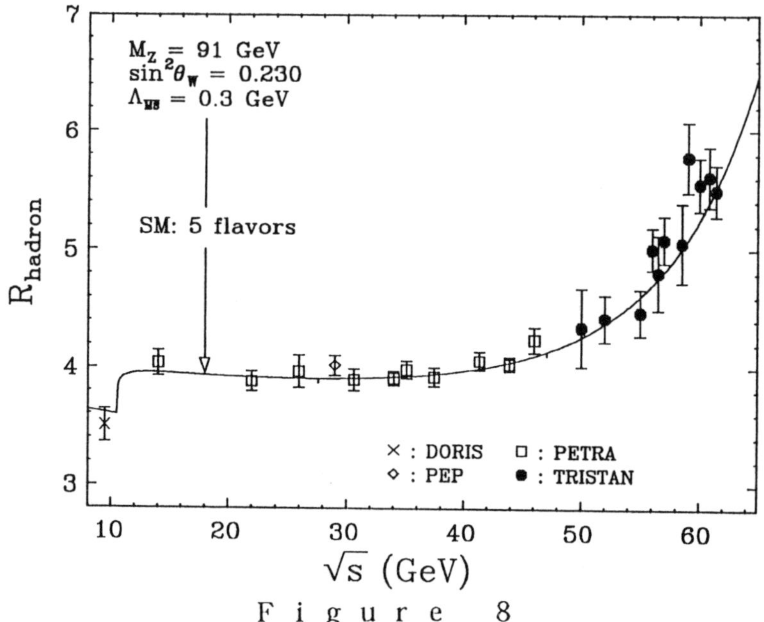

F i g u r e   8

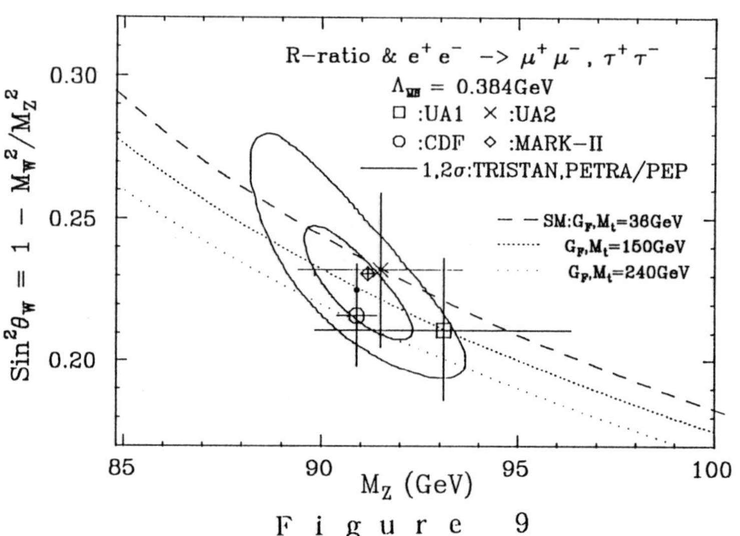

F i g u r e   9

the Born term contribution, we have used a full E-W correction up to $O(\alpha^3)$ [18] and $O(\alpha_s^3)$ QCD correction [19]. Basic parameters which enter the calculation are $\sin^2\theta_W$, M_Z, $\Lambda_{\overline{MS}}$, top mass M_{top}, and Higgs mass M_H. M_{top} and M_H enter through higher order radiative correction. The effect of top mass on δ is from ~ 0 to -1.3 % for $M_{top} =$ 50 $\sim$ 200 GeV while that of Higgs is from -0.2 to +0.4 % for $M_H = 10 \sim 1000$ GeV. As the effects are small, we may fix the values of M_{top} and M_H at 40 and 100 GeV respectively and include the effect of mass variation into systematic error (fit 1). The best fit gives the results [17],

$$
\begin{aligned}
M_Z &= 89.2 \; {}^{+2.1}_{-1.8} \; \text{GeV,} \\
\sin^2\theta_W &= 0.233 \; {}^{+0.035}_{-0.025}, \\
\Lambda_{\overline{MS}} &= 0.327 \; {}^{+0.275}_{-0.206} \; \text{GeV.}
\end{aligned}
$$

Note that the value of M_Z is somewhat lower than those obtained by Mark-II, CDF or LEP [20, 2] because of 1–2σ higher R_{had} values. If we use lepton data R_{ll} and A_{ll} together with R_{had} from TRISTAN, PEP, and PETRA, the best values [21] are (Figure 9),

$$
\begin{aligned}
M_Z &= 90.9 \; {}^{+1.6}_{-1.8} \; \text{GeV,} \\
\sin^2\theta_W &= 0.225 \; {}^{+0.021}_{-0.021}, \\
\Lambda_{\overline{MS}} &= 0.384 \; {}^{+0.285}_{-0.235} \; \text{GeV.}
\end{aligned}
$$

We may conclude that the results are consistent with SM. Note, the conclusion, is solely based on e^+e^- reaction and is independent of other reactions.

An alternative way is to fix the value of M_Z at 91.1 GeV as determined from Z^0 decay and calculate $\sin^2\theta_W$ using the definition,

$$
\sin^2\theta_W = \frac{1}{2}\left(1 - \sqrt{1 - \frac{4\pi\alpha}{\sqrt{2}G_F\,(1-\Delta_r)\,M_Z^2}}\right),
$$

where Δ_r is the weak radiative correction, and calculate $\Lambda_{\overline{MS}}$, M_{top}, and M_H (fit 2). Using Hioki's radiative calculation [22], TOPAZ obtained [17],

$$
\begin{aligned}
M_{top} &= 157 \pm 115 \text{ GeV} \quad \text{or} \quad < 366 \text{ GeV with 90 \% C. L.,} \\
\Lambda_{\overline{MS}} &= 0.417 \; {}^{+0.261}_{-0.224} \text{ GeV.}
\end{aligned}
$$

Here, M_H is fixed at 100 GeV since its logarithmic dependence gives very small change to the result. The upper bound on the top mass is similar to what had already been obtained from the neutrino deep inelastic data [23], and serves as an independent check.

2.4 Physics Beyond the Standard Model

So far we have only looked for consistency with SM. Now we may reverse the attitude, take the data seriously and see if there is a deviation from SM. R_{ll} data from PEP and PETRA are known to be about 2σ below SM prediction whereas R_{had} of AMY and TOPAZ group at TRISTAN may be 1–2σ high. We can interpret these small deviation as the evidence for the existence of an extra Z' and try to derive its mass value. The group structure is extended from $SU(2)\times U(1)_Y$ to $SU(2)\times U(1)_Y\times U(1)_{Y'}$. The corresponding interaction Lagrangian becomes,

$$-\pounds_{NC} = eJ_{EM}{}^\mu A_\mu + g_1 J_1{}^\mu Z_{1\mu} + g_2 J_2{}^\mu Z_{2\mu}.$$

The known Z^0 and the new extra Z' are mixed states of Z_1 and Z_2 with mixing angle θ_E. The mixing angle and mass values M_1, M_2 of Z^0, Z' are related by the equation $\tan\theta_E = (M_0{}^2-M_1{}^2)/(M_2{}^2-M_0{}^2)$, where M_0 is $M / \cos\theta_W$. Addition of Z' changes the values of vector coupling constants v_l and v_d. The effect is to lower R_{ll} and increase R_{had} as desired. We may consider two cases. In case 1, we make a simplest assumption, $i.\,e.$, put $\theta_E = 0$ and $g_2 = \lambda\, g_{EW}$.

Figure 10 (a) shows excluded region in λ-$M_{Z'}$ plane obtained from VENUS data [24] with previous results obtained by UA1 and UA2 [25]. VENUS result expands the excluded region considerably. A similar result has been obtained by CDF, too [26]. In case two, we rely on more model-dependent analysis. Hagiwara $et\,al.$ [27] used super string inspired E_6 model. In this case, the extra Z' is a combination of two neutral bosons, Z_ψ and Z_χ with mixing angle β. For β value of $\tan^{-1}\sqrt{3/5}$, it is given name of Z_η. In obtaining the allowed value of M_2 vs β, all the data from PEP, PETRA, and TRISTAN together with most recent mass values of Z^0 and W were used. Figure 10 (b) shows the original fitting by Hagiwara $et\,al.$ (dashed line), and update with recent VENUS data [28] (dotted line). Also shown is the improvement if we include v_μ-e data from CHARM [29]. The mass value is also constrained from above. However, for the mass value around and beyond 800 GeV, the model with the extra Z' is virtually indistinguishable from SM at TRISTAN energy. The evidence for the existence of extra Z', therefore, is not strong. The improvement of the fit obtained by inclusion of the Z' is shown in Figure 11.

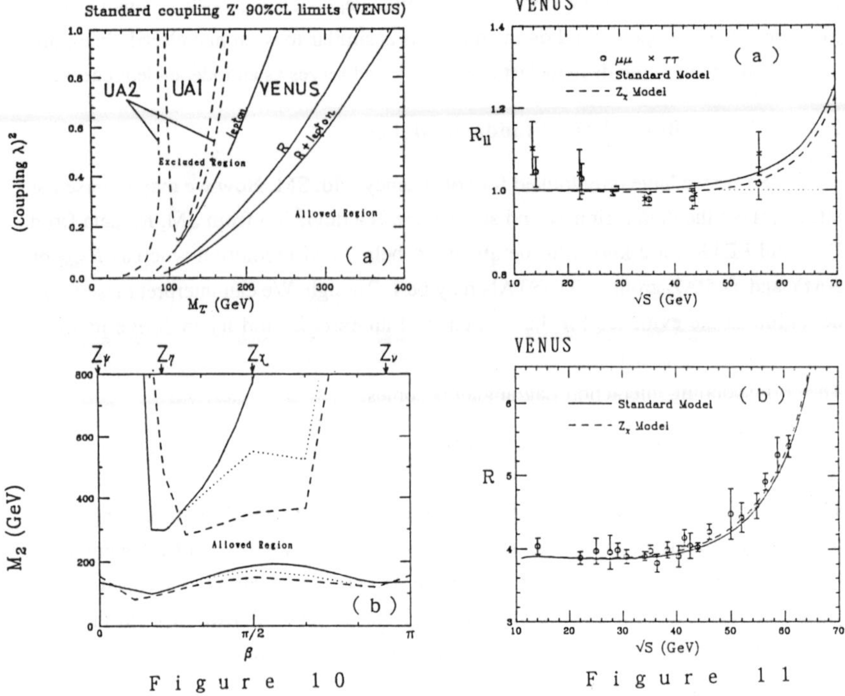

F i g u r e   1 0   F i g u r e   1 1

3.   STUDY OF QCD/JETS

3.1   Introduction

Interactions of quarks and their gauge field, gluon, are described by Quantum Chromo Dynamics, an exact gauge theory based on color SU(3). Because of its asymptotic behavior, perturbative method is applicable to treat partons (quarks and gluons) which are generated by high energy e^+e^- reactions. Generally two approaches are used to compare theory with experiment. In the parton shower approach (PS scheme, Figure 12 (a)), one generates gluon vertices in cascade according to Altarelli-Parisi formula, until parton virtuality Q^2 reaches its cutoff value which is usually set at ~ 1 $(GeV/c)^2$. In the analytic approach (ME scheme, Figure 12 (b)), one calculates exact matrix elements up to n-th order of QCD coupling constant α_S. In both approaches, however, partons have to be converted to hadrons to compare with the experiment. The hadronization is a non-perturbative process and its calculation relies heavily on models (Figure 12 (c)). ME scheme has an advantage of being exact up to $O(\alpha_S{}^n)$, but the number of final partons is limited to four at present ($n = 2$). In addition, the virtuality when the hadronization commences is usually very high ($\sim 0.04\sqrt{s}$) and makes the final result highly dependent on the model. PS scheme has advantages in that it can generate any number of partons and that the virtuality cutoff is low enough so that it depends less on hadronization models. However, the vertex treatment is approximate (LLA or NLLA) and may not describe certain hard processes as well as ME scheme.

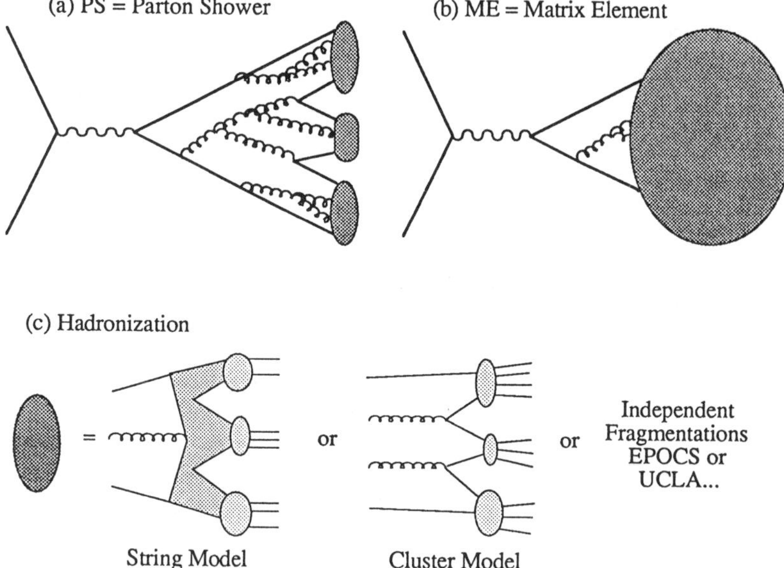

Figure 12

Among the hadronization models [30], the string model [31] has acquired the most popularity, although independent fragmentation models [32] have also been used extensively in the past. In most of the following treatments, data are treated either by ME+string or PS+string scheme.

One conspicuous feature of QCD is running nature of the coupling constant, namely $\alpha_s = \alpha_s(\mu)$ where μ is the renormalization scale usually taken to be $\sqrt{s}$ in the e^+e^- reaction. In the so-called modified minimal subtraction ($\overline{MS}$) scheme, an explicit formula for α_s up to the next-to-leading-log-approximation (NLLA) is given by [8],

$$\alpha_s(\mu) = \frac{4\pi}{\beta_0 L}\left(1 - \frac{\beta_1}{\beta_0^2}\frac{\ln L}{L}\right),$$

where, $L = \ln(\mu/\Lambda_{\overline{MS}})^2$, $\beta_0 = 11-N_f$, $\beta_1 = 102-(38/3)N_f$, and N_f is number of quark flavors which is 5 at PETRA and TRISTAN energy ($\Lambda_{\overline{MS}} = \Lambda^{(5)}_{\overline{MS}}$). Note there is one to one correspondence between α_s and $\Lambda_{\overline{MS}}$, but in the leading-log-approximation (LLA) which uses only the first term, $\Lambda_{\overline{MS}}$ cannot be determined from α_s or vice versa unless one specifies the renormalization scale μ.

3.2 Jet Production

<u>Running coupling constant</u>   Fractional n-parton production rate $R_n = (\sigma_{n\text{-}parton}/\sigma_{tot})$ for resolvable partons with relative minimum jet mass cutoff y_{min} ($m_{ij}^2/s \geq y_{min}$) is given by QCD as,

$$R_2 = 1 + C_{21}(y_{min})\,\alpha_s(s) + C_{22}(y_{min})\,\alpha_s(s)^2,$$
$$R_3 = C_{31}(y_{min})\,\alpha_s(s) + C_{32}(y_{min})\,\alpha_s(s)^2,$$
$$R_4 = C_{42}(y_{min})\,\alpha_s(s)^2.$$

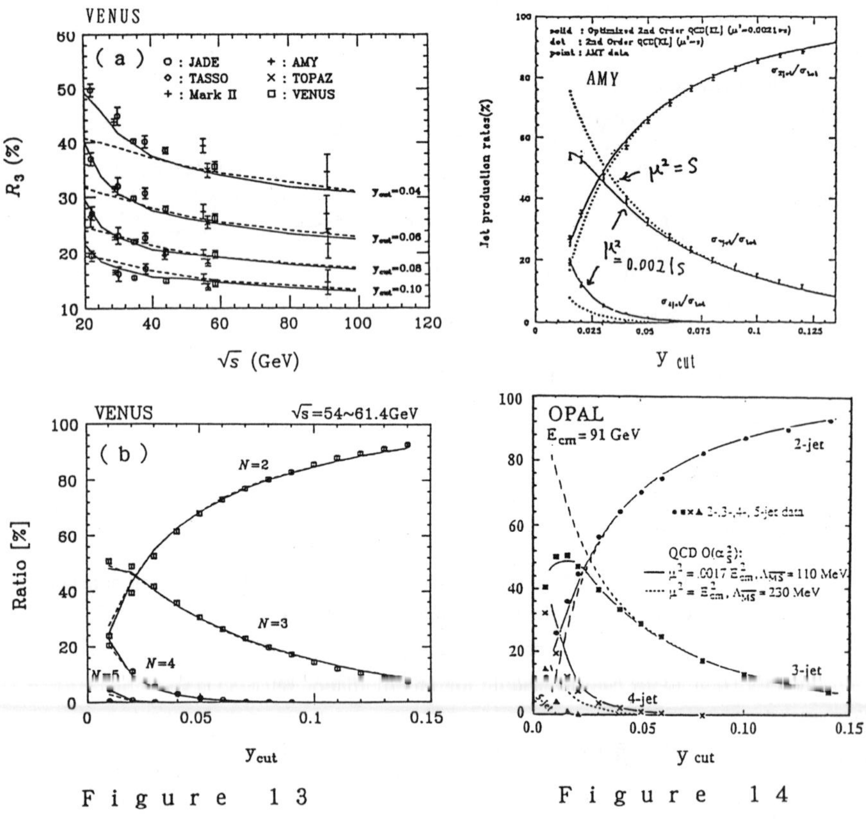

F i g u r e   1 3   F i g u r e   1 4

Experimentally, jet clustering is performed a la JADE [33]. For a hadron pair i, j relative mass is defined as $y_{ij} = 2E_iE_j (1-\cos\theta_{ij})/E_{vis}^2$. If minimum($y_{ij}$) is less than y_{cut}, combine i and j to form a pseudo-particle k, $p_k = p_i + p_j$ and repeat the procedure until all $y_{ij} \geq y_{cut}$. Monte Carlo shows that $R_{n\text{-}jet} \sim R_{n\text{-}parton}$ independent of models, if $y_{min} = y_{cut} > 0.04$ and $\sqrt{s} > 25$ GeV. As can be seen, s-dependence is in $\alpha_s(s)$, therefore $\alpha_s(s)$ runs if R_3 runs. Figure 13 (a) shows R_3 data plotted against s [33, 34, 35, 36]. Shown together are PS calculation (NLLA only (dotted line) and NLLA+EPOCS (solid line)) [35]. Running nature of α_s is excluded with significance of 4.6σ below TRISTAN energy and 5.7σ with Z^0 data [36].

<u>Determination of α_s or $\Lambda_{\overline{MS}}$</u>   To determine α_s or $\Lambda_{\overline{MS}}$ to second order, we need observables known to $O(\alpha_s^2)$. Usually, however, there is a big discrepancy between $O(\alpha_s^2)$ matrix element and hadron level observables and hence α_s value thus determined is model dependent. PS treatment reproduces observed R_n (Figure 13 (b) [36]) and is less model-dependent. However, the most popular PS calculation LUND 6.3 [31] uses only LLA hence $\Lambda_{\overline{MS}}$ is undetermined. In order to determine $\Lambda_{\overline{MS}}$ unambiguously, NLLA has to be used. VENUS group used a PS program which utilizes NLLA [37] combined with EPOCS hadronization model and obtained,

$$\Lambda_{\overline{MS}} = 254\ ^{+55}_{-47}\ \text{(stat.)} \pm 56\ \text{(syst.)}\ \text{MeV.}$$

Using α_s-$\Lambda_{\overline{MS}}$ relation, this corresponds to,

$$\alpha_s = 0.127 \pm 0.007 \quad \text{at } \sqrt{s} = 58.5\ \text{GeV.}$$

<u>Renormalization scale dependence</u>   Recently it is argued that proper choice of renormalization scale improves fit with observed data [38]. Theoretically, if added to all orders, any observable should not depend on the choice of renormalization scale or x ($\mu^2 = xs$). In finite order, choice of μ^2 affects $\Lambda_{\overline{MS}}$.

Using $O(\alpha_s^2)$ matrix element combined with hadronization model, AMY [39] adjusted $x = \mu^2/s$ to obtain a best fit with R_3 (Figure 14). The resultant $x = 0.0021$, $\Lambda_{\overline{MS}} = 173 \pm 60$ MeV are similar to $x = 0.0017$, $\Lambda_{\overline{MS}} = 95$ MeV obtained with Mark-II data [38]. Note, here xs is smaller than 5 quarks threshold hence $\Lambda_{\overline{MS}} = \Lambda^{(4)}_{\overline{MS}}$.

The procedure is observable dependent. If higher order correction is large, there is not much point of deciding proper scale. As an example, examine QCD correction to R_{had} which is the only observable known to $O(\alpha_s^3)$ [40].

$$r_{QCD} = \quad r_1\left(\frac{\alpha_s}{\pi}\right) \quad\quad + r_2\left(\frac{\alpha_s}{\pi}\right)^2 \quad + r_3\left(\frac{\alpha_s}{\pi}\right)^3 + ...,$$

$$\sim 0.04 \quad\quad\quad \sim 0.002 \quad\quad \sim 0.004$$

for $\alpha_s = 0.12$ with $r_1 = 1$, $r_2 = 1.411$, and $r_3 = 64.804$.

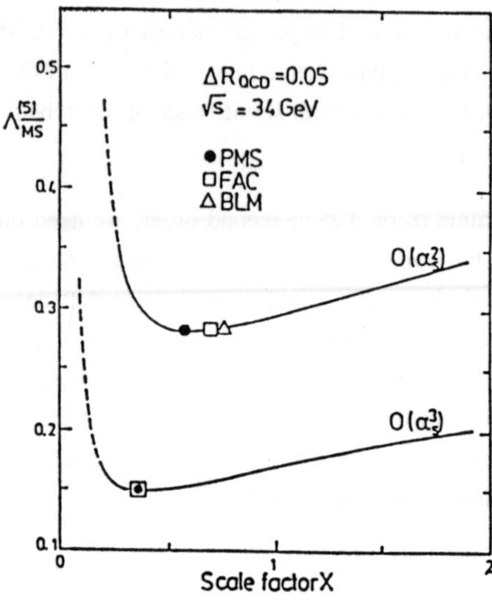

F i g u r e   1 5

Change of the scale $\mu^2 \to x\mu^2$ modifies $r_1' = r_1$, $r_2' = r_2 + r_1(\beta_0/2)\ln x$ etc.. In this example, $O(\alpha_s^3)$ term is larger than $O(\alpha_s^2)$ term. The corresponding change in $\Lambda_{\overline{MS}}$ is illustrated in Figure 15. $O(\alpha_s^3)$ $\Lambda_{\overline{MS}}$ is far from $O(\alpha_s^2)$ $\Lambda_{\overline{MS}}$ of any theoretically suggested choices (PMS, point of minimum sensitivity [41], FAC, fast apparent convergence [42], or BLM [43]).

The fact remains that small x does reproduce R_4 as well as other R_n's at small y_{cut} value if ME is used. This is also true at Z^0 energy (Figure 14 (b) [36]) Apparently, for small value of jet mass, small value of μ^2 has to be used. In PS treatment, this is built in as illustrated in Figure 13 (b). In summary, NLLA-PS model not only determines $\Lambda_{\overline{MS}}$ unambiguously but also gives best overall description of jet data.

3.3 More on $\Lambda_{\overline{MS}}$

<u>Asymmetry of energy-energy correlation (AEEC)</u>   Difference of $\Lambda_{\overline{MS}}$ by different treatment of QCD (ME and PS) may be avoided by using AEEC.

$$EEC(\chi) = \frac{1}{N_{ev}} \sum_{ev} \sum_i \sum_j \frac{E_i E_j}{E_{CM}^2} \delta(\chi - \chi_{ij}),$$

$$AEEC(\chi) = EEC(\pi - \chi) - EEC(\chi).$$

AEEC enhances non-collinear $q\bar{q}g$ events. Many fragmentation effects in collinear events cancel out. For AEEC, hadronization effect is known to be small and $O(\alpha_s^2)$ ME gives a good approximation. TOPAZ group carried out AEEC analysis (Figure 16) and obtained [44],

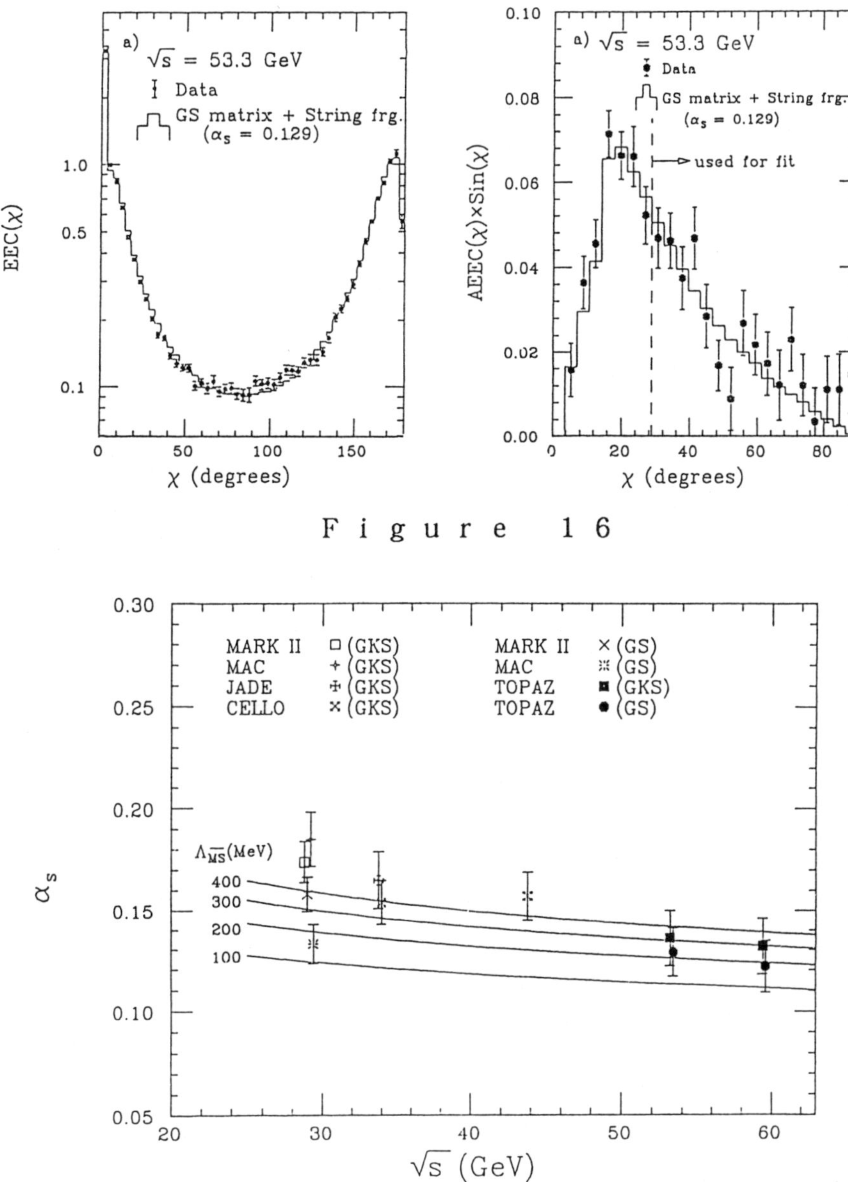

Figure 16

Figure 17

60

$$\alpha_s = 0.129 \pm 0.007 \pm 0.010 \qquad \text{at } \sqrt{s}=53.3 \text{ GeV},$$
$$\alpha_s = 0.122 \pm 0.008 \pm 0.010 \qquad \text{at } \sqrt{s}=59.5 \text{ GeV}.$$

Here GS [45] matrix element is used. Figure 17 shows s dependence of α_s as determined from AEEC. To compare with other data GKS [46] calculation are also shown.

Table 2 summarizes recent $\Lambda_{\overline{MS}}$ values obtained from various observables. Table 3 shows scale dependence of $\Lambda_{\overline{MS}}$, and table 4 compares $\Lambda_{\overline{MS}}$ of different order of α_s.

<u>Model independent limits on $\Lambda_{\overline{MS}}$ [47]</u>   Observables $V(s)$ may be expressed as,

$$V(s)=C_1\alpha_s + C_2\alpha_s^2 + \ldots\ldots + H(s).$$

The first two terms represent perturbative QCD calculation and $H(s)$ describes hadronization effect. Now choose observables such that $H(s)$ is always positive or negative regardless of models. Then put $H = 0$ and solve the equation. This gives model independent limit of $\Lambda_{\overline{MS}}$. The observables chosen to give lower limit of $\Lambda_{\overline{MS}}$ is partially integrated AEEC ($H < 0$) from 30 to 90 degrees and average value of the heavier jet mass of two jet production ($H > 0$) normalized by s. Models used are Ali IF (independent fragmentation model), Hoyer IF, LUND string model and LUND PS model. The obtained limits are,

TABLE 2

Summary of QCD Parameters

Exp.	$\sqrt{s}$ (GeV)	QCD M.E.	Fragmentation Model	Observable	$\alpha_s\,(\mu^2{=}s)$	$\Lambda_{\overline{MS}}$ (MeV)
Mark–II	29.0	GS	---	R_3	0.144	210 ±13
MAC	29.0	GS	Lund/ST	$\langle E_T^{in}\rangle$	0.112 ±0.008 ±0.007	
	29.0	GS	Ali/IF	$\langle E_T^{in}\rangle$	0.133 ±0.005 ±0.009	152 ±34
JADE	22 – 44	GS	---	R_3		210 ±13
	22 – 44	KL	---	R_3		205 ±13
	44	KL	---	$\dfrac{R_3(\sqrt{s_1})}{R_3(\sqrt{s_2})}$	0.154 ±0.038	$450\,^{+700}_{-370}$
TASSO	43.5	ERT	Lund/ST	$\langle M_h^2/s\rangle$	0.158 ±0.002 ±0.009	500 ±24 ±135
	43.5	ERT	Ali/IF	$\langle M_h^2/s\rangle$	0.134 ±0.002 ±0.007	210 ±14 ±55
	43.5	ERT	Lund/ST	$\langle (M_h^2-M_l^2)/s\rangle$	0.147 ±0.003 ±0.009	340 ±32 ±100
	43.5	ERT	Ali/IF	$\langle (M_h^2-M_l^2)/s\rangle$	0.120 ±0.002 ±0.007	100 ±12 ±30
	43.5	ERT	---	$\int dx_h \dfrac{d\sigma}{dx_h}$ Thrust $(x_h{=}p_h/\sqrt{s})$	0.129 ±0.006	170 ±47
TOPAZ	53.5	GS	Lund/ST	AEEC	0.129 ±0.007 ±0.008	$209\,^{+104}_{-78}$
	59.5	GS	Lund/ST	AEEC	0.122 ±0.008 ±0.010	
VENUS	58.5	PS (NLL)	EPOCS	R_3	$0.129 \pm 0.004\,^{+0.004}_{-0.005}$	$254\,^{+55}_{-47} \pm 56$
OPAL	91	KL	---	R_3	0.124 ±0.008	200 – 450

TABLE 3

Scale ($x=\mu^2/s$) Dependent Analysis of Rn

Exp.	$\sqrt{s}$ (GeV)	QCD M.E.	x	$\Lambda^{(5)}$ (MeV)
Mark–II (Bethke)	29.0	KL	0.0017 (0.001 ~ 0.02)	95 ±30
AMY	55	KL	$0.0021\,^{+0.0015}_{-0.0006}$	109 ($\Lambda^{(4)}$=173 ±22 ±55)
OPAL	91	KL	0.001 – 0.003	80 – 180

TBALE 4

Determination of $\Lambda_{\overline{MS}}$ from R

Author / Exp.	$\sqrt{s}$ (GeV)	$\Lambda_{\overline{MS}}$ (MeV) $O(\alpha_s^2)$	$\Lambda_{\overline{MS}}$ (MeV) $O(\alpha_s^3)$
[illegible] PEP,PETRA,TOPAZ	[illegible]	[illegible]	[illegible]
(W. de Boer) PEP,PETRA,TRISTAN	14 – 61	$650\,^{+450}_{-340}$	$370\,^{+230}_{-190}$
CESR,DORIS,PEP PETRA,TRISTAN	7 – 61	$460\,^{+290}_{-230}$	$260\,^{+160}_{-130}$

$$79 < \Lambda_{\overline{MS}} < 628 \text{ MeV} \qquad \text{at 90 \% confidence level.}$$

3.4 Gluon Self Coupling

The non-Abelian nature of QCD is best illustrated by self coupling of the gluon. Four jet production reaction contains contribution from a graph of two gluon emission by a gluon (Figure 18 (a)). Other contributions are $q\bar{q}$ production by a gluon (Figure 18 (b)) and two independent gluon emission by primary $q\bar{q}$ (Figure 18 (c)).

AMY found by Monte Carlo calculation [48], that contribution from (a), (b), (c) are (66) 0, (4) 51, (30) 44 % respectively for (non-) Abelian case. Unlike graph (b), graph (a) contains infrared singularity and the relative contribution of (a) and (b) are different for the two cases. Two distributions which are sensitive to the non-Abelian coupling are considered. The variables are θ_{BZ} and θ_{NR} defined as,

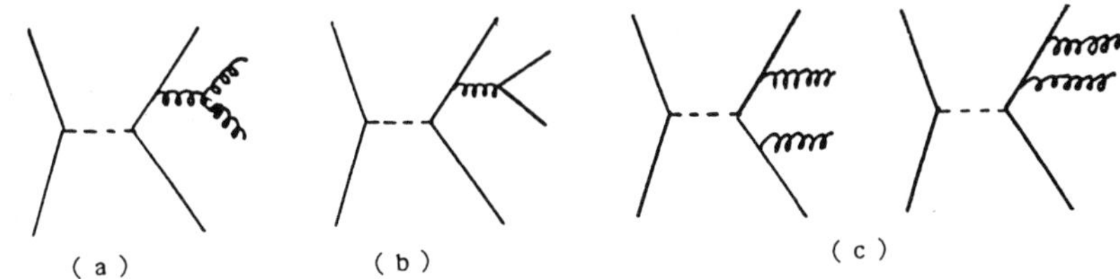

Figure 18

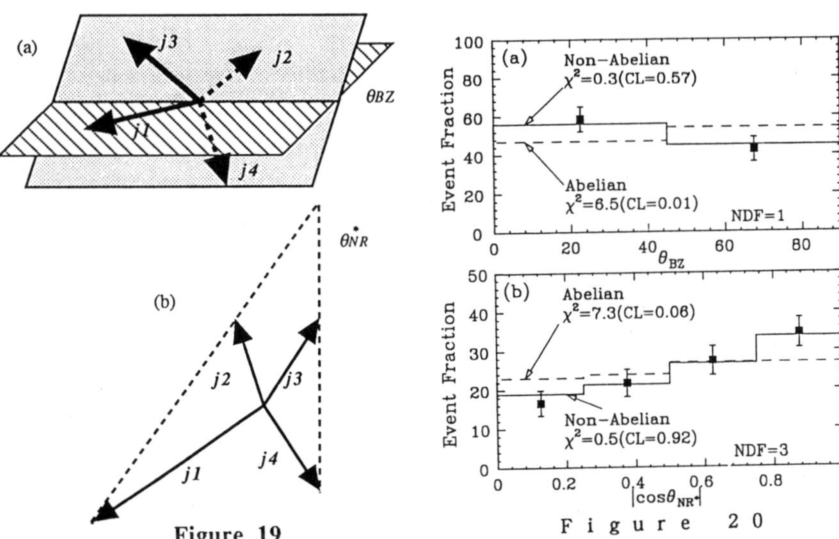

Figure 19   F i g u r e   2 0

$$\cos\theta_{BZ} = \frac{(P_1\times P_2)\cdot(P_3\times P_4)}{|P_1\times P_2|\,|P_3\times P_4|},$$

$$\cos\theta_{NR} = \frac{(P_1-P_2)\cdot(P_3-P_4)}{|P_1-P_2|\,|P_3-P_4|},$$
(Figure 19).

where, P_i's are momentum vectors of four jets ordered by their energy $E_1 > E_2 > E_3 > E_4$. The distributions are shown in Figure 20. Though the difference is small, it is statistically significant and confirms the self coupling nature of the gluon.

3.5 Model consideration

<u>Evidence of soft and collinear gluon emission</u>

TPC/2γ group picked up low sphericity (2-jet like) events and looked if soft and collinear pions accompany them [49]. Since soft gluons are dragged by their parent partons, they tend to produce a dip

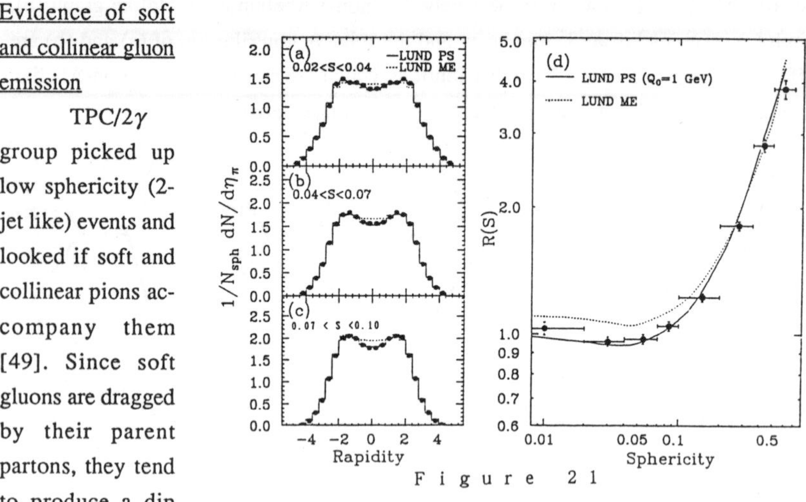

Figure 21

around zero in their rapidity distribution. Their particle identification capability was instrumental because mixture of heavy hadrons obscures it. Figure 21 (a)–(c) show pion rapidity distributions for different sphericity ranges and (d) shows relative yield of the pion in different rapidity regions with low sphericity. They are compared to ME+SF (string fragmentation) (dotted line) and PS+SF model with virtuality cut at ~1 (GeV/c)^2 (solid line). Clearly the data favors the latter. If the virtuality cut is set at ~4 (GeV/c)^2, the latter gives a similar result as the former.

<u>Strange and Charmed baryon production</u>   If high statistics data and particle spectra are available, fine tuning of the hadronization model is possible. As an example, we consider recent PEP4/9 result [50]. A number of strange baryon and a charmed baryon Λ_c^+ are observed and their fractional yields per event are determined (Figure 22). Models compared are two string models (Lund and UCLA [51]) and a cluster model (Webber [52]). Lund is the most popular of hadronization models, is known to

Previously Observed Signals (numerical results essentially final)

New Signals (numerical results preliminary)

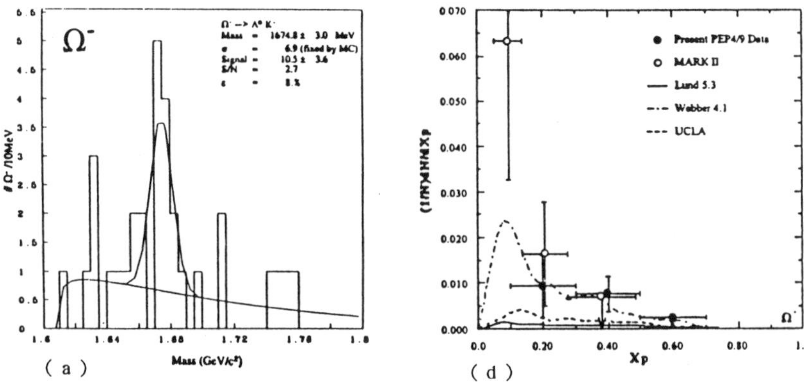

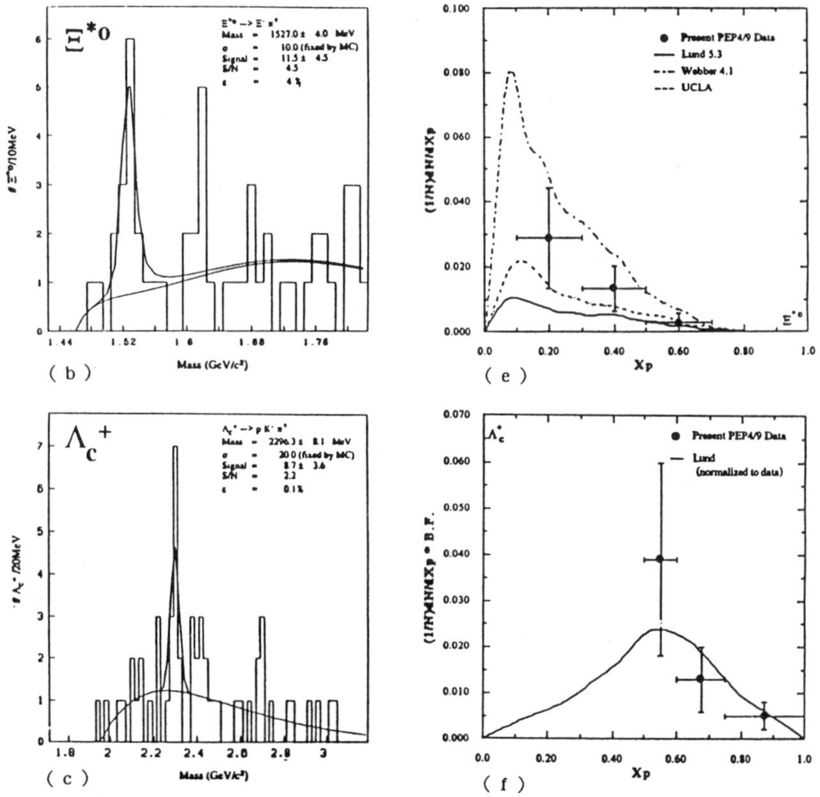

Figure 22 (a)-(c) Mass spectrum of Ω^-, Ξ^{*0}, and Λ_c^+, (d)-(f) comparison of particle spectra with models, Lund 5.3(solid line), Webber 4.1(dot dash line) and UCLA(dashed line).

Table 5

baryon	results	model		
	N/event	Lund 5.3	Webber4.1	UCLA
A^0	0.211±0.017	0.228	0.236	0.210
Ξ^-	0.020±0.005	0.015	0.036	0.016
Ω^-	0.0043±0.0026	0.0005	0.0053	0.0014
Ξ^{*0}	0.0091±0.0055	0.0038	0.0216	0.0070
Λ_c^+ [1]	0.14±0.08 [2]	0.055	0.042	0.030

[1] Observed in $pK^-\pi^+$

[2] $BR(\Lambda_c^+ \to pK^{-1}p^+) \sim 0.06$ is assumed. (0.3±0.2 if BR = 0.26 as in P. D. T.) Observed quantity is $n^*BR = 0.008 \pm 0.004$ (stat.) ± 0.002 (syst.).

reproduce well a variety of data but has many parameters to tune. With good physics insight one may hope to reduce the number of parameters as is the case with the other two models. UCLA model retains basic features of the string model but reduces the number of parameters considerably by taking account of transverse mass in longitudinal phase space. Webber model has the least number of parameters. Figure 22 (d)–(f) shows comparisons of the PEP4/9 scale variable x_p and Table 5 summarizes the result for the integrated value.

Observations are [53]:

1) String models (Lund, UCLA) were previously jeopardized by Ω multiplicity (models ~2.2σ low) but with new data, this discrepancy is resolved (models ~ 1.3σ low) and have now no substantial disagreement with light and starnge meson and baryon production rates at E_{cm}~30 GeV.

2) There are no distinction between Lund and UCLA.

3) Webber model is OK with Ω^- but continues to be ~2–4σ high on Λ, Ξ^-, Ξ^{*0}.

4) Λ_c production rate may possibly jeopardize the model but it depends on determining branch fraction $\Lambda_c \to pK^-\pi^+$.

A lot more data is necessary to distinguish the models.

4. Summary

In the realm of the electro-weak interaction, there are no evidences which contradict the standard model predictions. The number of neutrino is determined to be three and all properties of the quarks and leptons are as they should be. The top and tau neutrino must exist. The value of M_Z, determined by low energy e^+e^- reaction data agrees with that obtained at SLC and LEP. The mass of extra gauge boson Z', if it exists is larger than 400 GeV/c^2.

All jets behave more or less as expected from non-Abelian QCD. α_s runs. Self coupling nature of the gluon is observed. Rescaling of renormalization scale μ^2 remedies soft part of the jet behavior in second order matrix element treatment. However, higher order QCD calculation is needed for its significance to be understood. Parton shower coupled with string fragmentation model reproduces best overall behavior of jets.

The author wishes to thank Dr. K. Fujii and T. Tauchi for allowing him to use their compilation and fit of the data.

REFERENCES

[1] P. Langacker, Comments Nucl. Part. 19 (1989) 1.

[2] OPAL, Phys. Lett. B235 (1990) 389,
ALEPH, Phys. Lett. B234 (1990) 399, B235 (1990) 399,
L3, Phys. Lett. B237 (1990) 136.

[3] A. Maki, Talk, Int. Symp. on Lepton and Photon Int. at High Energies, Aug. 7-12, 1989, SLAC, Stanford, Calif., USA..

[4] Mark-II, Phys. Rev. Lett. 63 (1989) 2447,
OPAL, Phys. Lett. B236 (1990) 364,
ALEPH, Phys. Lett. B236 (1990) 511,
CDF, Phys. Rev. Lett. 64 (1990) 142, 148 and 152.

[5] AMY, Phys. Lett. B218 (1989) 112,
TOPAZ, Phys. Lett. B208 (1989) 319,
VENUS, to be published in Z. Phys. C.

[6] HRS, Phys. Rev. D40 (1989) 902,
CELLO, Phys. Lett. B222 (1989) 163,
TASSO, DESY 89-035,
AMY, Phys. Lett. B218 (1989) 112.

[7] K. Fujii, KEK 89-182, Talk, Int. EPS Conf., Madrid, Sept. 6-13, 1989.

[8] Particle Data Group, Phys. Lett. B204 (1988).

[9] R. Felst, Proc. Phys. in Collision, Tsukuba, 1987, p87.

[10] TASSO, Z. Phys. C44 (1989) 365.

[11] JADE, Z. Phys. C44 (1989) 567

[12] AMY, Phys. Rev. Lett. 63 (1989) 2341.

[13] ARGUS, Phys. Lett. B192 (1987) 245,
CLEO, Phys. Rev. Lett. 62 (1989) 2233.

[14] CELLO, DESY 89-125,
MAC, Phys. Lett. B218 (1989) 369,
TASSO, Z. Phys. C42 (1989) 17.

[15] P. Langacker and D. London, Phys. Rev. D38 (1988) 886.

[16] AMY, Phys. Lett. B218 (1989) 499,
VENUS, Phys. Lett. B234 (1990) 382.

[17] TOPAZ, Phys. Lett. B234(1990) 525.

[18] J. Fujimoto and Y. Shimizu, Mod. Phys. Lett. A3 (1988) 583.

[19] S. G. Gorishny, A. L. Kataev and S. A. Larin, Phys. Lett. B212 (1988) 238.

[20] Mark-II, Phys. Rev. Lett. 63 (1989) 724,
CDF, Phys. Rev. Lett. 63 (1989) 720.

[21] T. Tauchi, KEK 88-39 and private communication.

[22] Z. Hioki, Prog. Theor. Phys. 68 (1982) 2134.

[23] U. Amaldi et al., Phys. Rev. D36 (1987) 1385.

[24] VENUS, in preparation.

[25] UA1, Z. Phys. C44 (1989) 15,
UA2, Phys. Lett. B195 (1987) 613,

[26] CDF, S. Geer, Talk at Erice, Italy, July, 1989.

[27] K. Hagiwara, R. Najima, M. Sakuda and N. Terunuma, KEK 89-57.

[28] M. Sakuda, private communication.

[29] CHARM, Phys. Lett. B232 (1989) 539.

[30] For a review, see T. Sjörstrand, Int. J. Mod. Phys. A3 (1988) 751.

[31] T. Sjöstrand and M. Bengtsson, Comput. Phys. Commun. 43 (1987) 367.

[32] A. Ali et al., Phys. Lett. B93 (1980) 155,
P. Hoyer et al., Phys. Lett. B161 (1979) 349.

[33] JADE, Phys. Lett. B213 (1988) 235.

[34] TASSO, Phys. Lett. B214 (1988) 8,
Mark-II, Z. Phys. C43 (1989) 325.

[35] VENUS, to be published in Phys. Lett. B.

[36] OPAL, Phys. Lett. B235 (1990) 389.

[37] K. Kato and T. Munehisa, Phys. Rev. D36 (1987) 61,
T. Kamae, Proc. 24th HEP Conf. at Hamburg, 1988, p156.

[38] S. Bethke, Z. Phys. C43 (1989) 3311.

[39] AMY, KEK 89-53.

[40] W. de Boer, DESY 89-130, Talk, Phys. in Collision, Jerusalem, June, 1989.

[41] P. M. Stevenson, Phys. Rev. D23 (1981) 2916.

[42] G. Grunberg, Phys. Lett. B95 (1980) 70.

[43] S. Brodsky, G. P. Lapage and P. Mackenzie, Phys. Rev. D28 (1983) 493.

[44] TOPAZ, Phys. Lett. B227 (1989) 495.

[45] D. Gottschalk and M. P. Satz, Phys. Lett. B150 (1985) 451, Z. Phys. C34 (1987) 497.

[46] F. Gotbrod, G. Kramer and G. Shierholz, Z. Phys. C21 (1984) 235.

[47] CELLO, Z. Phys. C44 (1989) 63,
TASSO, DESY 89-069.

[48] AMY, Phys. Rev. Lett. 62 (1989) 1713.

[49] TPC/2γ, Z. Phys. C44 (1989) 357.

[50] C. D. Buchanan et al., PEP4/9 contribution to this conference.

[51] UCLA-HEP-89-003, C. D. Buchanan, Talk, KEK Topical Conf., Tsukuba, May 1989.

[52] G. Marchesini and B. R. Webber, Nucl. Phys. $\underline{B310}$ (1988) 461.

[53] The author is grateful for Professor C. D. Buchanan for many useful comments.

HEAVY QUARK PHYSICS*

M.S. Witherell

University of California, Santa Barbara

ABSTRACT

I give a review of selected topics in Heavy Quark Physics. The major part of the paper concerns semileptonic decays of heavy quarks, an area in which there are many new developments. I also give brief summaries of charm spectroscopy and hadronic decays of charmed and bottom mesons.

CHARM SPECTROSCOPY

It took about 10 years to complete the observation of the 6 $\ell = 0$ charmed mesons (D, D*, D_s, D_s^*). More recently, there has been surprisingly rapid progress in the last few years on the $\ell = 1$ charmed mesons, which are broad states that decay strongly. There are 4 such states expected for each of the three $c\bar{q}$ combinations ($q = u, d, s$). The mass splittings give information on the spin-orbit interaction of the $q\bar{q}$ potential at a radius larger than that probed by charmonium spectroscopy.[1]

Table 1 shows the states that have been observed by CLEO, ARGUS, and E691. Five of the 12 states have been seen, and probable spin-parity assignments have been made. The level of sophistication achieved in charmed meson spectroscopy is demonstrated by Fig. 1, which shows the ARGUS signal for the $D_s^{**}(2536)$.[2] The $I = 0$ meson cannot decay into $D_s + \pi$, which leads to the very distinctive $D^* + K^0$ decay signature.

There has been comparable progress in the spectroscopy of charmed baryons. The Λ_c, Σ_c^{++} and Σ_c^0 are well measured, and recently the Ξ_c baryons joined the group of baryons measured with good statistics. Figure 2 shows the Ξ_c^0 signal in the decay mode $\Xi^- \pi^+$ from CLEO.[3] The upsilon energy region has proved to be a remarkably productive place to study charmed baryons for both spectroscopy and decays. One can look forward to the time when high precision vertex detectors will be used with signals larger than those now available to measure lifetimes of the Ξ_c baryons accurately.

J^P	$M(c\bar{u})$	$M(c\bar{d})$	$M(c\bar{s})$
2^+	2459	2460	—
1^+	2420	2420	2536
1^+	—	—	—
0^+	—	—	—

Table 1. Masses of P-wave $c\bar{q}$ mesons.

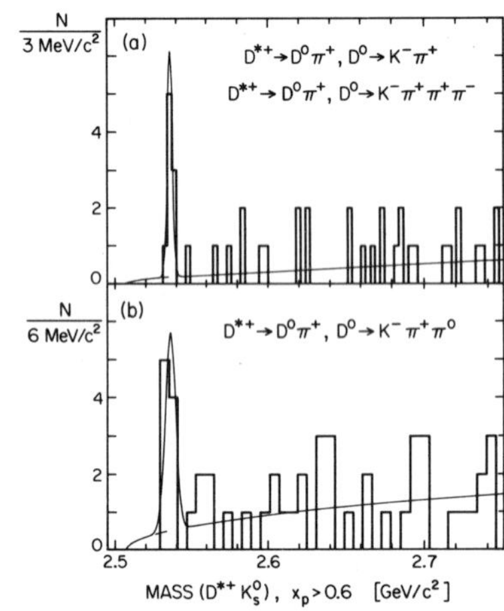

Fig. 1. The D^*K_s mass spectrum from ARGUS, showing the $D_s^{**}(2536)$.

SEMILEPTONIC DECAYS: THEORETICAL BACKGROUND

Special attention is given to the semileptonic decays of heavy quarks because they are the easiest to interpret. Figure 3 shows the diagram describing the process $M \to m\ell^- \bar{\nu}$, in which the parent

---

* Talk presented at DPF '90, Houston, Texas, January 3, 1990.

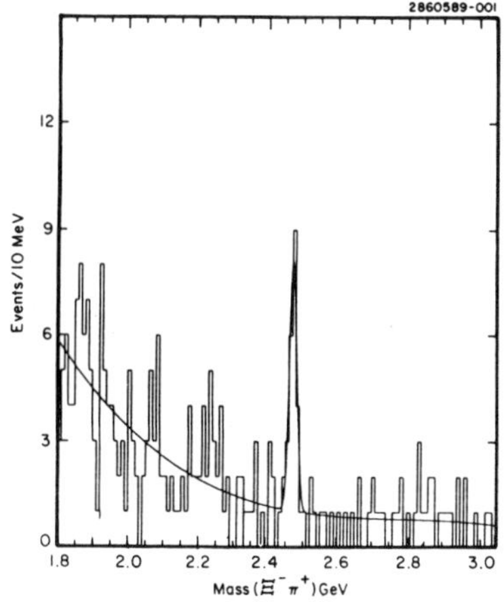

Fig. 2. The $\Xi^-\pi^+$ mass spectrum from CLEO, showing the Ξ_c^0.

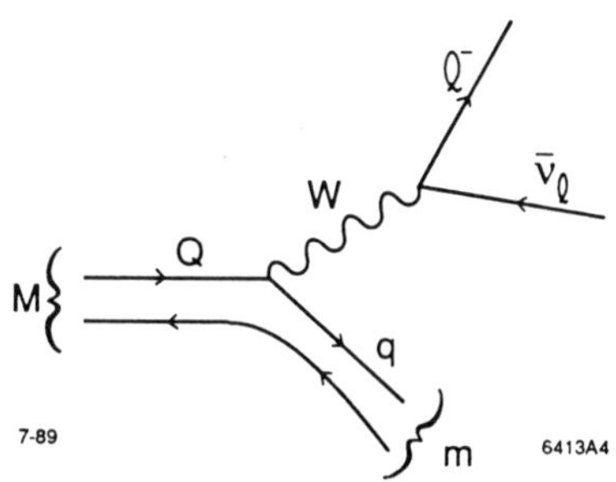

Fig. 3. The semileptonic decay of a heavy quark Q into a lighter quark, q, and a virtual w, which becomes a lepton and neutrino.

quark undergoes a muon-like decay to the final-state quark. The light antiquark in the initial meson M is a spectator quark, not involved in the weak decay.

We can cleanly separate this process into its weak interaction part, characterized by the Kobayashi-Maskawa matrix element V_{Qq}, and the strong interaction part. The strong interaction is completely described by the form factors, which give the amplitudes for the $c\bar{q}$ final state to turn into a pseudoscalar or vector meson. Many models have been constructed to predict these form factors, which depend on the variable $q^2 = M_{\ell\nu}^2$, the mass-squared of the virtual W boson. Independent of model, isospin gives the important relation: $\Gamma(\overline{B^0} \to D^+\ell^-\bar{\nu}) = \Gamma(B^- \to D^0\ell^-\bar{\nu})$, and similar equalities for all final state mesons.

The free quark model[4] should give a good description of inclusive semileptonic decays, in the limit $M(Q) \gg \infty$. For many reasons, however, it is often necessary to use models which calculate the specific exclusive final states. The experimental observations are sometimes more precise in specific exclusive modes, such as $\overline{B^0} \to D^{*+}\ell^-\bar{\nu}$. In the case of $b \to u\ell^-\bar{\nu}$, the only signal which has been isolated is at the endpoint of the electron energy spectrum, where the final state hadronic mass must be very small, and is therefore in the resonance region. Finally, in the case of charm decay, the energy release in the decay is not sufficient to insure success of the free quark model. For all of these cases, it is important to develop reliable models of exclusive decays of heavy quark mesons, and to verify them with experimental measurements. They are needed not only to extract the weak interactions of the bare quarks, but also to study the hadronic structure of these mesons for their own sake.

There are many models, all of which fall into three broad categories. Some models match wave functions near $q^2 = 0$, using an infinite-momentum frame picture.[5,6] Others estimate the wave functions when the final state meson is at rest in the frame of the initial meson, corresponding to $q^2 = q_{\max}^2$.[7,8] A third approach is to calculate the form factors on the lattice.[9,10] There are also other methods used.[11,12]

There is also a test that can be used to compare the exclusive models to the free quark model.

In an important paper,[13] Shifman and Voloshin have shown that

$$\Gamma(Q \to q\ell\bar{\nu}) = \Gamma(M \to P_q\ell\bar{\nu}) + \Gamma(M \to V_q\ell\bar{\nu}),$$

where P_q and V_q are the lowest-lying pseudoscalar and vector mesons, if 1) M_Q, $M_q \gg \Lambda_{\mathrm{QCD}}$ and 2) $(M_Q^2 - M_q^2)/2M_Q^2 \gg (\Lambda_{\mathrm{QCD}}/M_Q)^{1/2}$. In addition, Isgur has shown that this is true everywhere on the Dalitz plot.[14] This limit gives an important check of exclusive models, but only for $b \to c\ell\bar{\nu}$ decays. For the other three cases ($b \to u$, $c \to d$, and $c \to s$), the Shifman-Voloshin limit does not apply because the final state quark is too light.

Based on these arguments, I can give a consumer's guide to information on exclusive semileptonic decays, shown in Table 2.

The $B \to D\ell\bar{\nu}$ form factors are fairly reliable because of the Shifman-Voloshin limit, which allows them to be compared to a reliable quark model calculation. The reliability does not extend to $B \to \pi\ell\bar{\nu}$ and the related extraction of V_{ub}, however. Thus, one cannot use the success of describing experimental information on $B \to D\ell\bar{\nu}$ to certify the models used to extract V_{ub}.

	Theoretical $V_{Qq}{}^{a)}Z$	Theoretical form factors	Experimental measurements
$D \to K\ell\bar{\nu}$	* * *	*	* * *
$D \to \pi\ell\bar{\nu}$	* * *.	*	*
$B \to D\ell\bar{\nu}$		* *	* *
$B \to \pi\ell\bar{\nu}$		•	•

a) Knowledge of V_{Qq} from other sources, notably the constraints of three generation unitarity.

Table 2. Consumer's guide to semileptonic decays of heavy quarks. (* * * = most reliable; • = least.) In all cases, the rating is an average over the pseudoscalar (D) and vector (D*) mesons.

Another technique for obtaining reliable form factors for exclusive $b \to u$ decays has been emphasized by Isgur and Wise.[15] They argue that the $b \to u$ form factors near q_{max}^2 are related closely to the $c \to d$ form factors. Thus if one can measure $D \to \pi\ell\bar{\nu}$ and $\rho\ell\bar{\nu}$ over the entire range, one has a model-independent determination of the form factors needed to extract V_{ub}. This will be difficult, since there are only a few events in $D \to \pi\ell\bar{\nu}$ at the moment. On the other hand, these form factors are closely related to those for $D \to K\ell\bar{\nu}$ and $K^*\ell\bar{\nu}$, about which we have fairly good information.

The decay rate for semileptonic decay is calculated from the formula

$$d\Gamma = \tfrac{1}{2\mu}\,|(G_F/\sqrt{2})\,V_{Qq}\,L^\mu H_\mu|^2 d\pi_3$$

where $d\pi_3$ is the element of Lorentz invariant phase space and $L^\mu(H_\mu)$ contains the leptonic (hadronic) currents. The entire effect of the strong interaction is contained in H_μ. In the pseudoscalar final state (e.g., $\bar{B} \to D\ell\bar{\nu}$)

$$\frac{d\Gamma}{dq^2} = \frac{G^2|V_{Qq}|^2 K^3}{24\pi^3}|f_+(q^2)|^2,$$

where K is the momentum of the final state meson. There is only one form factor, a vector form factor, and the decay rate measures it directly.

The situation is more complex for the decay to the final state vector meson ($\bar{B} \to D^*\ell\bar{\nu}$):

$$\frac{d\Gamma}{dq^2} = \frac{G^2|V_{Qq}|^2 Kq^2}{96\pi^3 M^2}\left[|H_+|^2 + |H_-|^2 + |H_0|^2\right]$$

where the three helicity amplitudes H_+, H_-, H_0, correspond to the three helicity states of the virtual W. These amplitudes are linear combinations of three form factors:

$$H_0 = \tfrac{1}{2mq}\left[(M^2 - m^2 - q^2)(M + m)A_1(q^2) - \tfrac{4M^2K^2}{M+m}A_2(q^2)\right]$$

$$H_\pm = (M + m)A_1(q^2) \pm \left[2MK/(M + m)\right]V(q^2)$$

The three form factors include one vector form factor, $V(q^2)$, and two axial vector form factors, $A_1(q^2)$ and $A_2(q^2)$. Thus the decay rate is a rather complicated function of the form factors.

The theoretical models calculate the form factors, $f_+(q^2)$, $A_1(q^2)$, $A_2(q^2)$, and $V(q^2)$. The q^2-dependence of these form factors is usually assumed to have a single pole form:

$$f_+(q^2) = \tfrac{f_1(0)}{1 - q^2/M_V^2}.$$

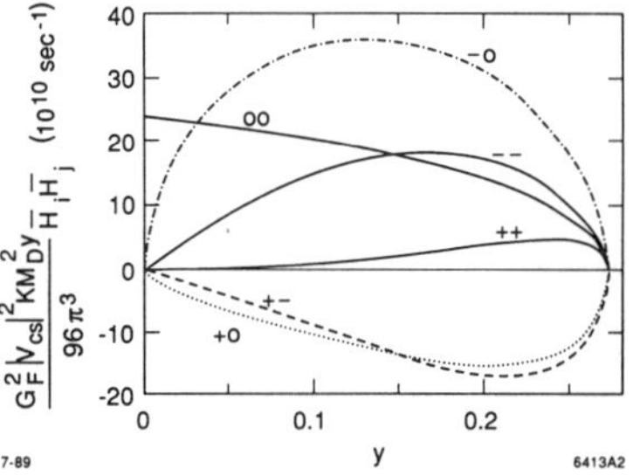

Fig. 4. The contribution to the rate according to a quark model, from Gilman and Singleton. The $|H^+|^2$ term is denoted $++$; the H^+H^0 interference term by $+0$; etc.

This is only an approximation, but in the one case ($D \to Ke\nu$) with experimental information, it works pretty well, with large errors. In $B \to De\nu$, the ratio $|f(q^2_{max})|^2/|f(0)|^2 \simeq 2$, so the exact form is not so important. The q^2 variation is similar in charm decay. On the other hand, in $B \to \pi e\nu$, $|f(q^2_{max})|^2/|f(0)|^2 \simeq 140$, and the exact form of this dependence must be measured to interpret experimental results.

For $\bar{B} \to D^*e\nu$ ($D \to \overline{K}^*e\nu$), then, the decay rate depends on the three form factors, each of which is calculated by the models. For this reason it is much more reliable to *compare form factors rather than rates* with theoretical models. As an intermediate step, the rate is often separated into longitudinal and transverse parts:

$$\frac{\Gamma_L}{\Gamma_T} = \frac{\int |H_0|^2 K q^2 dq^2}{\int (|H_+|^2 + |H_-|^2) K q^2 dq^2} \qquad (\simeq 0.5 \text{ if unpolarized})$$

The angular distribution for $\bar{B} \to D^*\ell\bar{\nu}$ is used to extract the longitudinal and transverse components. The relevant variables in the angular distribution are 1) θ_v, the angle of the π with respect to the virtual W in the D^* rest frame; 2) θ_ℓ, the angle of the lepton with respect to the D^* in the W rest frame; 3) ϕ, the azimuthal angle between the D^* decay plane and the W decay plane; and 4) $q^2 = M_{\ell\nu}^2$. The angular distribution for the three parts are

$$
\begin{array}{lll}
|H_+|^2 : & q^2(1 + \cos\theta_\ell)^2 \sin^2\theta_v & \text{(transverse)} \\
|H_-|^2 : & q^2(1 - \cos\theta_\ell)^2 \sin^2\theta_v & \text{(transverse)} \\
\text{and } |H_0|^2 : & (1 - \cos^2\theta_\ell)\cos^2\theta_v & \text{(longitudinal)}.
\end{array}
$$

Thus one can measure Γ_L/Γ_T by measuring the $\cos\theta_v$ distribution alone. Similar expressions hold for $D \to \overline{K}^*\ell\bar{\nu}$, but with the substitution $H_+ \leftrightarrow H_-$.

Figure 4, taken from the phenomenological work of Gilman and Singleton,[16] shows the contribution to the decay rate from the various helicity amplitudes as a function of q^2. The solid lines give the diagonal contributions for $|H_+|^2$ ($++$), $|H_-|^2$ ($--$) and $|H_0|^2$ (00), which are the only terms I am considering here. This shows that besides the distinctive $\cos\theta_v$ distributions that separate longitudinal from transverse components, there are important differences in the q^2 dependence. The longitudinal term approaches a constant value at $q^2 = 0$, but the transverse terms all vanish in this limit.

CHARM SEMILEPTONIC DECAYS

Although the K-M matrix elements for charm are well known from unitarity, the semileptonic decays are crucial in understanding the hadronic structure. The experimental results provide a direct measurement of the form factors, which can be compared with models. These comparisons can be used to select models, or to tune them, so that they can be applied more reliably to bottom decay, in which the K-M matrix element in unknown. In the alternative approach, suggested by Wise and Isgur, the charm form factors provide a direct measurement of the bottom form factors, which is more reliable than any particular model.

Mode	Branching Ratio (%)	Transition Rate (10^{10} sec^{-1})	Experiment
	Inclusive		
$D^0 \rightarrow e^+ x$	$7.5 \pm 1.1 \pm 0.4$	$17.4 \pm 2.5 \pm 0.9$	Mark III[a]
$D^+ \rightarrow e^+ x$	$17.0 \pm 1.9 \pm 0.7$	$15.7 \pm 1.8 \pm 0.6$	Mark III[a]
$D^0 \rightarrow K^- e^+ \nu_e$	$3.8 \pm 0.5 \pm 0.6$	$8.8 \pm 1.2 \pm 1.4$	E691[b]
	$3.4 \pm 0.5 \pm 0.4$	$7.8 \pm 1.2 \pm 0.9$	Mark III[c]
	3.5 ± 0.5	8.2 ± 1.2	Mean Value
$D^0 \rightarrow \pi^- e^+ \nu_e$	$0.39^{+0.23}_{-0.11} \pm 0.04$	$0.9^{+0.5}_{-0.3} \pm 0.1$	Mark III[c]
$D^+ \rightarrow \overline{K^{*0}} e^+ \nu_e$	$4.5 \pm 0.7 \pm 0.5$	$4.2 \pm 0.6 \pm 0.5$	E691[d]

a. From Reference 19.
b. From Reference 17.
c. From Reference 18.
d. From Reference 20.

Table 3. Semileptonic Decays.

The decay rate to the pseudoscalar meson, $D \rightarrow \bar{K}\ell\nu$, has been measured by two groups,[17,18] and they are shown in Table 3. Because the decay rate is proportional to the square of the single vector form factor, the average value of the decay rate, $(8.2 \pm 1.2) \times 10^{10}$ sec^{-1}, can be used to extract the form factor intercept at $q^2 = 0$, $f_+^K(0) = 0.75 \pm 0.05$. This is in agreement with theoretical predictions.[6,7,9,11] In addition, E691 makes a rough measurement of the q^2-dependence, and finds a good fit to the single pole form with $M_v = 2.1^{+0.4}_{-0.2}$ GeV/c^2, consistent with the D_s^* mass. There is also a measurement by Mark III of $D \rightarrow \pi\ell\nu$. Although this rate is consistent with the expected value $|V_{cd}|/|V_{cs}| = 0.22$, the sample of seven events does not allow a useful form factor measurement.

The $D^+ \rightarrow \overline{K^{*0}} e^+ \nu_e$ decay has been studied in some detail by the E691 collaboration.[20] The results on the decay rate is shown in Table 3. The first surprising result is that the $D \rightarrow K^*$ decay rate is roughly 1/2 of the $D \rightarrow K$ rate, although the models predict this number to be about 1.1. This is a serious disagreement which needs to be understood. Because of the complicated relationship between decay rates and form factors, one needs to look at the angular distributions to investigate the source of this discrepancy.

Figure 5 shows the $\cos\theta_v$ distribution from E691 in the $D \rightarrow K^* e\nu$ decay. The distribution shows a marked $\cos^2\theta_v$ behavior, in spite of a flat background contribution. The resulting value of $\Gamma_L/\Gamma_T = 2.4^{+1.7}_{-0.9} \pm 0.2$, should be compared with the unpolarized value of 0.5 and a value expected from models of about 1. Thus there is clearer evidence that the models do not describe the $D \rightarrow K^* e\nu$ decay well.

To get to the source of the problem, we have recently analyzed the E691 sample, using the full information of the angular distribution. The goal is to measure the form factors directly, for a proper comparison with models. Figure 6 shows the q^2 distribution for two regions, one in which

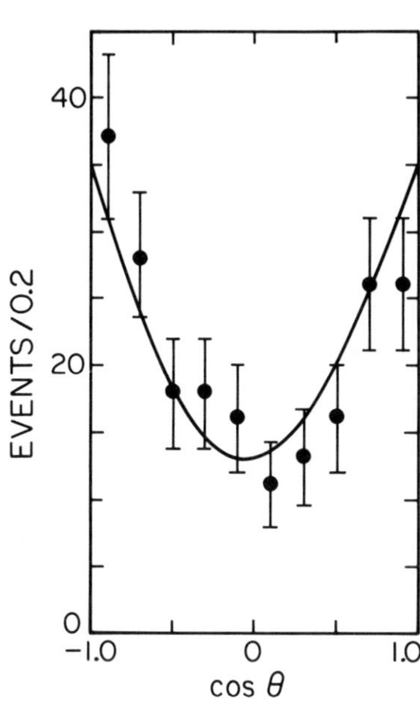

Fig. 5. The angular distribution of the K* decay into Kπ in D $\to$ K*$e\nu$ decay, from E691. The angle is the angle $\theta\nu$ defined in the text.

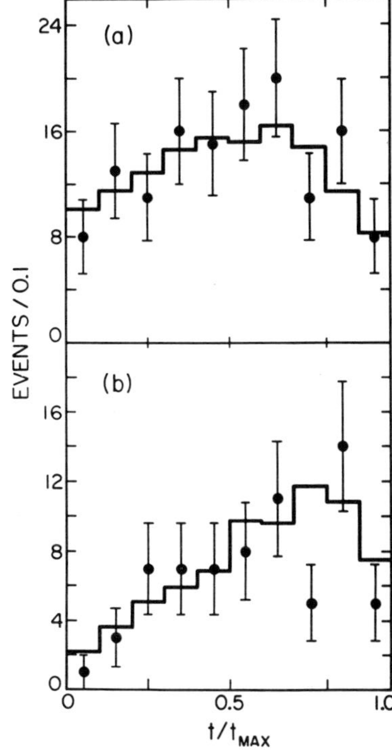

Fig. 6. The distribution of the momentum variable, $t = q^2 = M^2 e\nu$ for the E691 D $\to$ K*$e\nu$ sample: (a) $|\cos\theta_\nu| > 0.5$ and (b) $|\cos\theta_\nu| < 0.5$.

the longitudinal component is enhanced with cuts on the decay angles. One can see the expected difference in behavior at low q^2, which shows that a much improved measurement of Γ_L/Γ_T is possible using this information. It also shows that the best fit distribution agrees well with the data.

The results from this analysis are <u>preliminary</u>, and will change slightly, but not dramatically. The experimental measurements of the form factors are shown in Table 4, along with the theoretical predictions. The form factor A_1 is about 0.6 of the prediction with quite

	E691	ISGW[7]	WSB[6]	GS[16]	KS[5]
$A_1(0)$	$0.46 \pm 0.05 \pm 0.05$	0.8	0.9	0.8	1.0
$A_2(0)$	$0.0 \pm 0.2 \pm 0.1$	0.9	1.2	0.6	1.0
$V(0)$	$0.9 \pm 0.3 \pm 0.1$	1.1	1.3	1.5	1.0
$\frac{\Gamma_L}{\Gamma_T}$	$1.8^{+0.6}_{-0.4} \pm 0.3$	1.1	0.9	1.2	1.2

Table 4.

good errors, while V is also small. Most dramatic, however, is the result that A_2 is consistent with zero, and well below all predictions. Finally, the derived ratio $\Gamma_L/\Gamma_T = 1.8^{+0.6}_{-0.4} \pm 0.3$ has much better errors than the previous number, reflecting the power of including the q^2 dependence in the fit.

Thus the source of the discrepancy in $D \to K^* e\nu$ has been located, although it is not yet explained. The discrepancy is much larger than differences between models, so it is dangerous to use that model difference as a measure of systematic error in the theory. The exclusive prediction based on quark model ideas are suspect for the case of a light quark in the final state, and the models for $B \to \pi\ell\bar{\nu}$ and $\rho\ell\bar{\nu}$ are therefore unreliable.

BOTTOM TO CHARMED QUARK DECAYS

The decay of $b \to c\ell\bar{\nu}$ is used to determine V_{cb} in the K-M matrix. The exclusive decays $\bar{B} \to D\ell\bar{\nu}$ and $\bar{B} \to D^*\ell\bar{\nu}$ dominate the inclusive semileptonic decay. To study these, the CLEO and ARGUS collaborations apply cuts on minimum lepton momentum (1.0–1.4 GeV/c) and combine these leptons with D (or D^*) candidates. In the analysis, one calculates $M_m^2 = (M_B - E_D - E_\ell)^2 - (P_D + P_\ell)^2$, which ignores $\vec{P}_B = 330$ MeV/c because the direction is unknown. This leads to an error $\delta(M_m^2) \simeq 0.4$ GeV^2, which is not sufficient to separate D from D^* cleanly.

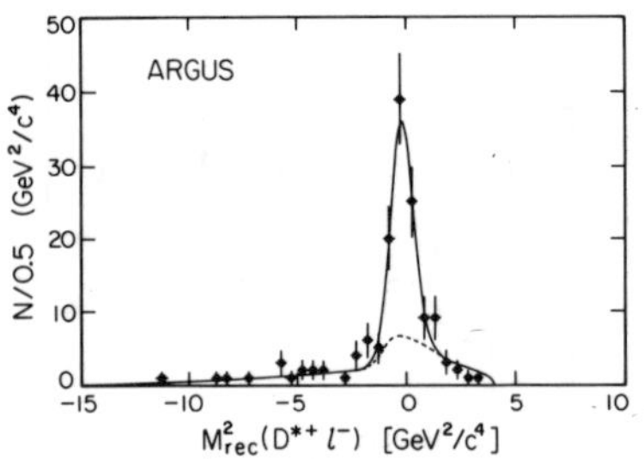

Fig. 7. The missing mass squared distribution for $D^{*+}\ell^-$ events from ARGUS.

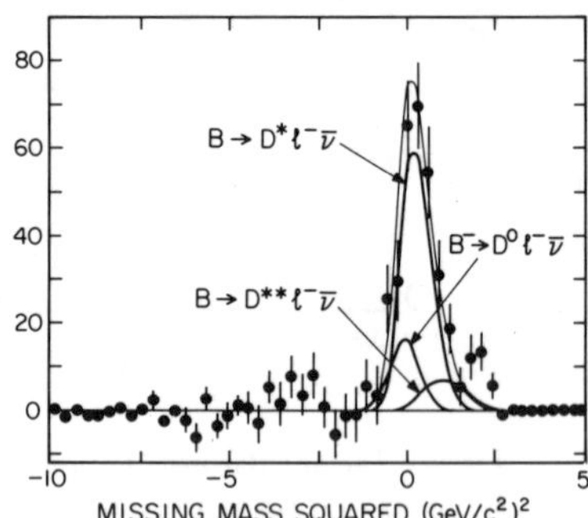

Fig. 8. The missing mass squared distribution for $D^0\ell^-$ events from CLEO.

As we saw earlier, the interpretation of the $D\ell\bar{\nu}$ mode is easier than $D^*\ell\bar{\nu}$ because there is only one form factor. Unfortunately, the measurement of $D\ell\bar{\nu}$ is hard because it is difficult to separate the feedthrough from $D^*\ell\bar{\nu}$ decays. Figure 7 shows the M_m^2 distribution for $D^{*+}\ell^-$ from ARGUS.[21] (There is similar information from CLEO.)[22] There is a clean peak at $M_m^2 = 0$ which is used to measure $B(\bar{B}^0 \to D^{*+}\ell^-\bar{\nu})$. Figure 8 shows a similar plot, from CLEO, of the $D^0\ell^-$ combinations.[23] Here one expects about 70% of the events to come from $D^{*+}\ell\bar{\nu}$ or $D^{*0}\ell\bar{\nu}$ final states, about 10% from $D^{**}\ell\bar{\nu}$, and only about 10% from $D^0\ell\bar{\nu}$ exclusive. This is why the measurement of the $D\ell\bar{\nu}$ branching ratios is so difficult.

The various measurements of V_{cb} from semileptonic decays are shown in Table 5. The value of $V_{cb} = 0.044 \pm 0.006$ is consistent with all inclusive and exclusive measurements. One expects this consistency, since for $b \to c\ell\bar{\nu}$ decay the Shifman-Voloshin limit is satisfied, and the exclusive models should agree well with the free quark model. There is an error of about 10% in extracting V_{cb} from the models (whether exclusive or inclusive). There is an additional error on V_{cb} due to the lifetime error, which is $\pm 6\%$ for modes using all B mesons, but is $\pm 10\%$ for the $\bar{B}^0 \to D^{*+}\ell\bar{\nu}$ measurement, since there is a large error on $\tau^+/\tau^0 = \tau(B^+)/\tau(B^0)$. To do better one needs better models and better measurement of the lifetime ratio.

MODE	BR (%)		V_{cb}	
	CLEO	ARGUS	CLEO	ARGUS
$\bar{B} \to X\ell\bar{\nu}$	$10.1 \pm 0.3 \pm 0.7$	$10.3 \pm 0.7 \pm 0.2$	$.046 \pm .006$	$.046 \pm .006$
$\bar{B}D\ell\bar{\nu}$	$2.4 \pm 0.8^{+0.7}_{-0.8}$	$1.8 \pm 0.6 \pm 0.5$	$.044 \pm .009$	$.044 \pm .009$
$\overline{B^0} \to D^{*+}\ell^-\bar{\nu}$	$4.6 \pm 0.5 \pm 0.7$	$5.4 \pm 0.8 \pm 1.3$	$.039 \pm .006$	$.046 \pm .008$

Table 5. Various measurements of V_{cb}.

The main check of the exclusive models is the D^* polarization. The analysis is identical to the $D \to K^*\ell\bar{\nu}$ case mentioned earlier, using only $\cos\theta_v$. The experimental results are $\Gamma_L/\Gamma_T = 0.83 \pm 0.36$ (CLEO) and 0.85 ± 0.45 (ARGUS). These values agree with the models which range from 0.9–1.1. As in the $D \to K^*\ell\bar{\nu}$ case, it should be possible to improve this measurement substantially with the present data by adding q^2 to the fit. To reiterate the earlier point, form factors should be extracted and compared with theory, rather than rates.

It is important to measure the lifetime ratio well for a number of reasons. As we have seen, a better measurement of this ratio is necessary to improve the error on V_{cb}. In addition, the present results on B_d mixing are sensitive to the lifetime ratio. Finally, the ratio has physics interest of its own, since present models of hadronic weak decay should predict τ^+/τ^0 quite precisely. In fact, τ^+/τ^0 is not expected to be exactly 1. We know that in the D system, the ratio of lifetimes $\tau(D^+)/\tau(D^0) = 2.5$, and we have a pretty good understanding of what causes the large difference. Scaling these effects to the B system, the expected values for τ^+/τ^0 range from 1.0 to 1.2.

The measurement of τ^+/τ^0 makes use of the fact that $\Gamma(B^- \to X^0\ell\bar{\nu}) = \Gamma(\overline{B^0} \to X^+\ell\bar{\nu})$ isospin. Thus measuring the ratio of semileptonic branching ratio determines the lifetime ratio. The results are $\tau^+/\tau^0 = 0.85 \pm 0.20^{+0.21}_{-0.16}$ (CLEO)[23] and $1.00 \pm 0.23 \pm 0.14$ (ARGUS).[24] There is rough agreement with expectations, but the errors are $\pm 30\%$, and need to be ± 5–10% to see the expected effects.

THE CRUCIAL QUESTION: WHAT IS V_{ub}?

In the 3-generation Cabibbo-Kobayashi-Maskawa model of quark couplings, the greatest uncertainties that can be resolved experimentally are those involving the top quark mass and V_{ub}. The ratio V_{ub}/V_{cb} is expected to be of order 0.1, simply from the constraints applied by other measurements. It is important to measure this ratio to an accuracy of, say, 20%.

This is a very difficult number to measure. The associated hadronic decay modes such as $B^0 \to \pi^+\pi^-$ should have branching ratios around 2×10^{-5}, well below present limits. The exclusive semileptonic decays, such as $B \to \pi\ell\bar{\nu}$ and $\rho\ell\bar{\nu}$, have larger branching ratios, but there is too much background to observe them in present experiments. The only way to search sensitively for $b \to u$ transitions is to look at the endpoint of the lepton spectrum, beyond the point where $b \to c\ell\bar{\nu}$ decays dominate. Unfortunately, calculation of V_{ub} from the rate in this region is very difficult, and present errors on extracting V_{ub} are large.

It is easiest to show the constraints imposed by unitarity of the CKM matrix by showing the triangle in Figure 9. If one divides all sides by the length of the base of the triangle, $|V_{cd}V_{cb}|$, one gets a rescaled triangle with vertices at $(0,0)$, $(1,0)$, and (ρ,η). In Figure 10 are a series of plots showing the allowed range in (ρ,η) space of the third vertex of this triangle. The dotted circles represent the constraints of V_{ub}/V_{cb}, which was assumed to lie in the range 0.04–0.16.

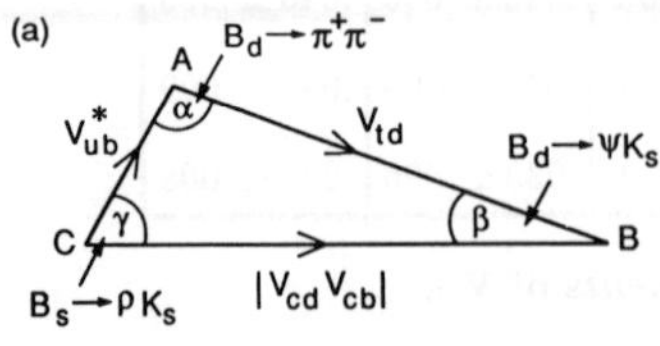
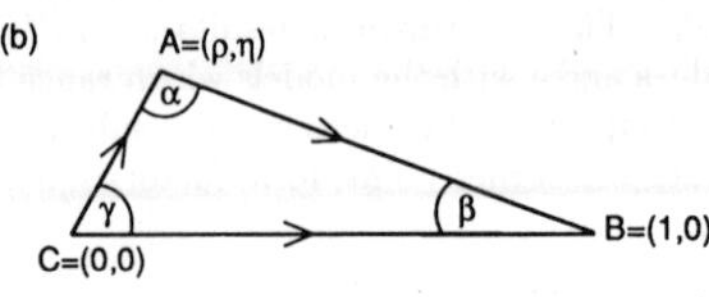

Fig. 9. Representation in the complex plane of the triangle formed by (a) the K-M matrix elements V_{ub}^*, V_{cd}, V_{cb}^* and V_{td}; and (b) the rescaled triangle with the bare set to unit length.

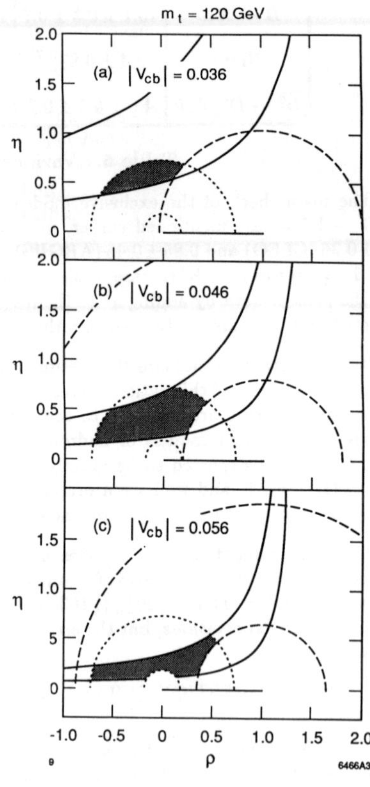

Fig. 10. Constraints from $|V_{ub}/V_{cb}|$ (dotted circles), B_d mixing (dashed circles), and $\mathcal{E}$ in K^0 decay (solid curves) on the rescaled unitarity triangle, for $m_t = 120$ GeV. The shaded region is that allowed by the vertex $A(\rho, \eta)$, under various assumptions for $|V_{cb}|$.

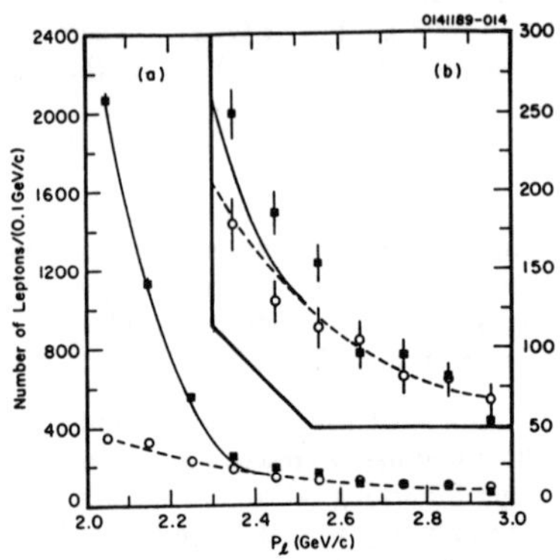

Fig. 11. The CLEO lepton spectrum near the B endpoint for on resonance data (filled squares) and scaled off resonance data (open circles). The curves are fits to the off resonance data, and off resonance plus $b \to c\ell\bar{\nu}$.

In analyzing the lepton spectrum, a series of cuts are applied to reduce the continuum background. The resulting spectra are shown in Fig. 11 for CLEO,[25] and Fig. 12 for ARGUS.[26] In Figure 11, the dashed curve shows the expected continuum contribution, and the solid curve also includes the $b \rightarrow c\ell\bar{\nu}$ component. In both cases there is an excess of events in the range 2.3–2.6 GeV/c, which is interpreted as evidence for $b \rightarrow u\ell\bar{\nu}$ decays.

In the CLEO analysis, the endpoint region is divided into two momentum bins, 2.2–2.4 GeV/c and 2.4–2.6 GeV/c. The net yields in these two bins are $62 \pm 28 \pm 28$ and $76 \pm 18 \pm 8$, based on raw data of 813 and 350 events in the two bins. In the ARGUS experiment, the event sample divided into one-lepton and two-lepton events, and the one-lepton events are required to show missing momentum due to the neutrino. The net yields are 27 ± 9 and 14 ± 5, based on 60 and 21 events, for 1- and 2-lepton events, respectively. Each experiment sees an excess of about four standard deviations.

Thus each experiment sees a significant signal, and they agree quite well in the derived values of $B(b \rightarrow u\ell\bar{\nu})$. Still, there is a large error in extracting V_{ub} because of theoretical uncertainties. Table 6 shows the derived value of $(V_{ub}/V_{cb})^2$ for each experiment, using four different models. Comparison of the two columns demonstrates that the experiments are in agreement. On the other hand, there is a range of a factor of 4 if one includes the differences among the models.

In fact, a conservative person would place an even larger uncertainty on the derived value of V_{ub}/V_{cb}. The free quark model might be expected to give a fairly good calculation for the rate integrated over the entire Dalitz plot. The measurements only cover a small region, however, and one that restricts the mass of the final state hadrons to be a π or ρ. Thus the measured rate depends critically on the details of how the particular hadronic states are populated.

Figure 13 shows the Dalitz plot for $B \rightarrow \pi\ell\bar{\nu}$, and the region accepted by CLEO and ARGUS. It demonstrates many of the problems faced by the exclusive models. The B pole lies just beyond the edge of the Dalitz plot, so that its influence dominates the upper corner. The form factor required, $|f(q^2)|^2$, varies by a factor of 100 over the Dalitz plot, so the electron energy spectrum depends critically on the details of this q^2 dependence. In addition, the reliability of the exclusive form factor models can not be insured by using the Shifman-Voloshin limit, because of the light quark in the final state. More importantly, the $D \rightarrow K^*e\nu$ results indicate that the models are not reliable, and that the uncertainty is underestimated by the differences among models.

Thus the normalization of the form factors in the exclusive models are uncertain by a large factor. There is a separate large uncertainty from the unknown q^2 dependence. Since the effect can only be observed in one tiny corner of the Dalitz plot, these effects can not be studied to improve

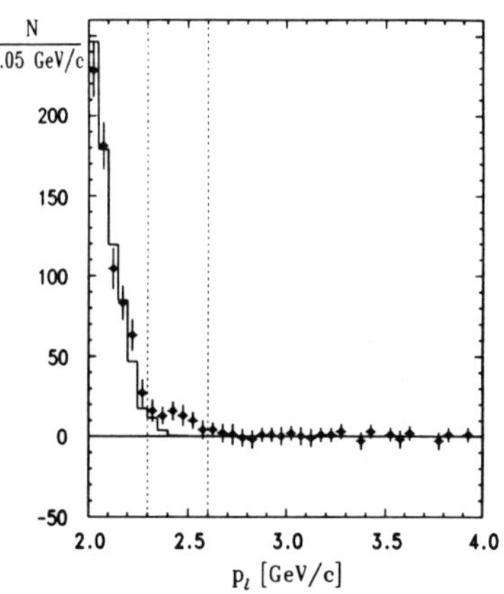

Fig. 12. The lepton spectrum from ARGUS near the B decay endpoint region, after subtraction of scaled off-resonanced data. The histogram shows the best fit to the $b \rightarrow c\ell\bar{\nu}$ contribution.

MODEL	$|V_{ub}/V_{cb}|^2$	$(\times 10^2)$
	CLEO	ARGUS
ISGW[7]	2.2 ± 0.6	3.2 ± 0.7
ACM[4]	0.8 ± 0.2	1.0 ± 0.2
WSB[6]	1.3 ± 0.4	1.4 ± 0.4
KS[5]	0.9 ± 0.2	0.8 ± 0.2

Table 6. Values of $|V_{ub}/V_{cb}|^2$ for using various theoretical models.

the model dependence. Given these uncertainties, the conservative judgment about the range of $(V_{ub}/V_{cb})^2$ is about an order of magnitude, 0.004–0.04.

To summarize the situation, there is strong evidence for $b \to u$ decays from CLEO and ARGUS, at about the four standard deviation level. On the other hand, it is difficult to extract the K-M matrix element reliably because of model uncertainties. The allowed range of 0.004–0.04 lies squarely in the middle of the range expected, given a top quark mass of 120–200 GeV.

Determination of V_{ub} with better precision is one of the most important goals for heavy quark physics. We could do this by measuring $b \to u\ell\bar{\nu}$ over the entire Dalitz plot and use the free quark model to extract V_{ub}, but it seems very difficult to reduce the background enough to measure the exclusive rate beyond the narrow endpoint region. The approach which is somewhat more practical experimentally is to measure B $\to \pi\ell\bar{\nu}$ and $\rho\ell\bar{\nu}$ over most of the Dalitz plot, which will probably require using doubly-tagged events. In addition, we need better measurement of D $\to$ K$\ell\bar{\nu}$ and K*$\ell\bar{\nu}$ to reduce the model uncertainty, or to determine the B form factors by scaling arguments. We also need additional guidance from the models, especially in understanding the discrepancy seen in charm decay.

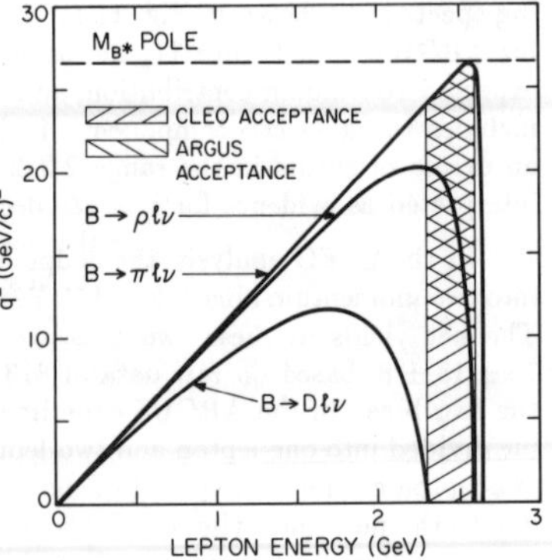

Fig. 13. Dalitz plot for B semileptonic decay. The kinematic boundaries are shown for $B \to \pi\ell\bar{\nu}$, $\rho\ell\bar{\nu}$, and D$\ell\bar{\nu}$. The <u>approximate</u> acceptance regions are shown for ARGUS and CLEO. The B* pole mass is also shown, which is relevant for $B \to \rho\ell\bar{\nu}$ and $\pi\ell\bar{\nu}$ decays.

HADRONIC CHARM DECAYS

The one topic in hadronic charm decay that has dominated recent attention is the measurement of the major D_s^+ decays. According to the general pattern, the D_s decays with an s$\bar{\text{s}}$ combination in the final state ($\phi\pi^+$, $\overline{K^{*0}}K^+$, $\overline{K^0}K^+$, $\phi\pi^+\pi^0$) are large and the ones without s$\bar{\text{s}}$ ($\rho^0\pi^+$, $\omega^0\pi^+$) are small. This is consistent with the picture that spectator decays dominate, so there usually is an s-quark from the $c \to su\bar{d}$ decay, in addition to the $\bar{\text{s}}$ spectator quark.

The only measurements that did not fit in well with the pattern expected were measurements on $D_s^+ \to \eta\pi^+$ and $\eta'\pi^+$. One expects these to be reasonably large, because of the s$\bar{\text{s}}$ content of the η or η'; typically, the models predict B$'(D_s^+ \to \eta [\eta']\pi^+) = 1.0$ [0.6], where I use B$'$ to denote the ratio of the branching ratio to that for the mode $D_s^+ \to \phi\pi^+$. The results which attracted some interest a year or more ago are the ones by Mark II[27] and NA14[28] in Table 7. For the Mark II measurement of $\eta\pi^+$, and even more for the two measurements of $\eta'\pi$, the branching ratios are many times that expected. This provoked a great deal of speculation on the source of the anomaly.

	MarkII[27]	NA14[28]	E691[29]	Mark III[30]
$\frac{B(D_s^+ \to \eta\pi^+)}{B(D_s^+ \to \phi\pi^+)}$	3.0 ± 1.3		< 1.5	< 2.5
$\frac{B(D_s^+ \to \eta'\pi^+)}{B(D_s^+ \to \phi\pi^+)}$	4.8 ± 2.1	$6.9 \pm 2.4 \pm 1.4$	< 1.6	< 1.9

Table 7. Branching ratio measurements for $D_s^+ \to \eta\pi^+$, $\eta'\pi^+$ relative to $D_s^+ \to \phi\pi^+$.

Recently, more sensitive searches by E691[29] and Mark III[30] show no evidence of either $\eta\pi^+$ or $\eta'\pi^+$ decay of the D_s^+. In each case the branching ratios are less than about $1.5 \times \phi\pi^+$ at the 90% confidence level. This agrees with the expected level, but disagrees slightly with the $\eta\pi^+$ measurement and seriously with the $\eta'\pi^+$ observations. Thus the previous indications of anomalously large branching ratios for these modes are not confirmed.

What we would like to do is to test that the pattern of major D_s^+ branching ratios is similar to that seen in D^0 decay, as suggested by the near equality of lifetimes. Unfortunately the lack of a well-measured absolute branching ratio for any mode prevents us from doing that. All measurements are quoted relative to the mode $D_s^+ \to \phi\pi^+$, which is the easiest to measure.

It is important to measure the absolute scale of branching ratios for two reasons. We would like to make detailed comparison of absolute branching ratios with those for D^0 and D^+. We would also like to use the D_s^+ branching ratios to measure the branching ratios for the important modes $\bar{B} \to D_s^- + D + X$. Simple spectator diagrams for B decays indicate that such decays should be very common.

There are four methods for measuring $B(D_s^+ \to \phi\pi^+)$, which are shown in Table 8. For a long time, people have assumed a D_s^+ cross section above threshold to extract the branching ratio, but it suffers from a large systematic error due to the uncertainty in the cross section. CLEO has presented a result based on starting with the total charm cross section, and subtracting the non-D_s components.[31] This is rather sensitive to, for example, the Λ_c^+ branching ratio. Mark III has published an upper limit on $B(D_s^+ \to \phi\pi^+)$ based on looking for double tagged events.[32] The limiting uncertainty is due to the large error on the dominant modes expected, expecially $D_s^+ \to \phi\pi^+\pi^0$ and $D_s^+ \to \overline{K^{*0}}K^{*+}$.

Method	Group	$B(D_s^+ \to \phi\pi^+)$ (%)
$e^+e^- \to D_s^+ \bar{D} X$, assume $\sigma(D_s^+)$	average	3.5 ± 1.5
$e^+e^- \to D_s^+ \bar{D} X$, assume $\sigma(c\bar{c})$	CLEO[31]	2 ± 1
$e^+e^- \to D_s^+ D_s^{*-}$, double-tagged	Mark III[32]	< 4.1
$\gamma N \to D_s^+ X,\ D_s^+ \to \phi e^+\nu_e$	E691[33]	> 3.4

Table 8. Measurements of the absolute branching ratio $B(D_s^+ \to \phi\pi^+)$.

There is a new result from E691 which implies a limit on $B(D_s^+ \to \phi\pi^+)$.[33] We find an upper limit on the ratio $B(D_s^+ \to \phi e^+\nu_e)/B(D_s^+ \to \phi\pi^+) < 0.45$ at 90% c.l. The absolute branching ratio of $D_s^+ \to \phi e^+\nu_e$ is closely related to that for $D^+ \to \overline{K^{*0}}e^+\nu_e$, however, since they are both spectator decays. The prediction is $B(D_s^+ \to \phi e^+\nu_e) = 1.7 \pm 0.3\%$, where the error is dominated by the uncertainty in the ratio of form factors, which are quite model independent. The derived limit, $B(D_s^+ \to \phi\pi^+) > 3.4\%$ at 90% c.l., is fairly insensitive to the exact error assigned to this prediction.

Although the different measurements give rather different pictures, the four measurements in Table 8 are not inconsistent. A simple weighted average gives a $\chi^2 = 4$ for 2 degrees of freedom. If one scales up the errors in the way made popular by the particle data book, then one gets a world average of $B(D_s^+ \to \phi\pi^+) = 3.1 \pm 0.7\%$.

There has been real progress in measuring the simplest exclusive hadronic B decays, and one can begin to ask whether the results agree with simple models. To have some framework for judging the measurements, I will use the simple model of Bauer, Stech, and Wirbel,[34] without any corrections. It assumes that the two spectator decays of Fig. 14 are dominant, with amplitudes a_1 and a_2, respectively, and that W-exchange and final state interactions are negligible. These neglected contributions change the predictions substantially in charm decay, about 30% for the dominant branching ratios, and somewhat worse for modes with small predictions. One could expect these corrections to be smaller for bottom decay.

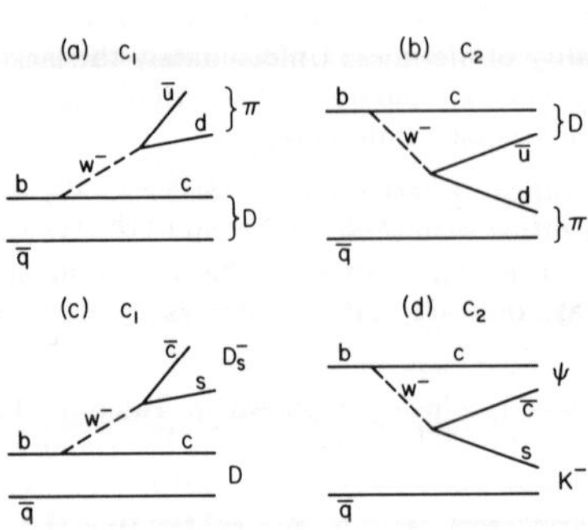

Fig. 14. Quark level diagrams for B decays. It is expected that $c_2/c_1 \simeq -0.2$, scaling from D decay.

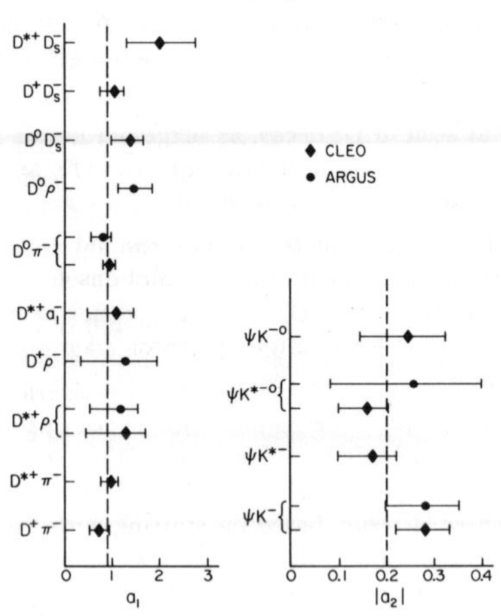

Fig. 15. Comparison of hadronic B decay branching ratios with the model of Bauer, Stech and Wirbel. The modes with D_s are calculated assuming $B(D_s \to \phi\pi) = 4.1\%$.

In his talk at the Madrid EPS conference, Morrison[35] used this model to compare a series of experimental measurements. He did this by evaluating a_1 (or a_2) for each mode. The results are shown in Fig. 15. All of the results proportional to $|a_1^2|$ that do not include a D_s are quite consistent with a central value of $a_1 = 0.9$. In addition, the a_2 modes, all of which contain a ψ in the final state, are consistent with $a_2 = -0.2$, but with large errors. This comparison clearly shows that the simple model works pretty well, and that to observe more detailed effects one needs to have more precise measurements.

The modes containing a D_s are plotted using the assumption $B(D_s^+ \to \phi\pi^+) = 4.1\%$. With this value they are also consistent with $a_1 \simeq 0.9$, and would still be with $B(D_s^+ \to \phi\pi^+) \simeq 3\%$. It is inconsistent with this value for $B(D_s^+ \to \phi\pi^+)$ less than about 2.5%, however. Thus if the absolute branching were lower than this, it would imply that the effective decay constant used to calculate two-body decays containing a D_s^+ is incorrect.

AGENDA FOR THE FUTURE

Looking over the development in heavy quark physics over the last five years or so, it is remarkable how much more detailed questions we are now addressing. We are measuring detailed hadronic structure of mesons containing heavy quarks, in the manner that kaons were studied some years ago. We can look into the near future and enumerate the most important issues to pursue. Let us look at what information is necessary to significantly improve the definition of the unitarity triangle, as seen in Figure 0.

The top quark mass is known to be greater than 80 and smaller than about 250 GeV. There is only one way to narrow this range, and that is to increase the luminosity at hadron colliders and look more sensitively for top quarks. The plots in Fig. 10 also show the effect of the range of $|V_{cb}|$, from 0.036 to 0.056. This will be improved by measuring the separate B^0 and B^+ lifetimes.

The dotted line in Figure 10 corresponds to a range of 0.04–0.16 for $|V_{ub}/V_{cb}|$. With the information presented earlier, I concluded that the allowed range is closer to 0.06–0.20. As I discussed in some detail, one should measure B $\rightarrow \pi\ell\bar{\nu}$ and B $\rightarrow \rho\ell\bar{\nu}$ over the entire range of q^2 to improve this. In addition, one needs experimental help from charm form factors and theoretical help from the models to handle the case of light quarks in the final state.

The other two constraints used are limited by theory, and not experiment. The range $0.50 \leq X_d \leq 0.78$ is already quite good, but its usefulness is weak because of the uncertainty in calculating $B_B f_B^2$, which are relevant in interpreting the mixing parameter. Similarly, the CP-violation parameter ϵ from K $\rightarrow \pi\pi$ decays is measured very well, but the calculations of B_K are still imprecise. It is hoped that lattice gauge calculations will improve these parameters in the near future.

This program of experimental and theoretical work will insure an active period in heavy quark physics over the next few years. With a much improved picture of the unitarity triangle, the stage will be set for the main act, the investigation of CP violation in B decays. It is a reasonable goal that at DPF2000 we should be discussing the measurement of one or more angles in the unitarity triangle. These angles can only be measured by measuring the CP asymmetries in B decays.

ACKNOWLEDGEMENTS

I have benefited from Rollin Morrison's Madrid paper, as well as from many discussions with the author. Fred Gilman has patiently explained many of the fine points of semileptonic decays.

REFERENCES

1. J.L. Rosner, *Comments Nucl. Part. Phys.* **16**, 109 (1986), and references therein.

2. H. Albrecht et al., *Phys. Lett.* **B230**, 162 (1989).

3. M.S. Alam et al., *Phys. Rev. Lett.* **62**, 863 (1989).

4. G. Altarelli et al., *Nucl. Phys.* **B208**, 365 (1982).

5. J.G. Körner and G.A. Schuler, *Z. Phys.* **C38**, 511 (1988); erratum **C41**, 690 (1989).

6. M. Wirbel et al., *Z. Phys.* **C29**, 637 (1985).

7. B. Grinstein et al., *Phys. Rev. Lett.* **56**, 298 (1986); and N. Isgur et al., *Phys. Rev. D* **39**, 799 (1989).

8. J. Altomari and L. Wolfenstein, *Phys. Rev. Lett.* **58**, 1583 (1987) and *Phys. Rev. D* **37**, 681 (1988).

9. C. Bernard et al., Brookhaven preprint BNL-45321 (1990)

10. M. Crisafulli et al., *Phys. Lett.* **B223**, 90 (1989).

11. C.A. Dominguez and N. Paver, *Phys. Lett.* **B207**, 499 (1988); erratum **B211**, 500 (1988).

12. J.M. Cline et al., *Phys. Rev. D* **40**, 793 (1989).

13. M.A. Shifman and M.B. Voloshin, *Sov. J. Nucl. Phys.* **47**, 511 (1988).

14. N. Isgur, *Phys. Rev. D* **40**, 101 (1989).

15. N. Isgur and M.B. Wise, Caltech preprints CALT-68-1585 and CALT-68-1608.

16. F.J. Gilman and R.L. Singleton, *Phys. Rev. D* **41**, 142 (1990).

17. J.C. Anjos et al., *Phys. Rev. Lett.* **62**, 1587 (1989).

18. J. Adler et al., *Phys. Rev. Lett.* **62**, 1821 (1989).

19. R.M. Baltrusaitis et al., *Phys. Rev. Lett.* **54**, 1976 (1985).

20. J.C. Anjos et al., *Phys. Rev. Lett.* **62**, 722 (1989).

21. H. Albrecht et al., *Phys. Lett.* **B219**, 121 (1989).

22. D. Bortoletto et al., *Phys. Rev. Lett.* **63**, 1667 (1989).

23. S. Stone, *Nucl. Phys.* **B13 (Proc Suppl.)**, 286 (1990).

24. H. Albrecht et al., *Phys. Lett.* **B232**, 554 (1989).

25. R. Fulton et al., *Phys. Rev. Lett.* **64**, 16 (1990).

26. H. Albrecht et al., *Phys. Lett.* **B234**, 409 (1990).

27. G. Wormser et al., *Phys. Rev. Lett.* **61**, 1057 (1989).

28. G. Wormser, in Proceedings at the Heavy Quark Symposium, AIP Conference Proceedings **196**, 219 (AIP, New York, 1989).

29. J.C. Anjos et al., *Phys. Lett.* **B223**, 267 (1989), and paper to this conference.

30. T.E. Browder, SLAC preprint SLAC-PUB-5118.

31. W.Y. Chen et al., *Phys. Lett.* **B226**, 192 (1989).

32. J. Adler et al., *Phys. Rev. Lett.* **64**, 169 (1990).

33. J.C. Anjos et al., submitted to Phys. Rev. Lett.

34. M. Bauer et al., *Z. Phys.* **C34**, 103 (1987).

35. R.J. Morrison, in Proceedings of the European Physical Society Meeting (1989).

Fermion Number Violation in the Standard Model

R. D. Peccei

Department of Physics, UCLA, Los Angeles, California 90024

ABSTRACT

The chiral structure of the electroweak theory causes the fermion number current to have an anomalous divergence. I review here how this anomalous violation of fermion number can become important at temperatures near the electroweak scale, leading to the erasure of any preexisting fermion asymmetry, late in the universe's expansion. I also discuss how this same phenomena may lead to catastrophic fermion number violation at high energies. Here, however, issues connected with the proper unitarization of the theory are still unclear. Thus the question of whether abundant fermion number violation occurs at high energy still remains open.

I. Fermion Number Violation and the Universe's Baryon Asymmetry

In the standard electroweak theory fermion number (baryon plus lepton number) is classically conserved, but is violated at the quantum level [1] due to the presence of an Adler, Bell, Jackiw [2] anomaly. Even though the fermion number current is vectorial

$$J^\mu_{B+L} = \Sigma_i \bar{f}_i \gamma^\mu f_i = \Sigma_i \bar{f}_{iL} \gamma^\mu f_{iL} + \Sigma_i \bar{f}_{iR} \gamma^\mu f_{iR}, \tag{1}$$

because the left handed fermion fields f_{iL} are part of $SU(2)$ doublets, while the right handed fields f_{iR} are singlets, the divergence of J^μ_{B+L} does not vanish. The triangle graph of Fig. 1, involving the $SU(2)$ gauge fields, only gets contributions from f_{iL} and as a result one finds [1]:

$$\partial_\mu J^\mu_{B+L} = \frac{\alpha_w}{4\pi} N_g W^{\mu\nu}_a \widetilde{W}_{a\mu\nu} \tag{2}$$

Here $\alpha_w = \frac{g_2^2}{4\pi} \simeq \frac{1}{30}$, N_g is the number of generations and $\widetilde{W}_{a\mu\nu}$ is the dual of the $SU(2)$ field strength.

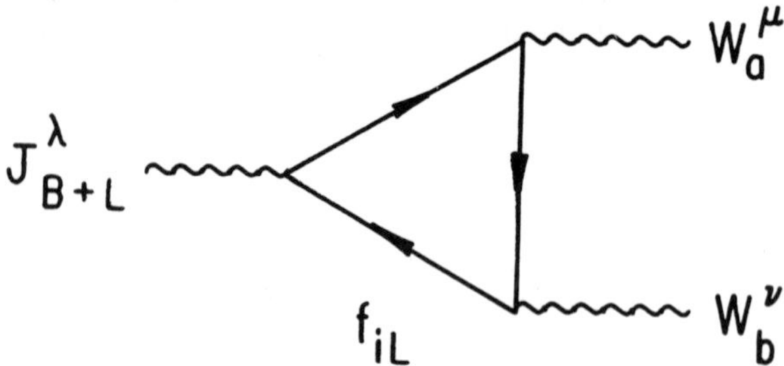

Fig. 1: Triangle graph responsible for fermion number non conservation in the electroweak theory.

Even though Eq. (2) shows that fermion number is not conserved in the standard model, the original estimate by 't Hooft [1] of the rate of occurrence of processes involving fermion violation were rather discouraging. In his seminal paper, where he first pointed out the possibility of violating fermion number in the standard model, 't Hooft also showed that these fermion number violating processes were always associated with gauge field configurations with non trivial Pontryagin number

$$\nu = \frac{\alpha_w}{8\pi} \int d^4x \, W_a^{\mu\nu} \widetilde{W}_{a\mu\nu} \quad , \tag{3}$$

so that the total fermion number change involved is

$$\Delta(B+L) = 2N_g\nu \qquad \nu = \pm 1, \pm 2, \; \cdots \tag{4}$$

To estimate the probability amplitude for such a transition to happen 't Hooft used the weight associated with instanton configurations for $\nu = 1$ found by **Belavin** et al [3], since these configurations give the minimum action in Euclidean space for $\nu = 1$. His estimate for these processes was then simply:

$$P_{B+L\text{viol}} \sim |e^{-S^{Eucl\,min\,(\nu=1)}}|^2 = e^{-\frac{4\pi}{\alpha_w}} \tag{5}$$

Clearly for the electroweak theory, where $\alpha_w \approx \frac{1}{30}$, Eq. (5) gives a tiny number and appears to make the whole issue of fermion number violation in the standard model just an academic curiosity.

About 5 years ago Kuzmin, Rubakov and Shaposhnikov [4] revived interest in this question by considering fermion violating processes at high temperatures. The zero temperature result of Eq. (5) can be interpreted as a tunnelling process between inequivalent gauge vacua in the $A_0 = 0$ gauge [5]. The suggestion of the authors of [4] is that, at finite temperature, rather than tunnelling between the two vacuum states, it is more profitable to actually get over the barrier by a thermal fluctuation. Thus, at finite T, they arrived at a Boltzman-like expression for $P_{B+L\text{viol}}$:

$$P_{B+L\text{viol}} \sim e^{-\frac{V_0}{T}} \tag{6}$$

where V_0 is the height of the barrier separating the inequivalent gauge vacua. Furthermore, they associated V_0 with the minimum energy of a static spherically symmetric solution of the electroweak theory, obtained previously by Klinkhamer and Manton [6] - the, so called, sphaleron.

The sphaleron is characterized by gauge and Higgs configurations

$$\vec{W} = \frac{v}{g_2} \frac{f(\xi)}{\xi} \hat{r} \times \vec{\tau}; \quad \Phi = \frac{v}{\sqrt{2}} h(\xi) \hat{r} \cdot \vec{\tau} \binom{1}{0} \tag{7}$$

Here v is the Higgs vacuum expectation value, $v \simeq 250\ GeV$, and the functions $f(\xi)$ and $h(\xi)$, with $\xi = 2M_W r$, are fixed by minimizing the energy and have a range $\xi \sim 0(1)$. In view of (7), not surprisingly, the energy of the sphaleron is of order

$$E_{sph} \sim \frac{M_W}{\alpha_w} \sim 10\ TeV \qquad (8)$$

Since at finite temperature $M_W = M_W(T)$, the height of the barrier $V_0 \simeq E_{sph}(T)$ actually decreases as T goes towards the electroweak phase transition. So the probability of having fermion number violating processes via thermal fluctuations is, essentially, not exponentially damped at temperatures near the electroweak phase transition and is very much above the $T = 0$ estimate of Eq. (5).

The observation of Kuzmin et al [4] that the rates for fermion number violating processes are large near $T = T_{EW}$ - the electroweak phase transition temperature - has important cosmological consequences. Indeed, using recent calculations of Arnold and McLerran [7], where also the nonexponential prefactors in the rate are included, one finds that, near T_{EW}, $\Gamma_{B+L \text{viol}} >> H$ where H is the Hubble parameter at this temperature. It follows, therefore, that fermion violating processes are in **equilibrium** near T_{EW}, and hence that any asymmetry established at an earlier epoch between the number of fermions and antifermions gets erased. Since we do observe a non zero asymmetry between baryons and antibaryons at present:

$$\eta = \frac{n_B - n_{\bar{B}}}{n_\gamma} \simeq \frac{n_B}{n_\gamma} \simeq 10^{-10} \qquad (9)$$

these results suggest two alternatives:

i) The Universe is not $B \leftrightarrow L$ symmetric. So, although anomaly related electroweak processes erase all the existing $(B+L)$ asymmetry at $T \sim T_{EW}$, the observed baryon asymmetry is connected to a $(B - L)$ asymmetry established at an earlier epoch. This is a perfectly viable possibility since there exist GUTs which allow for the generation of both a $(B - L)$ and a $(B + L)$ asymmetry [8]

ii) The Universe is $B \leftrightarrow L$ symmetric and all the fermion number asymmetry is erased at $T \sim T_{EW}$. However, as the Universe cools further the fermion number violating processes go again out of equilibrium and an asymmetry is established at temperatures below the electroweak scale.

The first possibility above is interesting since, if true, we are learning indirectly from low energy physics that the baryon asymmetry in the universe is really a $B - L$ asymmetry. Thus we are putting constraints on **what** possible GUTs are allowed. For instance, $SU(5)$ is disfavored since all $SU(3) \times SU(2) \times U(1)$ conserving, dimension six operators also conserve $(B-L)$ [9]. The second option is even more interesting, since it implies that η is really a number which should be calculable from the low energy theory. In particular, the fact that there are more baryons than antibaryons should be directly connected with the sign of the CP violating parameter $Re \ \epsilon$ in K decays! Unfortunately, early attempts to generate the Universe's baryon asymmetry below T_{EW} have not been successful, principally because there appears to be not enough CP violation in the low energy theory to generate a large enough asymmetry [10]. *

In the literature another possibility, besides the two above, was aired. Namely, that the sphaleron estimate of Eq. (6) is just not correct [11] [12]! However, careful recent discussions of this problem by Arnold and McLerran [13], Khlebnikov and Shaposhnikov [14] Grigoriev, Rubakov and Shaposhnikov [15] and Dine, Lechtenfeld, Sakita, Fischler and Polchinski [16] have done much to dispel these doubts. Although it is **a straw** man, it is useful for me to indicate one such objection here since it will directly bring me to reexamine in a different light the $T = 0$ estimate of 't Hooft. Ellis et al [11] objected to the sphaleron estimate for $P_{B+L\text{viol}}$ by arguing that even at finite temperature instantons provide the lowest action configuration in a $\nu = 1$ sector [17]. Thus, irrespective of how the fermion violating transition occurred, $P_{B+L\text{viol}}$ according to the authors of [11] must always be bounded by the tunnelling factor $e^{-\frac{4\pi}{\alpha_w}}$, with $\alpha_w = \alpha_w(T)$, the coupling constant at the scale set by T. It is fairly obvious that in a heat bath, however, such reasoning can be falsified by appropriate entropy factors [16], so that this objection is really not tenable. More interestingly, however, Arnold and McLerran [13] suggested another qualitative way in which such a tunnelling factor could be obviated. They argued, by means of physical examples, that only for a few particle processes one ought to expect the presence of a tunnelling factor. For the case of many quanta - of which the sphaleron

---

* I do not believe, however, that the GIM suppression adduced in [10] for the $B + L$ violating transitions. obtains. Indeed, as I will discuss below, for $B + L$ violation to take place **all** fermions must participate. Hence I see no reason why there should be a GIM suppression.

configuration is a prototype - such tunnelling factors can be **compensated** by other factors related to the multiplicity of the event. These arguments are interesting since these considerations should apply irrespective of the ambient temperature. Thus they naturally lead one to examine anew 't Hooft's $T = 0$ estimate for $P_{B+L\text{viol}}$.

II. Catastrophic Fermion Number Violation at 100 TeV?

A. Ringwald [18] and later, but independently, O. Espinosa [19] performed an explicit $T = 0$ calculation for fermion number violating processes, which explicitly **vindicated** the qualitative picture suggested by Arnold and McLerran [13]. What Ringwald [18] and Espinosa [19] showed, effectively, was that the tunneling factor in $B + L$ violating processes involving **many quanta**, in the approximation of neglecting any back reaction, at high energy gets fully compensated, leading to cross sections of the $0(1\ pb)$ for these processes at $\sqrt{\hat{s}} \simeq 50\ TeV$. This is a remarkable result and one that is pregnant with experimental consequences, particularly for the SSC! However, the coherent effects which kill the tunnelling factor at high energies (and which presumably account for how the sphaleron approximation works at high temperatures) in fact lead to a cross section which grows exponentially with energy and thus violates unitarity! Obviously, before one can take these results as indications that the dominant electroweak processes at high energy involve $B + L$ violations, one must first understand how - and when - unitarity gets restored.

The calculation of Ringwald and Espinosa are based on a formula for fermion number violating Green's functions which was derived a long time ago by Helen Quinn and myself [20]. Our paper, which was written in the early days of instanton physics, was primarily concerned with the question of how it was at all possible to have a Green's function which carried non zero fermion number. As a result, after we understood how the existence of chirally asymmetric zero modes in $\nu \neq 0$ sectors of the theory can give rise to a $B + L$ violating Green's function, we were content just to write down a formula for these Green's functions, but never examined this formula in any detail! Thus, unfortunately, (or perhaps fortunately, since I am sure that if we had realized the implications of our formulas, it would have confused us terribly then), we missed the important point discovered by Ringwald and Espinosa. Namely, that in the semiclassical approximation one is employing, these Green's functions are effectively factorizable, giving rise as a consequence to scattering cross sections which grow extremely rapidly with energy.

I will illustrate what is going on by considering, for simplicity, the case of one generation ($N_g = 1$). The relevant Green's function to calculate in this case involves two u-quarks one d–quark and an electron field * What Quinn and I showed [20] is that this Green's function does not vanish in a $\nu = 1$ sector because the (Euclidean) Dirac operator in the presence of background fields of non vanishing index ν have an asymmetric zero mode spectrum. In fact, we showed that, at the semiclassical level, the Green's function G_{uude} was just given by the product of the (left handed) zero mode solutions of

$$\gamma \cdot D[W]\psi_0 = 0 \tag{10}$$

for the quark and lepton fields in question. **

Ringwald [18] and Espinosa [19] rederived independently our formula for G_{uude}, but they noticed a conundrum connected with this formula which we missed. The background field W in the Dirac operator of Eq. (10) - or if there are also Higgs fields in the problem, the background gauge and Higgs fields - at the semiclassical level depends on collective coordinate parameters which describe various features of these solutions. For instance, these background fields may depend on a size parameter ρ, a location z^μ or some appropriate gauge degree of freedom. It follows, therefore, that also the zero mode solution ψ_0 depends on these parameters, e.g. $\psi_0 = \psi_0[x - z; \ \rho]$. In the formula for the Green's function G_{uude}, which involves the product of the zero modes for all the fermion fields participating in the process, one must integrate over these collective coordinates. Hence, schematically, one has

$$G_{uude} \sim \epsilon_{\alpha\beta\gamma} \int d^4z d\rho \psi_0^{u\,\alpha}[x_1 - z; \ \rho]\psi_0^{u\,\beta}[x_2 - z; \rho]$$
$$\psi_0^{d\,\gamma}[x_3 - z; \rho]\psi_0^e[x_4 - z; \rho]e^{-\frac{2\pi}{\alpha_w(\rho)}} \tag{11}$$

Since one is dealing with a spontaneously broken theory, the integral over ρ is

---

* The quark field's color indices are appropriately antisymmetrized to give a color singlet.

** The Atiyah Singer theorem [21] relates ν to the number of left-handed and right-handed zero modes of Eq. (10): $\nu = n_L - n_R$. Thus for $\nu = 1$ one expects to have only one left-handed zero mode for each fermion species, but no right-handed zero modes.

cut off at a scale $\rho \sim \frac{1}{v}$ *. The integral over the location z of the background field restores the translational invariance in G_{uude}. However, translating all the space time points x_i by z in (11) leads to an amplitude where the contributions of each fermion totally **factorize** from each other. Thus the result (11) is equivalent to having **a local effective Lagrangian** for fermion number violation:

$$\mathcal{L}_{\text{B+L viol}} \sim \frac{e^{-\frac{2\pi}{\alpha_w}}}{v^2} \epsilon_{\alpha\beta\gamma} u_{\alpha L}^T C d_{\gamma L} u_{\beta L}^T C e_L \tag{12}$$

Eq. (12) displays clearly the conundrum uncovered by Ringwald and Espinosa. If one takes (12) at face value then the resulting cross section for fermion number violating processes violates unitarity since σ grows linearly with s

$$\sigma \sim e^{-\frac{4\pi}{\alpha_w}} \frac{s}{v^4} \tag{13}$$

Of course, because of the tiny tunnelling factor, unitarity is only violated at stupendously large values of s : $s \sim v^2 e^{\frac{2\pi}{\alpha_w}}$. It is certainly questionable if at these energies the semiclassical calculation which leads to the Green's function (11) really applies. The factorizability present in Eq. (11), which is at the root of the rapid growth with energy in (13), arises because one treats each fermion in the process independently, assigning to each an identical zero mode wavefunction ψ_0. In this approximation no back reaction is accounted for. However, the back reaction should give some damping, since it correlates the fermions.

Once one understands what is going on in the case of one generation, the extension to three generations is trivial. The Green's function which is non vanishing in the $\nu = 1$ sector in this case involves 12 fermion fields. Semiclassically one again has a factorizable amplitude, which is equivalent to a local effective Lagrangian of the schematic form

$$\mathcal{L}_{\text{B+L viol}} \sim \frac{e^{-\frac{2\pi}{\alpha_w}}}{v^{14}} \psi^{12} \tag{14}$$

---

* To be more precise, for $v \neq 0$ there does not actually exist a finite action solution of the Euclidean equations of motion [1]. However, for $\rho v << 1$ one recovers the usual instanton solution, with an additional piece in the action due to the Higgs field given, approximately, by $\pi^2 v^2 \rho^2$.

Hence the cross section for the B + L violating process $q + q \to 7\bar{q} + 3\bar{\ell}$ grows even more dramatically with energy

$$\sigma \sim e^{-\frac{4\pi}{\alpha_w}} \frac{s^{13}}{v^{28}} \tag{15}$$

Unitarity is now violated at lower energies ($s \sim 10^8\ GeV$), but energies still very far away from the scale of the weak interactions.

Ringwald [18] and Espinosa [19], however, made the crucial observation that the **inclusive** fermion number violating cross section grows even faster. Indeed, as we shall see, one arrives at an exponential growth of σ with s, using the same semiclassical approximation of Eq. (11). For the process $q+q \to 7\bar{q}+3\bar{\ell}+n_w W + n_h H$ one needs to calculate the Green's function $G \sim\ < T(\psi^{12} W^{n_w} H^{n_h}) >$. Semiclassically ψ gets replaced by the zero mode solution $\psi_0\left[W_{\mathrm{back}}, H_{\mathrm{back}}\right]$ in the presence of the $\nu = 1$ background fields, while $W \to W_{\mathrm{back}}$ and $H \to H_{\mathrm{back}}$. Since the background fields are localized at z^μ, it follows that again one gets an effective local form of this Green's function, equivalent to *

$$\mathcal{L}_{B+L\ \mathrm{viol}} \sim \frac{e^{-\frac{2\pi}{\alpha_w}} \psi^{12} W^{n_w} H^{n_h}}{v^{14+n_w+n_h}\ g_2^{n_w}} \tag{16}$$

Pictorially, one gets the highly local particle production process shown in Fig. 2.

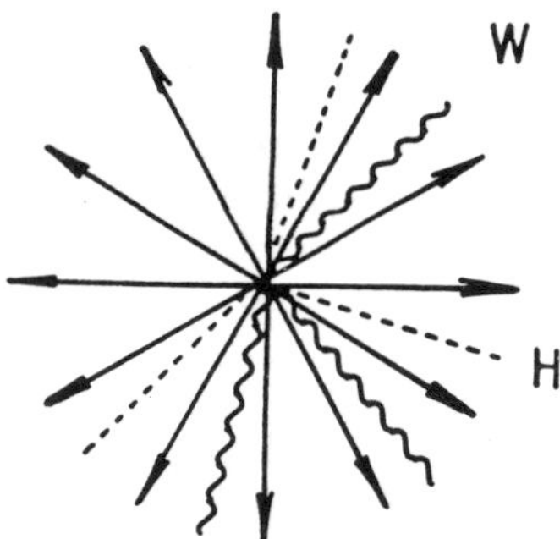

Fig. 2: Local form of the $B + L$ violating interaction in the semiclassical

---

* Recall from Eq. (7) that each vector classical field is inversely proportional to g_2.

92

approximation which leads to factorization.

The effective inclusive cross section for this process grows as $s^{13+n_w+n_h}$. Indeed, if one takes into account of appropriate factorial factors associated with phase space and the permutation of the $n_w W's$ and $n_h H's$, one finds for large n [18, 19]

$$\sigma^n_{B+L \; \text{viol}} \sim \frac{s^{n-1}}{v^{2n} n!} e^{-\frac{4\pi}{\alpha_w}} \tag{17}$$

Thus, for the total inclusive cross section the semiclassical approximation leads to an exponentially growing cross section with energy

$$\sigma^{\text{total}}_{B+L \; \text{viol}} \sim e^{-\frac{4\pi}{\alpha_w}} \frac{1}{s} \, e^{\frac{s}{v^2}} \tag{18}$$

Obviously, such a formula violates unitarity and indicates clearly the limitation of the semiclassical approximation employed. One can estimate roughly at what energy one begins to be in trouble. This will occur when the exponentially growing factor in (18) compensates for the tunnelling factor, corresponding to an energy

$$\sqrt{s} \sim \frac{M_w}{\alpha_w} \sim E_{sph} \tag{19}$$

Doing a more precise calculation, Ringwald [18] showed that the $B+L$ violating cross sections get to be of the order of the $B + L$ conserving cross sections for $\sqrt{\hat{s}} \simeq 50 \; TeV$, where the average number of produced bosons is of the order 50-60! However, precisely at these energies one is beginning to get in trouble with unitarity and one cannot really trust the calculation.

McLerran, Vainshtein and Voloshin [28] examined the results of Ringwald [18] and Espinosa [19] and concluded that, although these results require extrapolation beyond the validity of the constrained instanton approximation [22] used, the corrections to their results, in fact, seemed to indicate that the $B + L$ violating cross section was **bigger** rather than smaller than that obtained in [18, 19]! Effectively the amplitude becomes large in the Ringwald-Espinosa calculation near $\sqrt{\hat{s}} \simeq E_b^{\text{const inst}} \simeq 40 - 50 \; TeV$, while the McLerran et al approach seemed to indicate that this already occurred at $\sqrt{\hat{s}} \approx E_{sph} \approx 10 \; TeV$. One can understand this behavior roughly as follows [23]. In the $W_0 = 0$ gauge, the constrained instanton interpolates between the $\nu = 0$ and $\nu = 1$ vacua with the Euclidean time τ acting as a path parameter, with the background fields passing through the sphaleron configuration (7) at $\tau = 0$. The barrier height at $\tau = 0$ is just the minimum of the static energy [24]:

$$E_b(\rho) \simeq \frac{3\pi}{4\alpha_w}\frac{1}{\rho} + \frac{3}{8}\pi^2 v^2 \rho^2 \qquad (20)$$

This minimum occurs at $\rho \sim \frac{1}{M_W}$ and $E_b^{Min} \simeq E_{\rm sph}$. To minimize the Euclidean action

$$S_{\rm Euclid} = \frac{2\pi}{\alpha_w} + \pi^2 v^2 \rho^2 \qquad (21)$$

one needs $\rho \simeq \rho_{\rm inst} \simeq \frac{1}{\pi v} << \frac{1}{M_W}$. At this radius indeed the barrier height $E_b^{\rm const\ inst} \simeq 40\ TeV$. What McLerran et al [21] argue is that the corrections to the constrained instanton calculation push one to configurations of size $\rho_{\rm min} \sim \frac{1}{M_W}$ and hence that the $B + L$ violating processes become important when $\sqrt{\hat{s}} \simeq E_{\rm sph} = 10\ TeV$.

In the picture of McLerran et al [21], unitarity is restored through multi instanton repetitions, which become important only as one approaches the unitarity limit. So $\sigma_{B+L\ \rm viol}$ grows with energy essentially up to the unitarity limit and then one gets a "black disk" cross section

$$\sigma_{B+L\ \rm viol} \simeq \pi R^2 \qquad (22)$$

with $R \sim \frac{1}{E_{\rm sph}}$. Although this would be wonderful super high energy physics for the electroweak sector, the question, however, is if this is true or not! In fact, there is also a quite different way to restore unitarity than via multi instanton iterations at the last minute. Namely, one can conceive that correction to the semiclassical approximation of (11) destroy the factorization property, which is responsible for the rapid growth with energy of the $B + L$ violating amplitudes. These corrections correspond to taking into account of the back reaction on the fermions and, intuitively, they should provide some softening already at the one instanton level [25]. McLerran et al [21], roughly speaking, tried to take into account of some corrections to the factorizability approximation by considering the effect of the fermions as being equivalent to having a constant source. They found, however, in this approximation an enhancement, not a decrease in the size of the amplitude. As I will argue below, based on the work of my UCLA colleagues, Aoki, Cornwall and Mazur [26, 27, 28], the space-time properties of the effective correlating source are important and can lead to qualitatively different phenomena.

The locality of the interactions responsible for the rapid growth in energy of the $B + L$ violating cross section can be traced to the use of **independent** zero mode eigensolutions of the Euclidean Dirac equation (10) for each fermion field. One can try to take into account of the interdependence of these solutions by writing [26] the zero mode solution in the presence of the background field W as

$$\psi_0^i(x_i) = e^{J[W(x_i)]}\eta \tag{23}$$

where η is some constant spinor. Then formally one may account for the correlations in a Green's function containing n fermions by considering the modified action

$$S_{eff}[W] = S[W] - \Sigma_{j=1}^n J[W(x_j)] \tag{24}$$

Stationary solutions of (24), rather than those of the action $S[W]$, will take into account of the fact that there are n fermions at the locations $\{x_j\}$.

The presence of the "external" source J has an important technical consequence. The field equations for the gauge fields * now read

$$D^\mu W^{\mu\nu}(x) = \Sigma_j^n \frac{\delta J[W(x_j)]}{\delta W^\nu(x)} \equiv J^\nu(x; \{x_j\}) \tag{25}$$

Because of the presence of the effective current J^ν on the RHS, it is no longer true that the classical background field solutions are characterized by collective coordinates z^μ, ρ, etc., which are arbitrary. If we denote by W_0^ν the solutions of $D^\mu W^{\mu\nu} = 0$, then Eq. (25) has **only** a solution if J^ν is orthogonal to W_0^ν

$$\int dx\, J^\nu(x; \{x_j\})W_0^\nu(x) = 0 \tag{26}$$

---

* I neglect here the ordinary fermionic current on the RHS of (25), which is irrelevant for the argument at hand.

Eq. (26) puts a constraint on the collective coordinates, z_0^μ, ρ_0, etc, characterizing the classical solution $W_0^\nu(x) = W_0^\nu(x - z_0; \rho_0)$. Indeed, Aoki and Mazur [26] show that Eq. (26) is equivalent to requiring that the effective action (24) be stationary with respect to variation in the collective coordinates.

Taking (26) into account, one can write the solution of (25) as

$$W^\mu(x) = W_0^\mu[x - z_0; \rho_0] + \int dhy \tilde{G}^{\mu\nu}(x|y) J^\nu(y) \tag{27}$$

where $\tilde{G}$ is the Green's function for the problem at hand, with the zero modes removed. Eq. (27) contains two effects: (i) a classical solution W_0^μ where, however, because of the constraint (26) the collective coordinates are fixed; (ii) corrections to this classical solution due to the back reaction. The second effect, since it correlates all the various points $\{x_j\}$ where the fermions are located, will give non factorizable corrections to the zeroth order factorizable contribution [27]. However, the detail import of these effects is nontrivial to estimate. I will concentrate therefore on the first effect, which is also important since it implicitly fixes the relevant collective coordinates in terms of the process one wants to compute.

To illustrate what is going on consider the zero mode solution in the presence of instantons, which can be written as [1]

$$\psi_0[x_i] = \frac{\rho_0}{\left[(x_i - z_0)^2 + \rho_0^2\right]^{3/2}} \eta \tag{28}$$

Thus

$$S_{\text{eff}} = S[W] - n\ell n\rho_0 - \frac{3}{2}\Sigma_i \ell n[(x_i - z_0)^2 + \rho_0^2] \tag{29}$$

The orthogonality relation (26) is satisfied by minimizing S_{eff} with respect to ρ_0 and z_0^μ, which implies the following constraints [27]

$$z_0^\mu = \frac{\Sigma_i \frac{x_i^\mu}{(x_i - z_0)^2 + \rho_0^2}}{\Sigma_i [(x_i - z_0)^2 + \rho_0^2]^{-1}} \quad ; \quad \rho_0^2 = \frac{n}{3\Sigma_i [(x_i - z_0)^2 + \rho_0^2]^{-1}} \tag{30}$$

These constraints must be introduced into the semiclassical formula (11) and will directly affect the result. In particular, it is no longer true that the origin z_0^μ of the background field can be moved anywhere, since (30) relates z_0^μ to the locations $\{x_j\}$ of the fermions. Thus factorizability will no longer hold exactly. Furthermore the scale ρ_0 is no longer only fixed by the Higgs vacuum expectation value v, but it depends also via (30) on the coordinates.

In practice it does not appear to be too easy to incorporate these constraints, although as Cornwall emphasizes [28], qualitatively the constraint on ρ_0 appears to be the most important. One can see this as follows. Neglecting this constraint, the size of the classical solution is set by $\rho_0 \sim \frac{1}{v}$ (or $\rho_0 \sim \frac{1}{M_W}$, if one follows the arguments of [21]). However, incorporating the constraint (30), ρ_0 is of the order of the average distance d between the fermions. If one produces n quanta from an initial energy $\sqrt{s}$, one expects [28]

$$\rho_0 \simeq d \simeq \frac{n}{\sqrt{s}} \tag{31}$$

If this reasoning is correct, it is clear that the back reaction constraints automatically unitarize the Ringwald-Espinosa amplitude since, effectively, $v \to \frac{n}{\sqrt{s}}$. Thus the partial cross sections (17) really do not grow with s, but they may still be large for $n > \frac{1}{\alpha_w}$ [28]:

$$\sigma^n_{B+L \text{ viol}} \sim \frac{1}{s} \Big[\frac{n^{2n}}{n!} e^{-\frac{4\pi}{\alpha_w}} \Big] \tag{32}$$

III. Postscript and Concluding Remarks

Let me summarize some of the principal points discussed. Because of the asymmetry of the $SU(2)$ assignments for the left and right handed fermions, the standard model does not conserve total fermion number. Although at the semiclassical level the $B + L$ violating amplitudes are damped by a factor of $e^{-\frac{2\pi}{\alpha_w}}$, coherent effects, mediated by sphalerons, effectively overwhelm this damping at temperatures near that of the electroweak phase transition. Furthermore, the rate for $B + L$ violation is sufficiently large that possible asymmetries in the total fermion number, established earlier in the Universe, are erased. This suggests that the observed baryon asymmetry in the Universe is either due to $B - L$ violation at the GUT scale, or is generated at temperatures below those of the electroweak phase transition.

In the semiclassical approximation one finds that the amplitudes for fermion number violation factorize in momentum space, allowing one to develop large coherent effects not only at high temperatures but also at high energies. Indeed, in this approximation, the tunnelling factor of $e^{-\frac{4\pi}{\alpha_w}}$ in the total cross section for fermion number violating processes is overwhelmed at energies of the order of $E \sim E_{sph}$. However, the exponential growth of the cross section with energy also indicates that this approximation violates unitarity and is unreliable at precisely energies of this order of magnitude. Thus to ascertain whether $B + L$ violation is an important phenomena or not at high energy requires first understanding how, and at what energy, corrections to the semiclassical calculation employed restore unitarity.

At the time when this meeting was held two points of view appeared tenable. The more conservative point of view held that, after incorporating the fermion back reaction properly and taking into account of quantum corrections to the semiclassical calculation, one would find that $B + L$ violating processes are still small due to the tunnelling factor. The rapid rise in the $B + L$ violating cross section is only true for $E << E_{sph}$. Soon thereafter, damping effects effectively render these processes phenomenologically irrelevant. However also the diametrically opposed point of view had some credence. Namely that the corrections to the semiclassical calculation in the presence of just one $\nu = 1$ background field, do not substantially alter the rapid growth with energy of the inclusive cross section. In this view unitarity is restored only by including additional instanton-antistanton configurations, which become important at energies where the tunnelling factor is essentially overwhelmed. Due to $B + L$ violation. electroweak processes become strong at high energies.

This interesting dichotomy, and the obvious physical interest of the question, have spurred considerable activity in this topic since the Rice meeting. It is obvious beyond my purview here to enter into any detail on these very recent contributions. However, I would like to make three observations which I think are pertinent:

> i) The basic dichotomy outlined above has not really been resolved although, as one has understood more about this problem, it appears now perhaps less likely that $B + L$ violation will be the dominant electroweak phenomena at high energy.

> ii) The study of model field theories, like the Abelian Higgs model in 2 dimensions [29], or a weakly coupled $SU(2)$ supersymmetric Yang Mills

model [30], can be very useful to elucidate what goes on in the electroweak theory. So does the intuition that one can build-up in even simpler quantum mechanical analog models [31] [32], or by studying properties of large order perturbation theory [28] [33].

iii) Some general, and rather interesting, properties of the total $B + L$ violating inclusive cross section can be established [34] [35]. In particular, one can show that the corrections to the semiclassical calculation **also** exponentiate [36] [37] [38], so that

$$\sigma_{B+L \text{ viol}} = \frac{1}{s} \exp \frac{4\pi}{\alpha} \, F \left[\frac{s}{E^2_{sph}} \right] \tag{33}$$

One knows that $F(0) = -1$ from the work of 'Hooft [1]. From the Ringwald-Espinosa calculation one learned that the slope is positive there: $F'(0) > 0$. However. the real question, which has not been answered yet, is if there is some energy $\sqrt{s}$ where F itself becomes positive, or at least takes a value very close to zero.

Acknowledgements

I am grateful to Ken Aoki, Mike Cornwall, Pawel Mazur, Larry McLerran, Shmuel Nussinov, Andreas Ringwald, Valery Rubakov, Misha Shaposhnikov, Arkady Vainshtein, Misha Voloshin, Larry Yaffe and Valya Zakharov for illuminating discussions on these matters.

References

[1] G 't Hooft, Phys. Rev. Lett. **37** (1976) 8; Phys. Rev. D14 (1976) 3422

[2] S. L. Adler, Phys. Rev. 177 (1969) 2926; J. S. Bell and R. Jackiw, Nuovo Cimento 60 (1969) 47

[3] A. Belavin, A. Polyakov, A. S. Schartz and Yu. Tyupkin, Phys. Lett. 59B (1975) 85

[4] V. Kuzmin, V. Rubakov and M. Shaposhnikov, Phys. Lett. 155B (1985) 36

[5] C. G. Callan, R. Dashen and D. Gross, Phys. Lett. 63B (1976) 334; R. Jackiw and E. Rebbi, Phys. Rev. Lett 37 (1976) 172

[6] F. Klinkhamer and N. Manton, Phys. Rev. D30 (1984) 2212

[7] P. Arnold and L. McLerran, Phys. Rev. D36 (1987) 581

[8] M. Fukugita and T. Yanagida, Phys. Lett. 174B (1986) 45

[9] S. Weinberg, Phys. Rev. Lett 43 (1979) 1566; F. Wilczek and A. Zee, Phys. Rev. Lett. 43 (1979) 1571

[10] M. Shaposhnikov, Nucl. Phys. B287 (1987) 1757; B299 (1988) 797

[11] J. Ellis, R. Flores, S. Rudaz and D. Seckel, Phys. Lett. 194B (1987) 241

[12] A. Cohen, M Dugan and A. Manohar, Phys. Lett. 222B (1989) 91

[13] P. Arnold and L. McLerran, Phys. Rev. D37 (1988) 1020

[14] S. Yu. Khlebnikov and M. Shaposhnikov, Nucl. Phys. B308 (1988) 885

[15] D. Yu. Grigoriev, V. A. Rubakov and M. Shaposhnikov, Phys. Lett. 216B (1989) 172

[16] M. Dine, O. Lechtenfeld, B. Sakita, W. Fischler and J. Polchinski, CCNY-HEP- 89/18, December 1989

[17] D. Gross, R. Pisarski and L. Yaffe, Rev. Mod. Phys. 53 (1981) 43

[18] A. Ringwald, Nucl. Phys. B 330 (1990) 1

[19] O. Espinosa, Caltech Preprint, CALT-68-1586, November 1989

[20] R. D. Peccei and H. R. Quinn, Nuovo Cimento A41 (1977) 1309

[21] L. McLerran, A. Vainshtein and M. Voloshin, Minnesota preprint TPI-MINN 89/36T

[22] I. Affleck, Nucl. Phys. B 191 (1981) 445

[23] A. Ringwald, DESY 89-153, in Proceedings of the 3rd Hellenic School on Elementary Particle Physics, Corfu, Greece, 1989

[24] V. Rubakov, Nucl. Phys. B256 (1986) 509

[25] R. D. Peccei, UCLA/89/TEP/65, in Proceedings of the Workshop on Theoretical and Phenomenological Aspects of Underground Physics, L'Aquila, Italy, 1989

[26] K. Aoki and P. Mazur, UCLA/89/TEP/67

[27] K. Aoki, UCLA/90/TEP/7, Mod. Phys. Lett. A. to be published

[28] J. M. Cornwall, UCLA/90/TEP/2, Phys. Lett. B. to be published

[29] S. Yu. Khlebnikov, V. A. Rubakov and P. G. Tinyakov, Institute of Nuclear Research, Moscow preprint, 1990

[30] L. McLerran, A. Vainshtein and M. Voloshin, Minnesota preprint, TPI-MINN 90/8-T and TPI-MINN 90/27-T

[31] R. Singleton Jr., L. Susskind, L. Thorlacius, SLAC preprint, SLAC-Pub-5225

[32] T. Banks, G. Farrar, M. Dine, D. Karabali and B. Sakita, Rutgers preprint, RU-90-20

[33] A. Casher and S. Nussinov, Tel Aviv preprint, 1990

[34] V. Zakharov, Minnesota preprint, TPI-MINN 90/7-T

[35] M. Porrati, **Berkeley** preprint, LBL-28980

[36] P. Arnold and M. Mattis Argonne preprint ANL-HEP-PR-90-27

[37] L. Yaffe, private communication, and to be published

[38] S. Yu. Khlebnikov, V. A. Rubakov and P. G. Tinyakov, Institute for Nuclear Research, Moscow preprint, 1990

ELECTRON-POSITRON COLLIDERS AT INP: STATUS AND PROSPECTS

A. N. SKRINSKY
Institute of Nuclear Physics
630090 Novosibirsk, USSR

Experiments with electron-positron colliding beams long ago became an essential part of high energy physics. Fig. 1 and Table 1 show evidently two main directions in the development of e^+e^- colliders - energy and luminosity increase. Such experiments started in 1967 in Novosibirsk on the storage ring VEPP-2, where the excitation curve of the ρ-meson resonance in the reaction $e^+e^- \to \pi^+\pi^-$ was obtained. The maximum luminosity achieved on that collider was $5 \cdot 10^{28}$ cm^{-2} s^{-1} only. For comparison, $1 \cdot 10^{32}$ cm^{-2} s^{-1} has been already achieved on CESR at Cornell, whereas at future electron-positron factories one can hope to approach the value $1 \cdot 10^{34}$ cm^{-2} s^{-1}. The rate of energy increase is also high. This year a very important stage has been passed — SLC and LEP have observed the first production of Z^0 bosons in e^+e^- collisions. On behalf of the whole accelerator community I'd like to congratulate the SLAC and CERN laboratories for their successes. Further considerable energy increase is inevitably related to a transfer from circular electron-positron colliders to linear ones. The obvious need in international collaboration on development of the linear colliders begins to acquire practical put-lines and the first joint step will be construction of the experimental installation for operation with beams of sub-micron transverse size. Such an installation is being planned by the collaboration SLAC-INP-KEK-CERN using SLC as a base.

One should keep in mind that increase in energy and luminosity of e^+e^- colliders does not exhaust all the problems of their development. The use of transversely polarized beams enabled one to improve the accuracy of particle mass measurement by orders of magnitude,[1] see Table 2. Longitudinally polarized colliding beams will make the physics potential of e^+e^- experiments considerably higher. Of great interest is also the improvement of monochromaticity of the colliding beams,[2] which is rather high even without special efforts.

Consider now the status of e^+e^- facilities now available at the Institute of Nuclear Physics (INP). For over 15 years the VEPP-2M collider has provided luminosity by more than one order of magnitude higher than total luminosity of all other colliders in this energy range (up to 1.4 GeV in total). Start of operation of a new booster BEP will enable a further several time increase in the efficient luminosity (Fig. 3). Our experience with the VEPP-2M facility allowed us to find a way of increasing luminosity in the region of the ϕ-meson resonance by two more orders. The Novosibirsk ϕ factory (Fig. 4) will provide about 10^{11} pairs of $K_S K_L$ mesons per year.[3] This will ensure the improvement of the accuracy of the most important parameters of CP-violation in the K_L decays by one order and K_S decays by five orders of magnitude. This very problem of studying CP-violation is a main stimulus for the development of the ϕ-factory, but of course, new possibilities will open for investigation of rare and, at the same time, interesting decay modes of ϕ and K mesons.

The special features of the ϕ-factory under development are:

•the orbit shape resembling digit 8 enables one to combine both interaction points of each pair of e^+e^- bunches in the region of one detector;

•focusing in the interaction region using superconducting solenoids providing symmetric focusing with β-functions less than 1 cm;

•use of superconducting bending magnets ensures low damping time;

•the choice of the focusing structure providing equal beam size independently excited by quantum fluctuations in both transverse degrees of freedom;

•possibility of operation with 2 or 3 bunches in each beam with electrostatic separation in all undesirable interaction points. The main parameters of the ϕ-factory compared to VEPP-2M are presented in Table 3.

Of special interest is the operation mode with two bunches per beam having identical parameters. If trajectories are accurately combined, this mode provides precise compensation of coherent electromagnetic bunch fields before the collision. One should of course study possible instabilities, but it seems quite probable to surpass several times a tune shift limit because of beam-beam interactions and have a quadratic gain in luminosity (if one can provide a necessary number of particles stably stored in each beam).

To attain the maximum luminosity of the ϕ-factory one should provide a high injection rate of positrons (and correspondingly electrons) directly at the energy of experiment. One should keep in mind that the process of single bremsstrahlung will result in losing about 10^9 e^+/sec at the luminosity of $3 \cdot 10^{33}$ cm^{-2} s^{-1}. It is desirable to have injection by small portions at ultimately low perturbations of already stored particles. The losses during injection must be small and short in time to provide nearly continuous detector operation. This results in serious requirements for the injection systems and we, therefore, plan to make a ϕ-factory a part of the facility in design now, providing injection of high quality intense beams of electrons and positrons to the existing VEPP-4M collider and the B-factory VEPP-5.

At present, large modernization of the VEPP-4M (Fig. 5 and Table 4) is near completion which pursues several purposes. First, the luminosity of the storage ring will be considerably higher: instead of $6 \cdot 10^{30}$ cm^{-2} s^{-1} achieved earlier it should exceed $1 \cdot 10^{32}$ cm^{-2} s^{-1} (Fig. 6).

However, even after such modernization the luminosity of this collider will not exceed that already achieved at CESR. Therefore much attention is paid to the construction of the KEDR detector developed together with Italian physicists. Some of its parameters, we believe, will surpass those of other detectors, mainly due to the use of liquid Kr for the electromagnetic calorimeter.

There is one field in which the VEPP-4M collider is beyond competition — that of two-photon physics (in the range of two photon mass up to 4 GeV). Special structure of the entire interaction region (Fig. 7) allows detection of electrons and positrons moving at small angles with respect to the initial directions in the two-photon reaction:

$$e^+ + e^- \rightarrow e^+ + e^- + X \tag{1}$$

with energy and angular resolution providing X-mass measurement with an accuracy of about 10 MeV, close to a natural energy spread of the beams. Detecting also the produced system X in the central detector one obtains the picture of hadronic production with photon-photon colliding beams similar to that in Fig. 8.

Further relatively small changes in the interaction region (Fig. 9) will allow experiments with longitudinally polarized e^+e^-. If one studies e^+ and e^- helicity dependence of hadronic cross sections giving information about weak interactions of b-quarks, one of the main problems is to decrease systematic uncertainties. The total contribution of helicities to the cross section is about 1% and one should measure it with the accuracy not worse than, say, 20%. Therefore, one should provide identical conditions for the orbital motion of the e^+ and e^- bunches for events with different given helicity (parallel data taking, constant external fields, equalized number of particles in the bunches). Favourable conditions for such experiments will arise after construction of the new injection facility when electrons and positrons will be injected just at the energy of experiment.

The whole facility will look like in Fig. 10. The injector part consists of two linacs with a wavelength of 10 cm and the energy of 500 MeV (or one 500 MeV linac, passing e^- and then e^+) as well as the cooling storage ring for the energy of 510 MeV. It will provide low-emittance electron and positron bunches at a rate of 10^{10} e^+, e^- per second.

A part of cycles will end by injection into the ϕ-factory, in other cycles particles will be accelerated in the main linac with a wavelength of about 2 cm, acceleration rate of 100 MeV/m and a total energy of up to 8 GeV. The units developed for the linear collider VLEPP will serve as linac elements. After acceleration of up to the energy of experiment bunches will be injected into VEPP-4M or VEPP-5.

VEPP-5 is a B-factory with a center-of-mass energy of up to 11-13 GeV and luminosity close to $1 \cdot 10^{34}$ cm^{-2} s^{-1}. To achieve such a luminosity one should have two separate rings for electrons and positrons. The required values of bunch parameters can be roughly estimated from the usual formula

$$\mathcal{L}_{max} = \frac{c}{r_e} \cdot \frac{N_b \xi_{max}}{L_b \sigma_{||}} \tag{2}$$

where N_b is the number of particles in each bunch, $\sigma_{||}$ is a bunch halflength, L_b is a distance between bunches, ξ_{max} is a maximum allowed tune shift.

In the accepted variant, beam energies differ from each other , e.g. at $\Upsilon(4S)$ as the positron and electron energies are 7 and 4 GeV, respectively. Beam separation is performed by the magnetic field in a horizontal plane, and the low energy ring is installed inside (or below) the high energy one in the same tunnel.[5] In this case the distance between the bunches can be made less than 4 m with good separation of the bunches in all extra interaction points. Correspondingly, it is much easier to obtain the required values of bunch parameters. A possible lattice of the interaction region is shown in Fig. 11. Different beam energies are also attractive for studies of CP-violation in some processes with B-mesons.

Serious attention is also drawn to the development of such collider structures which would make higher the effective monochromatization of the colliding beams.[4] In B-meson studies this helps to select interesting events. Operating at narrow resonances $\Upsilon(1S)$, $\Upsilon(2S)$, $\Upsilon(3S)$ and at not yet discovered D-states, this provides better signal to noise ratio as well as higher collider productivity. A special operation mode at low energy will allow additional monochromatization to be achieved making possible a search for irregularities in the structure of Ψ and Ψ' resonances.

Main parameters of the VEPP-5 project are presented in Table 5.

But the major direction of our Institute activities in the field of electron-positron colliders is design and construction of the linear collider VLEPP. Already in the end of sixties it has been clear for us in Novosibirsk that to reach energies of hundreds GeV one should proceed to linear colliders. In 1971, at the International Seminar on perspectives of high energy physics in Morges (Switzerland) we have reported on these considerations. At that time use of the superconducting linacs with energy recuperation at the second part of the linac seemed more prospective. However, later we moved to using pulsed linacs with the normal conductivity (this variant has been also discussed by us in Morges). In 1978, at the International Seminar in Novosibirsk, devoted to the memory of A. M. Budker, we presented the selfconsistent physical project of the linear collider.[6,7] Its main points were:

•acceleration of a single bunch with correct matching of the number of particles, bunch length, wavelength and acceleration rate allows a high efficiency (dozens per cent) in transforming the energy accumulated in the accelerating structure into the beam energy at a good monochromaticity of about 1%;

•acceleration rate can be as high as 100 MeV/m which was several times larger than usual at that time;

•transverse instability, arising during the acceleration of even a single bunch in linac, can be efficiently suppressed by introducing of correct energy gradient along the bunch, this effect now referred to as BNS damping,[8] with its compensation at the end of acceleration;

•during the collision of intensive bunches coherent "two-beam" instability can arise; this effect limits the density in the colliding bunches at given energy;

•the fields in the intensive bunches are so high at the required micron cross sections of the beams in the interaction region that the radiation in the coherent field of the counter bunch (beamstrahlung) can become large and require using flat bunches;

•required intensive and super-low-emittance bunches can really be obtained in cooling storage rings at the energy about 1 GeV;

•used bunches of high energy electrons and positrons can provide regeneration of the bunches for the next cycle via conversion.

Further development of this project kept these main features, adding to them a possibility of obtaining polarized electrons and positrons, refining details and optimizing the whole project. The cucrrent scheme of VLEPP is shown in Fig. 12, main parameters are presented in Table 6.

In 1987, an important event occured: the state decision was taken about construction of the electron-positron linear collider VLEPP for the energy of 1000 GeV X 1000 GeV. This facility will be constructed at Protvino near the accelerator UNK which is under construction now. To this aim a Branch of our Institute of Nuclear Physics has been created. The decision to construct VLEPP there is related to the possibility of using construction facilities already developed for UNK. Besides that one can try in future use the co-operation of UNK and VLEPP for creating electron-proton colliding beams with the energy up to 2 TeV x 3 TeV and high enough luminosity.

However main efforts on development of the VLEPP components as well as technology of their serial production will be concentrated in Novosibirsk.

All electron-positron facilities of the INP under construction and design are oriented at the wide collaborations both inside the country and with abroad. We invite our colleagues from all the interested centres and laboratories to participate in this work at all possible stages: collider development and construction, construction of detectors, data taking, and analysis.

REFERENCES

[1] A. A. Zholents, I. Ya. Protopopov and A. N. Skrinsky, Proc. of the VIth All-Union Conference on Charged Particle Accelerators, Vol. 1, p. 132, Dubna 1979.

[2] A. N. Skrinsky and Yu. M. Shatunov, Sov. Phys. Usrekhi 158, N. 2, p. 315, 1989.

[3] L. M. Barkov et al., "Phi-factory Project in Novosibirsk," presented at this conference.

[4] A. N. Dubrovin et al.,, Proc. of the EPAC Conf., Rome 1988.

[5] A. N. Dubrovin et al., Proc. of ϕ-Factory Workshop, Blois, France, 1989.

[6] V. E. Balakin, G. I. Budker, and A. N. Skrinsky, "Problems of High Energy Physics and Controlled Fusion," Moscow, "Nauka," 1981, p. 11; V. E. Balakin, G. I. Budker, and A. N. Skrinsky, Proc. of the VIth All-Union Conf. on Charged Particle Accelerators, Vol. 1, p. 27, Dubna, 1979.

[7] V. E. Balakin and A. N. Skrinsky, Proc. of the IInd ICFA Workshop, Les Diablerets, Switzerland, 1979, p. 31.

[8] V. E. Balakin et al., Proc. of the VIth All-Union Conference on Charged Particle Accelerators, Vol. 1, p. 143, Dubna, 1979.

L, cm^{-2}, s^{-1}
ВЛЭПП
CLIC
LLC
LEP
TRISTAN
SLC
CESR
DORIS
PEP
PETRA
ВЭПП-5
ВЭПП-4M
ВЭПП-4
b
c
τ
φ-factory
BEPC
SPEAR
ADONE
DCI
ВЭПП-2
ACO
ВЭПП-2M
ВЭПП-2MM
ρ,ω
φ
ψ
ψ'
Y
Y'
Y''
z_0
E_{tot}, GeV
0.1
1
10
100
1000
10000
10^{27}
10^{28}
10^{29}
10^{30}
10^{31}
10^{32}
10^{33}
10^{34}
10^{35}

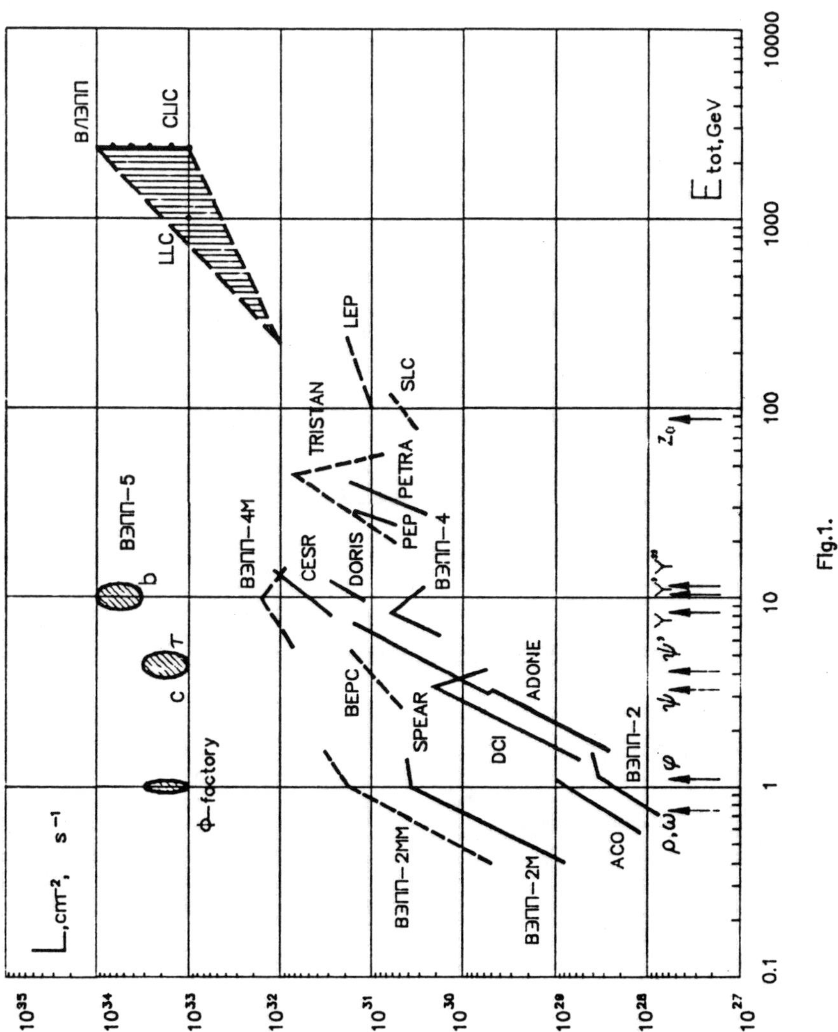

Fig.1.

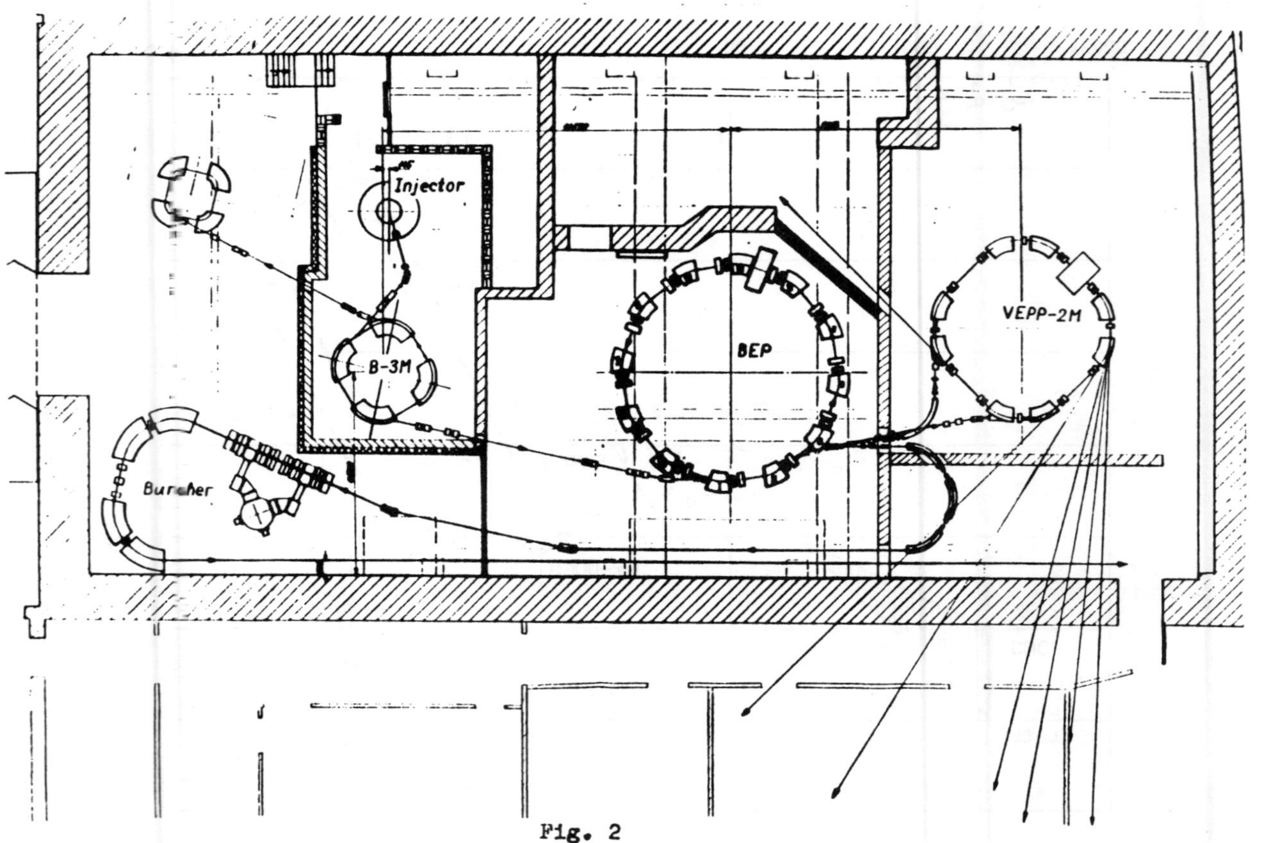

Fig. 2

L
cm⁻² s⁻¹
10³⁴
10³³
φ-factory
10³²
1989
VEPP-2M
10³¹
1986
10³⁰
1984
10²⁹
10²⁸
200 300 400 500 600 700
E
MeV

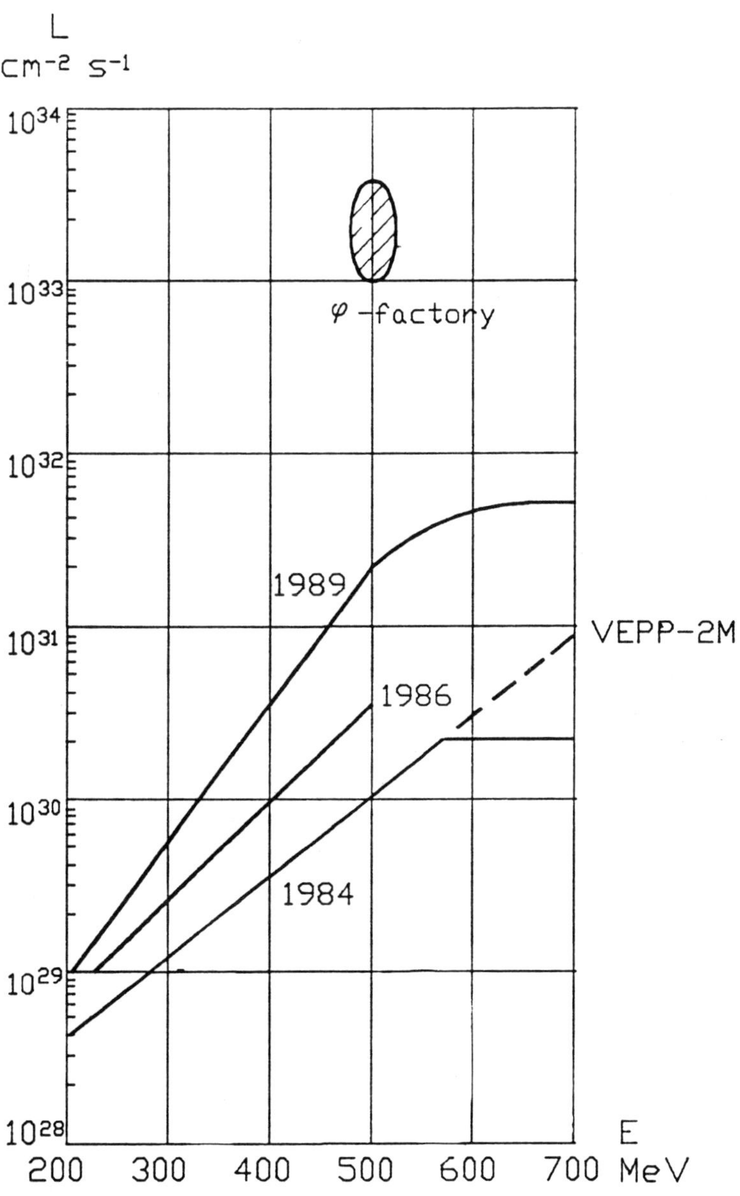

Fig.3.

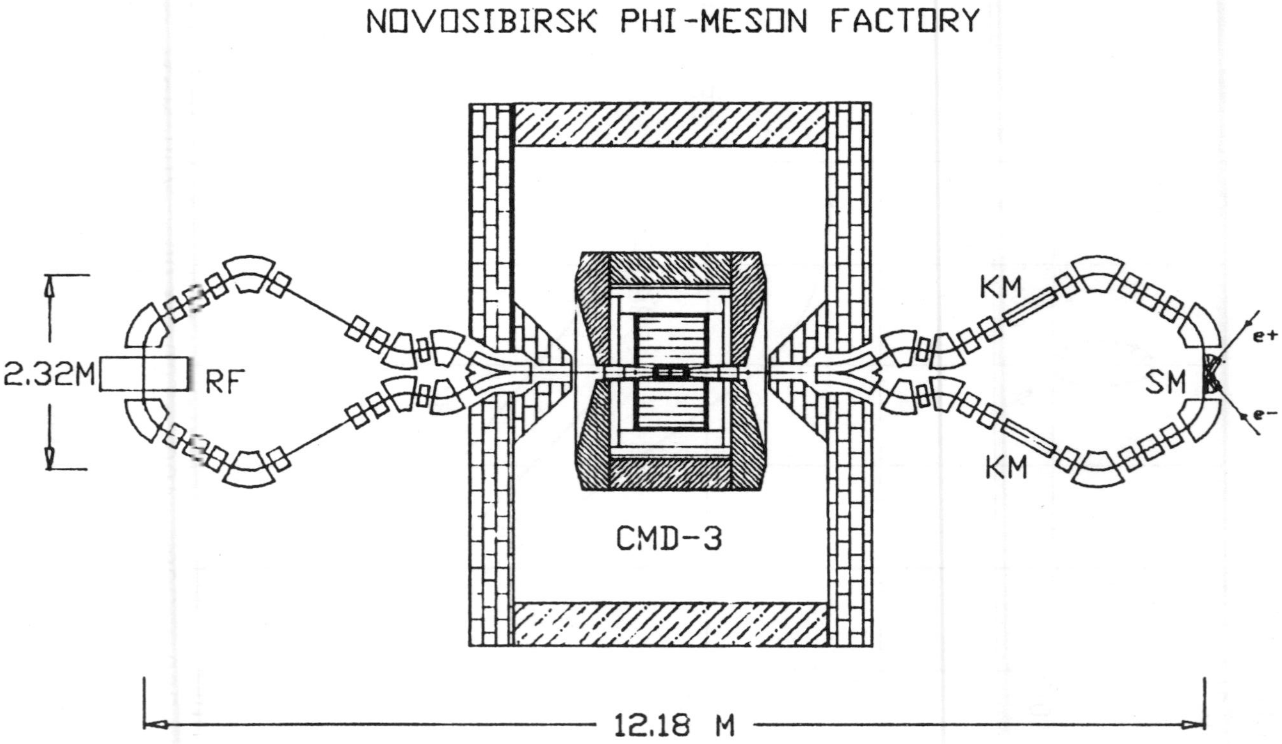

NOVOSIBIRSK PHI-MESON FACTORY

Fig.4.

VEPP-4M
SR
RF
VEPP-3
18°
8m
45m
B-4
LINAC
e+
e-
KEDR
Novosibirsk-Italy
colaboration.

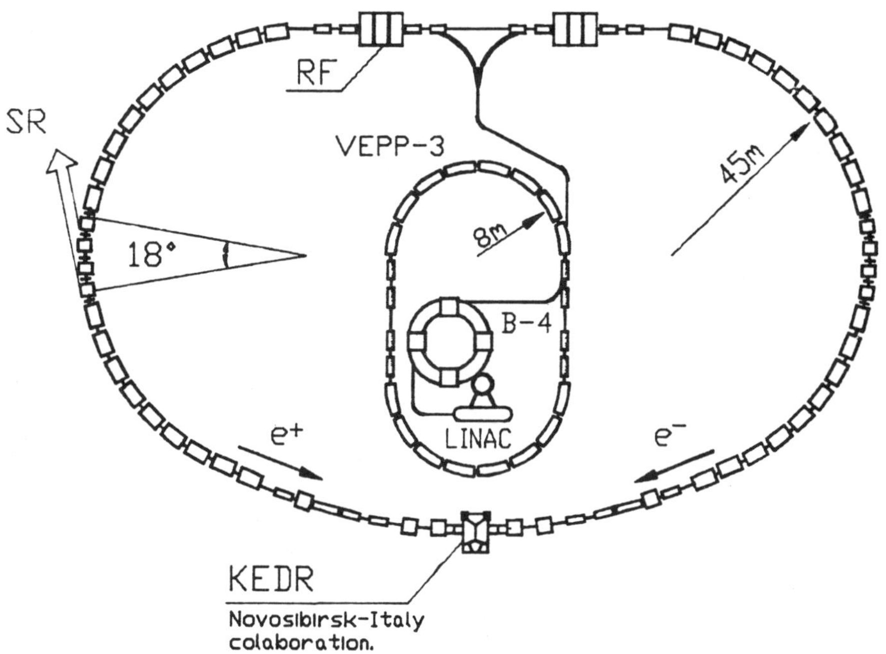

Fig.5.

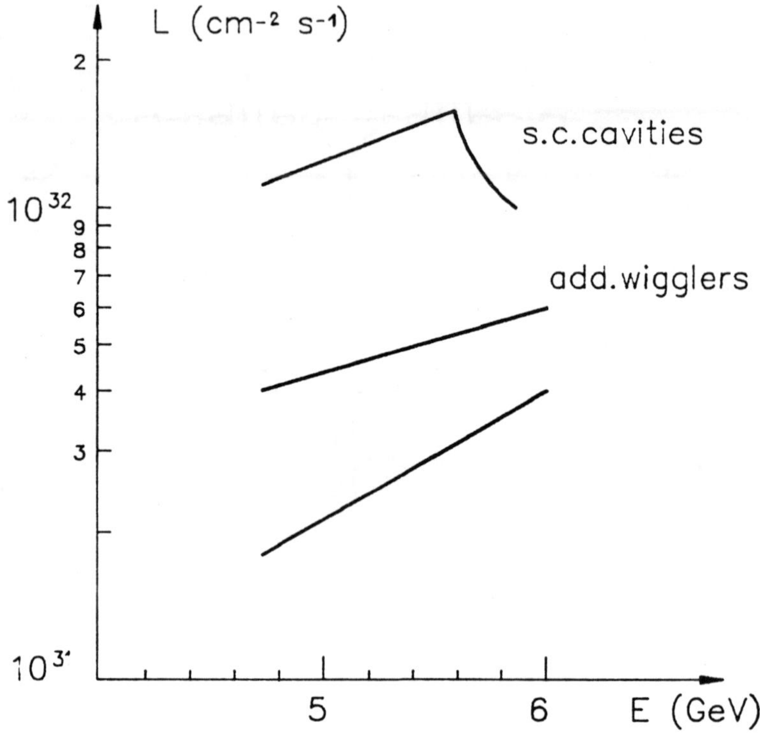

Fig.6.

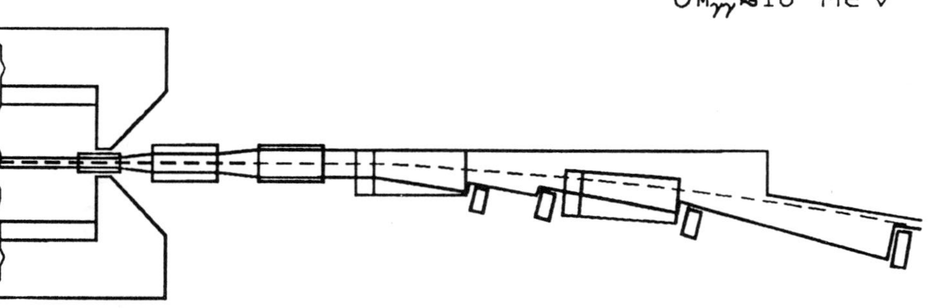

Lay-out of the tagging sistem for VEPP-4M.
Electrons are detected in the energy range
$0,4E_0 < E < 0,97E_0$, $\theta < 10\,\text{mrad}$.
Energy resolution $\sigma \frac{E}{E} \approx 2\ 10^{-3}$
$\sigma_{M_{\gamma\gamma}} \approx 10$ MeV

Fig.7.

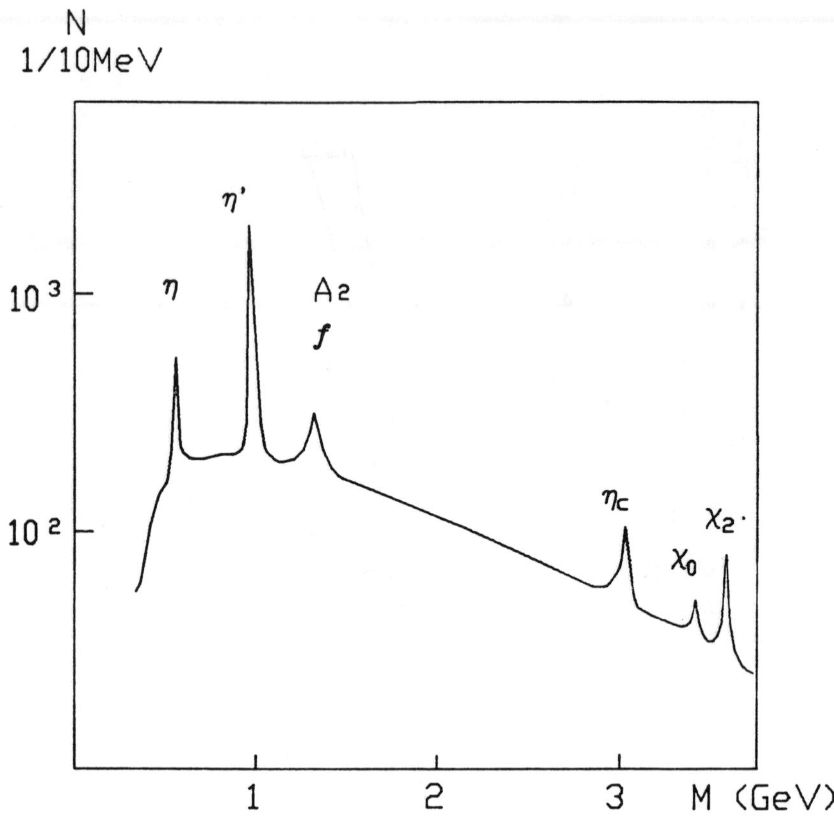

The number of events expected at
the integrated luminosity of 100 pb⁻¹

Fig.8.

Longitudinal polarization
on VEPP-4M

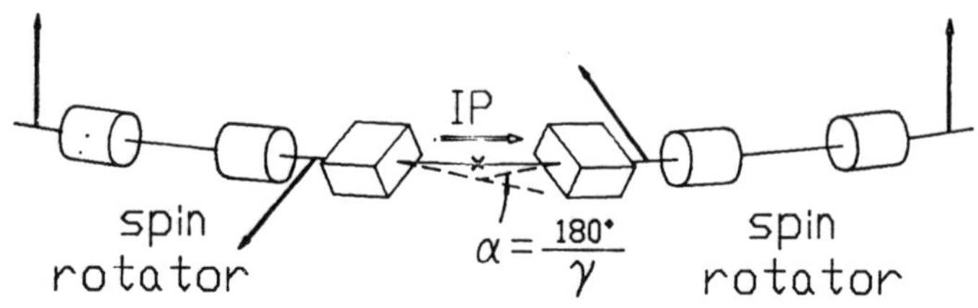

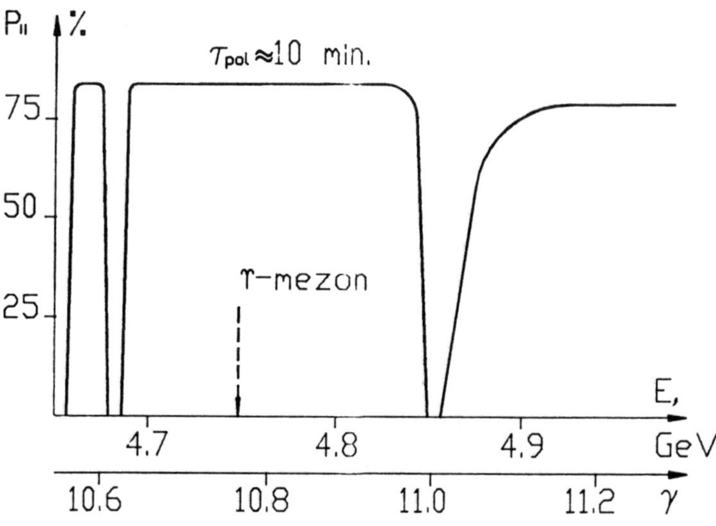

Fig.9.

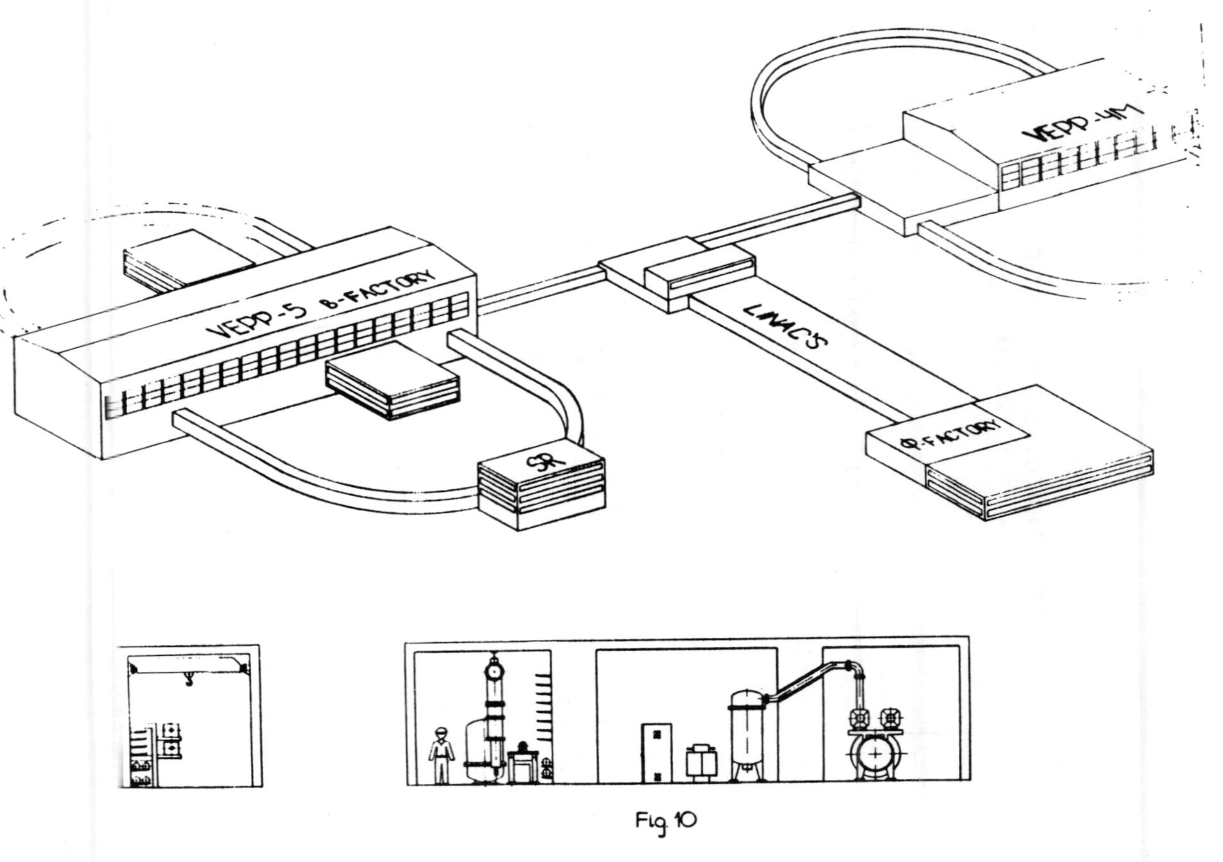

Fig. 10

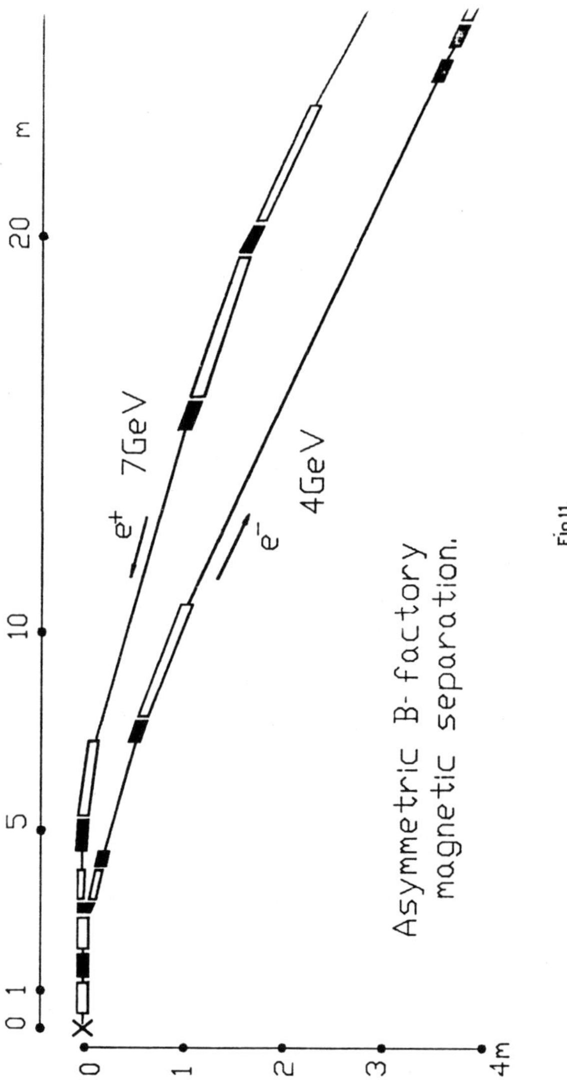

Fig.11. Asymmetric B-factory magnetic separation.

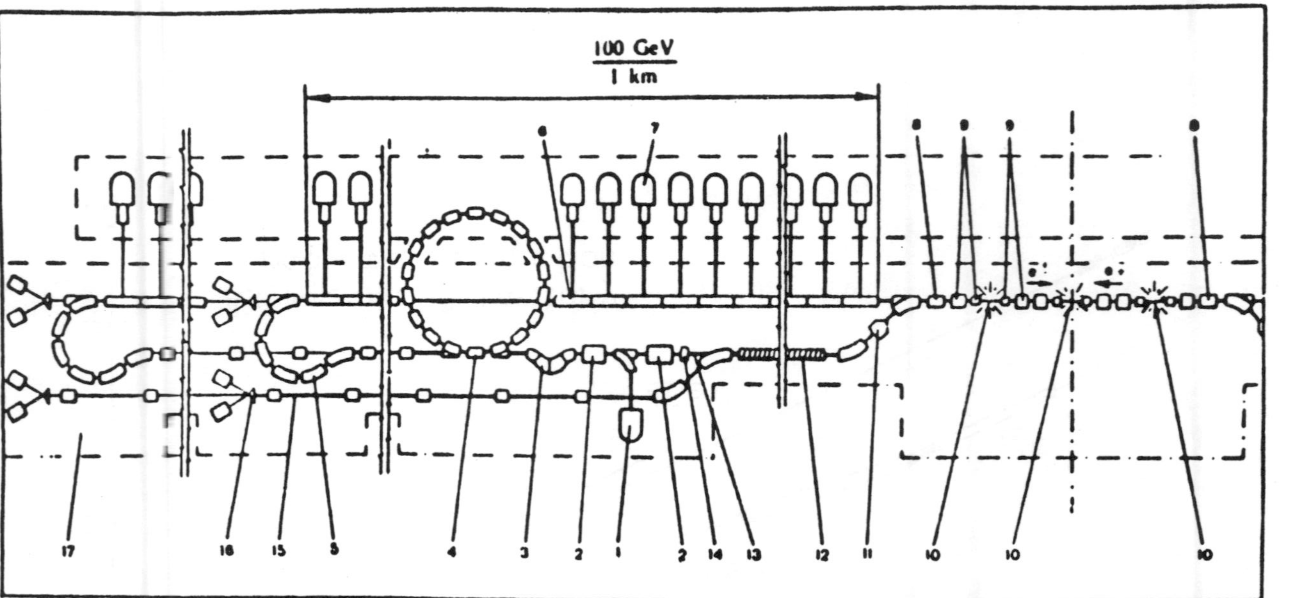

The general lay-out of the VLEPP facility:

1—initial injector; 2—intermediate accelerator; 3—debuncher-monochromalizer; 4—storage ring; 5—buncher; 6—accelerating sections; 7—RF-generators; 8—pulse deflector; 9—focusing lenses; 10—collision points; 11—spectrometer; 12—helical ondulator; 13—the beam of γ-quanta; 14—conversion target; 15—residual electron (positron) beam; 16—electron (positron) beam experiments; 17—the second stage.

Fig. 12,

		Total Energy, GeV	Luminosity $cm^{-2}s^{-1}$	First experiment	Shut down
VEPP-2	USSR Novosibirsk	1,35	$3*10^{28}$	1967	1970
ACO	France	1,1	$1*10^{29}$	1968	1974
ADONE	Italy	3,2	$6*10^{29}$	1969	1975
SPEAR	USA	8,4	$2*10^{31}$	1973	
VEPP-2M	USSR Novosibirsk	1,4	$4*10^{30}$	1974	1987
DORIS	BRD	10,2	$3*10^{31}$	1974	
DCI	France	3,1	$2*10^{30}$	1976	1984
PETRA	BRD	46	$2*10^{31}$	1978	1987
CESR	USA	11,2	$1*10^{32}$	1979	
VEPP-4	USSR Novosibirsk	10,2	$6*10^{30}$	1980	1985
PEP	USA	30	$2*10^{31}$	1980	
TRISTAN	Japan	60	$(8*10^{31})$	1987	
SLC	USA	100	$(6*10^{30})$	1989	
BEPC	China	5,6	$(2*10^{31})$	1988	
VEPP-2MM	USSR Novosibirsk	1,4	$(3*10^{31})$	(1990)	
LEP-1	CERN	100	$(1*10^{31})$	1989	
VEPP-4M	USSR Novosibirsk	12	$(1*10^{32})$	(1990)	
LEP-2	CERN	200	$(2*10^{31})$	(1992)	
VEPP-5	USSR Novosibirsk	14	$(1*10^{34})$	(1994)	
VLEPP	USSR Protvino	1000 2000	$(1*10^{33})$ $(1*10^{33})$	(1996) (1998)	
TLC	USA	1000	$(1*10^{33})$		
CLIC	CERN	2000	$(1*10^{33})$		

Tab.1.

Particle mass, MeV

Particle	World average value	Experimental results	Accuracy improvement
K^-	493.657 ± 0.020		
K^+	493.84 ± 0.13	493.670 ± 0.029	5
K^0	497.67 ± 0.13	497.661 ± 0.033	4
ω	782.4 ± 0.2	781.78 ± 0.10	2
ϕ	1019.70 ± 0.24	1019.52 ± 0.13	2.5
ψ	3097.1 ± 0.9	3096.93 ± 0.09	10
ψ'	3685.3 ± 1.2	3686.00 ± 0.10	10
Υ	9456.2 ± 9.5	9460.59 ± 0.12	80
Υ'	$10'016.0\pm10.0$	$10'023.6\pm0.5$	20
Υ''	$10'347.0\pm10.0$	$10'355.3\pm0.5$	2

TABLE1. BASIC PARAMETERS OF THE PHI-FACTORY AND BEAMS

PARAMETERS		UNITS	PHI-FACTORY	VEPP-2M WITH WIGGLER
CIRCUMFERENCE	C	m	28.0	17.88
ACCELERATING VOLTAGE FREQUENCY	f_0	MHz	700	200
MOMENTUM COMPACTION FACTOR	α	---	0.003	0.167
EMITTANCES	ε_{xo}	cm*rad	$2.8*10^{-5}$	$4.6*10^{-5}$
	ε_{zo}	cm*rad	$2.8*10^{-5}$	$5.5*10^{-7}$
RADIATIVE ENERGY LOSS PER TURN	ΔE_0	keV	40	9.1
DIMENSIONLESS DAMPING DECREMENTS BETWEEN INTERACTION POINTS	δ_z	---	$2.3*10^{-5}$	$0.44*10^{-5}$
	δ_x	---	$2.3*10^{-5}$	$0.38*10^{-5}$
	δ_s	---	$4.7*10^{-5}$	$0.94*10^{-5}$
R.M.S. ENERGY SPREAD IN THE BEAM	$\sigma_{\frac{\Delta E}{E}}$	---	$5*10^{-4}$	$6*10^{-4}$
BETA-FUNCTIONS AT THE I.P.	β_z	cm	0.5	4.5
	β_x	cm	0.5	48
R.M.S. LONGITUDINAL SIZE	σ_s	cm	0.4	3.5
BETATRON TUNES	ν_z	---	6.05	3.09
	ν_x	---	6.05	3.06
NUMBER OF PARTICLES IN THE BUNCH	N	e^+, e^-	$8.9*10^{10}$	$3.7*10^{10}$
SPACE CHARGE PARAMETERS	ξ_z	---	0.07	0.05
	ξ_x	---	0.07	0.02
LUMINOSITY IN SINGLE-BUNCH MODE	L_{max}	$cm^{-2}*s^{-1}$	$1*10^{33}$	$1*10^{31}$
LUMINOSITY IN THREE-BUNCH MODE	L_{max}	$cm^{-2}*s^{-1}$	$3*10^{33}$	---

Tab.3.

VEPP-4M

ENERGY	4-6 GeV
CIRCUMFERENCE	36604 cm
REVOLUTION FREQUENCY	819.01 kHz
PERIODIC CELL	
FOCUSING STRUCTURE	FODO
MEAN RADIUS	4549. 587 cm
BENDING RADIUS	3453. 6 cm
HORIZONTAL APERTURE	6 cm
VERTICAL APERTURE	2.7 cm
HORIZONTAL BETA-FUNCTION	1105 cm
VERTICAL BETA-FUNCTION	1442 cm
DISPERSION FUNCTION	121 cm
WORKING POINT Q_x, Q_z	8.53, 7.57
MOMENTUM COMPACTION FACTOR	0. 017
FOR BEAM ENERGY 6 GeV	
SYNCHROTRON RADIATION LOSS PER TURN	4 MeV
HORIZONTAL EMITTANCE	$4 \cdot 10^{-5}$
VERTICAL EMITTANCE	10^{-7}
ENERGY SPREAD	$1.1 \ 10^{-3}$
DAMPING TIME	2 msec .
RF SYSTEM	
WAVELENGTH	165 cm
HARMONIC NUMBER	222
MAXIMUM CIRCUMFERENTIAL VOLTAGE	9 MV
SEPARATRISS ENERGY SIZE	2%
SYNCHROTRON TUNE	0.03
BUNCH LENGTH σ_l	4 cm
RF POWER FOR BEAMS	400 kWt
RF POWER IN THE CAVITIES	800 kWt
INTERACTION POINT	
HORIZONTAL BETA-FUNCTION	74.5 cm
VERTICAL BETA-FUNCTION	5 cm
DISPERSION FUNCTION	80 cm
BEAM SIZES σ_x σ_z	1040 mkm, 7 mkm
LINEAR BEAM-BEAM Q SHIFT ε_z ε_x	0.05, 0.005
NUMBER OF PARTICLES IN THE BUNCH	$2 \ 10^{11}$
NUMBER OF BUNCHES	2
BEAM CURRENT	50 mA
LUMINOSITY	$7 \cdot 10^{31}$

Table 4

PARAMETERS of ASYMMETRIC 4x7 GeV B-FACTORY

	7	4
Beam energy (GeV)	7	4
Circumference (m)	600	588 [600]
Emittance (m·rad) x 10^{-10} ε_h	250	80
ε_v	2,5	2,2
Energy spread (10^{-3})	0,9	0,9
Damping partition number Gs	1,8	1,3
Gh	1,2	1,7
Bending radius (m)	50	24
Betatron frequency Qv	24	25
Qh	25	26
Synchrotron frequency Qs	0,024	0,018
Momentum compaction (10^{-3})	2,8	2,1
Accelerating voltage (MV)	10	4
Bunch length (cm), σ	1	1
Synchrotron radiation power (MW)	3,2	0,9
Crossing parameters (cm) β_h	40	50
ψ	50	40
β_v	1	1
Particles per bunch (10^{10})	6	8
Number of bunches	150	147 [150]
Tune shift ζ_v	0,05	0,05
ζ_h	0,008	0,009

Center-of-mass energy (GeV)	10,6
Center-of-mass energy spread (MeV) σ_v	1,4
Peak luminosity ($cm^{-2} s^{-1}$) 10^{33}	5
Number of interaction regions	2

Table 5.

TABLE 6

V L E P P

	I st.	II st.
$2 \times E$ GeV	2×500	2×1000
L cm^{-2}sec^{-1}	10^{33}	10^{34}
$2 \times l$ km	2×6	2×12
$\bar{P}$ m w	100	200
λ cm	2.1	2.1
f_{Hz}	100	100
n^{+}, n^{-}	$(1 \div 2) \cdot 10^{11}$	$(1 \div 2) \cdot 10^{11}$
T	1996_{y}	1998_{y}

Final focus test facility -
collaboration: SLAC-KEK-INP (-CERN)

LEP RESULTS

J.J. Thresher
CERN
Geneva, Switzerland

1. Introduction

The first period of LEP operation ended on 22nd December 1989, rather less than two weeks before the start of this Conference. It brought to a close a most exciting time at CERN, starting in July 1989, when those involved in commissioning LEP attempted for the first time to start up the complete machine and then to give the experiments their first taste of what LEP had in store for them.

By July almost all sections of LEP had been individually checked out. In particular, the entire injection chain had been tested with positrons a year earlier when a very successful injection test into the first completed LEP octant was carried out. Also by July the LEP detectors had been installed and were ready to take data with at least the most important sub-detector systems able to operate. By way of introduction a brief history of these first months of LEP operation is given below.

The first steps in bringing LEP into operation started on 14th July 1989 when positrons were injected into the ring for the first time. After only 55 minutes of magnet adjustments they had completed a full turn at the injection energy of 20 GeV. Further commissioning with positrons at this energy then followed to establish a stable circulating beam and then on 25th July the first electrons were successfully injected into LEP. By 31st July, after much work on beam accumulation had been done, a current of some 250 μA of positrons, i.e. about 60 μA in each of the four bunches was reached at 20 GeV and four days later on 4th August positrons were successfully ramped up to 47.5 GeV. Finally, at 23.15 on 13th August, just less than one month after the start of LEP commissioning, electrons and positrons were brought into collision at an energy of 45.5 GeV per beam. Five minutes later the first Z° to be observed at LEP was recorded in the OPAL detector.

Although at this stage the luminosity in the four interaction regions was rather low, it was decided to continue with a short five day "pilot" run to obtain some experience of running both the machine and the experiments for physics. In this period there was a total of 15 hours operation with colliding beams. The vacuum was sufficiently good at this early stage to allow beam current lifetimes of four hours to be obtained. The average luminosity during the 15 hours of collisions was approximately 2×10^{28} cm^{-2} sec^{-1} giving an integrated luminosity of about 1 nb^{-1} at each interaction point. In all 58 Z° events decaying to hadrons, e^+e^-, $\mu^+\mu^-$ or $\tau^+\tau^-$, were observed by the four detectors as follows:

Experiment	ALEPH	DELPHI	L3	OPAL	TOTAL
No. of Z° events	15	6	16	21	58

Table 1

The first serious physics run started about one month later on 20th September after a short shutdown of the machine and a further period of commissioning. One week was devoted to running on the Z° peak and two weeks to doing two energy scans across the peak. The main goals were to measure the mass of the Z° and to determine the number of neutrino types by measuring the cross section at the peak energy.

For the last week of this run it was possible to bring the four sets of superconducting low β quadrupoles into operation which reduced the value of β_V at the interaction points from the 20 cm obtained without them to the nominal design value of 7 cm. Although LEP has since been shown to run successfully at $\beta_V = 4.3$ cm, the operating conditions of the machine were not stable and all subsequent running has been with $\beta_V = 7$ cm.

The rest of the year's running from 25th October to 22nd December was divided into three periods of between two and three weeks each which were shared between commissioning and machine development studies and operating for physics. Here the main physics goals were to obtain a good measurement of the width and an improved measurement of the mass. Much other physics, of course, emerged at the same time.

To conclude this introduction, figures for the integrated luminosity and for the numbers of Z° events which were observed to decay into hadrons are shown in Table 2 below.

Running period		ALEPH	DELPHI	L3	OPAL	TOTALS
1. 20th September to 9th October 1989	$\int Ldt$ (nb^{-1})	150	54	157	187	548
	No. of events	3300	1066	2538	4359	11263
2. 25th October to 6th Nov. 1989	$\int Ldt$ (nb^{-1})	400	230	280	320	1230
	No. of events	8500	4500	4800	7510	25310
3.&4. 15th Nov. to 22nd Dec. 1989	$\int Ldt$ (nb^{-1})	700	380	690	830	2600
	No. of events	16800	7400	11600	18600	54400
Totals for 1989	$\int Ldt$ (nb^{-1})	1250	660	1130	1340	4380
	No. of events	28600	13000	18900	30500	91000

Table 2

As can be seen a total of some 91,000 hadronic events from Z° decays were observed. Including decays to charged lepton pairs, the total number of Z° events observed by the four experiments amounted to about 103,000.

All experiments ran with reasonable efficiency during this first phase of LEP operation. DELPHI, more than the others, suffered from early teething troubles, largely to do with problems in its read-out and data acquisition systems which fortunately were identified during the course of running and will be put right before the machine starts up again in March 1990.

2.   Performance of the LEP Machine

It is appropriate to mention at this point that throughout the months of LEP operation the fixed target programme of the SPS at CERN continued uninterrupted. As can be seen in Figure 1, both the CERN PS and the SPS form part of the LEP injection chain. Even the periods of electron-positron accumulation in LEP can be carried out parasitically during normal fixed target operation of the SPS; only when the SPS is operated for proton-antiproton collider physics is this not possible.

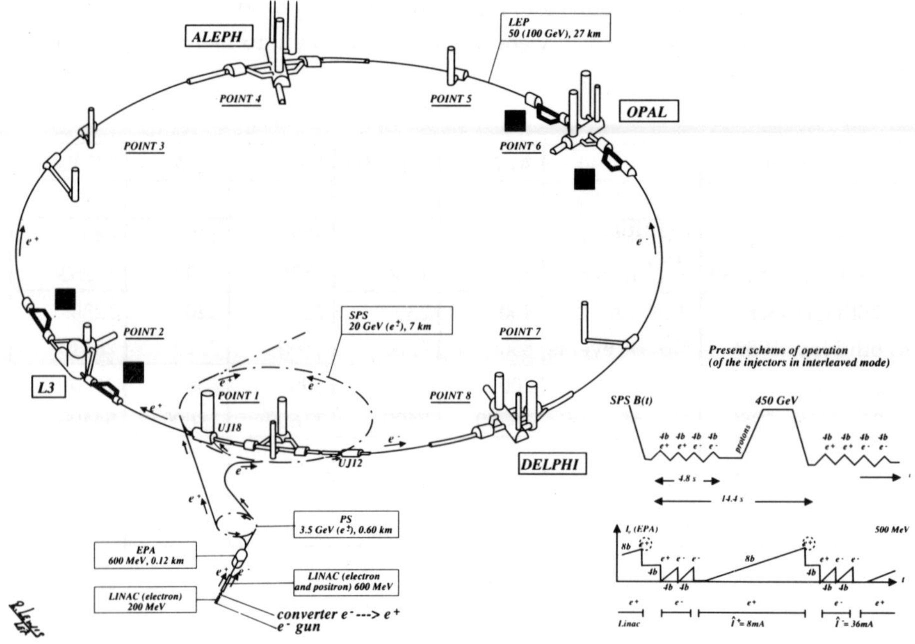

Figure 1   The Injection System for LEP.

Very soon after LEP came into operation it was discovered that there were serious effects due to coupling between the vertical and horizontal components of the beams. This was shown to vary approximately as $1/E$ and was distributed more or less uniformly around the ring. It caused the beam size to be larger vertically than it should which in turn limited the beam intensity at injection. It also introduced vertical dispersion around the ring, in particular in the region of the RF cavities which resulted in an increase in the vertical emittance and the appearance of synchrotron-betatron resonances in the machine.

Initially corrections for these effects could be made by adjusting the tune of the machine and by using eight rotated quadrupoles which were already in place to compensate for the effects of the solenoids in the detectors. Also, during a short shutdown in October sixteen more were installed. The outcome was that currents somewhat in excess of 1 mA per beam could be regularly obtained, to be compared with the design value of 3 mA. Although higher currents up to 2 mA in a single beam were often reached at the injection energy, conditions were not stable enough to enable them to be used in practice. A further eight more rotated quadrupoles will be added on either side of each of the 4 RF stations during the shutdown early in 1990 in an endeavour to solve the problems in these regions.

The existence of the large coupling was found to be due primarily to a significant transverse horizontal component of field in the dipole magnets of the LEP ring which, in part at least, is believed to be caused by a thin layer of nickel plated onto the aluminium vacuum vessel to help give good adhesion between it and the surrounding lead shield (see Figure 2). It appears that during the plating process the nickel had somehow become permanently magnetized. Further studies are in progress to confirm this diagnosis and to devise means of eliminating the effect.

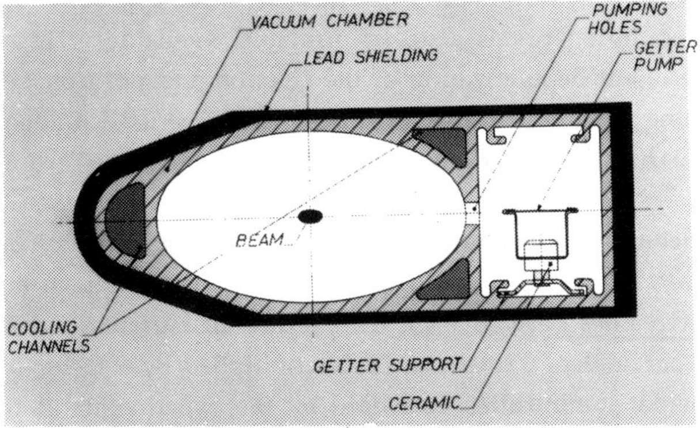

Figure 2   Cross-section of the LEP Vacuum Chamber.

On the more positive side the machine has a very low input impedance which may mean that the current per bunch could in the end be raised to about 1 mA as compared with the design figure of 0.75 mA which could give a sizeable increase in luminosity. Also it has been found that the machine conditions are reasonably reproducible from one fill of the ring to the next. One consequence of this is that the collimators in the machine which are used to reduce backgrounds in the experiments can, after each fill, be reset to standard settings for the start of the following run. Here it should be remarked that there are 140 collimator jaws which have to be readjusted after opening them out for injection and accumulation of electrons and positrons in LEP. Much time is therefore saved by the simple resetting procedures which are possible. The collimators reduce the backgrounds at the experiments by more than a factor of 1000 resulting in very clean running conditions which many, who have worked on other e^+e^- machines in the past, claim to be the best they have experienced.

130

Another very positive aspect is the accuracy to which the energy of the beams in the machine can be measured. So far two independent methods have been used. One of these is based on the measurement of the magnetic field integral around the ring and this gives the value of the energy to an accuracy $\Delta E/E$, of $\pm 5 \times 10^{-4}$. The other method involves a comparison at the injection energy of 20 GeV of the revolution frequencies of positron and protons in the LEP ring. Here an accuracy of $\Delta E/E = \pm 3 \times 10^{-4}$ has been achieved including uncertainties in extrapolating from 20 GeV to energies around 45 GeV. A large part of the uncertainty comes from the measurements which must be made to establish what differences, if any, there are between the positron and proton orbits. It is believed that with further work on the orbit measurements a level of $\pm 2 \times 10^{-4}$ for $\Delta E/E$ could ultimately be reached by this approach. A third scheme for measuring the energy is also, in principle, available. This relies on the method of resonant depolarization of the electron and positron beams with the goal of reaching a value of $\pm 10^{-5}$ for $\Delta E/E$. It will be used if adequate levels of transverse polarization are produced in LEP via the Sokolov-Tornov effect.

To conclude this section an indication of the performance of LEP since it started is given in Figures 3 to 8. The luminosity values shown in these figures are calculated values derived from the product of currents and the theoretical beam optics parameters. They should be multiplied by a factor of about 0.7 to obtain the actual luminosities measured by the experiments at the interaction points.

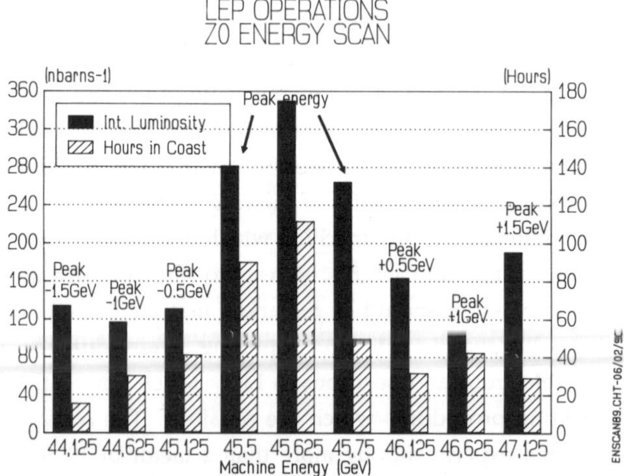

Figure 3   Integrated luminosities and the times spent with e^+e^- beams in collision at different LEP energies.

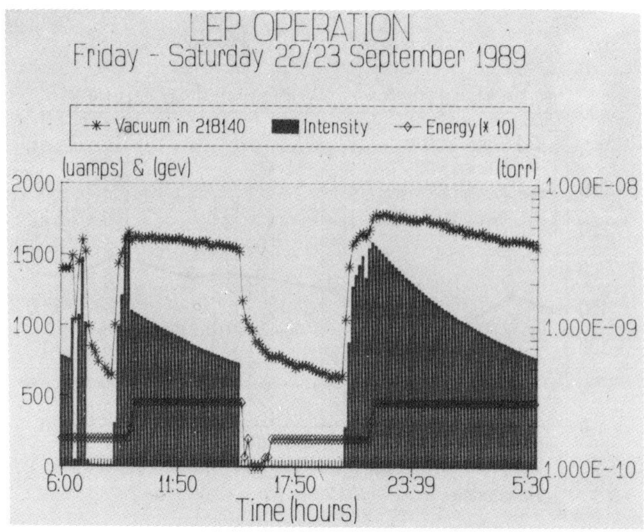

Figure 4   Some features of LEP performance.

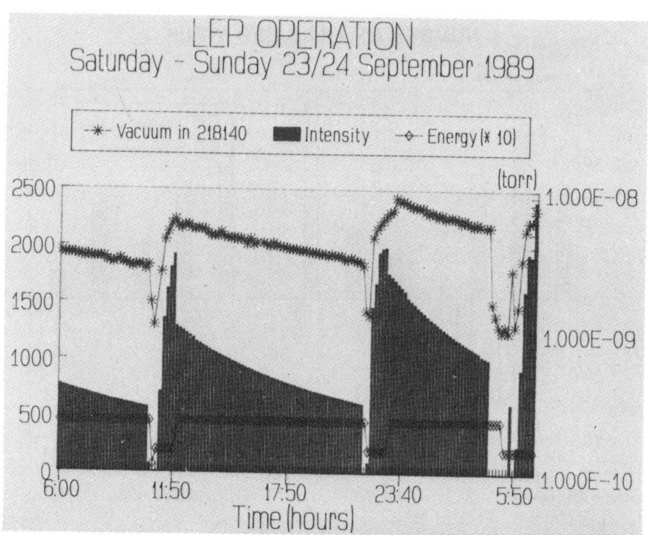

Figure 5   Some features of LEP performance.

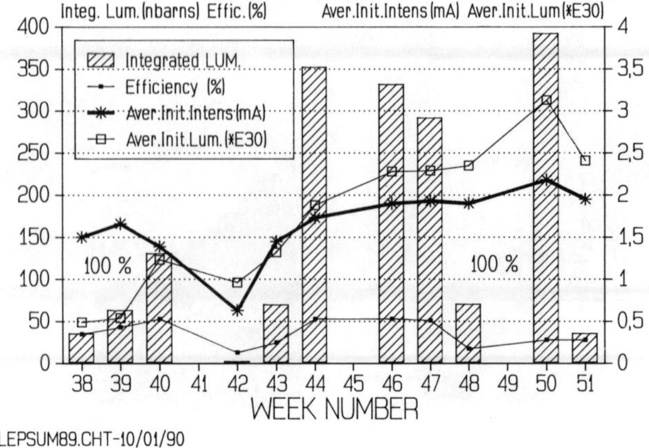

Figure 6 Information on the LEP luminosity delivered to each LEP experiment during 1989.

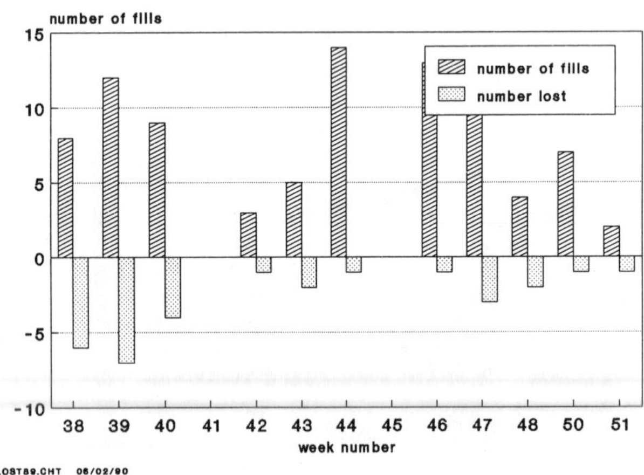

Figure 7 Statistics on the number of e^+e^- fills in LEP and the number lost through machine malfunctions or electricity supply failures in 1989.

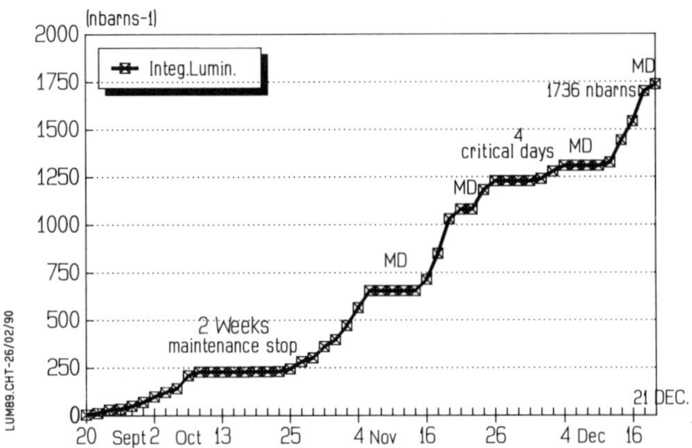

Figure 8   Evolution of the integrated luminosity delivered to each LEP experiment in 1989. As discussed in the text a normalization factor of approximately 0.7 should be applied to the luminosity scale shown.

The pattern of running so far has been to devote two days per week of scheduled LEP running to machine development and the remaining five days per week to running for the experiments. Table 3 gives information which

<u>Present LEP Operating Characteristics</u>

Initial Luminosity ($\beta_V = 7$ cm)	$\sim 2.5 \times 10^{30}$ cm^{-2} sec^{-1}
Total Beam Current (Initial)	~ 2 mA
Typical Run Duration	~ 6 hours
Typical time between Runs	3 - 6 hours
Typical Beam Current Lifetime (first 30 mins)	~ 13 hours
Efficiency = Physics running/scheduled time	$\sim 50\%$

<u>Design Figures</u>

Total Beam Current = 6 ma $\beta_V = 7$ cm;

Luminosity = 1.6×10^{31} cm^{-2} sec^{-1}

Table 3

illustrates some of the operating features of LEP during the latter part of the 1989 period of running.

3.   Physics Results from LEP

As mentioned in the introductory section of this paper, the main physics aims of LEP running in 1989 were to study features of the Z° boson such as its mass and width, to obtain a measure of the number of light neutrinos from the cross-section at the Z° peak and to determine the partial widths for decays to hadrons and lepton pairs including the so-called invisible width for decays to light neutrinos. It has also been possible to search for new particles and, since none has so far been found, to place lower mass limits on those such as the Higgs boson, the t and b' quarks, supersymmetric particles, etc. that could in principle have been produced should they exist with masses in the LEP energy range. Other results include tests of QCD by studying jet production in the large hadronic event samples obtained. The QCD results are discussed in the report presented to this meeting by G. Altarelli and will not be presented here.

Each of the four LEP experiments has results on most of these topics and as of 4th January 1990 about twenty papers have been published on the data that have been analysed so far.

3.1 The Z° Resonance Parameters

Data have been taken at 11 energies across the Z° peak from 88.28 GeV to 93.04 GeV[1]. Three of the energies close to the peak were separated by approximately 0.25 GeV while most of the other points were taken at 1 GeV intervals. The results for multihadron events have been assumed to have a Breit-Wigner shape and fitted to an expression which can take the form:

$$\sigma_{had} = \sigma^{o}_{had} \; \frac{s\Gamma_Z^2}{(s-M_Z^2)^2 + s^2 \, \Gamma_Z^2 / M_Z^2} \; (1 + \delta_{rad}(s)) \qquad (1)$$

$$\text{where} \quad \sigma^{o}_{had} = 12 \, \Pi \, \frac{\Gamma_{ee} \, \Gamma_{had}}{M_Z^2 \, \Gamma_Z^2}$$

Γ_{ee}, Γ_{had} are the partial widths for Z° decays to e^+e^- and $q\bar{q}$ respectively.

M_Z is the mass and Γ_Z the total width of the Z°.

$\delta_{rad}(s)$ incorporates corrections for the effects of initial state radiation which amount to about 30% but can be determined to be better than 0.5%.

Using equation (1) a three parameter fit to the data can be made to determine the Z° mass and width and the peak cross-section, which are virtually model independent. The results for M_Z and Γ_Z which were available at the time of this conference are shown in Table 4 where the errors on M_Z show, first the experimental error and second the error of 0.027 GeV due to the uncertainty in the LEP energy which is common to all measurements. The combined result for Γ_Z may be compared with the value of 2.487 GeV predicted by the Standard Model assuming $M_{top} = M_{Higgs} = 100$ GeV and $\alpha_s = 0.12$.

<u>M_Z, Γ_Z and N_ν from an analysis of multihadron</u>

<u>events in the LEP Experiments</u>

	M_Z(GeV)	Γ_Z(GeV)	N_ν
ALEPH	91.182±.026±.027	2.541±.056	3.01±.15
DELPHI	91.10±.04±.027		
L3	91.166±.026±.027	2.539±.055	3.32±.17
OPAL	91.132±.025±.027	2.571±.057	3.41±.20
Combined	91.152±.014±.027	2.549±.032	3.18±.10

Table 4

Also shown in Table 4 are the results derived for the number of light neutrino types by performing a two parameter fit to the data within the framework of the Standard Model with M_Z and N_ν as the free parameters. The combined result which ignores a small theoretical uncertainty of ± .05 is within two standard deviations of $N_\nu = 3$ and more than 5 standard deviations from $N_\nu = 4$. Figure 9 shows the data from which these results were derived.

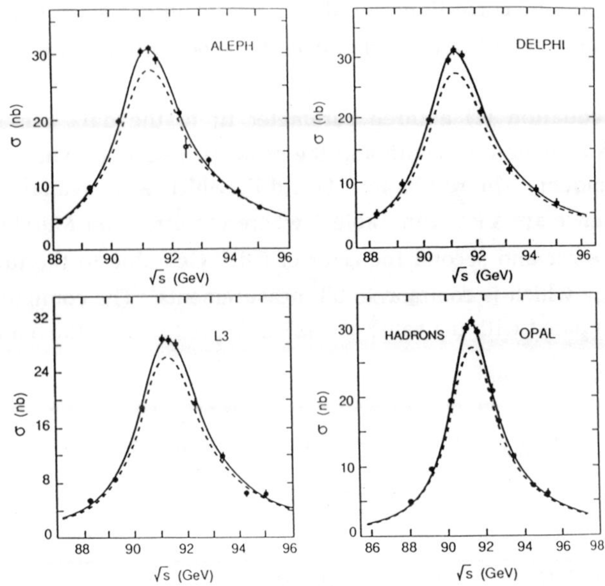

Figure 9   Cross-sections for $e^+e^- \to$ hadrons as measured in each of the four LEP experiments. Standard Model predictions at the values for M_Z given in Table 4 and for $N_\nu = 3$ are shown (solid lines). Also shown are the corresponding predictions for $N_\nu = 4$ (dashed lines).

3.2   Partial Widths and Asymmetries

The LEP experiments have also produced results on the partial widths for decays to charged leptons and hadrons. The results from the four experiments are given in Table 5.

(MeV)	ALEPH	DELPHI	L3	OPAL	Mean	St. Model
$\Gamma_{e^+e^-}$	82.1±3.4	80.5±6.8	79±2	79.6±3.0	79.8±1.5	
$\Gamma_{\mu^+\mu^-}$	87.9±6.0	83.5±11	83±2	82.8±7.0	83.4±1.8	
$\Gamma_{\tau^+\tau^-}$	86.1±5.6	76.5±11	84±5	80.5±7.3	83.4±3.2	
$\Gamma_{l^+l^-}$ (Av)	83.9±2.2	80.9±1.4	81.3±1.4	80.2±2.6	81.5±1.1	83.5
Γ_{hadr}	1804±44			1785±64		1737
Γ_{invis}	495±41			526±64		500

Table 5

All results are consistent with lepton universality and with the Standard Model.

For decays to e^+e^- major corrections were required for t-channel effects which were large in the forward direction as shown in Figure 10 for the ALEPH data.

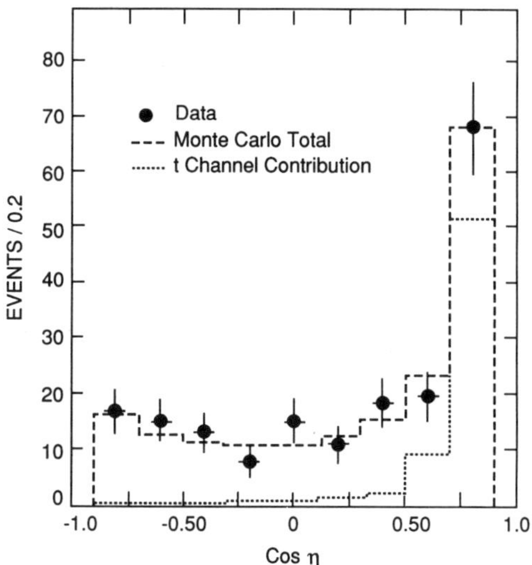

Figure 10   The angular distribution for $e^+e^- \rightarrow e^+e^-$ in the ALEPH experiment showing the t-channel peak in the forward direction.

The main background in a given leptonic channel was from misidentified events from other leptonic channels; in all cases corrections for this could be made without adding significantly to the errors.

The various experiments have treated their data in somewhat different ways. In particular ALEPH[2] and OPAL[3] have carried out model independent analyses. ALEPH has done this essentially by comparing the numbers of leptonic events in the different channels with the corresponding number of hadronic events and calculating the ratios while OPAL carried out a combined analysis of all their data by fitting the cross-sections as a function of energy across the Z° peak assuming it to have a modified Breit-Wigner shape. The results have been used to determine the values of Γ_{invis} shown in the table, which lead to a value for $N_\nu = 3.04 \pm 0.20$ from the combined ALEPH and OPAL results, consistent with the result given in Table 4 but less model dependent.

ALEPH[2] and L3 have used their data on $\Gamma_{l^+l^-}$, assuming lepton universality, to determine $\sin^2\Theta_W(M_Z)$. The results are

$$\sin^2\Theta_W(M_Z)$$

L3	0.239 ± 0.005
ALEPH	0.231 ± 0.008

Results of similar analyses to those mentioned above are expected to be available from all four experiments in the near future.

Finally two of the experiments have provided results on lepton forward-backward asymmetries, A_{FB}, which can be used to derive values for the product $g_A^2 \cdot g_V^2$ of the neutral current coupling constants. OPAL have looked at A_{FB} for decays to $\mu^+\mu^-$ and $\tau^+\tau^-$ and L3 for $\mu^+\mu^-$ alone[4]. The results from L3 are shown in Figures 11 and 12 where data from other experiments have been used to determine the signs of the coupling constants.

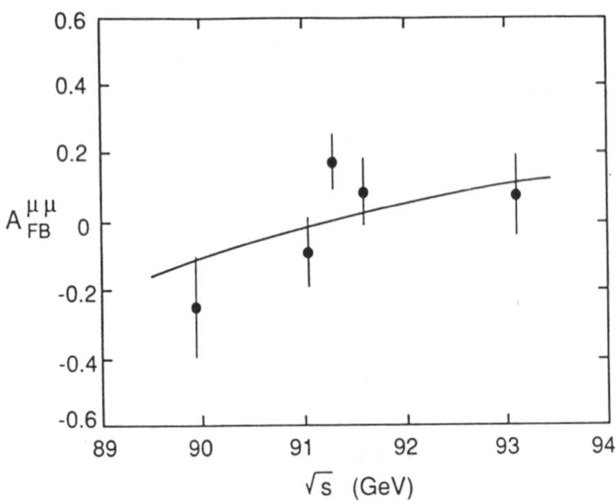

Figure 11   Measured forward-backward charge asymmetry, A_{FB}, as a function of the c.m. energy for the reaction $e^+e^- \rightarrow \mu^+\mu^-$. The curve is the prediction of the Standard Model. A_{FB} is fitted as a function of $\sqrt{s}$ by varying g_A and g_V with the constraint that $g_A^2 + g_V^2 = 0.250 \pm 0.007$ as described in Reference 4). The result of the fit gives $g_A = 0.495 \pm 0.007$ and $g_V = 0.066^{+0.046}_{-0.027}$. The data are from the L3 experiment.

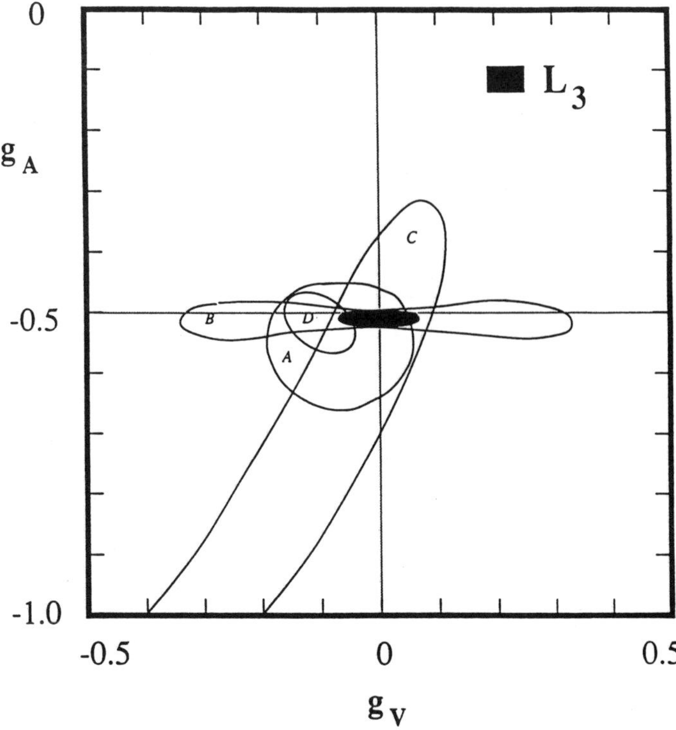

Figure 12   Results obtained from neutrino experiments expressed as 1σ limits on g_A and g_V. Area (A) is the result of the CHARM experiment, area (B) is the combined result from PETRA and PEP experiments, area (C) is the reactor $\bar{\nu}_e e$ result and area (D) is the BNL result. The black area represents the L3 measurement[4].

3.3   Search for New Particles

The LEP experiments have made wide ranging searches for new particles. As mentioned earlier, nothing new has been found below masses which approach very close to the limits kinematically possible for e^+e^- energies around the peak of the Z° cross-section. The methods are based on an analysis of observed event shapes (thrust, sphericity, acoplanarity etc.) compared with Monte Carlo predictions for the various topologies chosen. The results at the 95% C.L. limit are shown in Table 6.

<u>Search for New Particles</u>
(Mass Limits in GeV)

95% C.L.	ALEPH	OPAL	L3	DELPHI
M_t	> 45.8	> 44.5		
M_b'	> 46.2	> 45.2		
M_L	> 45.7	> 44.3		
$M_{\tilde{e}}$	> 44.2	> 43.4	> 41.0	
$M_{\tilde{\mu}}$	> 43.6	> 43.0	> 41.0	
$M_{\tilde{\tau}}$	> 42.3	> 43.0		
$M_{\tilde{W}}$	> 46.0	> 45.0	> 44.0	
M_H (SM)	> 24	> 19.3		
$M_H^\pm$				> 32.0
M_l^*	> 44.6			

Table 6

A selection of the data from which these results were derived is given in the following figures.

ALEPH

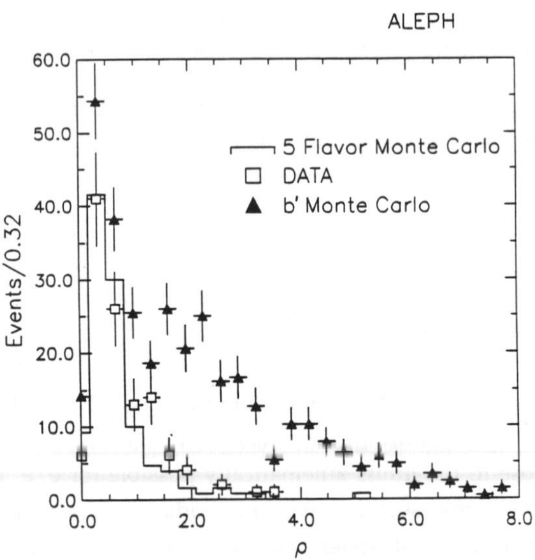

Figure 13   ALEPH data and Monte Carlo simulated events normalised to the same integrated luminosity for $Z^\circ \to b\bar{b}'$ at a mass of 40 GeV.

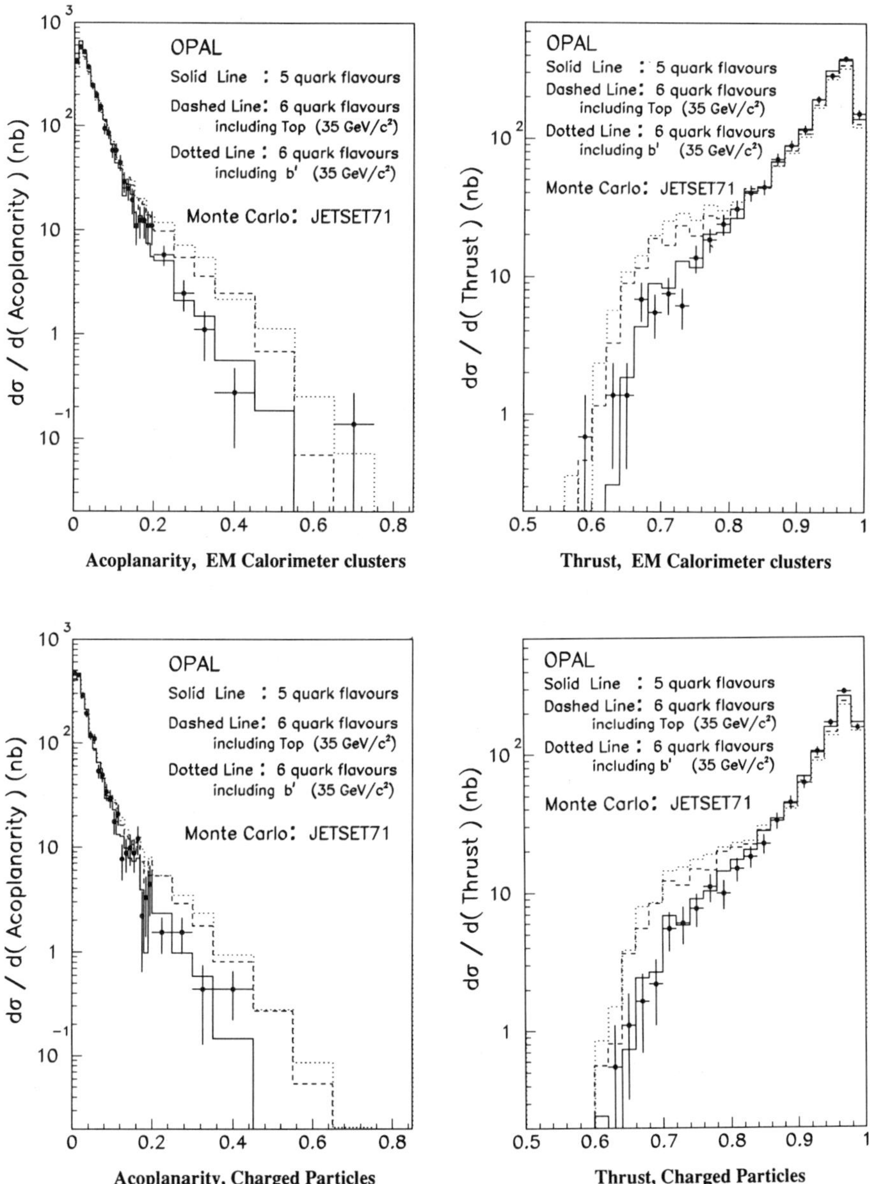

Figure 14 Acoplanarity and thrust distributions measured in OPAL. The data are compared with five flavour Monte Carlo distributions (solid histograms) and with the Monte Carlo predictions including a sixth quark with a mass of 35 GeV (dashed histograms for a top quark and dotted histograms for a b' quark).

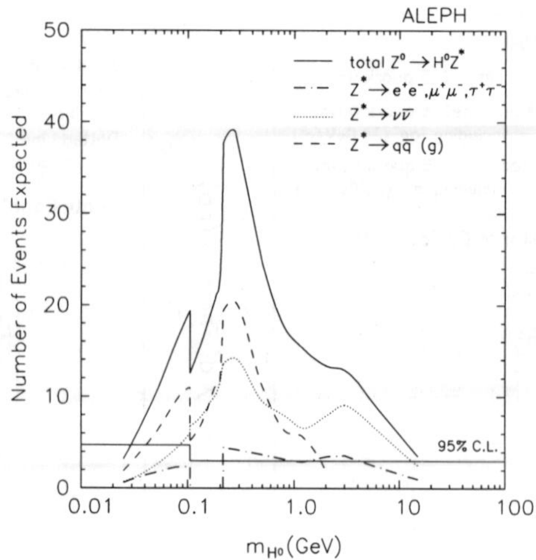

Figure 15　The number of events expected in the ALEPH data (540 nb^{-1}) for $Z^\circ \to H^\circ l^+l^-$, $Z^\circ \to H^\circ\, \nu\bar{\nu}$ and $Z^\circ \to H^\circ\, q\bar{q}$ (g) and the sum of all these channels as a function of the Higgs mass. The 95% CL limit excludes the region from 32 MeV to 15 GeV.

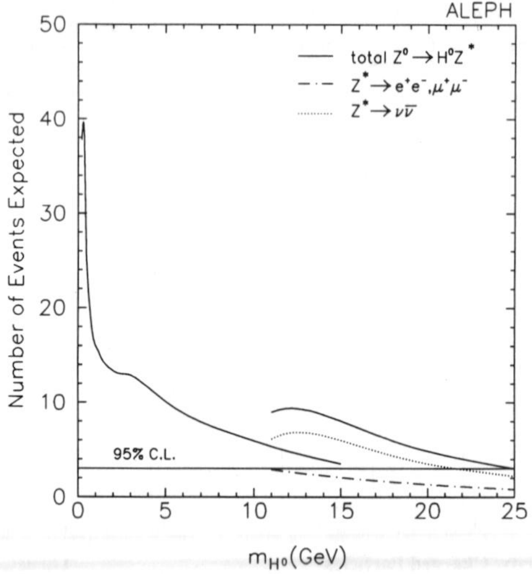

Figure 16　The number of events expected in the ALEPH data for $Z^\circ \to H^\circ l^+l^-$, $Z^\circ \to H^\circ\, \nu\bar{\nu}$ and $Z^\circ \to H^\circ\, q\bar{q}$ (g) and the sum of all these channels as a function of the Higgs Mass. The curve on the left is for 540 nb^{-1} (see also Figure 15) and that on the right for 1200 nb^{-1}. The 95% CL limit of the latter extends the excluded region up to a mass of 24 GeV.

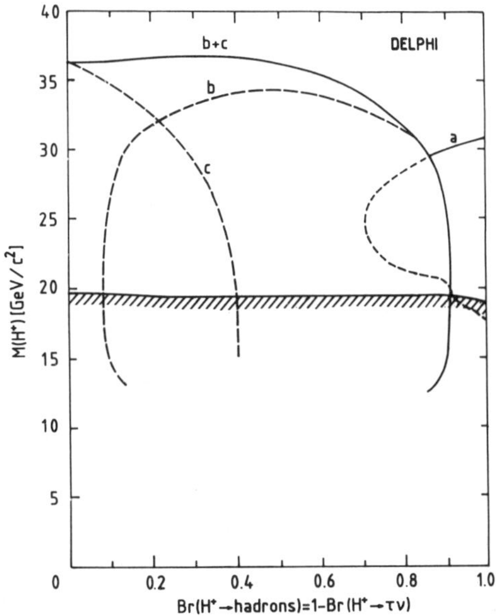

Figure 17   Exclusion contours, at 95% CL for the DELPHI experiment for $Z^\circ \to H^+H^-$ in terms of the hadronic branching ratio and of the charged Higgs mass. Curve (a) corresponds to the 4 jet channel, curve (b) to the $\tau\nu c\bar{s}$ channel and curve (c) to the $\tau\nu\tau\nu$ channel. The hatched area is from the Jade experiment at PETRA.

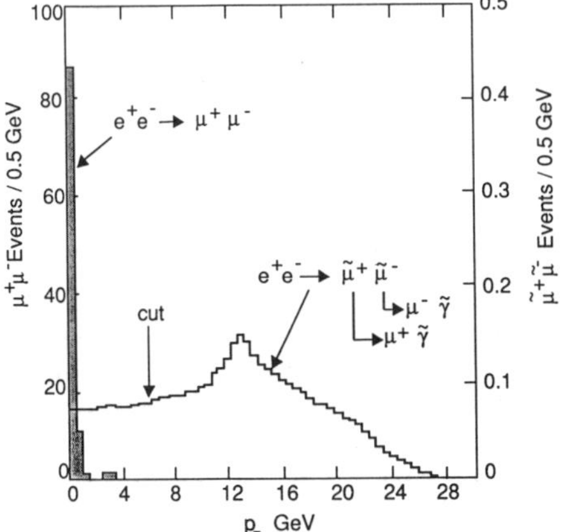

Figure 18   Missing P_T spectra for the reaction $e^+e^- \to Z^\circ \to \mu^+\mu^-$ from L3 experimental data (dark histogram) and for Monte Carlo predictions for dimuon events from the reaction $e^+e^- \to Z^\circ \to \tilde{\mu}^+\tilde{\mu}^- \to \mu^+\mu^-\tilde{\gamma}\tilde{\gamma}$ for a scalar muon mass of 41 GeV and a photino mass of 10 GeV.

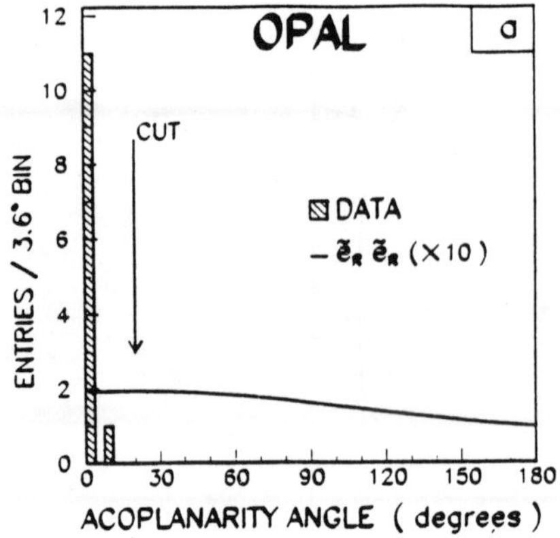

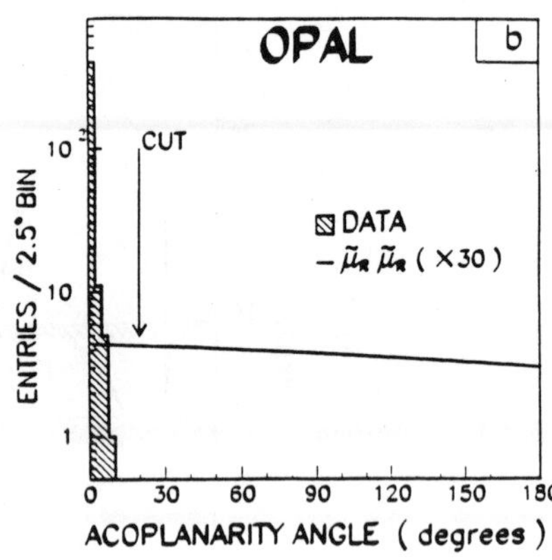

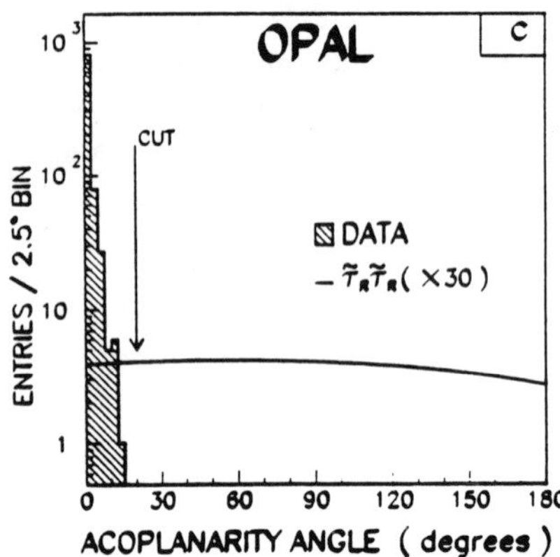

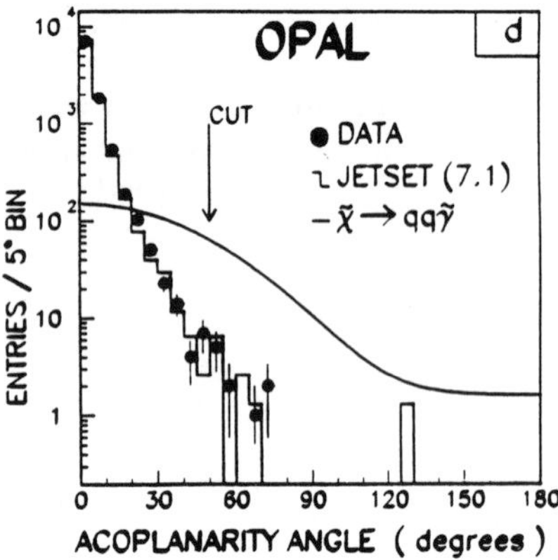

Figure 19   Acoplanarity distributions from the OPAL experiment for a) clusters in the electromagnetic calorimeter b) 2-track events with at least one μ candidate c) two low multiplicity jets d) hadronic jets. In Figs. a)-c) the data are represented by the hatched histograms and in Fig. d) by the points with error bars. The curves in a)-c) are the Monte Carlo expectations for $\tilde{l}_R$ with a mass of 40 GeV and a $\tilde{\gamma}$ mass of 10 GeV. In fig. d) the expectation for the $\tilde{\chi}^{\pm}$ with a mass of 43 GeV is given by the curve and the expectation for hadronic Z^0 decays, as simulated by the Jetset7.1 Monte Carlo, by the histogram (normalized to the number of events).

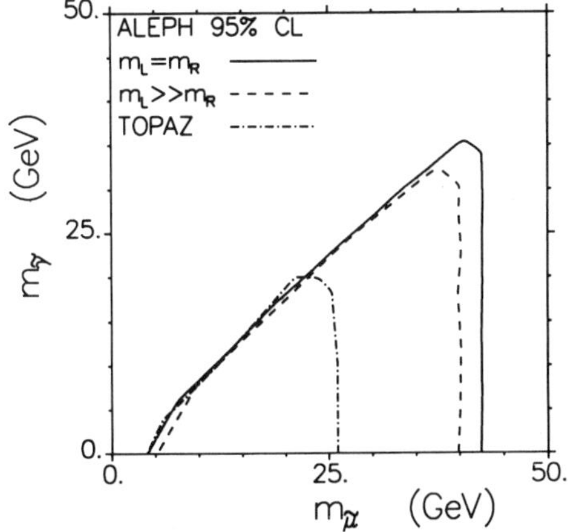

Figure 20   Region excluded at 95% CL by ALEPH data for scalar muon and photino masses for scalar muon pair production. Previous results from TOPAZ are also shown.

4.   Concluding Remarks

The future plans for LEP operation are to continue to run in the region of the Z° peak for another three or four years to accumulate better statistics. In the meantime, superconducting RF cavities will be added to the normal copper cavities already installed for LEP to enable the energy of the beams in LEP to be increased to the full design figure of 95-100 GeV per beam. The aim is to be ready for operation above the W^+W^- pair production threshold by 1994. Feasibility studies are also in progress aimed at extending the potential of the machine by increasing the luminosity of LEP at the energy of the Z° peak by at least an order of magnitude. This would be achieved by increasing the number of bunches in LEP and making improvements to the single bunch intensity and beam optics. The feasibility of providing longitudinally polarized beams in LEP is also being studied. At this stage, it is not clear what levels of polarization can be achieved but tests in LEP are planned for this year to see if measurable levels of transverse polarization can be produced.

By any reasonable standards it can surely be said that LEP and its experiments have made a good start. Certainly the amount of information that has emerged at this early stage has generally exceeded the expectations of a year ago. Nevertheless, it must be remembered that LEP was built to provide very high statistics, i.e. in the region of 5 to 10 x 10^6 events per experiment, at the Z° energy compared with ~ 25,000 so far and to enable physics at higher energies particularly above the W^+W^- pair production threshold to be studied in detail. These remain the primary aims; so for both the machine and the experiments there is still a long way to go.

5.   Acknowledgements

I should like to thank all four LEP experiments for making their results available to me prior to publication for inclusion in this report[6]. Also I am grateful to many members of the LEP and SPS Divisions at CERN for useful discussions and information including the provision of a large number of the figures contained in this report.

References

(1)   ALEPH Collaboration, D. Decamp et al., Phys. Lett. 231B (1989) 519
       DELPHI Collaboration, P. Aarnio et al., Phys. Lett. 231B (1989) 539
       L3 Collaboration, B. Adeva et al., Phys. Lett. 231B (1989) 509
       OPAL Collaboration, M.Z. Akrawy et al., Phys. Lett. 231B (1989) 530.

(2)   ALEPH Collaboration, D. Decamp et al., Phys. Lett. 235B (1990) 399.

(3)   OPAL Collaboration, M.Z. Akrawy et al., Phys. Lett. 240B (1990) 497.

(4)   L3 Collaboration, B. Adeva et al., Phys. Lett. 238B (1990) 122.

(5) For M_t, $M_{b'}$:

ALEPH Collaboration: D. Decamp et al., Phys. Lett. 236B (1990) 511
OPAL Collaboration: M.Z. Akrawy et al., Phys. Lett. 236B (1990) 364

For M_L:

ALEPH Collaboration: D. Decamp et al., Phys. Lett. 236B (1990) 511
OPAL Collaboration: M.Z. Akrawy et al., Phys. Lett. 240B (1990) 250

For $M_{\tilde{e}}$, $M_{\tilde{\mu}}$, $M_{\tilde{\tau}}$, $M_{\tilde{W}}$:

ALEPH Collaboration: D. Decamp et al., Phys. Lett. 236B (1990) 86
L3 Collaboration: B. Adeva et al., Phys. Lett. 233B (1989) 530
OPAL Collaboration: M.Z. Akrawy et al., Phys. Lett. 240B (1990) 261

For M_{H°:

ALEPH Collaboration: D. Decamp et al., Phys. Lett. 236B (1990) 233 and
CERN-EP/90-16
OPAL Collaboration: M.Z. Akrawy et al., Phys. Lett. 236B (1990) 224

For $M_{H^\pm}$:

DELPHI Collaboration: P. Abreu et al., Phys. Lett. 241B (1990) 449.

For $M_l{}^*$:

ALEPH Collaboration: D. Decamp et al., Phys. Lett. 236B (1990) 501.

(6) Most of the results presented here were preliminary at the time of the DPF 90 Meeting. Since then many of these results have been published and the final figures may differ somewhat from those in this report.

Some Remarks on String Theory and Conformal Field Theory[*]

John H. Schwarz

California Institute of Technology, Pasadena, CA 91125

Abstract

Some topics of current interest in string theory and conformal field theory are discussed. These include space-time duality, evidence for a minimum length, and issues in compactification and phenomenology.

---

[*] This work supported in part by the U.S. Department of Energy under Contract No. DE-AC0381-ER40050

Introduction

Since my last DPF report three years ago,[1] there has been steady progress in uncovering the problems and prospects for superstring theory as a framework for the ultimate unification of particles and forces. The most significant development, perhaps, has been the emergence of two-dimensional conformal field theory. It is relevant to the description of classical string solutions and to 2D critical phenomena. Most strikingly, it is a substantial new branch of mathematics that seems to have a life of its own quite apart from any physical applications.

Let me begin by briefly reviewing the general setup by which we hope to get from string theory to a description of nature. The starting point can be taken as the search for consistent relativistic quantum theories of one-dimensional extended objects (strings). This leads to just three possibilities, each of which is free from adjustable parameters or other arbitrariness. These are the type I, type II, and heterotic superstrings. They are called "superstrings" both because they can be formulated with supersymmetry on the two-dimensional world sheet and because they admit solutions with space-time supersymmetry. A fourth possibility, the original bosonic string, is not included in my list because it apparently does not have a stable ground state (*i.e.*, all solutions have one or more tachyons).

The standard approach to studying the physical consequences of a string theory is to look for classical solutions with a D-dimensional Minkowski space-time and additional "internal" degrees of freedom K. Sometimes, but not always, the internal degrees of freedom can be interpreted as corresponding to compactified spatial dimensions, à la Kaluza–Klein. There are an enormous number of classical solutions. Many are known, but many more remain to be discovered. However, there are a few general conclusions that can be drawn for all three theories: 1) The space-time dimension D consists of one time and $D-1$ spatial dimensions and $D \leq 10$. 2) Every classical solution has a particle spectrum that contains one (and only one) graviton in the D-dimensional sense. Its interactions at low energies (compared to the Planck scale) agree with the requirements of general relativity.[2] 3) The quantum perturbation expansion about a classical solution is well-behaved in those cases that satisfy a property called "modular invariance." This property ensures the cancellation of anomalies in all gauge symmetries — Yang–Mills, supersymmetry, and general coordinate invariance. It also ensures that perturbative corrections (*i.e.*, loop amplitudes) have no ultraviolet divergences. For flat $D = 10$ space-time, Mandelstam claims to have a complete proof of finiteness to all orders. Others are close to confirming and generalizing the conclusion.[3] The divergences that do occur are well understood and quite harmless. They correspond to mass shifts and the usual infrared phenomena. Actually, for amplitudes with massless external particles in $D = 10$ not even these occur. It is remarkable that in two of the theories (type II and heterotic), which contain only oriented closed strings, there is just one diagram at each order of the expansion! The myriad point-particle theory Feynman diagrams correspond to various degenerate limits of the world-sheet geometry. It should be emphasized, however, that there is no consistent point-particle field theory containing gravity that is ultraviolet finite or renormalizable. This fact is what hooked me on the string approach to unification some 16 years ago.[2]

The next question is whether any of the classical solutions is a plausible first approximation to a realistic quantum ground state. In the case of the heterotic string there are many possible choices of K that result in a low-energy theory that is essentially the minimal supersymmetric standard model. Thus it looks quite promising, assuming that there is low-energy supersymmetry. In the case of the type II superstring, there is a theorem to the effect that no choice of K can give a low-energy theory possessing quarks and leptons with the correct standard model quantum numbers,[4] although one can come tantalizingly close. There is some effort being made to find ways to circumvent this theorem, but the prospects are not bright. The type I superstring is more difficult to analyze than the others because it contains open strings as well as closed ones. Very little is known yet about the possibilities for solutions with $D = 4$.[5] This requires more study.

In the case of the heterotic string, crucial questions include the following: What are the mathematical possibilities for the internal space K? Is there a compelling theoretical basis for preferring some to others? Does any choice lead to realistic, and even usefully predictive, phenomenology? Are nonperturbative effects important? What ensures that the cosmological constant vanishes once supersymmetry is broken? These questions have been discussed at length by many authors. Here I will just make some relevant remarks.

Supersymmetry

As things stand today, it would *not* be fair to claim that low-energy supersymmetry is a first principles prediction of string theory, but it certainly has a natural home there. Moreover, it seems to be the most promising solution to the gauge-hierarchy and fine-tuning problems. It is true that some people (including myself) have suggested that a yet-to-be discovered 'string miracle' might ensure the ultraviolet cancellations needed to avert the hierarchy problem in the absence of space-time supersymmetry, but no convincing examples or mechanisms have been found. If supersymmetry is the right answer, then the characteristic mass scale must be the electroweak scale. This means that with a little luck it could be discovered within in the next few years.

It would be an enormous shot-in-the-arm for string theory if supersymmetry showed up experimentally. The best prospect on the immediate horizon seems to be in the study of Z decays.[6] The recent LEP and SLC experiments reported at this conference make it quite clear that there are only three neutrinos, and hence only three standard generations of quarks and leptons. However, it is conceivable that the experimental value of N_ν could exceed three by some fraction. In fact, the current numbers, while consistent with three, do seem to show such a tendency. Should this persist and become statistically significant, it would be a dramatic signal of new physics. The most likely possibility, in my opinion, is supersymmetry.

Many models suggest that the lightest supersymmetry particle (LSP) is likely to be a 'neutralino'. This is a spin 1/2 fermion that is a linear combination of the neutral gauginos and higgsinos, all of which are expected to mix in general. This particle seems to be the best candidate for the dark matter that provides a substantial fraction of the mass of the universe. (Another contender, the axion, looks less promising now than it did at one time.[7]) By assuming this to be the case, in the context of the minimal supersymmetric standard model, Ellis and Zwirner conclude that its mass should be about 4–20 GeV. Thus it could be pair produced in Z decays. They conclude at the same time that the top quark mass should be about 90 GeV. However, in order to get specific predictions they need to make some assumptions about the structure of supersymmetry breaking. This is a subject that has led many people astray in the past, and even today noone really knows for sure what is the right thing to do, so in my opinion all the numbers should be taken with a heavy dose of salt.

In supersymmetric models there need to be (at least) two Higgs doublet super-multiplets. One Higgs multiplet provides mass to up-type quarks, while the other one provides mass to down-type quarks and charged leptons. Thus after the W and Z particles eat three scalars, a charged Higgs $H^{\pm}$, two CP even neutral Higgs H_1 and H_2, and a CP odd neutral Higgs P survive. The supersymmetry partners $\tilde{H}_1$ and $\tilde{H}_2$ are the higgsinos referred to above. The lightest neutralino, $\tilde{N}_1$, is a linear combination

$$\tilde{N}_1 = \alpha \tilde{W}_3 + \beta \tilde{B} + \gamma \tilde{H}_2 + \delta \tilde{H}_1. \tag{1}$$

Since standard model couplings do not allow the Z to decay into gaugino pairs, the rate for decay into neutralino pairs depends crucially on the higgsino admixtures. Since neutralinos interact with weak interaction strength, direct observation would be difficult. But their existence could be inferred if the neutrino counting experiments were to end up with a result greater than three. In neutrino units the rate is given by

$$R = (\gamma^2 - \delta^2)^2 (1 - 4m^2/m_Z^2)^{3/2}. \tag{2}$$

The authors of [6] argue that there is a good chance that the mixing angles are such that this will be observable.

There are other supersymmetry particles that might also show up in Z decay. The sneutrinos, which are spin zero, would each contribute to the missing width an amount equivalent to half a neutrino times a phase space suppression factor. Thus the experiments can be interpreted as providing limits on their masses. The models allow their masses to be either above or below $m_Z/2$. Another possibility is 'charginos', which are linear combinations of winos and charged higgsinos. Being charged, they would be directly observable, of course. I could hardly overstate what the significance of the experimental discovery of supersymmetry would be for the future of particle physics.

Compactification and Phenomenology

In the case of the heterotic string, the degrees of freedom described by K split into ones that propagate in one direction around the string loop ('left-movers') and ones that propagate in the opposite direction ('right-movers'). The idea, roughly speaking, is that they satisfy 2D wave equations whose general solution has the form $f = f_L(\sigma + \tau) + f_R(\sigma - \tau)$. The string has an intrinsic orientation because the left-movers and right-movers have distinct properties. The most general classical solution that gives flat $D = 4$ Minkowski space-time is obtained by associating K with a conformal field theory that satisfies certain requirements, which we will now sketch.

The analysis is best understood by making a Wick rotation of the string world sheet ($\tau \rightarrow i\tau$), so that is has Euclidean metric, and introducing a complex coordinate $z = \sigma + i\tau$. (Strictly speaking z is only defined in local patches.) Then f is a sum of holomorphic and anti-holomorphic pieces, $f = f_L(z) + f_R(\bar{z})$. The full 2D theory that describes the classical string dynamics is required to be conformally invariant, meaning that is invariant under arbitrary analytic changes of variables ($z \rightarrow \tilde{z}(z)$. This infinite-dimensional symmetry group is generated by the infinitesimal transformations $z \rightarrow z + \epsilon z^{1-n}$, which are implemented by the operators $L_n = z^{1-n}\frac{d}{dz}$. These operators satisfy the Virasoro algebra $[L_m, L_n] = (m - n)L_{m+n}$. There is a separate algebra of this type for left-movers and right-movers. In general, 2D conformal theories have a quantum mechanical anomaly in the conformal symmetry, which leads to a central extension of the algebra having the form

$$[L_m, L_n] = (m - n)L_{m+n} + \frac{c}{12}(m^3 - m)\delta_{m+n,0}.$$

The value of the central charge (or conformal anomaly) c depends on the particular theory.

In the string theory context, the central charge receives several contributions, which are required to cancel in order that the theory really be conformally invariant. The cancellation occurs between the 4D space-time coordinates ($c_L = 4$ and $c_R = 6$), the Faddeev-Popov ghosts ($c_L = -26$ and $c_R = -15$) and the internal coordinates K ($c_L = 22$ and $c_R = 9$). There are some further requirements besides the cancellation of conformal anomalies. The right-moving modes must have at least $N = 1$ super-symmetry on the world-sheet, and the left-moving and right-moving modes must be pieced together to ensure modular invariance.

If one further requires a second supersymmetry for right-movers and suitable quantization of an associated U(1) charge, then the resulting theory has $N = 1$ space-time supersymmetry. Almost all studies concentrate on these supersymmetric models for several very good reasons: Nonsupersymmetric solutions usually have tachyons that cause divergence problems and destabilize the vacuum. They also develop a cosmological constant at one loop, which makes the study of multi-loop amplitudes problematical and frustrates phenomenological analysis. Space-time supersymmetry is easily (one might even say naturally) achieved and averts these difficulties.

The requirement of modular invariance is quite restrictive. Its meaning is that the amplitude associated with a topologically nontrivial world sheet (torus or higher genus) does not depend on the topological class of the coordinates used in parametriz-ing the surface. A complete classification of modular invariant solutions is certainly not known. However, an interesting class of solutions is achieved by identifying the right moving degrees of freedom with a $c = 9$ subset of the $c = 22$ left moving degrees of freedom. The complete prescription for doing this has been given by Gepner.[8] Models constructed in this way are called (2,2) models. (The numbers refer to the number of right-moving and left-moving world sheet supersymmetries.) This is the most studied and best understood case, although the space of such models presumably has lower dimension than that of the more general (2,0) models. The (2,2) models can be roughly understood as corresponding to compactification on six-dimensional Calabi–Yau spaces (Kähler manifolds with vanishing first Chern class), a possibility that was first proposed some five years ago.[9] However, there are (at least) two subtleties. First of all, the internal degrees of freedom really correspond to a certain conformal field theory, and the correspondence between geometric spaces and conformal field theories is many to one. Quite distinct compactification geometries can actually give the same physical theory! The reasons for this will be explained in the next section. Thus the abstract conformal field theory is in a sense more fundamental than any particular geometry that corresponds to it. The second subtlety is that singular spaces with corners or conical singularities (orbifolds) can occur. Calabi–Yau spaces are (by definition) smooth. In some cases the singular spaces can be interpreted as limiting cases of families of Calabi–Yau spaces.

One of the nice features of Calabi–Yau compactification — or more generally (2,2) models — is that the low-energy effective supersymmetric four-dimensional theory that results has E_6 gauge symmetry, one of the standard grand unification groups. There is an arsenal of tricks for breaking this to your favorite subgroup including the standard model gauge group $SU(3) \times SU(2) \times U(1)$. The massless spin 1/2 fermions fall into the standard E_6 matter representations — 27's (generations) and $\overline{27}$'s (anti-generations). The number of occurrences of each kind of multiplet, N_{27} and $N_{\overline{27}}$, is determined by two topological quantum numbers of the Calabi–Yau space. (These are readily generalized to the more abstract (2,2) models.) Yukawa couplings, which determine mass ratios among other things, are also determined topologically. The difference $N_{27} - N_{\overline{27}}$, which corresponds to the Euler number of the manifold, is of particular interest. It is tempting to suppose that the $\overline{27}$'s pair up with 27's to acquire a unification scale mass. If this is so, then the observed number of generations would equal half of the absolute value of the Euler number χ. In practice people are encountering difficulty in identifying the actual mechanisms that trigger the pairing.

For this and other reasons, it is probably desirable not only to have $|\chi| = 6$, but to have as few paired generations as possible.

A complete classification of Calabi–Yau spaces does not yet exist. Five years ago Yau estimated that there might be about 10,000 distinct simply-connected topologies. Given a simply-connected space with a discrete group of symmetry transformations G that acts freely, it is possible to construct a new space with fundamental group G by identifying points related by the symmetry. (More complicated constructions are also possible in the case of groups that do not act freely.) It is useful to have a space that is not simply connected, because then Wilson loops that wrap around noncontractible cycles can be used to break the E_6 gauge symmetry. When Yau made his estimate only a few simply-connected Calabi–Yau spaces were known. For a period of time only one space with $|\chi| = 6$ was known. It was first constructed by modding out a simply connected space with $|\chi| = 54$ by a discrete symmetry group with nine elements. Since it was the only example known that could give three generations of quarks and leptons, it was used in a number of "phenomenological" studies. Since the pairing of the extra generations and anti-generations, as well as the breaking of supersymmetry and a number of other desiderata, could not be derived from first principles, various interim measures needed to be taken. Thus arose the subject of 'superstring inspired phenomenology.' I think this was (and continues to be) a useful exercise, even though everyone realizes that the results cannot be taken too seriously at this stage.

A great deal of progress has been made recently in the construction and classification of Calabi–Yau spaces. Many new ones have been obtained recently,[10] and now the number in captivity almost matches Yau's estimate. Moreover about 25 of them have $|\chi| = 6$. They are simply connected and give lots of anti-generations that must pair up, so they are probably not of phenomenological interest. However, their existence does demonstrate that the particular space that was being studied earlier is not that special.

Space-Time Duality

Like many aspects of string theory, the use of the word "duality" has a curious history. Phenomenological studies of hadronic interactions in the 1960s led to the realization that high energy amplitudes could be approximated either by a sum of direct-channel resonance contributions or by a dual description in terms of crossed-channel Regge-pole exchanges. The desire to construct explicit mathematical examples with this property was a central motivation of Veneziano and others in the original papers on string theory — called "dual resonance models" at the time.[11] Re-expressed more abstractly in the language of two-dimensional conformal field theory, this property was called the "conformal bootstrap" by Belavin, Polyakov, and Zamolodchikov.[12] Since this phenomenon is associated to the string world sheet, and to avoid confusion with other uses of the word duality, I propose referring to it as "world-sheet duality."

A second use of the word duality has arisen in the description of string theories with toroidally compactified space-times. When one spatial coordinate forms a circle of radius R, the momentum component of the string in this dimension is quantized ($p = n/R$). This works the same for strings and point particles, but there is a second integer that only occurs for strings. A closed string can wrap around the circular dimension m times, where m is an arbitrary integer whose sign describes the orientation. Thus one can descibe string states by pairs of integers (m, n). As we will show, such a string state has a "zero-mode" contribution to its mass–squared proportional to $(n/R)^2 + (mR/\alpha')^2$, where α' is the Regge slope parameter which is inversely proportional to the string tension.[13] This formula exhibits a remarkable feature, which I believe is of rather deep significance. The formula is unchanged under the replacement $R \to \alpha'/R$, provided one simultaneously interchanges the momentum and winding numbers m and n.[14] Indeed the two theories are completely equivalent. Compactification on radius R and radius α'/R correspond to the same conformal field theory. It is the conformal field theory and not the choice of a particular spatial geometry for compactified dimensions that is critical in characterizing the theory.

As we will discuss, various generalizations of the Z_2 duality symmetry described above have been found. Clearly, this use of the word "duality" is quite different from the previous one. Since it concerns equivalences between seemingly distinct space-time manifolds, I propose to call it "space-time duality." Both types of duality have world-sheet and space-time aspects, but I think this terminology makes a useful distinction.

Depending on the choice of string theory, space-time duality is not always a symmetry. Sometimes a solution at radius R is equivalent to a different one at radius α'/R. One significant example relates $E_8 \times E_8$ to spin$(32)/Z_2$ heterotic strings.[15,16] Another relates type IIA and type IIB superstrings.[17] Incidentally, this gives compelling evidence in each case that one is dealing with distinct solutions of a single theory rather than two different theories.

Space-time duality seems to be trying to tell us something important. It suggests that there is effectively a minimum distance — the Planck length $\sqrt{\alpha'}$ — that can be probed in string theory.[18] This feature is expected in any sensible quantum theory of gravity, since there should be an uncertainty principle for the metric tensor itself, so it is pleasing to see it emerge. It also suggests a very amusing cosmological scenario if one imagines starting with a large contracting universe.[19] When the "radius" passes through the Planck length nothing very special happens except that gravity becomes strong, the temperature approaches a Planck scale value (called the Hagedorn temperature), and the detailed dynamics becomes rather complicated. Later, however, when $R << \sqrt{\alpha'}$, this can be reinterpreted (thanks to space-time duality) as an expanding universe of radius α'/R. Thus one has the possibility of effectively having a "bounce," without requiring any exotic forces to stop the collapse! The validity of such a picture depends, among other things, on the existence of a meaningful definition of "time" when $R \approx \sqrt{\alpha'}$.

A significant feature of space-time duality is that there is often enhanced gauge symmetry at fixed points of the transformation. This is generically the case for non-supersymmetric sectors (bosonic strings and heterotic left-movers) but not for supersymmetric ones (type II strings and heterotic right-movers). For example, in the case of circular compactification of a nonsupersymmetric dimension, the $U(1)$ gauge symmetry extends to $SU(2)$ at $R = \sqrt{\alpha'}$. The $U(1)$ has a simple Kaluza–Klein interpretation as the symmetry of a circle, but the occurrence of $SU(2)$ is a stringy phenomenon. It seems reasonable to suppose that for $R \neq \sqrt{\alpha'}$, there is still $SU(2)$ gauge symmetry, but it is spontaneously broken.[17] This viewpoint is supported by constructing low-energy effective Lagrangians containing (in addition to the gauge fields) appropriate scalar fields that act as order parameters.[20]

Let us now consider a more complex example than the simple circular compactification described above. It will yield an infinite group of generalized duality transformations. Suppose K consists of a d-dimensional torus. Denote the string components in the circular directions by X^a, $a = 1, ..., d$. We also allow constant values for the components of the metric and antisymmetric field, G_{ab} and B_{ab}, respectively, in these directions. Then the two-dimensional string action that describes the propagation of free strings in the compactified directions is[15] (setting $\alpha' = 1$)

$$\frac{1}{2} \int d\sigma d\tau \left[G_{ab} \partial^\alpha X^a \partial_\alpha X^b + \epsilon^{\alpha\beta} B_{ab} \partial_\alpha X^a \partial_\beta X^b \right] . \tag{3}$$

where $X^a \approx X^a + 2\pi m^a$, and $\{m^a\}$ are arbitrary integers. The angles and radii that characterize the d-dimensional toroidal space are encoded in the matrix G_{ab}. The coefficients $B_{ab} = -B_{ab}$ do not have a simple geometric interpretation. They are taken to be constants, which implies that the B term in the action is topological and does not modify the equation of motion $\partial_\alpha \partial^\alpha X^a = 0$. The world-sheet time is denoted by τ and the spatial coordinate by σ with period 2π. In particular, if $X^a(2\pi, \tau) = X^a(0, \tau) + 2\pi m^a$, then the string is wrapped m^a times around the ath circular dimension and m^a is called the winding number. In this case X^a can be written as the sum of $m^a \sigma$ and a periodic term.

As we have already remarked, the B term in the action does not affect the equations of motion. However, it does affect the quantum theory as a consequence of its appearance in the canonical momentum

$$P_a(\sigma, \tau) = G_{ab} \dot{X}^b + B_{ab} X'^b , \tag{4}$$

where, as usual, $\dot{X} = \frac{\partial}{\partial\tau}X$ and $X' = \frac{\partial}{\partial\sigma}X$. Single-valuedness of the wave function on the torus implies that the zero modes of P_a are integers n_a. Thus in terms of the winding numbers m^a and discrete momenta n_a, the zero mode part of the X^a coordinate is given by

$$X_0^a = x^a + m^a\sigma + G^{ab}(n_b - B_{bc}m^c)\tau\,, \tag{5}$$

where G^{ab} is the inverse of the matrix G_{ab}.

The parameters B_{ab} play a role analogous to the θ angle in QCD. Specifically, they are coefficients of a topological term in the action. As in the QCD case, this dependence is periodic. The dependence of the theory on each of the B_{ab} is periodic with unit period, since a shift $\Delta B_{ab} = N_{ab}$ for arbitrary integers N_{ab} can be compensated by a shift $N_{ab}m^b$ in the discrete momentum n_a. This provides equivalence relations on the space of theories parametrized by the moduli G_{ab}, B_{ab}. We will see that these are not the only ones.

Since $X^a(\sigma,\tau)$ satisfies a free wave equation, it has a "left–right" decomposition

$$X^a(\sigma,\tau) = X_L^a(\tau + \sigma) + X_R^a(\tau - \sigma)\,. \tag{6}$$

These in turn have mode expansions of the form

$$\begin{aligned}
X_L^a(\tau + \sigma) &= x_L^a + p_L^a(\tau + \sigma) + \frac{i}{\sqrt{2}}\sum_{m\neq 0}\frac{\tilde{\alpha}_m^a}{m}\,e^{-im(\tau+\sigma)} \\
X_R^a(\tau - \sigma) &= x_R^a + p_R^a(\tau - \sigma) + \frac{i}{\sqrt{2}}\sum_{m\neq 0}\frac{\alpha_m^a}{m}\,e^{-im(\tau-\sigma)}\,,
\end{aligned} \tag{7}$$

where $x^a = x_L^a + x_R^a$, and

$$\begin{aligned}
p_L^a &= \frac{1}{2}\left[m^a + G^{ab}(n_b - B_{bc}m^c)\right]\,, \\
p_R^a &= \frac{1}{2}\left[m^a + G^{ab}(n_b - B_{bc}m^c)\right]\,.
\end{aligned} \tag{8}$$

We are dealing with a simple unconstrained quantum system. Thus straightforward canonical quantization gives

$$\left[X^a(\sigma,\tau), P_b(\sigma',\tau)\right] \;=\; 2\pi i \delta^a_b \,\delta(\sigma-\sigma') \quad . \tag{9}$$

The Hamiltonian, constructed in the usual manner, is

$$\mathcal{H} = \frac{1}{2}\, G_{ab}\left(\dot{X}^a \dot{X}^b \,+\, X^{a'} X^{b'}\right) \propto G_{ab}\left(X_L^{\prime a} X_L^{\prime b} \,+\, X_R^{\prime a} X_R^{\prime b}\right) \quad . \tag{10}$$

This determines the mass-squared operator,

$$M^2 = \frac{1}{2} G_{ab}\, m^a m^b \,+\, \frac{1}{2} G^{ab}(n_a - B_{ac}m^c)(n_b - B_{bd}m^d)$$
$$+\; \text{oscillator terms} \;+\; \text{const.} \tag{11}$$

The "background fields" G_{ab} and B_{ab} only appear in the zero mode contributions. When all degrees of freedom are included, the left- and right-moving contributions to M^2 must agree (as a consequence of σ translational symmetry). The contribution of the zero modes to the difference is

$$G_{ab}(p_L^a p_L^b \,-\, p_R^a p_R^b) \;=\; m^a n_a \quad . \tag{12}$$

It is fortunate that this is an integer, so that it can be balanced by oscillator contributions.

It is striking that the equation above is invariant under interchange of the winding numbers and discrete momenta $m^a \leftrightarrow n_a$. This turns out to be a symmetry of the entire spectrum (and even the interacting theory) provided that one simultaneously redefines B_{ab} and G_{ab} in an appropriate way,[22] specifically $(G+B) \leftrightarrow (G+B)^{-1}$.

To understand the significance of the duality symmetry $(G+B) \leftrightarrow (G+B)^{-1}$, consider a single compactified dimension X. If it has radius R, $G_{11} = R^2/\alpha'$ and there is no B field. Thus the duality symmetry means that compactification on a circle of radius R is equivalent to compactification on a circle of radius α'/R, provided winding numbers and discrete momenta are interchanged. Both are summed over all integers in forming the partition function, which therefore is invariant under $R \leftrightarrow \alpha'/R$. In the general case it is invariant under $(G+B) \leftrightarrow (G+B)^{-1}$ as well as $B_{ab} \to B_{ab} + N_{ab}$, which we discussed earlier. The general duality symmetry also implies that the Lorentzian lattice spanned by vectors $\sqrt{2}(p_L^a, p_R^a)$ is even and self-dual.[21]

It is interesting to explore the "moduli space" of toroidal compactifications a little further. This space, which has dimension d^2, is parametrized by G_{ab} and B_{ab}. In the trivial one-dimensional case, after modding out by the Z_2 symmetry $G \leftrightarrow G^{-1}$, one is left with $G_{11} \geq 1$. In the case of a d-dimensional torus, the appearance of $p_L \cdot p_L$ and $p_R \cdot p_R$ in the M^2 formula and the left–right constraint show that *locally* the moduli space is $O(d,d)/O(d) \times O(d)$.

The global geometry is determined by the group of discrete symmetries generated by $B_{ab} \to B_{ab} + N_{ab}$ and $(G+B) \to (G+B)^{-1}$. Since these do not commute, they generate a rather large discrete group, which we would now like to determine. The principle that determines these symmetry transformations is that they preserve $p_L \cdot p_L \pm p_R \cdot p_R$. This implies that in terms of $2d \times 2d$ matrices written in $d \times d$ blocks, we have

$$\begin{pmatrix} m \\ n \end{pmatrix} \to \begin{pmatrix} m' \\ n' \end{pmatrix} = A \begin{pmatrix} m \\ n \end{pmatrix} \; , \tag{13}$$

where A has integral entries. Also

$$A^T \begin{pmatrix} 0 & 1 \\ 1 & 0 \end{pmatrix} A = \begin{pmatrix} 0 & 1 \\ 1 & 0 \end{pmatrix} \; , \tag{14}$$

which implies that $A \in O(d,d)$. The corresponding transformation of $G+B$ is given by

$$(m' \; n')M' \begin{pmatrix} m' \\ n' \end{pmatrix} = (m \; n)M \begin{pmatrix} m \\ n \end{pmatrix} \; . \tag{15}$$

where

$$M = \begin{pmatrix} G - B\,G^{-1}B & BG^{-1} \\ -G^{-1}\,B & G^{-1} \end{pmatrix} . \tag{16}$$

In other words, $A^T M' A = M$. For example, $A = \left(\begin{smallmatrix} 0 & 1 \\ 1 & 0 \end{smallmatrix}\right)$ corresponds to $(G + B) \to (G + B)^{-1}$.

Altogether we conclude that the moduli space of inequivalent toroidal compactifications of d dimensions, parametrized by inequivalent (G, B) is $O(d, d)/O(d) \times O(d)$ with real coefficients. Its global structure is specified by also modding out by $\tilde{O}(d, d; Z)$, as defined above. This group action has fixed points, which result in orbifold points of the resulting moduli space.

There is enhanced gauge symmetry at fixed points of the $\tilde{O}(d, d; Z)$ transformation group. A very simple example is $G = 1, B = 0$, which is a fixed point of $(G + B) \to (G+B)^{-1}$. $G = 1$ corresponds to all d compactified dimensions having radius $R = \sqrt{\alpha'}$ and being orthogonal. In this case there are $2d$ extra massless vector particles which extend the $[U(1)]^d$ gauge symmetry to $[SU(2)]^d$, a fact that is well-known.

The moduli space of possible string theory compactifications, characterized by the expectation values of scalar fields, has a rich structure. Special points with enhanced gauge symmetry are fixed points of discrete transformations that relate equivalent compactifications. Viewing the string theory as a gauge theory, the discrete transformations can be interpreted as finite gauge transformations. These features have been exhibited explicitly in a simple field theory model for the case of circular compactification.[20] The study of more complicated cases may be helpful for developing an understanding of string field theory.

Discussion

We have presented some ideas that are being developed as part of the program of understanding classical solutions of the heterotic string theory with an eye toward phenomenological application. Qualitatively, the results look very encouraging, especially if nature does really have low-energy supersymmetry. However, this approach has serious limitations due to its neglect of nonperturbative quantum effects. In the absence of these effects, space-time supersymmetry is unbroken and various scalars (including the notorious dilaton) are exactly massless. It is essential that nonperturbative effects break supersymmetry and give mass to all scalar fields. A really challenging question is how this can be achieved without generating a cosmological constant. As long as supersymmetry is unbroken, the vanishing of the cosmological constant is essentially a consequence of boson-fermion cancellations, although it is quite nontrivial to prove this. When supersymmetry is broken, there is no reason known why such cancellations should still be exact. Some people have suggested that the vanishing of the cosmological constant need not occur in the fundamental theory. Rather, they say, it is renormalized to zero by the effects of virtual wormholes in the topology of space-time.[23] I am not convinced that the reasoning used in these analyses is valid in the context of string theory — the only context in which quantum gravity effects are well-defined. I would find it much more satisfying to achieve an understanding in terms of microphysics.

The main reason nonperturbative quantum string effects have not been studied is because string theory has not yet been formulated nonperturbatively! Our current understanding of string theory is analogous to knowing the QCD perturbation expansion but not the QCD Lagrangian. With that amount of information it would be difficult to know about instantons — or even the existence of hadrons. Given the Lagrangian it is possible, at least in principle, to study the complete quantum theory numerically, for example by lattice Monte Carlo techniques. As yet there is no formulation of string theory with an equivalent logical status. This is not for lack of trying, however. The problem is very challenging, and the ultimate answer will undoubtedly be extremely profound.

One widely held hope is that even though it has an enormous number of classical solutions, the heterotic string has a unique quantum ground state. This is required if we hope to be able to derive all of particle physics from first principles. Of course, there is no a priori guarantee in this business, and maybe some parameters are just 'environmental' properties of our particular universe.

There are a number of interesting ideas being considered as possible ingredients of a complete formulation of quantum string theory. Superficially they seem quite unrelated, but they probably all have elements of truth and hidden connections. The conservative idea is to generalize what one does for point particle theories. This means formulating a field theory in terms of fields are functionals of loops, *i.e.*, they are operators that create or destroy entire strings. Such formulations do exist in special gauges and for special theories. However, what is required, and is still lacking, is a suitable functional generalization of general relativity that exhibits the entire symmetry structure of the theory. Part of the problem is that we don't know what the full symmetry is. There is some evidence that the entire spectrum of string excitations forms an irreducible representation of the unknown symmetry. The reason there is a spectrum of masses is because we always consider solutions in which the symmetry is spontaneously broken to a very small subgroup (corresponding to the massless particles). If I understand him correctly, Witten has suggested that a topological string field theory is required to exhibit the full symmetry.[24] Partly with this motivation he has been studying topological point particle theories as a warm-up exercise. This has turned out to be a remarkably rich subject of considerable interest in its own right.

The first step in formulating string field theory, be it topological or otherwise, is to identify the correct degrees of freedom, *i.e.*, the correct configuration space. This is the space on which the fields, from which the action is constructed, are defined. Classical string solutions correspond to field configurations for which the action is stationary. Now we know that there is a correspondence between classical string solutions and conformal field theories (subject to certain restrictions). This has motivated the proposal to identify the configuration space of string theory with the space of *all* 2D field theories, presumably subject to some restrictions that still need to be made precise. Then the action is defined on this theory space with the property that it is stationary at the conformal theories. This undoubtedly sounds very nebulous to anyone hearing it for the first time. However, an example of such an action is known! It is called the Zamolodchikov c function.[25] There are technical reasons why it cannot be the final answer, however. A modification appropriate to the closed bosonic string theory has been proposed by Shevitz and Strominger.[26] If this approach is eventually to work out, many further technical developments will be required.

One lesson seems to be that the more we can learn about two-dimensional field theory the better off we will be. A lot of effort is going into the problem of classifying 2D conformal field theories. Enormous progress has been made, especially for a subclass known as rational conformal field theories.[27] The 2D conformal theories are in turn a subclass of the set of integrable 2D theories, which are now also getting intense scrutiny. Substantial progress has been made on a class known as affine Toda models. These have potential terms that are sums of exponentials with parameters that are controlled by a Lie algebra. Some integrable models can be formulated (and solved) by perturbing conformal theories in suitable ways.

Another point of view from which to approach quantum string theory is suggested by the Polyakov path integral, which expresses string amplitudes as sums over two-dimensional 'random' surfaces. The perturbation expansion viewpoint breaks up this sum into pieces characterized by the genus of the surface. Thus the tree approximation is given by surfaces with the topology of a sphere, the one-loop correction corresponds to surfaces with the topology of a torus, and so forth. What this expansion misses is the surfaces of infinite genus, which (if they can be defined) ought to represent the nonperturbative contribution. Recently impressive progress has been made by formulating the sum in terms of triangulated surfaces rather by an expansion in genus. This seems to be yielding insights into nonperturbative aspects of the theory. So far this has only been done for certain highly oversimplified 'noncritical' string models.[28] Whether the techniques can be extended to theories of interest remains to be seen.

REFERENCES

1. J. H. Schwarz, pp. 2-20 in "Proceedings of the Salt Lake City Meeting," eds. C. DeTar and J. Ball (World Scientific, 1987).

2. J. Scherk and J. H. Schwarz, Nucl. Phys. **B81** (1974) 118.

3. E. D'Hoker, "Chiral Splitting and Unitarity of Closed Superstrings," preprint UCLA/89/TEP/54 and references therein.

4. L. Dixon, V. Kaplunovsky, and C. Vafa, Nucl. Phys. **B294** (1987) 43.

5. J. A. Harvey and J.A. Minahan, Phys. Lett. **188B** (1987)44; G. Pradisi and A. Sagnotti, Phys. Lett. **216B** (1989) 59; N. Ishibashi and T. Onogi, Nucl. Phys. **B318** (1989) 239; Z. Bern and D. C. Dunbar, "Open Superstrings in Four Dimensions," preprint LA-UR-89-3213.

6. J. Ellis and F. Zwirner, "Experimental Predictions from a Minimal Version of the Supersymmetric Standard Model," preprint UCB/PTH/89/15, LBL-27310,CERN-TH.5460/89; J. L. Lopez and D. V. Nanopoulos,"Three Generations: A Prediction Fulfilled," preprint CTP-TAMU-62/89 and references therein.

7. A. Dabholkar and J. M. Quashnock, "Pinning Down the Axion," preprint PUPT-1129 (1989).

8. D. Gepner, Nucl. Phys. **B296** (1988) 757.

9. P. Candelas, G. Horowitz, A. Strominger, and E. Witten, Nucl. Phys. **B258** (1985) 46,

10. M. Lynker and R. Schimmrigk, "A-D-E Quantum Calabi–Yau Manifolds," preprint UTTG-42-89, NSF-ITP-89-182.

11. M. Jacob, editor, "Dual Theory," Physics Reports Reprint Volume I (North-Holland, 1974).

12. A. A. Belavin, A. M. Polyakov, and A. B. Zamolodchikov, Nucl. Phys. **B241** (1984) 333.

13. M. B. Green, J. H. Schwarz, and L. Brink, Nucl. Phys. **B198** (1982) 474.

14. K. Kikkawa and M. Yamasaki, Phys. Lett. **149B** (1984) 357; N. Sakai and I. Senda, Prog. Theor. Phys. **75** (1984) 692.

15. K. S. Narain, M. H. Sarmadi, and E. Witten, Nucl. Phys. **B279** (1987) 369.

16. P. Ginsparg, Phys. Rev. **D35** (1987) 648.

17. M. Dine, P. Huet, and N. Seiberg, "Large and Small Radius in String Theory," preprint IASSNS-HEP-88/54 and CCNY-HEP-88/20.

18. D. J. Gross, "Superstrings and Unification, "Princeton preprint PUPT-1108 (1988); D. Amati, M. Ciafaloni, and G. Veneziano, Phys. Lett. **216B** (1989) 41.

19. R. Brandenberger and C. Vafa, "Superstrings in the Early Universe," Brown preprint HET-673 (1988).

20. J. H. Schwarz, "Spacetime Duality in String Theory," preprint CALT-68-1581 (1989).

21. K. S. Narain, Phys. Lett. **169B** (1986) 41.

22. A. Giveon, E. Rabinovici, and G. Veneziano, Nucl. Phys. **B322** (1989) 167; A. Shapere and F. Wilczek, Nucl. Phys. **B320** (1989) 669.

23. S. W. Hawking, Phys. Lett. **195B** (1987) 337; S. Giddings and A. Strominger, Nucl. Phys. **B307** (1988) 354; S. Coleman, Nucl. Phys. **B307** (1988) 967.

24. E. Witten, "The Search for Higher Symmetry in String Theory," preprint IASSNS-HEP-88/55.

25. A. B. Zamolodchikov, Pis'ma Zh. Eksp. Teor. Fiz. **43** (1986) 565; [JETP Lett. **43** (1986) 730.]

26. D. Shevitz and A. Strominger, "Two Dimensional Approach to String Field Theory," UC Santa Barbara preprint (1989).

27. G. Moore and N. Seiberg, "Lectures on RCFT," preprint RU-89-32, YCTP-P13-89 and references therein.

28. E. Brézin and V. Kazakov, ENS preprint, Oct. 1989; M. Douglas and S. Shenker, Rutgers preprint, Oct. 1989; D. Gross and A. Migdal, Princeton preprint, Oct. 1989.

QCD AND EXPERIMENT

G. Altarelli

CERN - Geneva

CONTENTS

1. - Introduction

2. - Experimental determination of $\alpha_s(Q)$
    2.1 - Total hadronic cross-section in e^+e^- annihilation
    2.2 - Scaling violations in deep inelastic leptoproduction
    2.3 - Quarkonic decays
    2.4 - $e^+e^- \to$ jets
    2.5 - Other processes
    2.6 - Summary and conclusions on α_s
    2.7 - First LEP results and QCD

Acknowledgements

1. - INTRODUCTION

Quantum Chromodynamics (QCD) (1), the SU(3) gauge theory of coloured quarks and gluons, is the present theory of strong interactions. Today it stands as a main building block of the standard model of the known interactions of fundamental particles (except gravity) based on the gauge group SU(3) × SU(2) × U(1). The most striking physical properties of QCD are asymptotic freedom and confinement. The first property (2) is rigorously established in perturbation theory and consists of the fact that the effective coupling decreases logarithmically at short distances. This is the basis for perturbative QCD which is relevant for processes involving large momentum transfers. Colour confinement is the property that the potential energy between coloured charges increases approximately linearly at large distances, so that only colour singlet states can be produced and observed. Originally a mere conjecture (3), with time the basis for colour confinement has become more solid partly from experiments on quarkonium spectroscopy (4) (which are consistent with a potential that increases at long distances) and especially from lattice simulations (5) [for example, the calculation (6) of the potential between static colour charges at large distances and the study (7) of the deconfinement phase transition]. Although not yet proved it is by now at least quite plausible that confinement is really implied by the theory.

Precise experimental tests of QCD are clearly as important as tests of the electroweak sector of the standard theory. Yet testing QCD is even more difficult than testing the electroweak theory. In fact the latter interactions are so weak that perturbation theory is always reliable at present energies. Moreover, leptons as well as photons, W's and Z's are at the same time the fields in the Lagrangian and the particles observed in our detectors. On the contrary, in QCD the perturbative approach, which is essentially the only viable method for extracting from the theory testable quantitative predictions, is only valid for processes with large momentum transfers. Even in the most favourable cases the expansion is only slowly converging because of the relatively large values of the QCD coupling $\alpha_s(Q)$ >> α_{QED}. Moreover QCD is ·a theory of confined quarks and gluons (the "partons") while only hadrons are actually observed. Except for the rather limited set of completely inclusive experiments, in general non-perturbative effects connected with soft parton cascades and hadronization tend to obscure the underlying simplicity of the parton dynamics. On one hand, in spite of its limitations perturbative QCD provides a rich testing ground for the theory. On the other hand, among non-perturbative methods, QCD on the lattice (5) is extremely promising for understanding crucial properties such as confinement, broken chiral symmetry, hadronic masses and matrix elements. However, computer limitations impose very drastic restrictions on the lattice size, on the accuracy of the extrapolation to the continuum limit and on the possibility of implementing the effects of quark loops on the various observables. As a consequence up to now lattice calculations have been more useful for adding important elements to our understanding of the theoretical foundations of QCD rather than for offering direct possibilities of comparison with experiment.

The difficulty of testing QCD is reflected in the fact that no single process or experiment by itself provides a clear-cut and precisely quantitative experimental proof of the theory, at least when practical limitations of feasible experiments are taken into account. There are no analogues in QCD of the g-2 experiment for the muon or the electron, of the Lamb shift and so on in QED. In view of the uncertainties connected with any given experiment, our confidence in QCD rests on the overall set of convergent positive indications which arise from an ever-increasing number of different experiments with all possible beams and targets in a wide range of large energies and momentum transfers. As a result the integrated experimental evidence in favour of QCD is quite impressive by now.

In particular, QCD provides a solid field theoretical basis for the parton approach. The beautiful "naive" parton model of Bjorken, Feynman and others (8) has evolved into the "QCD improved" parton model (9). This powerful approach has become such a familiar and widespread tool for everyday practice in high energy physics that one is led to take all its new successes as granted and, in a way, obvious. Actually the very definite and characteristic pattern of successful predictions given by the parton model already provides quite a solid ground of experimental evidence for QCD. In addition to inheriting the beautiful experimental score of the parton approach, QCD is also tested by an increasing list of phenomena which go beyond the naive parton model and therefore provide an additional and qualitatively different experimental basis for the specific dynamical structure of the theory.

The QCD theory (10-12) is specified by the Yang-Mills (13) Lagrangian corresponding to the unbroken colour gauge group SU(3) with quark matter fields in the fundamental three-dimensional representation of SU(3):

$$\mathcal{L}_{QCD} = -\frac{1}{4} \Sigma_{A=1}^{8} F_{\mu\nu}^{A} F^{A\mu\nu} + \Sigma_{j=1}^{f} \bar{q}_j (i\not{D}-m_j)q_j \tag{1.1}$$

with

$$F_{\mu\nu}^{A} = \partial_\mu g_\nu^{A} - \partial_\nu g_\mu^{A} - g_s c_{ABC} g_\mu^{B} g_\nu^{C}$$

$$D_\mu = \partial_\mu + ig_s \Sigma_{A=1}^{8} t^A g_\mu^{A} \tag{1.2}$$

where g_μ^{A} (A=1,...,8) are the gluon fields, t^A are the generators in the quark representation, normalized according to $\mathrm{Tr}(t^A t^B) = \frac{1}{2}\delta^{AB}$. In turn the normalization of t^A fixes the coupling g_s and the structure constants c_{ABC} given by $[t^A, t^B] = i c^{ABC} t^C$. Gauge fixing and ghost terms are to be added for a correct quantization of the theory (14), but are omitted here. Also omitted is the P, T and CP violating θ term (15)

$$\mathcal{L}_\theta = \theta \frac{g_s^2}{32\pi^2} \varepsilon_{\mu\nu\rho\sigma} \Sigma_{A=1}^{8} F^{A\mu\nu} F^{A\rho\sigma} \tag{1.3}$$

of instantonic origin (16) whose empirical smallness ($|\theta| \lesssim 10^{-8}$) (17) is not understood and represents an important conceptual problem for the standard model (18).

The selection of SU(3) as colour gauge group is unique in view of (a) the fact that the group must admit complex representations because it must be able to distinguish a quark from an antiquark (there are meson states made up of $q\bar{q}$ but not similar qq bound states). (b) There must be colour singlet (because we see no colour replicas of known hadrons), completely antisymmetric baryonic states made up of qqq in order to solve the statistics puzzle for the lowest-lying baryons of spin 1/2 and 3/2 in the

56 of (flavour) SU(6) (19). (c) As we shall see, the number of colours for each kind of quarks must be in agreement with the data on the total hadronic e^+e^- cross-section and on the $\pi^\circ \to 2\gamma$ rate and also, to some extent, on other processes like lepton pair production and the semi-leptonic branching ratio of the τ lepton. Within simple groups (a) restricts the choice to SU(N) with N $\geqslant$ 3, SO(4N+2) with N > 1 [SO(6) has the same algebra as SU(4)], and E6, and then (b) and (c) directly lead unambiguously to SU(3) with each flavour of quarks in a fundamental representation of the group. Too many coloured quarks not only would violate (c) but could also spoil asymptotic freedom (2). In particular the $\pi^\circ \to 2\gamma$ rate is fixed by the Adler-Bell-Jackiw anomaly (20) to the value (21):

$$\Gamma(\pi^0 \to 2\gamma)_{TH} = N_c^2 \ (Q_u^2 - Q_d^2)^2 \ \frac{\alpha^2 \ m_{\pi^0}^3}{32\pi^3 \ f_\pi^2} = (7.61 \pm 0.02) \ (\frac{N_c}{3})^2 \ \text{eV} \qquad (1.4)$$

obtained for f_π = 131.69$\pm$0.15 MeV (22). N_c is the number of quark colour replicas and $Q_{u,d}$ are the u,d quark charges. The theoretical error is expected to be of order m_π^2/M^2 (due to the soft-pion extrapolation from $q^2 = m_\pi^2$ down to $q^2=0$) where M is the scale of variation of the relevant form factor (M~1 GeV). The experimental value is given by (22):

$$\Gamma(\pi^0 \to 2\gamma)_{EXP} = (7.3 \pm 0.2) \ \text{eV} \qquad (1.5)$$

This beautiful result is a great achievement for the theory and a remarkable proof that $N_c = 3$.

$\alpha_s = (g_s^2/4\pi)$, the analogue in QCD of the QED fine structure constant, is widely used. The precise definition of the renormalized coupling must be specified. Technically the most practical definition is in the context of dimensional regularization (23,24) through the procedures of minimal subtraction (MS) (25) or of modified minimal subtraction ($\overline{MS}$) (26). Unless explicitly stated, we shall always refer to the $\overline{MS}$ definition of α_s in the following. Note that in the perturbative region the light quark masses can be neglected in most cases. Thus the definition of α_s refers to the massless theory. The renormalization group formalism (27) leads to the concept of running coupling. The running coupling $\alpha_s(Q)$ is a function of the energy scale Q. Loosely speaking $\alpha_s(Q)$ acts as an effective coupling for QCD interactions at distances of order 1/Q. The running coupling is defined by the relation

$$t = \int_{\alpha_s(\mu)}^{\alpha_s(Q)} \frac{d\alpha}{\beta(\alpha)} \qquad (1.6)$$

where $t = \ln Q^2/\mu^2$ and the initial value $\alpha_s \equiv \alpha_s(\mu)$ has been specified. The β function, obeying the relation

$$\frac{\partial}{\partial t} \alpha_s(Q) = \beta[\alpha_s(Q)] \qquad (1.7)$$

can be computed in perturbation theory (2,28,29)

174

$$\beta(\alpha) = -b_f \alpha^2 (1 + b'_f \alpha + b''_f \alpha^2 + \dots)$$

$$= -\left(11 - \frac{2}{3}f\right) \frac{\alpha^2}{4\pi} \left[1 + \frac{153-19f}{2(33-2f)} \frac{\alpha}{\pi} + \right.$$

$$\left. + \frac{3}{32(33-2f)} \left(2857 - \frac{5033}{9}f + \frac{325}{27}f^2\right) \left(\frac{\alpha}{\pi}\right)^2 + \dots \right] \tag{1.8}$$

The values of b_f (2) and b'_f (28) do not depend on the renormalization prescription, while b''_f (29) and the following terms change with the definition of α_s [however, b''_f as given in Eq. (1.8) is the same (30) in the MS and $\overline{\text{MS}}$ prescriptions]. Numerically, one obtains for $f = 5$:

$$\beta(\alpha) = -0.610 \, \alpha^2 \left[1 + 1.261 \frac{\alpha}{\pi} + 1.475 \left(\frac{\alpha}{\pi}\right)^2 + \dots \right] \tag{1.9}$$

When the first two terms in the expansion of $\beta(\alpha)$ are included in Eq. (1.6) defining $\alpha_s(Q)$, the integral can be performed and one obtains:

$$\frac{1}{\alpha_s(Q)} = \frac{1}{\alpha_s} + b_f \, t - b'_f \, \ell n \, \frac{\alpha_s(Q)}{\alpha_s} \tag{1.10}$$

To the required accuracy this equation can be solved in the form:

$$\alpha_s(Q) = \alpha_{os}(Q) \left[1 - b'_f \, \alpha_{os}(Q) \, \ell n \, \ell n \, Q^2/\Lambda^2 + 0(\alpha_{os}^2(Q))\right] \tag{1.11}$$

where

$$\alpha_{os}(Q) = \left(b_f \, \ell n \, Q^2/\Lambda^2\right)^{-1} \tag{1.12}$$

and the parameter μ appearing in $\alpha_s \equiv \alpha_s(\mu)$ has been traded for the parameter Λ:

$$\ell n \, \frac{\mu^2}{\Lambda^2} = \frac{1}{b_f \alpha_s} + \frac{b'_f}{b_f} \, \ell n \, b_f \, \alpha_s \tag{1.13}$$

As b_f and b'_f do not depend on the renormalization prescription the functional form of the running coupling is universal up to terms of order $\alpha_s^2(Q)$ included. Different definitions of the coupling imply different values of $\alpha_s(\mu)$, hence of Λ. Thus one talks of Λ_{MS}, $\Lambda_{\overline{\text{MS}}}$, etc., and for us $\Lambda \equiv \Lambda_{\overline{\text{MS}}}$. Clearly the difference only appears at next-to-leading order, because the variation of α_s appears at order α_s^2. Thus only sufficiently precise measurements analyzed in terms of theoretical formulae developed up to non-leading accuracy can really lead to a determination of $\Lambda_{\overline{\text{MS}}}$.

Note that b_f, b'_f,... depend on the number of excited quark flavours. In fact when working at large but fixed Q, with Q much larger than the masses of some light quarks and much smaller than the masses of some heavy quarks, then according to an intuitive decoupling theorem (31), the relevant number of flavours f is that of light quarks. In fact in QCD the theory obtained by dropping out heavy quarks is still renormalizable and the couplings do not grow with masses (unlike in spontaneously broken gauge theories where the couplings of Higgses and of longitudinal gauge boson modes increase with masses). Then the behaviour of amplitudes with light external particles is dictated in this range of Q by the set of diagrams where all internal lines are also light. For example, for $Q < am_b$ with $a \sim 1$-2 the large Q behaviour of $\alpha_s(Q)$ is determined by b_5 and b'_5, while b_4 and b'_4 are relevant for $\alpha_s(Q)$ at $Q \gg am_b$. Of course, the physical coupling is continuous so that some matching prescription must be specified. Experiments on e^+e^- annihilation at PETRA or TRISTAN work at $Q \sim 30$-50 GeV well above the b threshold. Their results are naturally compared with the asymptotic formula for f = 5. One can extrapolate their determination of $\alpha_s(Q)$ below the b threshold by setting, for example, (we only consider Q values above c threshold for simplicity)

$$\alpha_s(Q) = \alpha_{s,5}(Q) \cdot \theta(Q-am_b) + \alpha_{s,4}(Q) \cdot \theta(am_b - Q) \qquad (1.14)$$

with

$$\alpha_{s,5}(Q) = \alpha_s(Q,5) \qquad (1.15)$$

$$\frac{1}{\alpha_{s,4}(Q)} = \frac{1}{\alpha_s(Q,4)} + \frac{1}{\alpha_s(am_b,5)} - \frac{1}{\alpha_s(am_b,4)} \qquad (1.16)$$

where

$$\alpha_s(Q,f) = \frac{1}{b_f \ln Q^2/\Lambda_5^2} \left[1 - \frac{b'_f \ln \ln Q^2/\Lambda_5^2}{b_f \ln Q^2/\Lambda_5^2}\right] \qquad (1.17)$$

Note that the constant addition in Eq. (1.16) is subleading with respect to the logarithmically growing terms in $1/\alpha_s(Q,4)$ so that the correct asymptotic behaviour is not spoiled. This constant is fixed so that

$$\alpha_{s,4}(am_b) = \alpha_{s,5}(am_b) \qquad (1.18)$$

On the other hand, the results of experiments on deep inelastic scattering or on Υ decays are naturally expressed in terms of the asymptotic form for $\alpha_{s,4}(Q)$:

$$\alpha_{s,4}(Q) = \frac{1}{b_4 \ln Q^2/\Lambda^2_4} \left[1 - \frac{b'_4 \ln \ln Q^2/\Lambda^2_4}{b_4 \ln Q^2/\Lambda^2_4} \right] \qquad (1.19)$$

In this case it is $\alpha_{s,5}(Q)$ which is defined with a constant term in complete analogy with Eq. (1.16). In the range of Q of practical interest the values of Λ_4 and Λ_5 are quite different, being approximately related by:

$$\Lambda_5 \simeq 0.65 \ \Lambda_4 \qquad (1.20)$$

One sees that for practically the same $\alpha_s(Q)$ the corresponding values of Λ_4

and Λ_5 are quite different. On the other hand, it turns out that there is little numerical difference when the matching parameter a is moved in the interval $1 < a < 2$, or when other similar matching procedures are applied.

In conclusion, when a measurement of $\alpha_s(Q)$ is translated into a value of Λ or vice versa, it is essential that the exact functional form of $\alpha_s(Q)$ in terms of Λ is specified (whether one-loop or two-loop accuracy was assumed, which of the possible forms differing at $O(\alpha_s^3)$ has been adopted, the relevant number of excited flavours, the procedure for making α_s continuous at thresholds).

2. – EXPERIMENTAL DETERMINATIONS OF $\alpha_s(Q)$

The most direct set of quantitative tests of QCD consists of the comparison of several measurements of α_s in different processes. As already stated, for a meaningful determination of $\alpha_s(Q)$ at a well-defined scale Q, or equivalently for a measurement of Λ, in a given process, one needs a perturbative calculation of the corresponding observable at least at next-to-leading accuracy in α_s. In fact a change of scale Q or a change of definition of α_s only affects the result at the level of subleading terms. In the following we shall review and discuss the most significant determinations of α_s from the available data [we follow and update Ref. (32)].

2.1 Total Hadronic Cross-Section in e^+e^- Annihilation

Neglecting quark masses $\sigma \equiv \sigma(e^+e^- \to \text{hadrons})$ is given by (9):

$$\sigma = \frac{4\pi\alpha^2}{3Q^2} (1 + Z) R_{e^+e^-} \tag{2.1}$$

where $Q = \sqrt{s}$, Z is the effect of the weak neutral gauge boson, which can be explicitly computed in terms of $\sin^2\theta_W$ and M_Z, and $R_{e^+e^-}$ is given by (33,34,35):

$$R_{e^+e^-} = 3\Sigma_f \ Q_f^2 \ \{1 + \frac{\alpha_s(Q)}{\pi} + (1.986 - 0.115 \ f) \ [\frac{\alpha_s(Q)}{\pi}]^2 +$$

$$+ (70.985 - 1.200 \ f - 0.005 \ f^2) \ [\frac{\alpha_s(Q)}{\pi}]^3 - (\Sigma_f \ Q_f)^2 \ 1.679 \ [\frac{\alpha_s(Q)}{\pi}]^3 =$$

$$\stackrel{f=5}{=} \frac{11}{3} \ \{1 + \frac{\alpha_s(Q)}{\pi} + 1.411 \ [\frac{\alpha_s(Q)}{\pi}]^2 + 64.81 \ [\frac{\alpha_s(Q)}{\pi}]^3\} \tag{2.2}$$

where in the last line the small term proportional to $(\Sigma_f Q_f)^2$ has been reabsorbed for $f = 5$ into the main correction of order α_s^2. Note that $R_{e^+e^-}$ is (essentially) proportional to N_c, so that the value of this quantity is an important direct test of the existence of three colour replicas of quarks. The correction of order α_s was computed long ago (33) and is clearly independent of the renormalization scheme; that of order α_s^2 was computed in the $\overline{MS}$ scheme by three independent collaborations (34). Finally the calculation of $R_{e^+e^-}$ at three loops in the $\overline{MS}$ scheme was recently completed in Ref. (35). The precise expression of R given in Eq. (2.2) only applies to the photon exchange term. Differences due to the axial Z couplings are neglected in Eqs. (2.1) and (2.2).

Two different groups have analyzed all available data in the range 7 GeV < Q < 56 GeV in order to extract α_s. De Boer et al. (36) fix $\sin^2\theta_W = 0.23$ (the world average value) and obtain

$$\alpha_s(Q = 34 \text{ GeV}) = 0.140 \pm 0.016 \qquad (2.3)$$

Marshall (37) presents a combined fit of α_s and $\sin^2\theta_W$. Together with the fitted result $\sin^2\theta_W = 0.242\pm0.017$ the corresponding value of α_s is given by:

$$\alpha_s(Q = 34 \text{ GeV}) = 0.135 \pm 0.016 \qquad (2.4)$$

Note that the effect of the order α_s^3 correction to $R_{e^+e^-}$ is large enough to change the value of α_s by ~10% (the central value decreases from 0.155 down to 0.140). Although this variation of α_s is within the overall error still the size of the coefficient of $(\alpha_s/\pi)^3$ in the expansion of $R_{e^+e^-}$ is somewhat disturbing. In fact with the value of α_s given by Eq. (2.3) the expansion of $R_{e^+e^-}$ reads:

$$R_{e^+e^-} = \frac{11}{3} [1 + r_1 \frac{\alpha_s}{\pi} + r_2(\frac{\alpha_s}{\pi})^2 + r_3(\frac{\alpha_s}{\pi})^3 + \dots] \simeq$$

$$\simeq \frac{11}{3} [1 + 0.0446 + 0.0028 + 0.0057 + \dots] \qquad (2.5)$$

While it is true that, strictly speaking, beyond one loop the coefficients of the expansion depend both on the renormalization scheme and the choice of scale Q, however, the large ratio of r_3 with respect to r_1, r_2 makes it hopeless to improve the situation substantially by a reasonable reparametrization. Note that as far as one can tell from the three-loop result of Eq. (1.9), the β-function expansion in the MS or $\overline{MS}$ prescriptions is well behaved. The computation of the $0(\alpha_s^3)$ corrections to $R_{e^+e^-}$ dramatically shows the limitations of all the so-called optimization procedures (38). For whatever choice made at the two-loop level among these procedures, the result at three loops is always about as bad as in Eq. (2.5). This clearly shows that the optimization choices cannot pretend to reduce the theoretical error.

In conclusion, the determination of α_s through $R_{e^+e^-}$ is in principle very clear. Unfortunately this method has a limited sensitivity because the QCD correction is quite small. In spite of that the result obtained by combining all experiments together is amazingly precise. Rounding off the errors we can summarize the results as follows [recall Eq. (1.20)]

$$\alpha_s(Q = 34 \text{ GeV}) = 0.14 \pm 0.02 \qquad (2.6)$$

$$\Lambda_{\overline{MS}}^{(5)} \simeq 0.65 \; \Lambda_{\overline{MS}}^{(4)} \simeq (240 \pm {}^{230}_{140}) \; \text{MeV} \qquad\qquad (2.7)$$

2.2 Scaling Violations in Deep Inelastic Leptoproduction

In principle this is the most solid and powerful method for testing perturbative QCD and measuring α_s. As for the total hadronic cross-section in e^+e^- annihilation, also in this case the underlying theory is very well founded (9) [for example in terms of the light-cone operator expansion in (39)] and, the process being completely inclusive, there are no problems associated with the experimental definition of jets and its relation with theoretical partonic cross-sections. But with respect to the case of $\sigma(e^+e^- \rightarrow \text{hadrons})$ there are essential advantages. First, there are many independent structure functions (40) and all of them can be measured at different values of the Bjorken variable x for each given $q^2 = -q^2$ (where q_μ is the virtual γ or $W^\pm$ or Z four-momentum). Thus in this case one actually deals with a system of tests (typically, after a suitable binning, one can compare with theory a number of $\ell n Q^2$ slopes at different values of x). Second, for the structure functions the scaling violations are quantitatively more important than the small $O(\alpha_s)$ corrections to $\sigma(e^+e^- \rightarrow \text{hadrons})$ because they arise from the resummation of a series of logarithmically enhanced terms.

Over the last years an imposing experimental effort has been devoted to the measurement of scaling violations in deep inelastic scattering with electron or muon, neutrino or antineutrino beams (41) on hydrogen, deuterium and heavy nuclei. Recently a new generation of high precision experiments has been completed [EMC (42), BCDMS (43), CDHSW (44)]. Many important predictions of the theory have been confirmed. The existence of scaling violations is definitely established at Q^2 values large enough to support the prediction that their asymptotic decrease is only logarithmic. The observed pattern and magnitude of the scale breaking effects are in good agreement with the theoretical expectations. The values of Λ extracted from the data are consistent with other experimental derivations. However, at a closer look several serious problems remain in the comparison of the data among themselves and with the theory, so that unfortunately after so much theoretical and experimental work the situation is still not yet clear and satisfactory. There are experimental discrepancies that are certainly beyond the declared systematic and statistical errors among different measurements of the same observable. The most disturbing of these cases is perhaps the difference (45) between the BCDMS and the EMC measurements of the structure function $F_2(x,Q^2)$ with muon beams on hydrogen (see Fig. 1) (a priori the most accurate and comprehensive sets of hydrogen data). This discrepancy not only reflects itself in a sizeable difference on the values of the proton parton densities useful for physics at pp, $p\bar{p}$ and ep colliders (which are preferentially extracted from hydrogen data to avoid possible nuclear effects) but also leads to serious doubts on the whole analysis of scaling violations. In fact it turns out a posteriori that the amount of uncontrolled systematics present in at least one of the two experiments is large enough to considerably affect the measurement of α_s and Λ (45).

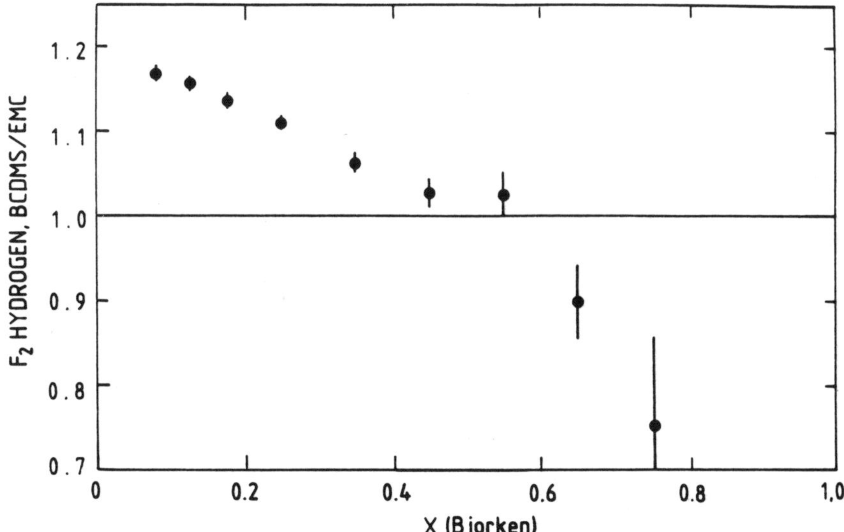

Fig. 1 : Ratio of F_2^p measured by BCDMS and EMC as function of x [Ref.(45)]. Only statistical errors are shown.

A different problem has to do with experiments on iron targets. In this case (45) there is reasonable consistency among different experiments but the data show deviations from the expected behaviour or at least complete agreement with the theory is only obtained at particularly large values of Q and W (the invariant mass of the produced hadronic system) where the statistical accuracy is however small.

We shall now concisely summarize the status of QCD tests based on scaling violations in leptoproduction. The Q^2 dependence of structure functions is dictated by the QCD evolution equations (46). In the non-singlet case these equations are simplest:

$$Q^2 \frac{\partial}{\partial Q^2} \ell n \, F_a^{NS}(x,Q^2) = \frac{\alpha_s(Q^2)}{2\pi} \frac{1}{F_a^{NS}(x,Q^2)} \int_x^1 \frac{dy}{y} \, F_a^{NS}(y,Q^2) \, \mathcal{P}_{qq}^{NS(a)}(\frac{x}{y},\alpha_s(Q^2))$$
$$(2.8)$$

where in leading approximation the kernel $\mathcal{P}_{qq}^{NS(a)}$ coincides with the simple distribution (46)

$$P_{qq}(z) = \frac{4}{3} \left[\frac{1+z^2}{(1-z)_+} + \frac{3}{2} \delta(1-z) \right] \tag{2.9}$$

Note that the lowest-order kernel P_{qq} is independent of the index a, i.e., it is the same for all non-singlet structure functions. The next-to-leading corrections to $\mathcal{P}_{qq}^{NS(a)}$ have been computed in Ref. (47).

The non-singlet evolution equations are exactly valid for the difference of any given structure function measured on proton and neutron targets, i.e., $F^{NS} = F^p - F^n$, or for the structure function F_3 which is given by the difference of neutrino and antineutrino scattering on a given target. Thus the analysis based on F_3 (or $F^p - F^n$) is particularly clear in principle but it suffers from the relatively large errors arising from taking a difference of cross-sections (in any case F_3 is not accessible to experiments with muon beams).

For a general structure function one can separate a non-singlet and a singlet part. For example, for $\mathcal{F}_2 = F_2/x$ one starts from:

$$\mathcal{F}_2(x,Q^2) = \sum_{i=1}^{2f} c_i q_i(x,Q^2) \tag{2.10}$$

where the sum runs over all active flavours of quarks <u>and</u> antiquarks and c_i are the relevant electroweak charges. Equation (2.10) is certainly valid to lowest order and can be taken as a redefinition of quark densities beyond leading order (48). Then

$$\mathcal{F}_2 = \mathcal{F}_2^{NS} + \mathcal{F}_2^S = \sum_{i=1}^{2f} (c_i - <c>) q_i + <c> \Sigma \tag{2.11}$$

where $<c> = 1/2f(\sum_{i=1}^{2f} c_i)$ and $\Sigma = \sum_{i=1}^{2f} q_i$. A gluon term is also present in the singlet evolution equation (46):

$$Q^2 \frac{\partial}{\partial Q^2} \mathcal{F}_2^S(x,Q^2) = \frac{\alpha_s(Q^2)}{2\pi} \int_x^1 \frac{dy}{y} [\mathcal{F}_2^S(y,Q^2) \mathcal{P}_{qq}^{S(2)}(\frac{x}{y},\alpha_s(Q^2)) +$$
$$+ 2f<c>g(y,Q^2) \mathcal{P}_{qg}^{(2)}(\frac{x}{y},\alpha_s(Q^2))] \tag{2.12}$$

The system is then closed by an analogous equation for the gluon density:

$$Q^2 \frac{\partial}{\partial Q^2} g(x,Q^2) = \frac{\alpha_s(Q^2)}{2\pi} \int_x^1 \frac{dy}{y} [\frac{\mathcal{F}_2^S(y,Q^2)}{<c>} \mathcal{P}_{gq}^{(2)}(\frac{x}{y},\alpha_s(Q^2)) +$$
$$+ g(y,Q^2) \mathcal{P}_{gg}^{(2)}(\frac{x}{y},\alpha_s(Q^2))] \tag{2.13}$$

The next-to-leading corrections to the splitting functions $\mathcal{P}$ for the singlet case have been computed in Refs. (49). While the lowest-order kernels are totally unambiguous the corresponding corrections of order α_s depend on the exact definition of quark and gluon densities beyond the leading order. For example, Eq. (2.10) provides a possible definition of quark densities to all orders (48).

An important feature of the QCD evolution equations, evident from Eqs. (2.8, 2.12, 2.13) is that the Q^2 derivative at x of a given structure function only depends on the quark and gluon densities at $y > x$. This allows us to predict the Q^2 evolution from the values of x actually measured. In fact, at fixed Q^2, in practice it is not possible to reach too small values of x. Furthermore as it is empirically true and theoretically reasonable that glue and sea densities are negligible with respect to valence quark densities at sufficiently large x, the gluon term in the singlet equation can correspondingly be omitted. Since the singlet kernel P_{qq}^S also approaches P_{qq}^{NS} at large x $[|P_{qq}^S - P_{qq}^{NS}| \sim (1-x)^5$ near $x \to 1]$ (9) one can approximately reobtain the much simpler non-singlet equation at $x > x_0$, with a suitable value of x_0. In practice, as we shall see, $x_0 \simeq 0.25$–0.30 is normally adopted.

In the evolution equations there are two variables x and Q^2. The shape in x of the structure functions at fixed $Q^2 = Q_0^2$ is not a prediction of perturbative QCD. But given the x dependence of the structure function at $Q^2 = Q_0^2$, then the evolution equations predict it at all Q^2 (in the singlet case the shape of the gluon density is also needed at $Q^2 = Q_0^2$). Although the x and Q^2 dependence are coupled by the evolution equations it is clear that for QCD tests what matters most is the Q^2 variation at fixed x rather than the x variation at fixed Q^2 (50). When the limited range in Q^2 and the experimental errors are taken into account one realizes that the QCD test in the non-singlet case essentially consists of checking that a single value of Λ can accommodate the measured logarithmic slopes $d\ell nF/d\ell nQ^2$ at a number of fixed values of x. It is in fact beyond the present possibilities to measure (within a single experiment) significant deviations from a straight line behaviour in ℓnQ^2. In the singlet case the gluon density, in addition to Λ, is also to be determined from the logarithmic slopes. In fact the gluons are not directly coupled and their distribution is also to be inferred from the scaling violations (or from processes other than leptoproduction, e.g., large p_T photons in p-$\bar{p}$ collisions).

After this concise theoretical summary we now review the data and the corresponding QCD analysis at next-to-leading accuracy.

The BCDMS collaboration (43) has measured F_2 with muon beams on carbon and hydrogen. This experiment has the largest statistics at large Q^2. For the carbon data $Q^2 > 25$ GeV^2 in the range $0.275 < x < 0.75$. The extrapolation to $x > 0.75$ does not introduce an important error because the structure functions are very small in this range. The data are analyzed in the non-singlet approximation. The results for the logarithmic slopes are shown in Fig. 2. The corresponding determination of $\Lambda_{\overline{MS}}^{(4)}$ for four flavours leads to the result

$$\Lambda_{\overline{MS}}^{(4)} = (230 \pm 20 \pm 60) \text{ MeV} \tag{2.14}$$

which corresponds to:

$$\alpha_s(Q = 10 \text{ GeV}) = 0.160 \pm 0.003 \pm 0.010 \tag{2.15}$$

The BCDMS collaboration (43) has also analyzed the data on hydrogen in the non-singlet approximation for $x \geqslant 0.275$ with $Q^2 > 20$ GeV^2. The logarithmic slopes on hydrogen are shown in Fig. 3 together with the QCD fit. In this case one finds:

$$\Lambda_{\overline{MS}}^{(4)} = (205 \pm 22 \pm 60) \text{ MeV} \tag{2.16}$$

or

$$\alpha_s(Q = 10 \text{ GeV}) = 0.156 \pm 0.004 \pm 0.011 \tag{2.17}$$

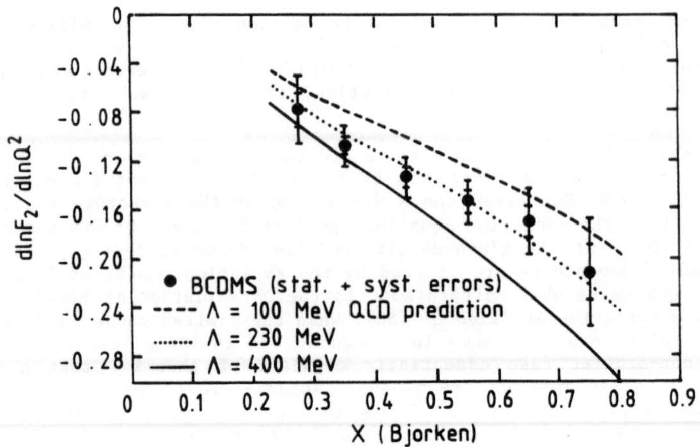

Fig. 2   :   The logarithmic slopes of F_2 measured on carbon by BCDMS (43) compared to the next-to-leading non-singlet QCD evolution for the indicated values of $\Lambda \equiv \Lambda_{\overline{MS}}^{(4)}$.

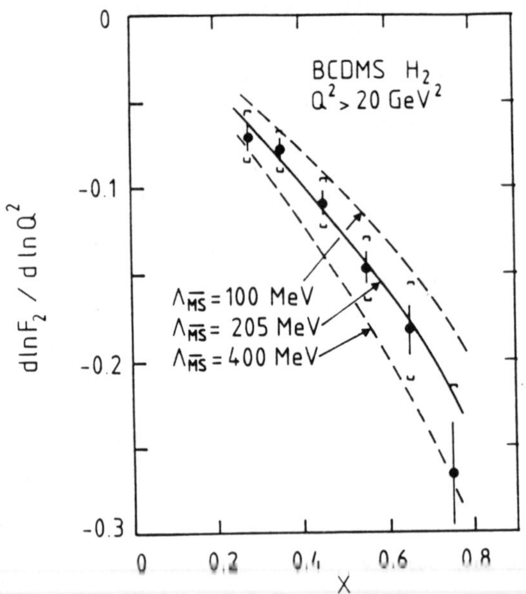

Fig. 3   :   The logarithmic slopes of F_2 measured on hydrogen by BCDMS (43) compared to the next-to-leading non-singlet QCD evolution for the indicated values of $\Lambda \equiv \Lambda_{\overline{MS}}^{(4)}$.

The hydrogen fit is in perfect agreement with the results on carbon. They can be combined to give

$$\Lambda_{\overline{MS}}^{(4)} \simeq 1.54 \quad \Lambda_{\overline{MS}}^{(5)} = (220 \pm 15 \pm 50) \text{ MeV} \tag{2.18}$$

The hydrogen data on F_2 by BCDMS are also available in the range $0.07 < x < 0.275$ (with $Q^2 > 8$ GeV^2 for $x < 0.16$, $Q^2 > 14$ GeV^2 for $0.16 < x < 0.25$) so that a singlet fit can be performed at small x (Fig. 4). In this region of x the logarithmic slopes require a sizeable gluon density. The gluon distribution is indeed concentrated at small values of x as demanded by consistency as the gluon term in the evolution equations was neglected for $x > 0.275$ in the non- singlet analysis.

In conclusion the BCDMS analysis presents a remarkable consistency among carbon data, hydrogen data at large x and hydrogen data at small x and a beautiful agreement with perturbative QCD predictions.

Unfortunately this idyllic picture is somewhat spoiled by the results from other experiments of a priori comparable precision. As already mentioned, there is a severe disagreement (45) (about three times larger than that allowed by the quoted systematic errors) between the BCDMS and the EMC data on F_2 for proton targets, shown in Fig. 1. Previous SLAC data (51) on F_2^p (up to $Q^2 \sim 20$ GeV^2) cannot resolve this discrepancy because there is essentially no overlap in x and Q^2. It is true that the discrepancy is mainly on the normalization of F_2 at different x and not on the logarithmic slopes (Fig. 5). The EMC data are indeed also consistent within errors with QCD. The non-singlet fit to F_2^p by EMC at $x > 0.35$ and $Q^2 > 8$ GeV^2 (with $<Q^2> \sim 22.5$ GeV^2) leads to:

$$\Lambda_{\overline{MS}}^{(4)} = (105 \pm \begin{smallmatrix} 55 \\ 45 \end{smallmatrix} \pm \begin{smallmatrix} 85 \\ 45 \end{smallmatrix}) \text{ MeV} \tag{2.19}$$

This value of $\Lambda_{\overline{MS}}$ is consistent with the corresponding BCDMS result in Eq. (2.16) although the EMC central value is smaller by a factor of two. The real problem is that the discrepancy indicates an uncontrolled systematics of large enough size (45) to make the agreement with QCD and the consistency of the fitted values of $\Lambda_{\overline{MS}}^{(4)}$ to some extent accidental.

The BCDMS results for $\Lambda_{\overline{MS}}$ obtained from the data on carbon are compatible with the CHARM collaboration (52) results obtained from neutrino scattering on marble ($CaCO_3$), a target not very much heavier than carbon ($A \sim 20$ vs. $A \sim 12$). The CHARM result, obtained from the non-singlet structure function F_3 at next-to-leading accuracy for $Q^2 \simeq 3$-78 GeV^2, is given by

$$\Lambda_{\overline{MS}}^{(4)} = (310 \pm 140 \pm 70) \text{ MeV} \tag{2.20}$$

There are many high statistics experiments on the iron structure functions ($A \sim 56$). F_2^{Fe} has been measured by CCFRR (53), CDHSW (44) with ν beams and by BFP (54) and EMC (42) with muon beams. The data are in reasonable agreement among them, within the stated uncertainties, although the EMC data are 5-10 % below the other data sets. Similarly the xF_3 measurements from CCFRR and CDHSW are also consistent. For iron structure functions the problem is that the logarithmic slopes in general show a steeper x dependence than expected from QCD with values of $\Lambda_{\overline{MS}}$ compatible with Eqs. (2.18, 2.19). For example, a comparison of the EMC data on F_2^{Fe} with the non-singlet QCD fit obtained at next-to-leading accuracy using the

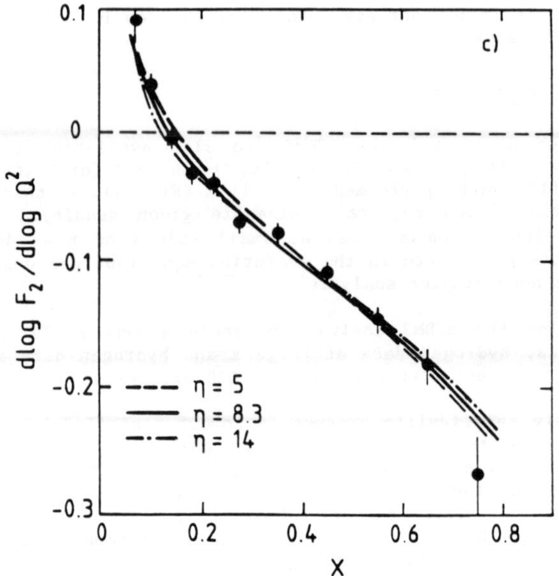

Fig. 4 : The logarithmic slopes of F_2 measured by BCDMS (43) on hydrogen compared to the next-to-leading singlet QCD evolution with $\Lambda_{\overline{MS}}^{(4)} \simeq 220$ MeV and a gluon density $xg(x,Q^2) \sim A(1-x)^\eta$ for $Q^2 = 5$ GeV^2.

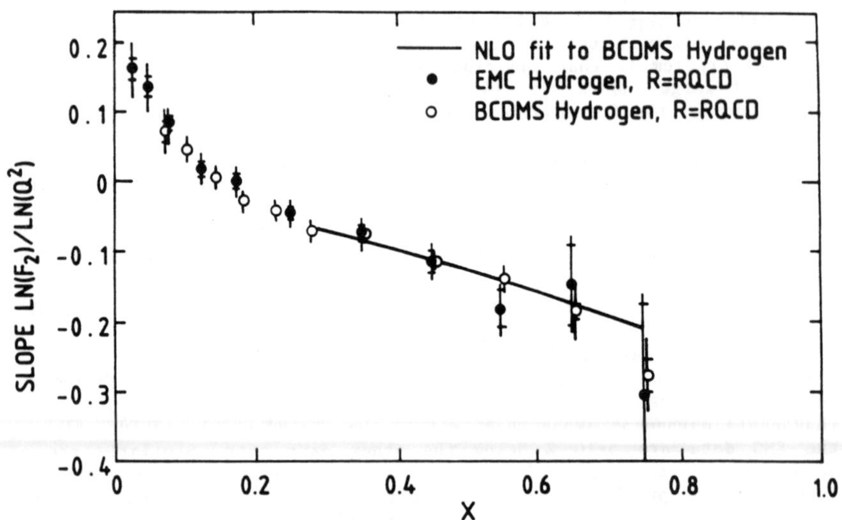

Fig. 5 : Comparison (45) of EMC (42) and BCDMS (43) logarithmic slopes of F_2 on hydrogen.

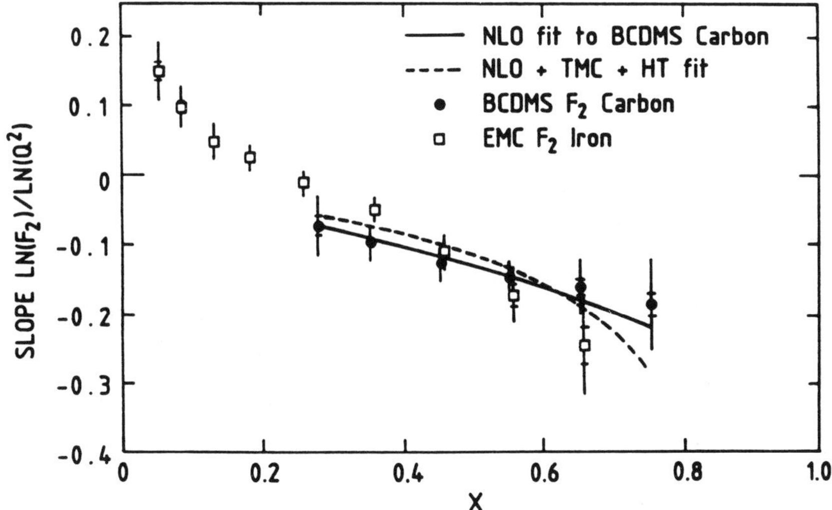

Fig. 6 : Comparison of data by EMC (42) on F_2 on iron and by BCDMS (43) on F_2 on carbon. The slopes on iron are apparently steeper than those on carbon.

value of $\Lambda_{\overline{MS}}$ measured by BCDMS on carbon is shown (45) in Fig. 6 (together with possible modifications induced by a model of higher twist effects and target mass-corrections). A quite similar behaviour is also observed in the logarithmic slopes of xF_3 recently measured by CDHSW (44). In this case agreement with QCD is reobtained for $Q^2 > 20$ GeV, within a more limited accuracy determined by the smaller statistics. From the iron data one derives the indication that pre-asymptotic or non-perturbative effects could be relatively more important in the case of heavy targets.

In conclusion, the BCDMS data on H and C have the highest statistics at the largest values of Q^2. These data are in beautiful agreement with QCD, and lead to the value of $\Lambda_{\overline{MS}}^{(4)}$ quoted in Eq. (2.18). This value of $\Lambda_{\overline{MS}}^{(4)}$ is in agreement with the values of $\Lambda_{\overline{MS}}^{(4)}$ quoted by EMC on H and CHARM on $CaCO_3$. However, the agreement with EMC is somewhat illusory in view of the large discrepancies between BCDMS and EMC on F_2^p. Finally, the data on Fe are unfortunately not very conclusive as tests of QCD and measurements of α_s.

2.3 Quarkonium Decays

The rates of quarkonium (especially the Υ) provide a nominally rather precise experimental determination of α_s and $\Lambda_{\overline{MS}}$. The problem in this case is the theoretical error. In the non-relativistic approximation the decay rates are proportional to the absolute square of the wave function at the origin. Thus, within the limits of this approximation, ratios of decay rates are independent of the wave function which is unknown. The most convenient ratios for the determination of α_s are $\cong \Gamma_{\gamma gg}/\Gamma_{ggg}$ and $\Gamma_{\mu\mu}/\Gamma_{ggg}$. Next-to-leading calculations of these ratios have been performed in the $\overline{MS}$ scheme (55). In the Υ case the results are:

$$\frac{\Gamma_{\mu\mu}}{\Gamma_{ggg}} = \frac{9\pi}{10(\pi^2-9)} \frac{\alpha^2}{\alpha_s^3(\mu)} \left\{ 1 - \frac{\alpha_s(\mu)}{\pi} \left(3\pi b_4 \, \ell n \, \frac{\mu^2}{m_Q^2} + 0.43 \right) + \dots \right\} \qquad (2.21)$$

$$\frac{\Gamma_{\gamma gg}}{\Gamma_{ggg}} = \frac{4}{5} \frac{\alpha}{\alpha_s(\mu)} \left\{ 1 - \frac{\alpha_s(\mu)}{\pi} \left(\pi b_4 \, \ell n \, \frac{\mu^2}{m_Q^2} + 2.6 \right) + \dots \right\} \qquad (2.22)$$

where $b_4 = (33-2f)/12\pi$ ($f=4$) and μ is an arbitrary scale of the order of the heavy quark mass $m_Q \sim (m_T/2)$.

The experimental value of $\Gamma_{\mu\mu}/\Gamma_{ggg}$ is known for T(1S), T(2S) and T(3S). It is obtained by the relation $\Gamma_{ggg}/\Gamma_{\mu\mu} = [1-B_{\gamma gg}-B_{\pi\pi}-B_{E1}]/B_{\mu\mu} - 3-R$ where $R \simeq 3.48$ and $B_{\pi\pi}$, B_{E1} are absent in the case of T(1S). Recent precise measurements (56) by CUSB and CLEO when combined lead to $B_{\mu\mu} = 2.61\pm0.14$, 1.38 ± 0.29, 1.73 ± 0.19 for T(1S), T(2S) and T(3S) respectively. These values lead to $\Gamma_{ggg}/\Gamma_{\mu\mu} = 31.0\pm2.0$, 34.0 ± 8.8, 30.8 ± 4.3 respectively. For $\mu = m_b \sim 4.9$ GeV one obtains:

$$1S : \alpha_s(m_b) = 0.1736 \pm 0.0037, \quad \Lambda_{\overline{MS}}^{(4)} = (157 \pm 13) \text{ MeV} \qquad (2.23)$$

$$2S : \alpha_s(m_b) = 0.176 \ \pm 0.016, \quad \Lambda_{\overline{MS}}^{(4)} = (167 \pm 58) \text{ MeV} \qquad (2.24)$$

$$3S : \alpha_s(m_b) = 0.1728 \pm 0.0083, \quad \Lambda_{\overline{MS}}^{(4)} = (154 \pm 29) \text{ MeV} \qquad (2.25)$$

The averages are $\alpha_s(m_b) = 0.1736\pm0.0033$ and $\Lambda_{\overline{MS}}^{(4)} = 157\pm12$ MeV.

Similarly the world average for $\Gamma_{\gamma gg}/\Gamma_{ggg}$ measured by CUSB, CLEO, ARGUS and CRYSTAL BALL is $(2.78\pm0.15)\%$ (57). From this value and Eq. (2.22) one obtains

$$\alpha_s(m_b) = 0.181 \pm 0.009 \qquad (2.26)$$

The agreement between the values of α_s derived from $\Gamma_{\mu\mu}/\Gamma_{ggg}$ for T(1S), T(2S) and T(3S) and from $\Gamma_{\gamma gg}/\Gamma_{ggg}$ for T(1S) is remarkable and is an experimental check of the wave function factorization. Actually consistent values of α_s are also obtained with more uncertainty from charmonium decays. An overall fit of the available data, also including a crude estimate of relativistic corrections was performed in Ref. (58) with the results $\alpha_s(m_c) \simeq 0.278 \pm 0.014$, $\alpha_s(m_b) = 0.185 \pm 0.006$ which correspond to $\Lambda_{\overline{MS}}^{(4)} = (199\pm22)$ MeV.

In the stated results for α_s and $\Lambda_{\overline{MS}}$ the error shown does not clearly include the theoretical error. This is certainly the largest source of uncertainty. Corrections to the non-relativistic approximation can still be sizeable in spite of the experimental success of factorization. The order of magnitude of v^2/c^2 is in fact ≈ 0.77 for charmonium and ≈ 0.10 for the T system. The effects of higher perturbative orders and of non-perturbative terms could be large because the energy scale is relatively small.

A large source of error is the dependence on the scale parameter μ. The expressions in Eqs. (2.21), (2.22) are functions of μ/Λ (through $\alpha_s(\mu)$) and of μ/m_b (explicitly appearing in the log). If one starts from a different μ one finds different $\Lambda_{\overline{MS}}^{(4)}$ and $\alpha_s(m_b)$. For example, the average derived from $\Gamma_{ggg}/\Gamma_{\mu\mu}$ becomes $\Lambda_{\overline{MS}}^{(4)} = 157\pm12\pm60$ MeV after inclusion of the error from the μ dependence (56).

In the case of $\Gamma_{\gamma gg}/\Gamma_{ggg}$ I see a further problem in the fact that the observed photon spectrum is not well understood in perturbation theory. The lowest-order spectrum is definitely too hard to accommodate the data (57). In Ref. (59) the observed soft γ spectrum is very convincingly explained as due to an effective mass of gluon jets. While the parton gluon is massless the physical gluon jet has a non-vanishing invariant mass. By a Monte Carlo simulation, Field (59) has shown that an average mass $<M> \sim 1.6$ GeV should be attributed to the gluon jet in order to reproduce the data. Perturbative effects (gluon splitting into $q\bar{q}$ or gg) should indeed induce an invariant mass of order $\alpha_s M_T$ which is not far from the observed value of $<M>$. However, the only existing calculation of the perturbative corrections (60) to the normalized spectrum gives a negligible improvement. This calculation could be wrong and in fact it is somewhat obscure in many respects. It is important to clarify this point because if the spectrum is indeed dominated by non-perturbative (or higher order) effects, then also the perturbative evaluation of the total width could be to some extent affected, even if inclusive quantities are usually more protected.

In conclusion, it appears difficult to me to compress the total theoretical error below the 10%-20% level. Actually, the theoretical error is so relatively small only because the different measurements on the Υ system are remarkably consistent among them. Thus I would tentatively conclude that:

$$\alpha_s(m_b) \simeq 0.175 \ (1 \pm 15\%) \tag{2.27}$$

Even with this enlarged error the resulting determination of $\Lambda_{\overline{MS}}^{(4)}$ is comparatively quite good:

$$\Lambda_{\overline{MS}}^{(4)} \simeq 1.54 \ \Lambda_{\overline{MS}}^{(5)} = (160^{+100}_{-80}) \ \text{MeV} \tag{2.28}$$

2.4 $e^+e^- \to$ Jets

All methods of measuring α_s described in the previous sections are based on totally inclusive processes. We now consider the determination of α_s from the observed properties of jets in the final state. The study of jets in e^+e^- annihilation has provided a formidable laboratory for QCD testing for about a decade. Many striking confirmations of the theory have been obtained (61): the observation of the predicted jet structure and the expected hierarchy of two, three, four... jets, evidence for gluons and their vector nature, the quantitative correspondence between the observed distributions in energy and angles and the QCD matrix elements. We concentrate here on the measurement of α_s from jets in e^+e^- annihilation.

The principle of the method is to measure a quantity which is zero in lowest order (corresponding only to a quark-antiquark pair in the final state) and starts at order α_s (quark-antiquark-gluon). For a meaningful determination of α_s it is necessary to know the same quantity at order α_s^2. This implies computing virtual corrections to three-parton amplitudes and real four-parton matrix elements. As is well known, the contribution to the rate of virtual and real diagrams is separately divergent while only the sum is finite for well-defined physical observables. The additional difficulty of jet physics consists of the obvious fact that the theory deals with partons and the physical observables are jets of hadrons. The relation between partons and the experimentally defined jets necessarily requires some model of non-perturbative fragmentation and hadronization.

While non-perturbative effects should asymptotically become negligible their influence on the extracted value of α_s is still sizeable at PEP/PETRA energies and contributes the main source of error.

For example, assume that one wants to compute some three-jet distribution $d\sigma$(three-jets). As we have seen the perturbative calculation at order α_s^2 needs "jet-dressing" to become finite. One must add to the contribution of three partons the integral over unresolved configurations from final states with four partons:

$$d\sigma \ (3\text{-jets}) = d\sigma \ (3\text{-partons}) + \int_{NR} d\sigma \ (4\text{-partons}) \qquad (2.29)$$

where NR indicates the non-resolved configurations, i.e., those where any two partons are too close to be separated so that the event is observed as a three-jet event. Clearly some jet resolution criterion is needed. Typically this is a cut on the invariant mass y_{ij} of a pair i,j of partons: below a given value of y_{ij} the pair is detected as a single jet. Evidently the presence of these cuts introduces a problem of cut dependence of the result. Not only that but the experimental jet identification criterion based on observed hadrons can only be translated into a cut on parton variables by a model of fragmentation and hadronization. With time there has been considerable progress (62) in the understanding of early discrepancies from different calculations (63,64) based on different four-parton resolution criteria. Also, for the determination of α_s, one now selects some global quantity [e.g., oblateness (9), energy-energy correlations and their asymmetry (65,66)] which are independent of, or less dependent on jet resolution criteria. Of course the dependence on fragmentation and hadronization effects always remains even if care is taken to concentrate on quantities which are invariant under collinear splitting of one into two massless particles. Finally one demands a good apparent convergence, i.e., that the resulting non-leading correction of order α_s^2 is not too large for a natural choice of the renormalization scale μ appearing in the leading term proportional to $\alpha_s(\mu)$.

Until recently the method for the determination of α_s which was used most was based on the asymmetry of energy-energy correlations (AEEC) (65,66). The energy-energy correlation (EEC) is defined by:

$$\frac{1}{\sigma}\frac{d\,\Sigma}{d\,\cos\chi} = \frac{1}{\sigma}\,\Sigma_{i,j}\int\frac{d\sigma}{dx_i\,dx_j\,d\,\cos\chi}\,x_ix_j\,dx_i\,dx_j \simeq$$
$$\simeq \frac{1}{N_{events}}\,\Sigma_{events}\,\Sigma_{i,j}\,x_ix_j\delta(\cos\theta_{ij}-\cos\chi) \qquad (2.30)$$

where χ is the fixed angle between two calorimeter cells, $x = 2E/\sqrt{s}$. Clearly the contribution at χ not too close to 0 and π arises from non-collinear events. The energy weights make EEC infra-red safe, and the linearity in x_i guarantees invariance under collinear splitting of particle i. The asymmetry AEEC is defined as

$$\frac{1}{\sigma}\frac{d\Sigma^{AEEC}}{d\,\cos\chi} = \frac{1}{\sigma}\frac{d\Sigma}{d\,\cos\chi}\,(\pi-\chi) - \frac{1}{\sigma}\frac{d\Sigma}{d\,\cos\chi}\,(\chi) \qquad (2.31)$$

The asymmetry is different from zero because in a typical three-jet event there is a slim jet in one hemisphere and a fat di-jet in the other one.

A purely perturbative calculation leads to a result of the form:

$$\frac{1}{\sigma}\frac{d\Sigma^{AEEC}}{d\cos\chi} = \frac{\alpha_s(Q)}{\pi} A (\cos\chi) \left[1 + \frac{\alpha_s(Q)}{\pi} R (\cos\chi) + \ldots\right] \qquad (2.32)$$

In the range $-0.95 < \cos\chi < 0.95$ the value of R (67) varies between 2.5 and 3.5, so that the expansion is apparently well behaved. For $|\cos\chi|$ near 1 the perturbative expansion should be improved by a resummation of the corresponding singularities.

Experimentally it is found that for $\chi > 30°$ the purely perturbative evaluation of the AEEC distribution at order α_s^2 provides an excellent fit to the data. The results on α_s determined from the purely perturbative fit are reproduced in Table 6 of Ref. (67). By taking the average one obtains:

$$\alpha_s(Q = 34 \text{ GeV}) = 0.121 \pm 0.003$$

(perturbative)   (2.33)

$$\Lambda_{\overline{MS}}^{(5)} = (100 \pm 15) \text{ MeV}$$

As usual, also in this case the problem is to estimate the theoretical error. The main source of error is expected to arise from fragmentation and hadronization effects. Thus the most natural thing to do is to compare the purely perturbative result with those obtained by including models of jet formation. Following Ref. (67) where a more complete discussion is given, we report the results based on two different Monte Carlo analyses including the perturbative calculations of order α_s^2 and a model of fragmentation: a model by Ali et al. (68) and a version of the Lund model (69). The results from these analyses are reported in Table 6.2 of Ref. (67). The average values are:

$$\alpha_s(Q = 34 \text{ GeV}) = 0.128 \pm 0.003$$

(Ali et al.)   (2.34)

$$\Lambda_{\overline{MS}}^{(5)} = (144 \pm 16) \text{ MeV}$$

$$\alpha_s(Q = 34 \text{ GeV}) = 0.148 \pm 0.002$$

(Lund)   (2.35)

$$\Lambda_{\overline{MS}}^{(5)} = (325 \pm 20) \text{ MeV}$$

These results are in agreement with some other existing measurements (61,67) based on oblateness or the planar triple energy correlation (70). The conclusion is that there is indeed a systematic difference in the results from the Ali et al. and the Lund model. Both models lead to an increase of α_s. But in the case of Ali et al. the change with respect to the perturbative result is much smaller than in the Lund case. The dispersion of the results can be taken as a indication of the theoretical error. Thus one concludes that from $e^+e^- \to$ jets the present result is something like

$$\alpha_s(Q = 34 \text{ GeV}) \simeq 0.135 \pm 0.015$$

(2.36)

$$\Lambda_{\overline{MS}}^{(5)} = (215 \pm 130) \text{ MeV}$$

We shall come back to the determination of α_s from data on $e^+e^- \to$ jets in Section 2.7 devoted to LEP results.

2.5 Other Processes

The main source of additional information on α_s is obtained from γ-γ reactions (71). The photon structure function F_2^γ measured in γ-γ collisions (one tagged photon of virtual squared mass $-Q^2$ on a quasi-real photon) is special because it is predicted to grow as $\ln Q^2$ (72). The logarithmic increase of F_2^γ is well supported by the data and is a nice confirmation of asymptotic freedom that preserves this prediction also in presence of QCD corrections that, however, very markedly modify the shape of the structure function (making it considerably softer).

The leading pointlike component is to a large extent computable (especially at relatively large x). However, early hopes of measuring α_s, free from hadronic non-perturbative unknowns, directly from the observed value of F_2^γ at sufficiently large Q^2 in some range of not too small x cannot be really fulfilled. It is by now generally recognized that the determination of $\Lambda_{\overline{MS}}$ from F_2^γ requires data at different values of Q^2 (71). In fact it is clear that in order to measure $\Lambda_{\overline{MS}}$ complete control of the next-to-leading terms is necessary. At that level (73) the complete separation of the computable pointlike-photon contribution from the hadronic terms becomes impossible. These terms arise from the hadronic component of the photon as visualized for example by vector-meson dominance (the virtual photon scatters on a ρ, ω, ϕ,... in the quasi-real photon). Actually spurious singularities in the space of moments (which affect the behaviour of F_2^γ at small x) are generated in the singlet sector if hadronic terms are not properly taken into account. It is therefore necessary to introduce a parametrization of the non- perturbative hadronic terms and to determine from the data the corresponding parameters together with $\Lambda_{\overline{MS}}$. This evidently generates some ambiguity on $\Lambda_{\overline{MS}}$ that the limitations of present data cannot eliminate.

Several theoretical approaches (74) have been advocated for a treatment of the hadronic component. Values of $\Lambda_{\overline{MS}}$ obtained by different experiments and procedures are listed in Table 1 (71,75). Note that most of the listed data were analyzed in terms of formulae for $f = 3$. Thus the reported values of $\Lambda_{\overline{MS}}$ should be mainly identified with $\Lambda_{\overline{MS}}^{(3)}$ which differs from $\Lambda_{\overline{MS}}^{(4)}$ by about 30% (because $\Lambda_{\overline{MS}}^{(3)} \simeq 1.3\ \Lambda_{\overline{MS}}^{(4)}$).

In conclusion, when translated in terms of $\Lambda_{\overline{MS}}^{(4)}$ the results on photon structure function lead to values of $\Lambda_{\overline{MS}}^{(4)}$ in the range 50-300 MeV which are perfectly consistent with other experiments.

There are many less precise or more ambiguous determinations of α_s from other experimental sources. For example, I can quote the determination of α_s at $\mu \simeq m_\tau$ from the leptonic branching ratio of the τ lepton. This branching ratio is $B_c = (18.3\pm0.3)\%$. The fact that it is close to 1/5 rather than to 1/3 is a proof that $N_c = 3$, because $B_c \simeq 1/(2+N_c)$. From the deviations from the value 1/5 a value of $\alpha_s(m_\tau)$ can be extracted (76) with some model dependence from the treatment of non-perturbative effects. At the other extreme a measurement of α_s at $\mu \simeq M_W$ has been attempted (77) by UA2 from W+jet production at the CERN $p\bar{p}$ collider. The results are in both cases consistent with the more precise methods already described.

2.6 Summary and Conclusion on α_s

We have tried to review and interpret the large amount of experimental information on α_s. A sample of the most significant determinations of α_s is reported in Table 2. In turn, each entry is a combination of different

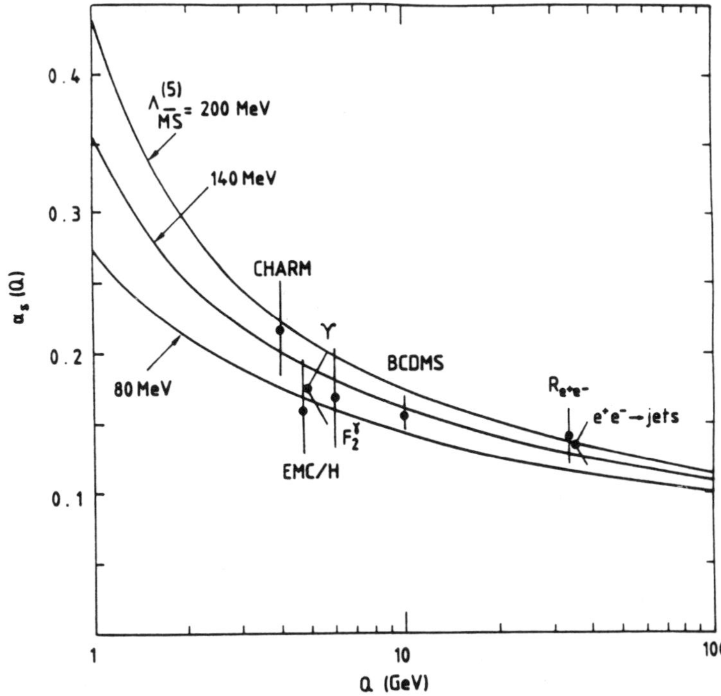

Fig. 7 : A summary of the determinations of α_s discussed in the text. The curves for $\Lambda_{\overline{MS}}^{(5)} = (140\pm60)$ MeV are obtained following the matching procedure at the b threshold explained in Eqs. (1.14)–(1.19) (with a $\simeq$ 1).

measurements. It is clearly remarkable that the final results on α_s from so many completely different sources are in very good agreement with each other. This is one of the most important quantitative tests of QCD. A plot of a more extended set of data is shown in Fig. 7 taken from Ref. (32). I do not think that it would be appropriate to combine the errors according to Gaussian statistics in order to derive an average value of Λ. In fact, the data on deep inelastic scattering are in part contradictory or not always in agreement with QCD; the error on the Υ entry is a personal estimate; the $e^+e^- \to$ jet result is obtained by a particular combination of the existing data and so on. However, I cannot avoid the task of providing the readers with a suggested set of values. In this spirit I propose:

$$\Lambda_{\overline{MS}}^{(4)} = (220 \pm 90) \text{ MeV} \tag{2.37}$$

or, equivalently:

$$\Lambda_{\overline{MS}}^{(5)} = (140 \pm 60) \text{ MeV} \tag{2.38}$$

where the reported error is about twice the value that would be obtained by combining the errors shown in Table 4.

An objection which is often made is that from the whole set of data we have discussed, once the errors are taken into account, the running of α_s cannot be clearly established in the range $Q \sim 3\text{--}50$ GeV. However, the fact that the values of α_s which are measured at large Q are so small, as shown in Table 2, is a very strong proof of the running of α_s, because such a relatively feeble strong force could not provide hadrons with the observed tight binding. A more significant test of QCD is evident from Fig. 7. A relatively loose determination of $\alpha_s(Q)$ at $Q \sim 1$ GeV leads to a very tight determination of $\alpha_s(Q)$ at large Q. For example, from the resulting value of $\Lambda_{\overline{MS}}^{(5)}$ given in Eq. (2.38) the prediction for the value of α_s to be measured at LEP (and HERA) is very precise:

$$\alpha_s(Q \simeq m_z) \simeq 0.11 \pm 0.01 \qquad (2.39)$$

Establishing that this prediction is experimentally true would be a very quantitative and accurate test of QCD, conceptually equivalent but more reasonable than trying to see the running in a given experiment.

2.7 First LEP results and QCD

We have seen that $\alpha_s(m_z)$ is predicted to lie in a very narrow range of values once the existing experiments are taken into account. It is an important goal of LEP to verify this prediction. There are already some interesting results which we now briefly describe.

The total and partial widths of the Z do not allow a precise determination of α_s. The averages of the 1989 data by the four LEP experiments for the hadronic and total widths of the Z are given by:

$$\Gamma_h = 1792 \pm 23 \text{ MeV}, \qquad \Gamma_T = 2539 \pm 26 \text{ MeV} \qquad (2.40)$$

The corresponding predictions of the standard model (78) for 60 GeV $< m_t <$ 200 GeV, $m_H = 10^{2\pm1}$ GeV and $\alpha_s(m_z) = 0.10\text{--}0.12$ are given by:

$$(\Gamma_h)_{SM} = 1739 \pm 24 \text{ MeV}, \qquad (\Gamma_T) = 2490 \pm 30 \text{ MeV} \qquad (2.41)$$

We see that the central values observed are a bit higher than the predictions. A fitted value of $\alpha_s(m_z)$ would lead to something like (79) $\alpha_s(m_z) = 0.18\pm0.08$. Note that the effect of α_s is to increase Γ_h by $\delta\Gamma_h \simeq 70\pm7$ MeV for $\alpha_s(m_z) = 0.10\text{--}0.12$.

While the data on the widths are not conclusive on α_s, all QCD tests so far performed on the final state are perfectly in agreement with expectations. In particular, DELPHI (80) and OPAL (81) present data that allow to measure $\alpha_s(m_z)$. We shall briefly discuss these results.

DELPHI (80) measures the average value of thrust:

$$\langle 1-T \rangle = 0.066 \pm 0.003 \qquad (2.42)$$

Before comparison with perturbative results derived at the parton level, this value has to be corrected for hadronization and b-decay effects. As discussed in Ref. (82) one estimates:

$$\frac{\delta\langle 1-T \rangle}{\langle 1-T \rangle} = -0.03 - 0.03 \qquad (2.43)$$

The first number is the hadronization correction estimated by the generator HERWIG, the second is the b-decay correction. The corrected value is then:

$$\langle 1-T\rangle_{corr} = 0.062 \pm 0.005 \tag{2.44}$$

where the error has been enlarged to take the uncertainties on the corrections into account. This corrected value can be compared with the perturbative prediction (82):

$$\langle 1-T\rangle_{GW} = A\,\frac{\alpha_s(\mu)}{\pi} + (A\pi b\,\ell n\,\frac{\mu^2}{m_Z^2} + B)\,(\frac{\alpha_s(\mu)}{\pi}) + \ldots \tag{2.45}$$

with $A = 1.05$, $b = 23/12\pi$ and $B = 10.1\pm0.08$. From this relation and the value in Eq. (2.44) one finds (83)

$$\alpha_s(m_Z) = 0.127 \pm 0.013 \tag{2.46}$$

where the quoted error also includes the scale ambiguity introduced by varying μ in the range $\frac{m_Z}{4} < \mu < m_Z$.

OPAL has presented a very impressive analysis of jet multiplicities which follows the same lines of earlier work (84) by MARK II, JADE, AMY, mainly prompted by S. Betke (85). A soft and infra-red safe procedure of jet counting is introduced. The observed numbers of jets are compared with perturbative calculations based on the same jet-finding algorithm in order to obtain a value for $\Lambda_{\overline{MS}}^{(5)}$. In a given event for each pair of particules k, ℓ one evaluates

$$y_{k\ell} = \frac{M_{k\ell}^2}{E_{vis}^2} = \frac{2E_k E_\ell(1-\cos\theta_{k\ell})}{E_{vis}^2} \tag{2.47}$$

The pair of smallest $M_{k\ell}^2$ is replaced by a single "particle" with $P_\mu = (p_k+p_\ell)_\mu$ until for all pairs $y_{k\ell} > y_{cut}$. The remaining clusters are called "jets". Note that in the evaluation of $y_{k\ell}$ by Eq. (2.47) all intermediate clusters are taken as massless, but masses are kept in the addition of four momenta. There are in principle many different procedures in treating particle and cluster masses. But the same procedure is adopted in the experiment and the theoretical calculations. The hadronization effects are studied by Monte Carlo simulations and are apparently small. It is found that QCD event generators reproduce the data very well. The corresponding calculations at the parton level were performed by Kramer and Lampe (86). The jet multiplicities are evidently of the form:

$$R_2 = \frac{\sigma_2}{\sigma_{TOT}} = 1 + c_{21}\,\alpha_s(\mu) + c_{22}(\mu)\,\alpha_s^2(\mu) + \ldots$$

$$R_3 = \frac{\sigma_3}{\sigma_{TOT}} = c_{31}\,\alpha_s(\mu) + c_{32}(\mu)\,\alpha_s^2(\mu) + \ldots$$

$$R_4 = \frac{\sigma_4}{\sigma_{TOT}} = c_{42}\,\alpha_s^2(\mu) + \ldots \tag{2.48}$$

194

All the coefficients c_{ij} shown here are known. Non-leading terms [necessary for a meaningful determination of $\Lambda_{\overline{MS}}^{(5)}$] were only computed for R_2 and R_3.

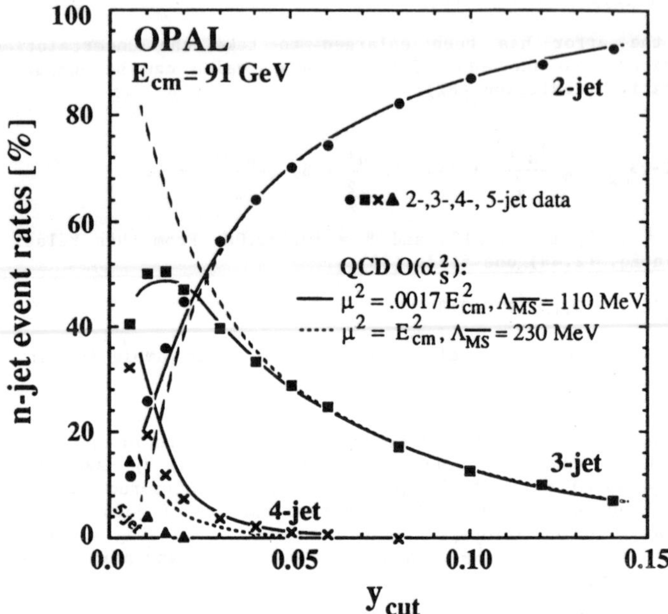

Fig. 8   :   The data on jet multiplicities compared with the $O(\alpha_s^2)$ QCD calculations by Kramer and Lampe with parameters as optimised at E_{cm} = 29 GeV.

A comparison of the data with the perturbative predictions is shown in Fig. 8. Note that in this plot the same data enter many times for different y_{cut}. The data can be fitted in terms of μ^2 and $\Lambda_{\overline{MS}}^{(5)}$. It is found that small values of μ give a better fit at small y_{cut}. But this is presumably only a reflection of the fact that the unknown non-leading corrections to R_4 are large and apparently positive. For a meaningful determination of $\Lambda_{\overline{MS}}^{(5)}$ one should actually take y_{cut} relatively large so that only R_2 and R_3 are important. By fitting both μ and $\Lambda_{\overline{MS}}^{(5)}$ one finds

$$\Lambda_{\overline{MS}}^{(5)} = (130 \pm 50) \text{ MeV}$$

$$\mu = (\ 4 \pm\ 1) \text{ GeV} \tag{2.49}$$

We see that the fitted value of μ is small but still in the perturbative region. By fixing $\mu = m_Z$ one obtains:

$$\Lambda_{\overline{MS}}^{(5)} = (325 \pm 125) \text{ MeV} \tag{2.50}$$

A conservative way to interpret this result is to include the μ dependence in the error on $\Lambda_{\overline{MS}}^{(5)}$. Even if one varies μ in the range from 3 to 11 GeV one has:

$$\Lambda_{\overline{MS}}^{(5)} = (265 \pm 185) \text{ MeV} \tag{2.51}$$

or

$$\alpha_s(m_Z) = 0.117 \pm 0.015 \tag{2.52}$$

If we compare this value with the result by DELPHI previously given in Eq. (2.46) we see that LEP indeed confirms that $\alpha_s(m_Z)$ is within the predicted range obtained from experiments at lower energies.

A combination of OPAL results on R_3/R_2 with similar measurements at PETRA, PEP and TRISTAN leads to a clear indication on the running of α_s as shown in Fig. 9. For that, however, one has to rely on a Monte Carlo simulation to prove that the correspondence between hadrons and partons is not different at different energies (84) (and similarly that the systematics of different experiments do not introduce large distortions).

In conclusion adequate theoretical calculations exist for many observables (which are soft and infra-red safe): line thrust, spherocity, oblateness, asymmetry in energy-energy correlations, jet multiplicities and so on (82). We expect from LEP a thorough study of all of them (not just those that appear to work better!) with the aim of disentangling in a systematic way parton effects from hadronization corrections. One hopes to measure α_s with proper consideration of all sorts of errors, also including μ-dependence. We have now already seen a promising beginning.

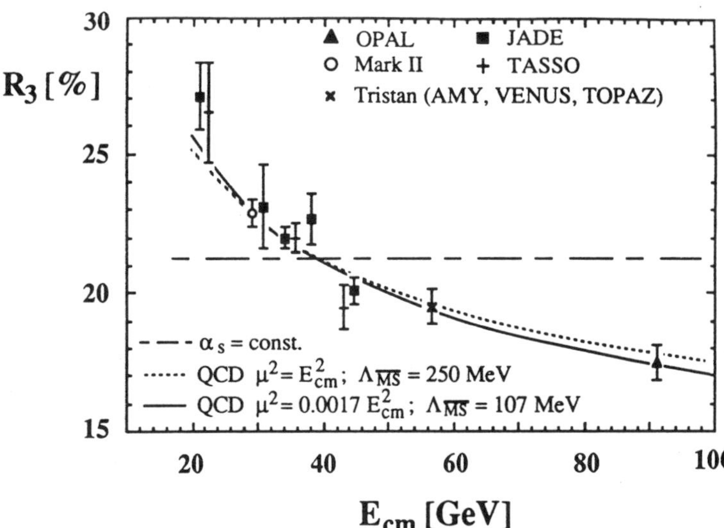

Fig. 9 : The energy dependence of the three-jet event production rates at $y_{cut} = 0.08$, compared with several assumptions about the energy dependence of α_s. The errors shown are statistical only. Since R_3 should only depend on energy through α_s this is an indication for the running of the QCD coupling according to asymptotic freedom.

ACKNOWLEDGEMENTS

It is a pleasure for me to thank the Organizers for their invitation and magnificent hospitality in Houston.

Table 1

Results on $\Lambda_{\overline{MS}}$ obtained by different experiments on the photon structure function. Entries quoted with the label (a), (b), ... for one given experiment differ by the assumed model for the hadronic component. The quoted value of $\Lambda_{\overline{MS}}$ in most cases should be read as $\Lambda_{\overline{MS}}^{(3)}$ because it is obtained from a fit to the (charm-subtracted) data done in terms of QCD formulae with f = 3. The first five entries were taken from Ref. (71), the TPC results from Ref. (75).

Experiment	Q^2 (GeV^2)	$\Lambda_{\overline{MS}}$ (MeV)
PLUTO (a)	3 – 100	$183 \pm {}^{80}_{54}$
PLUTO (b)	3 – 100	240 ± 90
PLUTO (c)	3 – 100	160 ± 60
JADE	10 – 220	250 ± 90
TASSO	7 – 70	$140 \pm {}^{190}_{60}$
TPC/2γ (a)	0.7 – 22	108 ± 32
TPC/2γ (b)	0.7 – 22	232 ± 59

<u>Table 2</u>

Summary of the most significant determinations of α_s and $\Lambda_{\overline{MS}}$. The values and the errors quoted are discussed in the text.

	α_s (34 GeV)	$\Lambda_{\overline{MS}}^{(4)}$ (MeV)	$\Lambda_{\overline{MS}}^{(5)}$ (MeV)
$R_{e^+e^-}$	0.14 ± 0.02	$370 \pm {}^{350}_{220}$	$240 \pm {}^{230}_{140}$
BCDMS	0.127 ± 0.006	220 ± 60	140 ± 40
T	0.123 ± 0.009	180 ± 80	120 ± 50
$e^+e^- \to$ jets	0.135 ± 0.015	330 ± 200	215 ± 130
γ-structure function	0.120 ± 0.016	175 ± 125	115 ± 80

REFERENCES

(1) Gell-Mann, M., Acta Phys. Austriaca, Suppl. IX:733 (1972);
Fritzsch, H., Gell-Mann, M., XVIth International Conference on High
Energy Physics, Batavia, II:135 (1972);
Fritzsch, H., Gell-Mann M., Leutwyler, H., Phys. Lett. 47B:365
(1973).

(2) Gross, D.J., Wilczek, F., Phys. Rev. Lett. 30:1343 (1973); Phys.
Rev. D8:3633 (1973);
Politzer, H.D., Phys. Rev. Lett. 30:1346 (1973).

(3) Weinberg, S., Phys. Rev. Lett. 31:494 (1973).

(4) See, for example:
Lee-Franzini, J., Nucl. Phys. B (Proc. Suppl.) 3:139 (1988).

(5) See, for example:
Petronzio, R., Proceedings of the XXIV International Conference on
High Energy Physics, Munich, Springer Verlag p. 136 (1988).

(6) Stack, J.D., Phys. Rev. D29:1213 (1984);
Barkai, B., Moriarty, K.J.M., Rebbi, C., Phys. Rev. D30:1283 (1984);
See also:
Hasenfratz, P., Proceedings of the XXIII International Conference on
High Energy Physics, Berkeley, World Scientific p. 169 (1986).

(7) See, for example:
Karsch, F., Z. Phys. C38:147 (1988) and the APE Collaboration,
Bacilieri, P., et al., Phys. Rev. Lett. 61:1545 (1988);
Brown,F.R., et al., Phys. Rev. Lett. 61:2058 (1988).

(8) See, for example:
Feynman, R.P., Photon Hadron Interactions, W.A. Benjamin, New York
(1972).

(9) See, for example:
Altarelli, G., Physics Reports 81:1 (1982).

(10) Yndurain, F.J., QCD: an Introduction to the Theory of Quarks and
Gluons, New York, Springer Verlag (1983);
Muta, T., Foundations of QCD, World Scientific, Lecture Notes in
Physics (1987).

(11) See the collection of articles in "Perturbative Quantum Chromodynamics", ed. by Jacob, M., Amsterdam, North Holland (1982).

(12) Collins, J.C., Soper, D.E., Ann. Rev. of Nucl. Part. Sci., to be
published.

(13) Yang, C.N., Mills, R.C., Phys. Rev. 96:191 (1954).

(14) See, for example:
Abers, E.S., Lee, B.W., Physics Reports 9:11 (1973).

(15) 't Hooft, G., Phys. Rev. Lett. 37:8 (1976); Phys. Rev. D14: 3432
(1976).

(16) Belavin, A.A., Polyakov, A.M., Schwartz, A.S., Tyupkin, Yu.S., Phys. Lett. 59B:85 (1975).

(17) Baluni, V., Phys. Rev. D19:2227 (1979);
Crewther, R.J., Di Vecchia, P., Veneziano G., Witten, E., Phys. Lett. 88B:123 (1979), E91B: 487 (1980).

(18) Peccei, R.D., in "CP violation", ed. by Jarlskog, C., World Scientific (1989).

(19) See, for example:
Greenberg, O.W., Ann. Rev. Nucl. Part. Sci. 28:327 (1978);

(20) Adler, S.L., Phys. Rev. 177:2426 (1969);
Bell, J.S., Jackiw, R., Nuov. Cim. A51:47 (1969).

(21) Fukuda H., Miyamoto, Y., Prog. Theor. Phys. 4:347 (1949);
Steinberger, J., Phys. Rev. 76:1180 (1969).

(22) Particle Data Group, Phys. Lett. B204:1 (1988).

(23) Ashmore, J.F., Nuov. Cim. Lett. 4:289 (1972);
Bollini, G.G., Giambiagi, J.J., Nuov. Cim. 12B:20 (1974).

(24) See, for example:
Itzykson, C., Zuber, J.B., Introduction to Quantum Field Theory, McGraw Hill (1980).

(25) 't Hooft, G., Nucl. Phys. B61:455 (1973).

(26) Bardeen, W.A., Buras, A.J., Duke, D.W., Muta, T., Phys. Rev. D18:3998 (1978).

(27) Stückelberg, E.C.G., Peterman, A., Helv. Phys. Acta 26:499 (1953);
Gell-Mann, M., Phys. Rev. 95:1300 (1954);
Bogoliubov, N.N., Shirkov, D.V., Dokl. Akad. Nauk. USSR 103: 206, 391 (1955);
Ovsiannikov, L.V., ibid. 109:1121 (1956);
Callan, G.C., Phys. Rev. D2:1541 (1970);
Symanzik, K., Comm. Math. Phys. 18:227 (1970).

(28) Caswell, W., Phys. Rev. Lett. 33:244 (1974);
Jones, D.R.T., Nucl. Phys. B75:531 (1974).

(29) Tarasov, O.V., Vladimirov, A.A., Zharkov, A.Yu., Phys. Lett. 93B:429 (1980).

(30) Braten E., Leveille, J.P., Phys. Rev. D24:1369 (1981);
Blumenfeld, A., Moshe, M., Phys. Rev. D26:648 (1982).

(31) Appelquist, T., Carazzone, J., Phys. Rev. D11:2856 (1975).

(32) Altarelli, G., Ann. Rev. Nucl. Part. Sci. 39:357 (1989).

(33) Jost R., Luttinger, J.M., Helv. Phys. Acta 23:201 (1950);
Appelquist, T., Georgi, H., Phys. Rev. D8:4000 (1973);

Zee, A., Phys. Rev. D8: 4038 (1973).

(34) Chetyrkin, K.G., Kataev, A.L., Tkachov, F.V., Phys. Lett. 85B: 277 (1979);
Dine, M., Sapirstein, J., Phys. Rev. Lett. 43:668 (1979);
Celmaster, W., Gonsalves, R.J., Phys. Rev. Lett. 44:560 (1979);
Phys. Rev. D21:3112 (1980).

(35) Gorishny, S.G., Kataev, A.L., Larin, S.A., Phys. Lett. B212:238 (1988).

(36) de Boer, W., Proceedings of the XXIV International Conference on High Energy Physics, Munich, Springer Verlag, p. 905 (1988).

(37) Marshall, R., ibid., p. 901 (1988).

(38) Stevenson, P.M., Phys. Rev. D23:2916 (1981);
Politzer, H.D., Nucl. Phys. B194:493 (1982);
Stevenson, P.M., Politzer, H.D., Nucl. Phys. B277:758 (1986);
Grunberg, G., Phys. Rev. D29:2315 (1984).

(39) Wilson, K., Phys. Rev. 179:1499 (1969);
Brandt, R., Preparata, G., Nucl. Phys. 27B:541 (1971).

(40) See, for example:
Altarelli, G., Riv. Nuov. Cim. 4:335 (1974).

(41) See, for example:
Sloan, T., Smadja, G., Voss, R., Physics Reports 162:45 (1988);
Diemoz, M., Ferroni, F., Longo, E., Physics Reports 130:293 (1986).

(42) EMC Collaboration, Aubert, J.J. et al., Nucl. Phys. B259:189 (1985);
B272:158 (1986); B293:740 (1987).

(43) BCDMS Collaboration, Benvenuti, A.C. et al., Phys. Lett. B195:91
1987); B195:97 (1987); CERN preprint EP/89-06 (1989) and EP/89-07 (1989).

(44) CDHSW Collaboration, Buchholz, P., Proceedings of the EPS International Conference on High Energy Physics, Bari, Laterra, p. 557 (1985);
Vallage, B., Thèse Université de Paris-Sud, Orsay (1986);
Perez, P., Proceedings of the International Conference on Neutrino Physics and Astrophysics, Sendai, p. 341 (1986).

(45) Mount, R., Proceedings of the XXIV International Conference on High Energy Physics, Munich, Springer Verlag, p. 1007 (1988).

(46) Altarelli, G., Parisi, G., Nucl. Phys. B126:298 (1977).

(47) Floratos, E.G., Ross, D.A., Sachrajda, C.T., Nucl. Phys. B129:66 (1977), EB139:545 (1978);
Curci, G., Furmanski, W., Petronzio, R., Nucl. Phys. B175:27 (1980);
Floratos, E.G., Lacaze, R., Kounnas, C., Phys. Lett. 98B:89 (1981);
Gonzales-Arroyo, A., Lopez, C., Yndurain, F.J., Nucl. Phys. B153:161 (1979).

(48) Altarelli, G., Ellis, R.K., Martinelli, G., Nucl. Phys. B157:461 (1979).

(49) Floratos, E.G., Ross, D.A., Sachrajda, C.T., Nucl. Phys. B152 (1979) 493;
Furmanski, W., Petronzio, R., Phys. Lett. 97B:438 (1980);
Floratos, E.G., Lacaze, R., Kounnas, C., Phys. Lett. 98B (1981) 285;
Herrod, R.T., Wada, S., Phys. Lett. 96B:195 (1981), Z. f. Phys. C9:351 (1981).

(50) For a recent discussion of this point, see the two theses by Ouraou, A. and Virchaux, M., Orsay (1988).

(51) SLAC-MIT Collaboration, Bodek A., et al., Phys. Rev D20:1471 (1979).

(52) CHARM Collaboration, Bergsma, F., et al., Phys. Lett. 153B:111 (1985).

(53) CCFRR Collaboration, McFarlane, D.B., et al., Z. Phys. C26:1 (1984).

(54) BFP Collaboration, Meyers, P.D., et al. Phys. Rev. D34:1265 (1986).

(55) Mackenzie, B.P., Lepage, G.P., Phys. Rev. Lett. 47:1244 (1981);
Barbieri, R., Gatto, R., Kögerler, R., Kunszt, Z., Phys. Lett. 57B:455 (1975);
Celmaster, W., Phys. Rev. D19:1517 (1979);
Barbieri, R., Caffo, M., Gatto, R., Remiddi, E., Phys. Lett. 95B:93 (1980).

(56) CUSB Collaboration, Kaarsberg, T.M. et al., Phys. Rev. Lett. 62:2077 (1989).

(57) Bizzeti, A., Proceedings of the XXIV International Conference on High Energy Physics, Munich, Springer Verlag, p. 909 (1988).

(58) Kwong, W., Mackenzie, P., Rosenfeld, R., Rosner, J.L., Phys. Rev. D37:3210 (1988).

(59) Field, R.D., Phys. Lett. 133B:248 (1983).

(60) Photiadis, D.M., Phys. Lett. 164B:160 (1985).

(61) See, for example:
Kramer, G., Springer Tracts in Mod. Phys. Vol. 102 (1984);
Wu, S.L., Physics Reports 107:59 (1984);
Adeva, B. et al., Physics Reports 109:131 (1984);
Naroska, B., Physics Reports 148:67 (1987);
Ali, A., Söding, P., Advanced Series on Directions in High Energy Physics, World Scientific, Vol. 1 (1988).

(62) Gottschalk, T.D., Phys. Lett. 109B:331 (1982);
Gottschalk, T.D., Shatz, M.P., Phys. Lett. 150B:451 (1985);
Sjöstrand, T., Z. Phys. C26:93 (1984).

(63) Ellis, R.K., Ross, D.A., Terrano, A.E., Phys. Rev. Lett. 45:1226 (1980); Nucl. Phys. B178:321 (1981);
Vermaseren, J.A.M., Gaemers, K.J.F., Oldham, S.J., Nucl. Phys. B187:301 (1981).

(64) Fabricius, K., Kramer, G., Schierholz, G., Schmitt, I., Z. Phys. C11:315 (1982); Gutbrod, F., Kramer, G., Schierholz, G., Z. Phys. C21:235 (1984).

(65) Basham, C., Brown, L., Ellis, S., Love, S., Phys. Rev. Lett. 41:1585 (1978); Phys. Rev. D19:2018 (1979); D24:2382 (1981).

(66) Ali, A., Barreiro, F., Phys. Lett. 118B:155 (1982); Nucl. Phys. B236:269 (1984); Richards, D.G., Stirling, W.J., Ellis, S.D., Phys. Lett. 119B:193 (1982); Nucl. Phys. B229:317 (1983).

(67) Ali, A., Barreiro, F., in the book quoted as the last of Refs. (61).

(68) Ali, A., Pietarinen, E., Kramer, G., Willrodt, J., Phys. Lett. 93B:155 (1980); See also: Hoyer, P. et al., Nucl. Phys. B161:349 (1979).

(69) Anderson, B., Gustafson, G., Ingelman, F., Sjöstrand, T., Physics Reports 97:31 (1983).

(70) Csikor F. et al., Phys. Rev. D31:1025 (1985); Phys. Rev. D34:129 (1986).

(71) For a recent review, see: Berger, C., Wagner, W., Physics Reports 146:1 (1987); Kolanoski, H., Zerwas, P., in the last of Refs. (61).

(72) Witten, E., Nucl. Phys. B120:189 (1977).

(73) Bardeen, W.A., Buras, A.J., Phys. Rev. D20:166 (1979).

(74) Antoniadis, I., Grunberg, G., Nucl. Phys. B213:445 (1983); Antoniadis, I., Marleau, L., Phys. Lett. B161:163 (1985); Field, J.H., Kapusta, F., Poggioli, L., Phys. Lett. 181B:362 (1986); Z. Phys. C36:121 (1987); Glück, M., Reya, E., Dortmund Univ. preprint (1988).

(75) Maxfield, S.J., Proceedings of the XXIV International Conference on High Energy Physics, Munich, Springer Verlag, p. 661 (1988).

(76) Braaten, E., Phys. Rev. Lett. 60:1606 (1988); Narison, S., Pich, A., Phys. Lett. B211:183 (1988).

(77) Meier, K., Proceedings of the XXIV International Conference on High Energy Physics, Munich, Springer Verlag, p. 729 (1988).

(78) Z Physics at LEP 1, ed. by Altarelli, G., Kleiss, R. and Verzegnassi, C., CERN Report 89-08 (1989).

(79) We thank L. Garrido for this evaluation which is obtained from the ALEPH data.

(80) DELPHI Collaboration, Aaaammmmmmio, P. et al, CERN-EP/90-19 (1990).

(81) OPAL Collaboration, Akrawy, M.Z. et al., CERN-EP/89-153 (1989).

(82) Kunszt, Z., Nason, P. et al., "QCD at LEP 1" in Ref. (78).

(83) Nason, P., Private communication

(84) JADE Collaboration, Bartel, W. et al., Z. Phys. C33:23 (1986);
     JADE Collaboration, Bethke, S. et al., Phys. Lett. B213:235 (1988);
     AMY Collaboration, Park, I. et al., Phys. Rev. Lett. 62:1713 (1989);
     MARK II Collaboration, Bethke, S. et al., Z. Phys. C43:325 (1989).

(85) Bethke, S., Z. Phys. C43:331 (1989).

(86) Kramer, G. and Lampe, B., Z. Phys. C39:101 (1988).

Z HIGHLIGHTS

PRECISION MEASUREMENT OF THE RATIO M_W/M_Z AT THE CERN $\bar{p}p$ COLLIDER

THE UA2 COLLABORATION
Bern - Cambridge - CERN - Heidelberg - Milano
Orsay (LAL) - Pavia - Perugia - Pisa - Saclay (CEN)

Presented by
J. R. INCANDELA*
EP Division, CERN
CH-1211 Geneva 23, Switzerland

ABSTRACT

In the 1988 and 1989 runs of the CERN SPS $\bar{p}p$ collider, large numbers of $W \rightarrow e\nu_e$ and $Z \rightarrow e^+e^-$ decays were collected by UA2. Subsamples of these events have been selected to determine individual boson masses, M_W and M_Z, for which systematic errors in energy scale are expected to be identical. As a result, a precise value for the ratio M_W/M_Z is obtained. This is used to determine $\sin^2\theta_W$, and it is combined with the recent, very precise measurements of M_Z at e^+e^- colliders to extract a precise value for M_W. The results are discussed within the context of the standard model.

---

* Present address: INFN Milano, Gruppo Alte Energie, Via Celoria 16, 20100 Milano, Italy

1. INTRODUCTION

The upgraded UA2 detector has operated successfully in the 1988 and 1989 runs of the CERN $\bar{p}p$ collider which attained peak luminosities of 3×10^{30} cm^{-2}s^{-1}. The data used for the study of the vector boson masses corresponds to an integrated luminosity of 7.4 pb^{-1}.

In our analysis[1] we select samples of $W \rightarrow e\nu$ and $Z \rightarrow e^+e^-$ events for which systematic errors in energy scale are believed to be identical allowing the cancellation of this contribution to the systematic error when the ratio of the masses is formed. The precise ratio is used to obtain a value of $\sin^2\theta_W$ and combined with the recent e^+e^- collider results for M_Z to obtain M_W to unprecedented precision.

2. DETECTOR AND DATA

The UA2 detector has no muon coverage so that only boson decays with electrons in the final state are studied. On-line selection relies upon calorimeter information only. Showers which are mostly contained in electromagnetic (EM) compartments are identified as electron candidates. The calorimeter EM (hadronic) compartments are made up of Pb (Fe) and scintillator layers extending to pseudo-rapidity $\eta = 2.5$ (3.0). The calorimeter is partitioned into projective towers of size $\Delta\phi \times \Delta\eta \approx 15° \times 0.2$. Electron triggers include an inclusive trigger which is fully efficient for $E_T(e) > 12$ GeV and a pair trigger efficent for $E_T(e_i) > 5.5$ GeV.

Calibration of the calorimeter is based upon test-beam studies and monitored via Co^{60} radioactive sources, analysis of transverse energy flow in minimum bias events, and additional periodic test-beam studies. Initially all modules were placed in test-beams. Electron response is studied via energy scans and by scanning in angle, impact point, and vertex offset. These data are used to construct a model of electron response for each cell-type. With regard to energy resolution the calorimeter can be divided into three parts; the endcaps, a central region, and an edge region in between, (see table I). The resolution in the edge region is poor because the EM cells have been shortened to make room for the central detectors.

The central detector is used for tracking and to identify electrons. Moving outward from the beam-pipe, there is a silicon layer (3024 pads with $\Delta\phi \times \Delta z = 30° \times 2$ mm), followed by a cylindrical drift chamber ($\Delta\phi = 22.5°$ sectors with 13 staggered sense wires per sector), followed by a second silicon layer (3024 pads with $\Delta\phi \times \Delta z = 15° \times 8.7$ mm). Beyond the last silicon layer there are two transition radiation detectors (320 and 432 sense wires, 1216 strips). The final layer of the central detector is made up of a scintillating fiber detector which incorporates 18 layers of 1 mm diameter fibers arranged into 6 stereo triplets for tracking, followed by a 1.5 X_o Pb converter and an additional 2 stereo triplets. Electrons initiate showers in the Pb which result in a large charge measurement in the two "preshower" triplets. This enables clear discrimination between electrons and charged hadrons. In addition, the position of the apex of the shower measured in the last two triplets can be matched with the track of an incoming electron to good precision enabling further rejection of backgrounds,(e.g. γ's from π^o decay overlapped by a $\pi^\pm$). The track-preshower match resolutions are $\sigma_{r\phi} = 0.4\ mm$ and $\sigma_z = 1.1\ mm$. For the endcap region, proportional tubes are arranged to perform similar matching with resolutions of $\sigma_x = \sigma_y = 5mm$. A matching quantity, $d^2 = (r\Delta\phi/\sigma_{r\phi})^2 + (\Delta z/\sigma_z)^2$ is defined and we require $d < 4$ for electrons. In the endcap regions $d < 5$ is required. We also compare shower profiles of candidate electrons to what we expect as a result of test-beam measurements. A χ^2-like factor is calculated and the requirement $P(\chi^2) > 10^{-4}$ is imposed.

Table I. Calorimeter Response to Electrons

	Central	Edge	Endcaps
Coverage	$\|\eta\| < 0.8$	$0.8 < \|\eta\| < 1.0$	$\|\eta\| > 1.0$
Resolution	$15\%/\sqrt{E}$	$30\%/\sqrt{E}$	$15\%/\sqrt{E} + 1.0\%$
Cell-Cell Variation	1.0%	5.0%	3.5%
Impact Correction	1.3%		2.0%
Energy Scale	1.0%	3.0%	1.0%

Reconstruction of Z decays requires only a pair of identified electrons. For reconstruction of the W decays only transverse quantities are well-defined since the neutrino is not seen. $P_T(\nu)$ depends upon momentum balance which includes the hadronic activity in the event. In $\bar{p}p$ collisions hadrons are produced via initial state gluon radiation and as a result of interactions of spectator partons. Thus $\vec{P}_T(\nu) \sim - \vec{P}_T(e) - \vec{P}_T(\text{hadrons})$ where the relationship is only approximate as a result of acceptance limitiations and detector effects which lead to a mis-measurement of the total hadron momentum, (see section 4). The electron measurement uses a minimum number of "core" cells to minimize the contribution of energy from hadrons from the underlying event. Typically 2 cells are used but the exact number depends upon where the electron track intersects the calorimeter.

For the mass analysis we exclude events with electrons in the edge region where the energy resolution is poor and also those which are within $\Delta\phi = 0.5^o$ or

$r\Delta\theta = 6.0$ *mm* of the cell boundaries where energy corrections are the most uncertain. W decays with P_T(hadrons) > 20 GeV (5% of total), are also cut since the large degree of hadronic activity increases the uncertainty on $P_T(\nu)$.

Finally, since the endcap and central calorimeters have different systematic uncertainties and the W and Z samples have different proportions of electrons in the two regions, we use only central electrons. The loss in statistics is more than compensated by the reduction in systematic uncertainty in the mass ratio. We are left with 1206 W decays but only 54 Z decays.

We introduce a second sample of Z's in which we require one and only one electron to be in the central fiducial region. Of the information available for the other electron we retain only the track. Its energy is obtained by requiring momentum balance in the transverse plane along an axis perpendicular to the bisector of the electron *directions*. (This is denoted the ξ axis and has the property that the component of total momentum along this axis is the most sensitive to the electron momenta.) The balance requirement is: $E_1\,\hat{n}_1 \cdot \hat{\xi} + \mathbf{E_2}\,\hat{n}_2 \cdot \hat{\xi} = \vec{P}_T(\text{hadrons}) \cdot \hat{\xi}$ and is solved for $\mathbf{E_2}$. An independent set of 94 events is found but with worse mass resolution. The mass resolution is $\delta M \sim 4.3$ GeV as compared with the central sample for which $\delta M \sim 1.9$ GeV. The energy scale of this sample is however inherited from the electron in the central fiducial region since $\vec{P}_T(\text{hadrons})$ is not directionally correlated with the ξ axis (i.e. $\langle \hat{\xi} \cdot \vec{P}_T(\text{hadrons}) \rangle \sim 0$.)

3. $\mathbf{M_Z}$

The data and mass fits are shown in Fig. 1. An unbinned maximum likelihood method [2] is used in the mass fits. Likelihood functions are compared with data to find the best values of the mass and width. A confidence level for the fit is determined by a Kolmogorov-Smirnov test [3] which is also bin-independent.

For the Z peak, there are several contributions to the shape of the mass spectrum: *(i)* the Breit-Wigner, *(ii)* smearing of the resonance as a result of the finite resolution of the electron energy, *(iii)* skewing of the peak as a result of the parton luminosity distribution, *(iv)* a shift of the peak toward higher mass due to inflation of the electron energies by hadrons from the underlying event, and *(v)* a shift downward in mass resulting from final state photon emission in which the photon energy is not added to the electron energy.

If the parton luminosity distribution as a function of effective mass is approximated by a decaying exponential, then it is possible to derive an analytic expression for the line-shape which takes into account contributions *(i)* - *(iii)*. This is done by convoluting a Breit-Wigner with a Gaussian distribution representing the mass

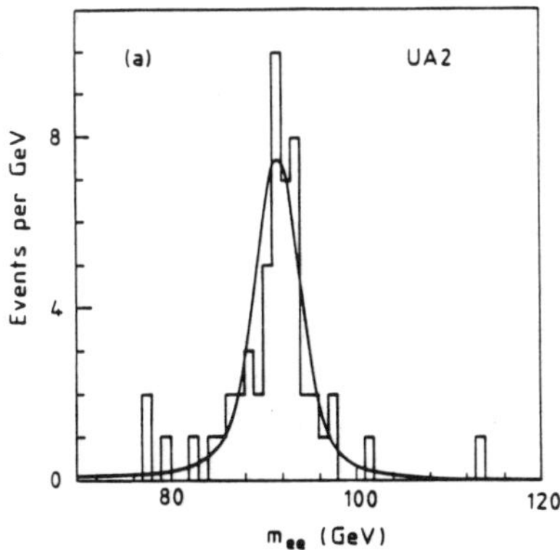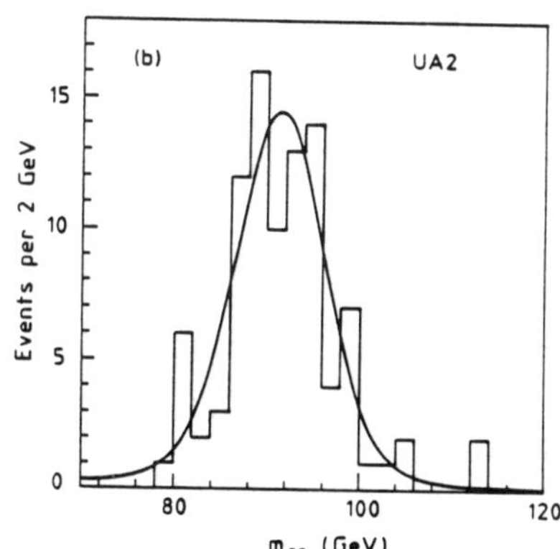

Fig. 1. M_Z Fits.
(a) Central Sample, (b) P_T Constrained Sample.

resolution and a decaying exponential. Letting m_Z and Γ_Z be the mass and width of the Z, m_{ee} the mass of the electron pair, σ_m the uncertainty on the mass, and $e^{-\beta m}$ the parameterization of the parton luminosity distribution, we obtain:

$$f(m_{ee}, \sigma_m; m_Z, \Gamma_Z) = \frac{1}{\sqrt{2\pi}\sigma_m} \times e^{(\beta^2 \sigma_m^2/2 - \beta m_{ee})} Re \left[\omega(x + iy)\right] \qquad (1)$$

Where $\omega(z)$ is the error function of complex argument, and x and y are given by:

$$x = \frac{\beta \sigma_m^2 - m_{ee} + m_Z}{\sqrt{2}\,\sigma_m} \qquad , \qquad y = \frac{\Gamma_Z}{2\sqrt{2}\sigma_m}$$

We find $\beta = .01980 \pm .00005$ by fitting to the parton luminosity distributions for various structure functions.

To take into account effects (iv) and (v) above, corrections are applied after fitting the data with the likelihood functions of eq. (1). The corrections are determined by comparing results obtained for fits to Monte Carlo (MC) data in which these effects have and have not been included. The correction for final state radiation is: $\Delta M = 0.10 \pm 0.10$ GeV. For the underlying event contribution, a study of W decays reveals that one-third of the electrons gain energy distributed as $e^{-2.8 E_T}$, (E_T in GeV). To determine the effect of this on M_Z, ensembles of MC data sets were used. Electron 4-vectors were generated taking into account all physics and detector considerations except final state radiation and underlying event energy. A corresponding pair mass was calculated. The electrons then had underlying event energy added at random and the pair mass was re-calculated. Correlated data sets of 54 events were then fit and the mass differences were histogrammed to determine fluctuations in the effect for a small statistical sample. We find that fit values must be corrected by $\Delta M = -0.24 \pm 0.05 \pm 0.05$ GeV, where the additional error corresponds to uncertainty in the parameterization.

Table II. M_Z Results After Corrections

Sample	Parameter	Results for Γ_Z Fixed	Results for Γ_Z Free
Central	M_Z	$91.55 \pm 0.43 \pm 0.12$	$91.56 \pm 0.45 \pm 0.12$
	Γ_Z	2.5	$2.96^{+0.98}_{-0.78}$
	C.L.	96%	99%
P_ξ Constrained	M_Z	$91.37 \pm 0.58 \pm 0.12$	$91.39 \pm 0.60 \pm 0.12$
	Γ_Z	2.5	$2.94^{+1.18}_{-.94}$
	C.L.	96%	97%

Fig. 1 shows the best-fitting likelihood functions superimposed on the data for the Z samples. Table II lists the results for fixed and free width. The latter are consistent with $\Gamma_Z = 2.5$ GeV as expected in the case of a heavy top quark. To obtain the final Z mass, corrections are applied and systematic uncertainties are added in quadrature. For the second Z sample the σ_m values used in eq (1) depend upon the response of the detector to hadrons, (section 4). This leads to an additional error of ±0.10 GeV. The final result is obtained as the weighted average of the results for the two samples: $M_Z = 91.51 \pm 0.36$ (stat) ± 0.12 (syst) ± 0.92(scale) GeV.

4. M_W

The W mass is determined by fitting of the transverse variables $P_T(e)$, $P_T(\nu)$ and $M_T(e\nu) = \sqrt{2P_T(e)P_T(\nu)(1 - \cos\phi_{e\nu})}$. These cannot be described by analytic functions. The likelihood functions are thus generated numerically by a Monte Carlo. The MC is used to parameterize the physics and the response of the detector in terms of a few variables such that the following global quantities can be reproduced: $\sum E_T \equiv \sum_{n=1}^{ncells}(E_n)_T$, $\vec{P}_T(\text{hadrons}) \equiv \left[\sum_{n=1}^{ncells} E_n\hat{u}_n\right]_T$, and $\vec{P}_T(e)$. The sums are over all calorimeter cells except those determining the electron energy, and $\hat{u}_n$ is a unit vector from the event vertex to the nth calorimeter cell with energy E_n. From studies of independent data sets and from MC we obtain conservative bounds and best values for the model parameters. Likelihood functions generated with the best values determine the central values of mass and width while those generated with extreme values of the parameters determine our systematic uncertainties.

Table III. M_W Fits Before Corrections and Systematic Errors

Transverse Variable	Parameter	Results for Γ_W Fixed	Results for Γ_W Free
$M_T(e\nu)$	M_W	80.75 ± 0.31	$80.78 \pm 0.31 \pm 0.12$
	Γ_W	2.5	$1.89^{+0.47}_{-0.40}$
	C.L.	84%	89%
$P_T(e)$	M_W	80.79 ± 0.38	80.83 ± 0.39
	Γ_W	2.5	$1.60^{+0.78}_{-0.68}$
	C.L.	95%	97%
$P_T(\nu)$	M_W	80.32 ± 0.41	80.33 ± 0.42
	Γ_W	2.5	$2.03^{+0.82}_{-0.72}$
	C.L.	83%	88%

The MC contains matrix elements for the decays $W \rightarrow e\nu_e$, $W \rightarrow \tau\nu_\tau$ with $\tau \rightarrow e\nu_e\bar{\nu}_\tau$, $Z \rightarrow e^+e^-$, and also interference with the Drell-Yan continuum; $\gamma^*/Z \rightarrow e^+e^-$. We study the effect of final state radiation with generators for the processes $W \rightarrow \gamma e\nu_e$ and $Z \rightarrow \gamma e^+e^-$ [4]. All of these processes can be simulated using any one of many parton distribution functions [5] evolved to $Q^2 = M_W^2$ or M_Z^2. The events are weighted according to a relativistic Breit-Wigner. $P_T(\text{boson})$ is obtained from calculations of Altarelli et al. [6]. We assume these give a good description of the data with uncertainties contained within the band of possibilities obtained by varying Λ_{QCD} for the parton distribution functions.

Once the kinematics are fixed, the event must be placed in the UA2 geometry. A vertex is generated as per our observed vertex distribution, ($\bar{z} \sim 0.0$, $\sigma_z \sim 13$ cm) and the electron trajectory and its impact point with the calorimeter are calculated. Fiducial cuts are applied as in the analysis of the data. Electron response is modelled via test-beam observations (section 2). Fluctuations in energy lost in material ahead of the calorimeter are simulated as are cell-to-cell gain variations. For final state radiation a simple algorithm is used in which the γ has its 4-vector added to the electron's if it is within a 20° cone about the electron track and it is included in the underlying event sums if outside the cone. The contribution to the electron energy from hadrons in the underlying event is also included.

The most difficult part of constructing the MC comes in modelling the underlying event which is essential to the determination of $P_T(\text{hadrons})$ and hence $P_T(\nu)$. Ultimately the model can be checked for consistency with Z data since there exists a relatively accurate measurement of $P_T(Z)$ from the electrons which can be compared with that from hadrons. The two basic parameters in the hadron response model are the resolution σ and the response correction Δ. The former accounts for the cumulative effect of fluctuations in hadronic energy measured in the calorimeter cells. The latter accounts for the systematic under-estimate of these quantities due to acceptance limitations, and calorimeter non-linearities and read-out thresholds.

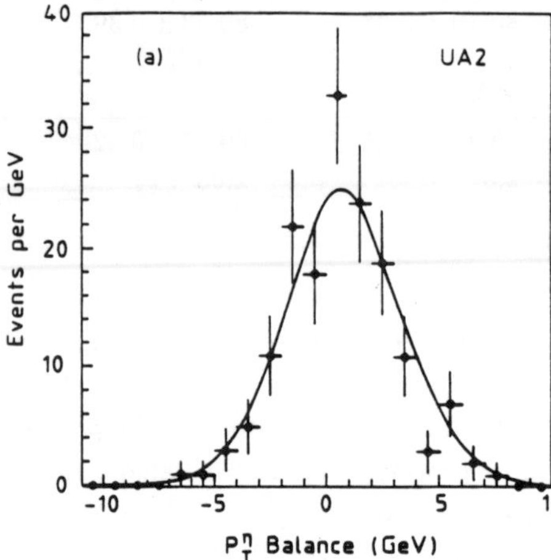

Fig. 2. Momentum Balance Along the η Axis.
The total Z sample is compared with the output of the MC with default values of σ and Δ.

To determine σ as a function of $\sum E_T$, minimum bias and inclusive two-jet data were studied. In minimum bias events there does not exist a significant physics source of missing momentum, (no hard ν), and consequently the net momentum observed is a measure of the hadronic momentum resolution of the detector. We take this to be a lower bound on what we expect for boson decays since the latter may also have a recoiling jet due to initial state radiation. For inclusive two-jet events the recoiling component will be larger than for W decays due to the dominance of glue-glue interactions, (at $\sqrt{s} = 630$ GeV, W and Z production is dominated by $q\bar{q}$ interactions [7]). Hence the momentum resolution determined with these data provides an upper limit for the resolution to be used in our model. The average of these parameterizations which are of the form $\sigma = a(\sum E_T)^b$, is taken to be our best guess for boson decays.

A lower bound for the response correction Δ was found via MC study of acceptance effects only. As upper bound we take the result for all effects as calculated with the PYTHIA MC which yields a larger correction than is necessary for our Z data. Our best guess is again taken to be intermediate between the bounds.

The hadron response model was used to calculate the component of total mo-

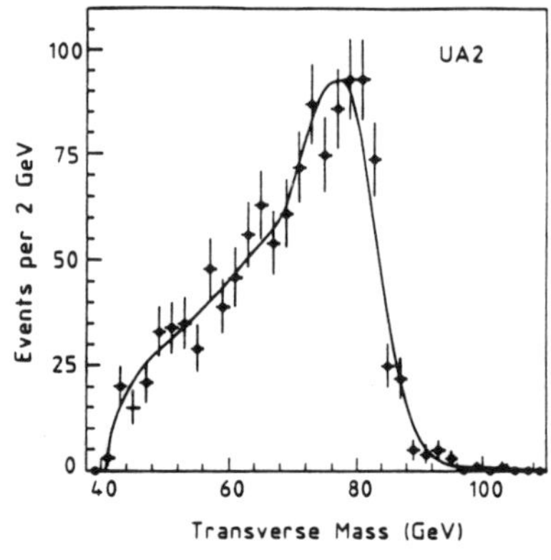

Fig. 3. $M_T(e\nu)$ Fit.
The $M_T(e\nu)$ distribution is shown with the best fit from the MC overlayed.

mentum in Z events along the positive η axis, (defined by the inner bisector of the two electron directions) and compared with data as seen in Fig. 2. This distribution is of particular interest since its width depends upon σ while its mean depends upon Δ. The prediction of the model is found to be in agreement with Z decay data within the limited statistics of the total Z sample. The model also reproduces the observed $P_T(W)$ spectrum when $\Lambda_{QCD} = 0.160$ GeV is chosen [8].

With a satisfactory hadron response model, likelihood functions can be generated and used to determine M_W. The fit to the $M_T(e\nu)$ distribution is shown in Fig. 3. Table III lists the results for all transverse variables for fixed width ($\Gamma = 2.1$ GeV, consistent with heavy top), and for floated width. The likelihood functions all contain a 3.8% contribution from $W \to \tau\nu_\tau$ with $\tau \to e\nu_e\bar{\nu}_\tau$.

Table IV. M_W Fit Corrections and Systematic Errors

Model Variation	M_T Fit	$P_T(e)$ Fit	$P_T(\nu)$ Fit
Hadron Model $(\sigma, \Delta, \Lambda_{QCD})$	±115	±215	±350
Parton Functions	±100	±160	±130
Neutrino Scale	±85		±170
E(e) Resolution	±40	±50	±60
Underlying Event	±30	±50	±20
Fit Procedure	±100	±100	±150
Radiative Decays	$+40 \pm 40$	$+60 \pm 60$	$+160 \pm 160$
Total	$+40 \pm 210$	$+60 \pm 300$	$+160 \pm 470$

Once again it is necessary to account for the effect of final state radiation and systematic errors must be determined. The latter are obtained by varying the

MC parameters within bounds to produce alternate likelihood functions. These are used to fit the data and a large MC control sample in order to determine systematic variations which are smaller than typical statistical uncertainties. Table IV lists the variations and the inferred uncertainties.

From tables II and IV it is seen that the most precise result is obtained in the M_T fit and it is taken as our final measurement of M_W:

$$M_W = 80.79 \pm 0.31(\text{stat}) \pm 0.21(\text{syst}) \pm 0.81(\text{scale})\,\text{GeV}$$

Fits are in agreement to within statistical uncertainties alone indicating that the hadron response model adequately accounts for any difference between hadron and electron energy scales.

5. M_W/M_Z AND THE STANDARD MODEL

Although the hadron response model adequately matches the neutrino energy scale to that of the electron, the energy scales of the W and Z electrons could still differ. In particular Z decay electrons have higher mean energy than those from W decays and hence deposit more energy on average in the hadronic compartments of the calorimeter which have different response to EM showers than do the EM compartments. Possible non-linearities in the electron energy reconstruction that could result from this were modelled, and test-beam measurements for electons with $E(e) = 40$ GeV, (intermediate between mean energies in W and Z decays), were used in interpolating to W and Z electron energies. The overall change in the mass difference $M_Z - M_W$ due to such an effect was found to be less than 100 MeV and is ignored. The final mass ratio is thus :

$$M_W/M_Z = 0.8831 \pm 0.0048(\text{stat}) \pm 0.0026(\text{syst})$$

This can be combined with the results from LEP and SLC [9] for M_Z to obtain:

$$M_W = 80.49 \pm 0.43(\text{stat}) \pm 0.24(\text{syst})\ \text{GeV}$$

This result can be compared with theoretical expectations. The standard model of electroweak interactions with a single complex Higg's doublet has three fundamental free parameters, (ignoring Higg's and fermion masses). It is now standard practice to choose these to be α, G_μ and M_Z which are all well-measured. To compute higher order corrections, a renormalization scheme must be chosen and it is natural for us to choose the scheme of Sirlin [10] for which $\sin^2\theta_W = 1 - (M_W/M_Z)^2$, which leads to the relations [11]:

$$M_W^2 = \frac{A^2}{(1 - \Delta r)\sin^2\theta_W}, \quad M_Z^2 = \frac{A^2}{(1 - \Delta r)\sin^2\theta_W \cos^2\theta_W}$$

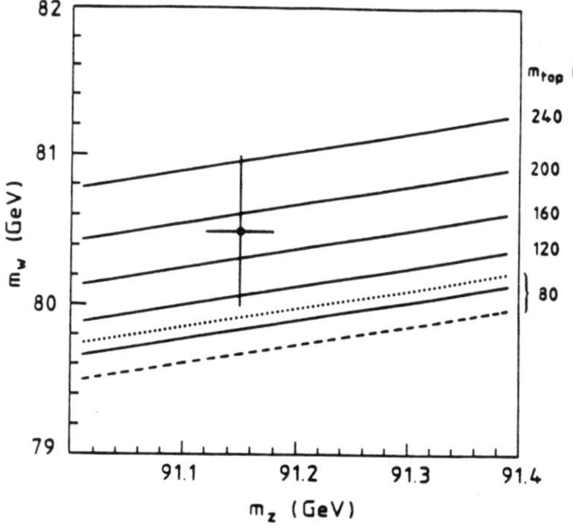

Fig. 4. Comparison with Standard Model.
Solid lines indicate allowed values of M_W and M_Z for a given M_t for $M_{Higgs} = 100$ GeV. The dotted (dashed) line indicates the expectations for $M_t = 80$ GeV and $M_{Higgs} = 10(1000)$ GeV. The data point is obtained from UA2 (M_W), and LEP/SLC (M_Z).

where $A \equiv [\pi\alpha/\sqrt{2}G_\mu]^2 = 37.2805 \pm 0.0003$ GeV. The quantity Δr represents the effect of the radiative corrections arising from virtual loops in the boson propagators. It depends on the square of top mass, M_t, and the logarithm of M_{Higgs}. The corrections can be calculated and an accurate numerical value is computed in the program of Hollik and Burgers [12]. Fig. 4 shows the measured boson masses as compared with expectations of the standard model for various combinations of M_t and M_{Higgs}. The value of M_W measured by UA2 is consistent with a heavy top quark. This can be quantified by determining confidence intervals for the value of Δr calculated using the measured boson masses and converting these to limits on M_t for conservative choices of M_{Higgs}. Assuming $M_{Higgs} > 24$ GeV as reported at LEP [13] and $M_{Higgs} < 1$ TeV, then:

$$\textbf{76 GeV [90\% C.L.]} < \textbf{M}_t < \textbf{272 (289) GeV [90 (95)\% C.L.]}$$

These limits do not, in general, apply for extensions of the standard model which have additional Higg's multiplets.

From the mass ratio we also obtain a value for the weak mixing angle:

$$\sin^2\theta_W = 0.2202 \pm 0.0084(\text{stat}) \pm 0.0045(\text{syst})$$

which is consistent with the world average from neutral current experiments [14]:

$$\sin^2\theta_W(\nu N) = 0.2309 \pm 0.0029(\text{stat}) \pm 0.0049(\text{syst}).$$

ACKNOWLEDGEMENTS

We gratefully acknowledge P. Darriulat for his contributions to the UA2 upgrade project. We also thank the technical staff of collaborating institutes. This experiment was made possible by the successful operation of the CERN $\bar{p}p$ Collider to whose staff we extend our appreciation.

Funding for this experirment was provided by Schweizerischen Nationalfonds zur Forderung der Wisenschaftlichen Forschung, the UK Science and Engineering Research council, the Bundesministerium fur Forschung und Technologie, the Institut National de Physique Nucléare et de Physique des Particules, the Istituto Nazionale della Fisica Nucleare and the Institut de Recherche Fondamentale (CEA).

REFERENCES

[1]   J. Alitti et al.: **A Precise Determination of the W and Z Masses at the CERN $\bar{p}p$ Collider**, CERN-EP/90-22, submitted to Phys. Lett. B.

[2]   V. Blobel, **Formulae and Methods in Experimental Data Evaluation Vol. 3**, (EPS - Computational Physics Group, 1984),p. L1

[3]   W.H.Press et al. **Numerical Recipes**, (Cambridge 1986), p.472

[4]   F.A. Berends,R. Kleiss : Z. Phys. C. - Particles and Fields **39** (1988) 24

[5]   -M. Diemoz, F. Feronni, E. Longo, G. Martinelli: Z. Phys. C. - Particles and Fields **27**(1985)617
-A.D. Martin, R.G. Roberts, W.J. Stirling : Mod. Phys. Lett. **A4**(1989)1135
-E. Eichten, I. Hinchkliffe, K. Lane, C. Quigg : Rev. Mod. Phys. **56** (1984) 579; erratum **58**(1986)1065
-D.W. Duke, J.F. Owens : Phys. Rev. **D30**(1984)49

[6]   G. Altarelli, R.K. Ellis, G. Martinelli : Z. Phys. C - Particles and Fields **27**(1985)617

[7]   P.B. Arnold, M. Hall Reno : Nucl. Phys. **B319** (1989)37

[8]   G. Blaylock, these proceedings.

[9]   -G. Abrams et al. (MARK II) : Phys. Rev. Lett. **63** (1989)2173
-D. Decamp et al. (ALEPH) : Phys. Lett. **231B** (1989)519 and CERN preprint CERN-EP/89-169
-P. Aarnio et al. (DELPHI) : Phys. Lett. **231B** (1989)539
-B.Adeva et al. (L3) : Phys. Lett. **231B**(1989)509
-M.Z.Akrawy et al. (OPAL) : Phys. Lett. **231B** (1989) 530

[10] A. Sirlin : Phys. Rev. **D22**(1980)971

[11] W.J. Marciano, A. Sirlin : Phys. Rev. **D29**(1984)945

[12] W. Hollik, G. Burgers : Z Physics at LEP, Vol. 1, CERN 89-08

[13] -D. Decamp et al. (ALEPH) : CERN-EP/90-16
-M.Z. Akrawy et al. (OPAL) : CERN-EP/89-174

[14] G.L. Fogli, D. Haidt : Z. Phys. C - Particles and Fields **40**(1988)379

Review of Results from the
Mark II Experiment at SLC

David P. Coupal
representing the Mark II Collaboration

Stanford Linear Accelerator Center
Stanford, California 94309

Abstract

This paper reviews results on Z° physics from the 1989 run of the Mark II experiment at the SLAC Linear Collider. Based on about 20 nb^{-1} we present results on the mass, width and branching ratios of the Z° boson, the number of light neutrino species, properties of hadronic decays and searches for new particles.

1. Introduction

In the Standard Model the Z° boson couples to all the fundamental fermions. Thus e^+e^- machines of sufficient energy to produce the Z° boson provide excellent laboratories for studying the Standard Model, both the electroweak sector and QCD, and for searching for new particles or physics. The Mark II experiment at the SLAC Linear Collider is the first experiment to look at decays of the Z° at one of these machines.

The SLAC Linear Collider is a single-pass collider that uses the linac to accelerate both electrons and positrons. Arcs at the end of the linac bend the beams around and a final set of optics focusses the beams down to a radius of 2-3 micron at the interaction point (IP). Most of the 1989 running was with a repetition rate of 60 Hz and bunch intensities of $1\text{-}2\times10^{10}$ particles per bunch resulting in typical instantaneous luminosities of 7×10^{27} cm^{-2} sec^{-1}.

The Mark II detector is shown in Figure 1 and described in detail elsewhere[1].

*Work supported in part by the Department of Energy contracts DE-AC03-81ER40050 (CIT), DE-AM03-76SF00010 (UCSC), DE-AC02-86ER40253 (Colorado), DE-AC03-83ER40103 (Hawaii), DE-AC02-84ER40125 (Indiana), DE-AC03-76SF00098 (LBL), DE-AC02-76ER01112 (Michigan), and DE-AC03-76SF00515 (SLAC), and by the National Science Foundation (Johns Hopkins).

The vertex detectors were not installed for the 1989 running. A cylindrical central drift chamber (CDC) in a 4.75 kG solenoidal field provides charged particle tracking and particle identification using dE/dx. Outside the CDC are TOF counters, the magnet coil, and a lead/liquid Argon barrel calorimeter. The endcap regions are covered by a lead/proportional tube calorimeter. Muon counters outside the barrel calorimeter cover 45% of 4π. A Small Angle Monitor (SAM) consisting of tracking and calorimetry provides low angle coverage and luminosity monitoring. A detector at small-

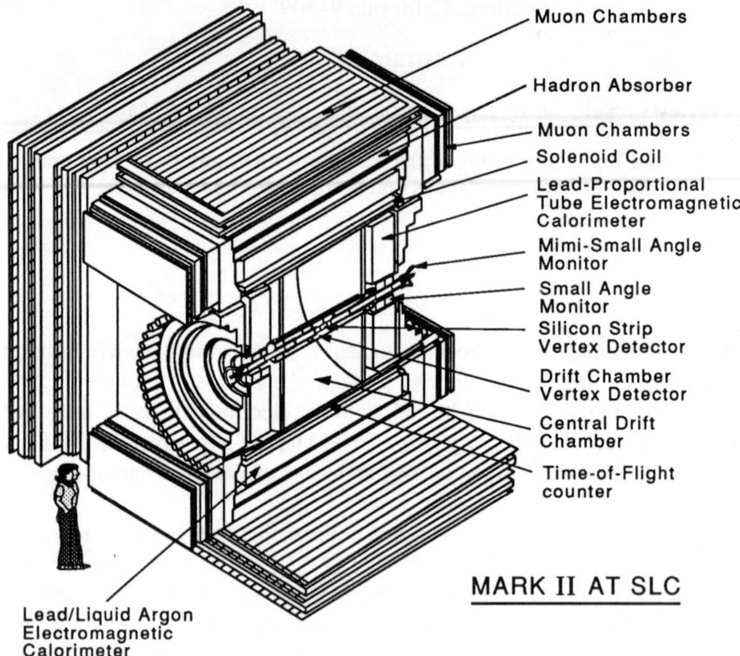

Figure 1. The Mark II Detector at SLC.

er angles (Mini-SAM) provides higher statistics luminosity monitoring. The central detector has a charged trigger that demands at least two tracks in the CDC and a calorimeter trigger that looks for clusters of energy deposition. Monte Carlo studies predict a combined trigger efficiency of over 99% for hadronic decays of the Z°.

A measurement of the energies of both beams is critical for the Z° mass and width determination. To this end, spectrometers measure the energy of each bunch on its way from the IP to their respective dumps. These spectrometers[7] measure the energies to an accuracy of 20 MeV. Possible position-energy correlations in the colliding beams add to the error in the center-of-mass energy (E_{cm}) giving a total error of 35 MeV.

In the following we review Mark II results based on about 20 nb^{-1} of data collected between May and October of 1989. This sample corresponds to roughly 450 hadronic decays of the Z°. The next section will briefly describe the data and subsequent sections present results on the Z° mass and width, branching ratios, hadronic decays and new particle searches.

2. Data

The 1989 data consists of a total of 20 nb^{-1} of data taken at 10 different values of E_{cm}. The sample consists of roughly 450 hadronic decays, 13 μ pairs, 21 τ pairs and 18 electron pairs.

Cuts on the data depend on the particular physics analysis but some of the same track quality cuts are used by most of the analyses presented here. Charged tracks are required to originate from a cylinder of radius 1 cm and half-length 3 cm centered on the IP. Charged tracking is efficient down to $|\cos\theta|<.90$ but analyses that require good momentum measurement require $|\cos\theta|<.82$. Energy clusters in the calorimeters are required to be at least 1 GeV.

In the Monte Carlo detector simulation, particles from the desired physics process are tracked through the detector and used to generate fake raw data. This data is then mixed with real data from random beam crossings to simulate the beam-related backgrounds in the detector. The final mixed Monte Carlo data is submitted to the same analysis as the real data.

3. Mass, Width and Number of Neutrinos

The mass of the Z° boson (M_Z) is a fundamental parameter of the Standard Model and its decay width (Γ_Z) contains information on the number of particles that couple to the Z° and the strength of those couplings. M_Z and Γ_Z can be determined by fitting the resonance in the decay rate $\Gamma(e^+e^- \to Z^\circ \to f)$ as a function of E_{cm}, where f is some final state.

Details of the Mark II analysis of the Z° resonance can be found elsewhere[3]. We use for the final state, f, all hadronic Z decays and decays into μ and τ pairs. Hadronic event selection requires at least 3 charged tracks with $|\cos\theta|<.90$ and at least $.05 \times E_{cm}$ of energy visible in both the forward and backward hemispheres. Backgrounds from beam gas and two photon interactions are negligible. The efficiency, including the trigger, for hadronic decays is 95% as determined from Monte Carlo. The μ and τ pair events are required to have $|\cos\theta_T|<.65$, where θ_T is the thrust angle. Restriction to this angular range insures high trigger efficiency and unambiguous identification.

The luminosity is measured at each value of E_{cm} using the SAM and Mini-SAM detectors described above. Bhabha scattering events into the SAM ($50<\theta<160$ mrad) are selected by requiring 40% of the beam energy in each SAM. The overall normalization is done to a smaller fiducial volume ($60<\theta<160$ mrad) with an accurately cal-

222

culable cross section. The Bhabha cross section into this region is 25.2 nb at 91.1 GeV. The estimated systematic errors are 2% from unknown higher order radiative corrections and 2% due to detector resolution and reconstruction. The selection of Bhabha events in the Mini-SAM involves looking for back-to-back showers in pairs of quadrants. The overall cross section is determinied by normalizing to the SAM giving 227 nb at E_{cm}=91.1 GeV (234 nb for the last 3 scan points following a detector realignment).

Figure 2 shows the resulting resonance shape in the cross section versus E_{cm}. The data are fitted to a relativistic Breit-Wigner of the form:

$$\sigma_Z(E) = \frac{12\pi}{M_Z^2}\frac{s\Gamma_e\Gamma_f}{\left(s-M_Z^2\right)^2 + s^2\Gamma^2/M_Z^2}(1+\delta(E))$$

where $s=E_{cm}^2$, δ is an analytical approximation of the radiative corrections[4], Γ_e is the electron partial width, and Γ_f is the partial width for decays into our fiducial volume. Γ_f is given in terms of the hadronic, μ and τ partial widths by $\Gamma_f=\Gamma_h+f\times(\Gamma_\mu+\Gamma_\tau)$ where $f=.556$ is the fraction of μ and τ events with $|\cos\theta_T|<.65$. The total Z width, Γ, is given by $\Gamma=\Gamma_h+\Gamma_e+\Gamma_\mu+\Gamma_\tau+N_v\times\Gamma_v$, where N_v is the number of light neutrino families. Three fits are done to the data in Figure 2. The first fit varies only the mass (M_Z), the second fit allows the mass and number of neutrinos to vary and the third fit allows the mass, total width and peak cross section (σ_0) to vary. The results are shown in Table 1. The result for N_v can be translated into a 95% CL upper limit of $N_v<3.9$, excluding to this level a

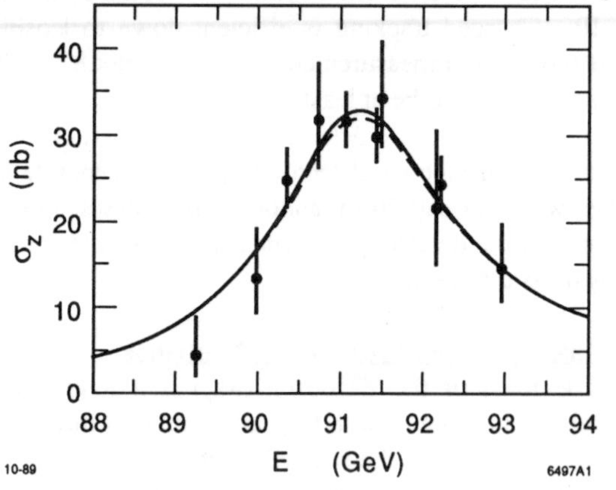

Figure 2. e^+e^- annihilation cross sections to hadronic events plus μ and τ pairs with $|\cos\theta_T|<.65$. The dashed curve represents the result of the first fit and the solid curve the second and third fits, which are indistinguishable.

	Table 1. Results of fit to Z° resonance.			
Fit	M_Z	N_v	Γ	σ_0
	GeV/c^2		GeV	nb
1	91.14±0.12	--	--	--
2	91.14±0.12	2.8±0.6	--	--
3	91.14±0.12	--	2.42+.45-.35	45±4

fourth light neutrino specie. The fits for Γ and σ_0 should be compared to the Standard Model predictions of 2.45 GeV and 43.6 nb respectively.

4. Branching Ratios

The Standard Model makes definite predictions of the couplings of the Z° to the fermions. It predicts for the ratio of the decay rate into a pair of leptons to the rate into hadrons to be $\Gamma(Z^\circ \to \ell^+\ell^-)/\Gamma(Z^\circ \to hadrons) = .048$. To separate the different leptonic decay modes, the calorimetry is used to distinguish electron and muon events and a cut on the minimum momentum of the 2 tracks is used to separate 1-1 topology τ decays from the electron and μ events. After correcting the electron sample for the presence of Bhabha events we get $\Gamma_{ee}/\Gamma_{had} = .037^{+.016}_{-.012}$, $\Gamma_{\mu\mu}/\Gamma_{had} = .053^{+.020}_{-.015}$, and $\Gamma_{\tau\tau}/\Gamma_{had} = .066^{+.021}_{-.017}$ in good agreement with the Standard Model. Details of this analysis can be found elsewhere[5].

We have also measured the branching ratio of Z° into bottom quarks[6]. In this analysis we use the semileptonic decay of the b quark. For electrons and muons with momentum greater than 2 GeV, Figure 3 shows the p_T relative to the closest jet. Electrons are identified in the calorimetry and muons in the muon counters. Counting the number of events containing a tagged lepton with p_T greater than 1.25 GeV/c and correcting for background and

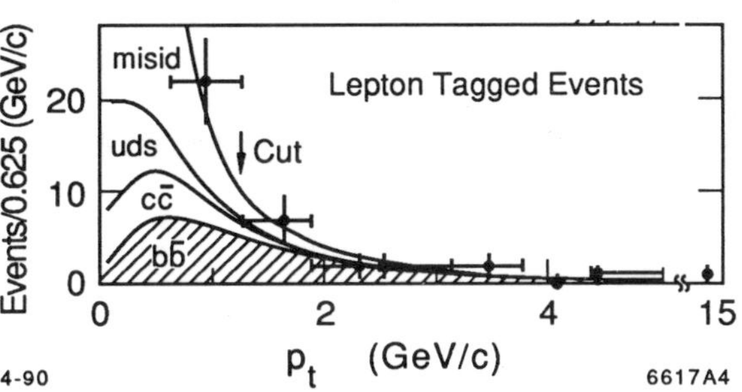

Figure 3. p_t distribution for tracks tagged as electrons or muons with the Monte Carlo prediction of the different contributions.

acceptance gives $\Gamma_{b\bar{b}}/\Gamma_{had} = .23^{+.11}_{-.09}$, in agreement with the Standard Model value of .22 .

5. Hadronic Decays

In the Standard Model hadronic decays are 70% of the available decay modes, providing a clean environment for high statistics studies of QCD. We review Mark II results on global shape parameters[7], which are sensitive to parton level processes, and inclusive particle distributions[8], which give insight into the fragmentation process. Fragmentation models used in the comparisons include the Lund 6.3[9], the Webber 4.1[10], and the Caltech II 86[11] parton shower models and a Lund model based on a second order QCD matrix element calulation by Gottschalk and Shatz[9,12]. The parameters of these models were tuned to fit the Mark II data taken at PEP at E_{cm}

= 29 GeV. We also present the first measurement of α_S at this higher q^2 and compare to a measurement done at PEP with the same detector andtechnique[13].

Event selection cuts include the demand that the event have at least 7 tracks and that the visible energy be at least 50% of E_{cm}. These cuts insure that the background is low and the event is well contained within the detector. The efficiency as determined from Monte Carlo is 78±2%.

Common global shape parameters used in many analyses are thrust, sphericity and aplanarity. Figure 4 shows one of these variables, sphericity, compared to the 4 models. The mean values of the quantities (corrected for detector effects) are shown in Figure 5 as a function of E_{cm} and compared to other measurements at lower energy and with the Lund parton shower prediction. Also shown in Figure 5 is the fraction of 3-jet events, where jets are found using the standard JADE algorithm with y_{cut} = .08.

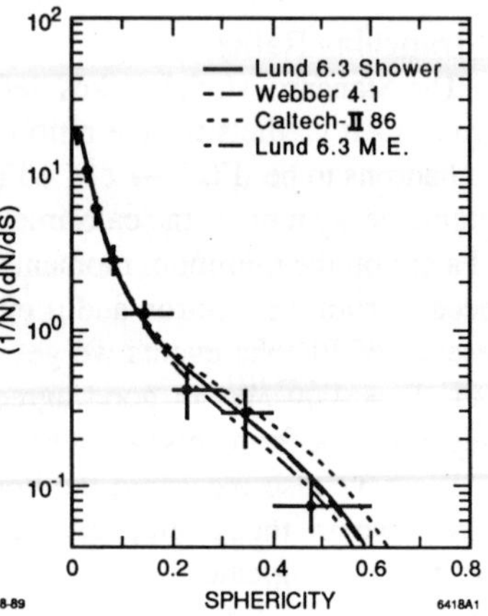

Figure 4. Sphericity distribution for hadronic decays of the Z°.

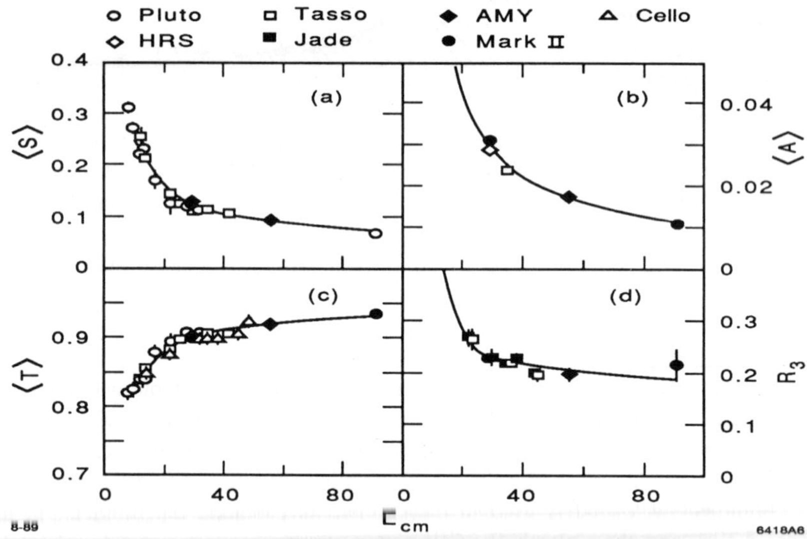

Figure 5. Corrected mean sphericity, aplanarity, thrust and 3-jet fraction versus E_{cm}. The solid curve is the prediction from the Lund parton shower model.

Fragmentation models can be studied by looking at charged particle inclusive distributions. The corrected mean charged multiplicity was found to be 20.1±1.2 . Figure 7(a) shows the corrected charged particle inclusive distribution in the scaled momentum $x = 2p/E_{cm}$. Figure 7(b) shows it versus E_{cm} for several x bins together

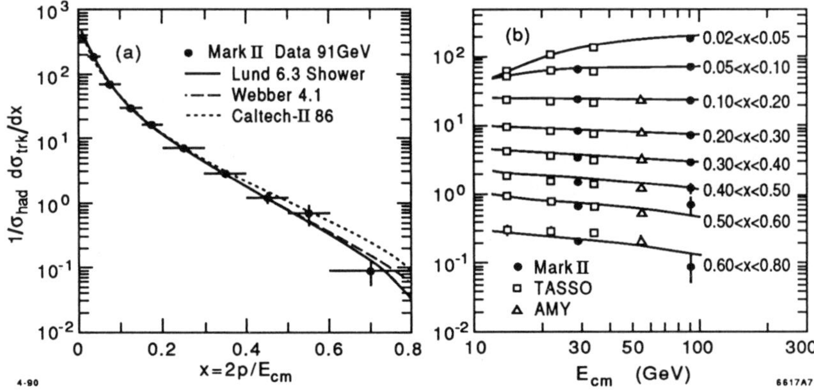

Figure 7. (a) Inclusive x distribution for hadronic decays compared to parton shower models. (b) Inclusive distribution versus E_{cm} in x bins. The solid lines are the Lund model prediction.

with $e^{+}e^{-}$ data at lower E_{cm}. The mean of the square of the momentum transverse to the sphericiy axis both in the event plane ($p_{\perp in}$) and out of the event plane ($p_{\perp out}$) is shown in Figure 8. These data are compared to Mark II data from PEP, data from other experiments and the Lund model.

The strong coupling constant, α_S, is a fundamental parameter of QCD. We have measured α_S by fitting the distrubtion of differential jet multiplicity versus the value of y_{cut}. This technique has been applied to Mark II PEP data taken at E_{cm} = 29 GeV and again at the SLC with E_{cm} = 91.1. Choosing the renormalization point at $q^2 = E_{cm}^2$ gives α_S=.123±.009±.005 at SLC and α_S=.149±.002±.007 at PEP in good agreement with the QCD prediction of the running of α_S with q^2.

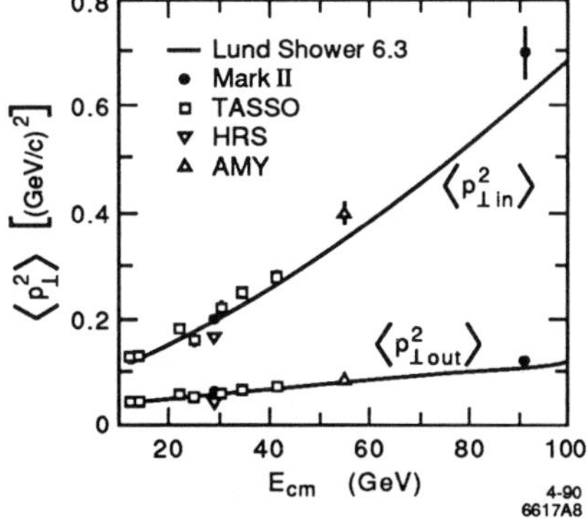

Figure 8. Mean values of $p_{\perp in}^2$ and $p_{\perp out}^2$ vs E_{cm} for various experiments.

6. Search For New Particles

The new energy regime opened up by the SLC makes it natural to look for the production of new particles, in particular heavier equivalents of the known fermions such as the top or b' quark or a heavy neutrino. A fourth light neutrino has already been eliminated by the measurement of the Z° resonance. We review below limits on the mass of the top and b' quarks using an isolated track and event shape analysis[14]. We also present limits on heavy neutrinos using the isolated track analysis[14], search for displaced vertices[15] and a search for jets containing 2 tracks consistent with a heavy neutrino decaying into $\tau \ell^\pm \nu_\ell$ [16].

The isolated track analysis uses the fact that semileptonic decays of heavy quarks will tend to produce tracks well isolated from the jet coming from the secondary quark. We define a parameter $\rho_{track} = \{2E(1-\cos\theta_{jet})\}^{1/2}$ where θ_{jet} is the angle to the nearest jet and define ρ for the event as the maximum of all the ρ_{track}. A cut of $\rho > 1.8$ is used to select top or b' events. We observe one event with $\rho > 1.8$ in our sample of Z° hadronic decays. The expected background varies from .9 to 1.8 events depending on the fragmentation model. To be conservative we assume the smallest value (.9 events) and can set limits of $M_{top} > 40.0$ and $M_{b'} > 44.7$ GeV/c^2.

A similar analysis can be done looking for the non-leptonic 4-jet decays for the heavy quarks. We define the quantity

$$M_{out} = \frac{E_{cm}}{E_{vis}} \frac{1}{c} \sum \left| p_T^{out} \right|$$

where p_t^{out} is the momentum transverse to the event plane. A cut of $M_{out} > 18$ GeV/c^2 is used to select heavy quark decays. This selection is also sensitive to heavy quark decays involving charged Higgs (t → H$^+$b, b' → H$^-$c). We find 6 events with $M_{out} > 18$ and expect 4.8-11.7 depending on the model. Again taking the most conservative (4.8 events) we can set the limits $M_t > 40.7$ and $M_{b'} > 44.2$ GeV/c^2. If the dominant decay mode is through a charged Higgs (M(H$^\pm$) > 25 GeV) then the limits are $M_t > 42.5$ and $M_{b'} > 45.2$ GeV/c^2.

As already discussed above the measurement of the Z° resonance shape rules out the existence of a light 4th generation neutrino. This limit excludes stable neutrino masses up to about 19 GeV/c^2. The phase-space suppression in the production of heavier neutrinos limits the sensitivity to higher masses. Direct searches are possible provided the heavy neutrino decays by some mechanism. Barring the existence of a charged partner lighter than the massive neutrino, a common mechanism of generating a massive unstable neutrino is through mixing with the lighter neutrinos. Just as in the quark sector, the mass eigenstate, ν_4, may be a mixture of the weak eigenstates, ν_ℓ (ℓ=e,μ,τ,L) so that

$$\nu_\ell = \sum_{i=1}^{4} U_{\ell i} \nu_i$$

leading to decay rates of $\nu_4 \to \ell^\pm + W^*$ dependent on $|U_{\ell 4}|^2$.

For short-lived decays (large $|U_{\ell 4}|^2$) the isolated track analysis can be used to set limits assuming Standard Model coupling to the Z°. Figure 9 shows the limits of

$M(\nu_4)$ vs $|U_{\ell 4}|^2$ from this analysis for $\ell = e$, μ and τ.

Smaller values of $|U_{\ell 4}|^2$ imply a long-lived neutrino and therefore the decays may be displaced from the interaction point. We looked for displaced decay vertices by looking at the impact parameter of the charged tracks in an event. Figure 9 shows

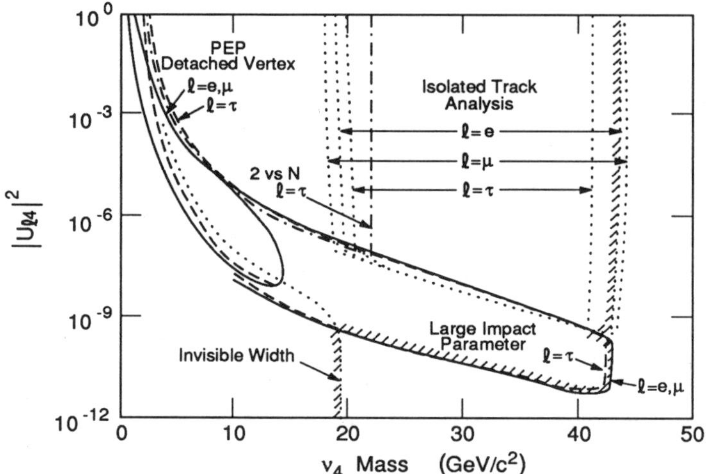

Figure 9. 95% C. L. excluded regions for a 4th generation massive Dirac neutrino as a function of mass and mixing. The analyses used in the different limits are described in the text. The hatched line shows the combined limit for $\ell = e$.

the limits from this analysis along with the limits from the search for detached vertices at PEP. Finally, Figure 9 also shows the limit between masses of 2.2 and 20 GeV/c^2 derived from a search for events where one heavy lepton decays via $\nu_4 \rightarrow \tau \ell^{\pm} \nu_{\ell}$ and the other decays hadronically (2 vs N).

7. Conclusions

Based on a sample of roughly 500 Z° decays, the Mark II experiment has been able to measure the mass of the Z° to 120 MeV, eliminate the existence of a 4th generation light neutrino at the 95%CL and confirm the Standard Model in the leptonic and hadronic decays. We have also set limits on the existence of the top and b′ quark and a massive neutrino decaying through mixing to lighter neutrinos.

The Mark II has installed silicon strip and drift chamber vertex detectors.We will take data again in the Summer of 1990 during which we hope to log an additional 3-4 k Z° decays.

228

REFERENCES

[1]   G. Abrams *et al.*, Nucl. Instrum. Methods **A 281**, 55 (1989).

[2]   J. Kent *et al.*, SLAC-PUB-4922 (1989); M. Levi, J. Nash, and S. Watson, Nucl. Instrum. Methods **A281**,265 (1989); M. Levi *et al.*, SLAC-PUB-4921 (1989).

[3]   B. Harral, these proceedings: G. S. Abrams *et al.*, Phys. Rev. Lett. **63** 2173 (1989).

[4]   R. N. Cahn, Phys. Rev. D **36**, 2666 (1987), Eqs. 4.4 and 3.1 .

[5]   M. Kuhlen, these proceedings; G. S. Abrams *et al.*. Phys. Rev. Lett. **63**, 2780 (1989).

[6]   J. F. Kral, these proceedings; J. F. Kral *et al.*, Phys. Rev. Lett. **64**, 1211 (1990).

[7]   E. Wicklund, these proceedings; G. S. Abrams *et al.*, Phys. Rev. Lett. **63** (1989).

[8]   K. O'Shaughnessy, these proceedings; G. S. Abrams *et al.*, Phys. Rev. Lett. **64** 1334 (1990).

[9]   T. Sjostrand, Comput. Phys. Commun. **39**, 347 (1986); T. Sjostrand and M Bengtsson, Comput. Phys. Commun. **43**, 367 (1987); M. Bengtsson and T. Sjostrand, Nucl. Phys. **B289**, 810 (1987).

[10] G. Marchesini and B. R. Webber, Nucl. Phys. **B238**, 1 (1984); B.R. Webber, Nucl Phys. **B238**, 492 (1984).

[11] T. D. Gottschalk and D. Morris, Nucl. Phys. **B288**, 729 (1987).

[12] T. D. Gottschalk and M. P. Shatz, Phys. Lett. **150B**, 451 (1985); CALTECH Report No. CALT-68-1172, -1173,-1199, 1985 (unpublished); T. D. Gottschalk (priviate communication).

[13] S. Komamiya, these proceedings; S. Komamiya, *et al.*, Phys. Rev. Lett. **64**, 987 (1990).

[14] R. Van Kooten, C. K. Jung, and S. Komamiya, these proceedings;G. S. Abrams *et al.*, Phys. Rev. Lett. **63**, 2447 (1989).

[15] R. Van Kooten, C. K. Jung, and S Komamiya, these preoceedings; C. K. Jung *et al.*, Phys. Rev. Lett. **64**, 1091 (1990).

[16] R. Van Kooten, C. K. Jung and S Komamiya, these proceddings; P. R. Burchat *et al.*, SLAC-PUB-5172, LBL-28458 (Jan 1990), (Submitted to Phys. Rev. Lett.)

Z Highlight - Results from ALEPH

Sau Lan Wu
Department of Physics, University of Wisconsin
Madison, Wisconsin 53706
Representing the ALEPH Collaboration

Section 1: INTRODUCTION

The LEP operation at CERN commenced successfully in July, 1990, and data taking for all four LEP experiments started one month later. I report here the physics results from the data of about 31,700 Z° decays into fermion pairs, collected with the ALEPH detector at LEP.

The topics covered in this paper are:

(1)   Measurements of Electroweak Parameters from Z° Decays into Fermion Pairs and the Number of Families with Light Neutrinos.

(2)   Search for Higgs Bosons.

(3)   Search for New Quarks and New Leptons.

(4)   Search for Supersymmetric Particles.

Since similar results by MARK II of SLC, DELPHI, L3 and OPAL of LEP are given in this proceeding, references to their results are not included in this report.

Section 2: MEASUREMENTS OF ELECTROWEAK PARAMETERS FROM Z° DECAY INTO FERMION PAIRS AND THE NUMBER OF FAMILIES WITH LIGHT NEUTRINOS

2.1 Parameters for Z° Resonance

Soon after the operation of LEP at CERN, a large number of Z° particles became available for precise measurements [1], [2]. The first publication of the ALEPH Collaboration in September 1989 on the basis of 3,700 hadronic Z° decays, and a systematically precise determination of the peak cross-section, was able to demonstrate that the number of neutrino families into which the Z° decays is 3.27 ± 0.30, and so with great likelihood just three [1]. With the October and November runs the statistics had increased fivefold and analysis of the leptonic in addition to the hadronic decays permitted a more precise determination, $N_\nu = 3.01 \pm 0.16$, as well as a first measurement of the leptonic branching ratios. The latter are interesting for two reasons. The experimental ratio $R = \Gamma_{had}/\Gamma_{\ell\ell}$ of the hadronic to leptonic partial width can be used in the determination of N_ν in place of the electroweak value, improving the autonomy of the experiment, and the experimental widths can be compared with the predictions of the electroweak theory and so to test the theory at a higher level of sensitivity than was previously possible.

In this section we present results of the Z° resonance based on 28,000 hadronic and 3,700 leptonic decays. This total data sample obtained by ALEPH in 1989 corresponds to an integrated luminosity of 1.2pb^{-1}.

The ALEPH detector is described in detail in ref. [3]. A display in the Time Projection Chamber of a computer constructed event taken during the pilot run in August 1989 is shown in Fig.1. The description of the measurements for the luminosity, the hadronic cross-section

($\sigma(Z^\circ \to$ hadrons)) and the leptonic cross-section ($\sigma(Z^\circ \to \ell^+ \ell^-)$), where $\ell^\pm = e^\pm, \mu^\pm$ and $\tau^\pm$, are given in the parallel session of this proceeding [4].

The peak cross-section for $e^+e^- \to f\bar{f}$ due to Z° exchange, when unfolded from initial state radiation, is:

$$\sigma^\circ_{f\bar{f}} = \frac{12\pi}{M_z^2} \frac{\Gamma_{ee}\Gamma_{f\bar{f}}}{\Gamma_z^2}$$

where Γ_{ee} and $\Gamma_{f\bar{f}}$ are the partial widths of Z° decay into e^+e^- or any fermion pair $f\bar{f}$ respectively, Γ_z is the total width and M_z the mass of the Z° boson. The invisible width is defined as $\Gamma_{inv} = \Gamma_z - \Gamma_{had} - \Gamma_{ee} - \Gamma_{\mu\mu} - \Gamma_{\tau\tau}$.

The resonance parameters have been determined by means of three different computer programs to calculate the initial state radiation and the contribution from Z°-photon interference terms. The first two programs, based on formulae by Borelli et al. [5] and by Berends et al. [6], use analytic expressions, while the third one following Ref. [7] uses a numerical integration. Differences in the results from the three programs are less than 0.1 of the experimental error. Results quoted are from the program using the formula of reference [7]. The cross-sections are fitted by minimizing χ^2 taking into account the various correlations between the different final states due to luminosity statistics and due to systematic errors in the luminosity and event selections.

Fitting the hadronic and three lepton pair cross-sections simultaneously one can determine six parameters: the Z° mass, the full width, the peak hadronic cross-section and the three lepton partial widths in units of the hadronic width. The fit result for the six parameters is:

$$M_z = (91.203 \pm 0.021_{exp.} \pm 0.030_{LEP}) \text{ GeV}$$
$$\Gamma_z = (2.558 \pm 0.046) \text{ GeV}$$
$$\sigma^\circ_{had} = (41.23 \pm 0.75) \text{ nb}$$
$$\Gamma_{ee}/\Gamma_{had} = 0.0474 \pm 0.0021$$
$$\Gamma_{\mu\mu}/\Gamma_{had} = 0.0463 \pm 0.0018$$
$$\Gamma_{\tau\tau}/\Gamma_{had} = 0.0474 \pm 0.0017$$

with $\chi^2 = 19.5$ for 28 degrees of freedom. These parameters are generally uncorrelated; the only important correlation is between the hadronic peak cross-section and the total width (-46%).

The second error on the Z° mass is due to the uncertainty in the LEP center-of-mass energy. The possible changes in the beam energy of a few times 10^{-5} from one run period to another are negligible.

The hadronic cross-section and the expectations for two, three and four neutrinos are shown in Fig.2. The cross-sections for the three lepton pair channels are shown in Fig 3a-c

Other parameters, including the partial widths and branching ratios, can be derived from these results. The three leptonic widths agree well with each other as expected from lepton universality.

Assuming lepton universality we repeat the fit for four parameters: M_z, σ°_{had}, Γ_z, $\Gamma_{\ell\ell}/\Gamma_{had}$ using the hadron sample and the lepton sample selected without distinction on the final state flavor. The result is:

$$M_z = (91.197 \pm 0.021_{exp.} \pm 0.030_{LEP}) \text{ GeV}$$

$$\sigma^\circ_{had} = (41.30 \pm 0.75) \text{ nb}$$

$$\Gamma_z = (2.562 \pm 0.047) \text{ GeV}$$

$$\Gamma_{\ell\ell}/\Gamma_{had} = 0.0473 \pm 0.0012$$

where $\chi^2 = 8.8$ for 12 degrees of freedom. Equivalently the results can be expressed by $\Gamma_{\ell\ell}$ and $\Gamma_{inv}/\Gamma_{\ell\ell}$ rather than by σ°_{had} and Γ_Z:

$$\Gamma_{\ell\ell} = (85.1 \pm 1.7) \text{ MeV}$$

$$\Gamma_{inv}/\Gamma_{\ell\ell} = 5.91 \pm 0.33$$

In Fig.3d the cross-sections for the lepton pairs are shown.

2.2 The Measurement of the Number of Light Neutrino Species

The number of light neutrino species can be obtained from the measured hadron peak cross-section assuming $\Gamma_{inv} = N_\nu \Gamma_\nu$. With

$$\Gamma_z = \Gamma_{had} + 3\Gamma_{\ell\ell} + N_\nu \Gamma_\nu$$

$$N_\nu = \frac{\Gamma_{\ell\ell}}{\Gamma_\nu} \left(\sqrt{\frac{12\pi}{M_z^2 \, \sigma^\circ_{had}} R} - R - 3 \right)$$

where R stands for $\Gamma_{had}/\Gamma_{\ell\ell}$.

Using the Standard Model prediction for $\Gamma_{\ell\ell}/\Gamma_\nu = 0.5010 \pm 0.0005$, $R = 20.93 \pm 0.21$ and our measured values of M_z and σ°_{had} one finds:

$$N_\nu = 2.99 \pm 0.14_{exp} \pm 0.03_{theory}.$$

The second error comes from the uncertainty in the prediction of R which is dominated by the uncertainty in the QCD correction. If we use instead our measured value of $R = 21.10 \pm 0.58$ we obtain:

$$N_\nu = 2.97 \pm 0.16_{exp} \pm 0.03_{theory}.$$

2.3 Comparison with the Standard Model

In the Standard Model both Γ_{had} and $\Gamma_{\ell\ell}$ depend on the top mass while the QCD correction only affects Γ_{had}. This prediction is shown in Fig.4 for top masses from 50 GeV to 250 GeV. Our measurements are consistent with the Standard Model favoring a larger value for the top mass.

The ratios $\Gamma_{had}/\Gamma_{\ell\ell}$ and $\Gamma_{inv}/\Gamma_{\ell\ell}$ are almost independent of the top and Higgs masses and therefore permit a very powerful test of the Standard Model. The only uncertainty in the prediction comes from the QCD correction to the hadronic cross-section. In Fig.5 the correlation between the two ratios $\Gamma_{had}/\Gamma_{\ell\ell}$ and $\Gamma_{inv}/\Gamma_{\ell\ell}$ is shown together with the prediction; again the data are well consistent with the Standard Model prediction for 3 neutrinos including the QCD corrections.

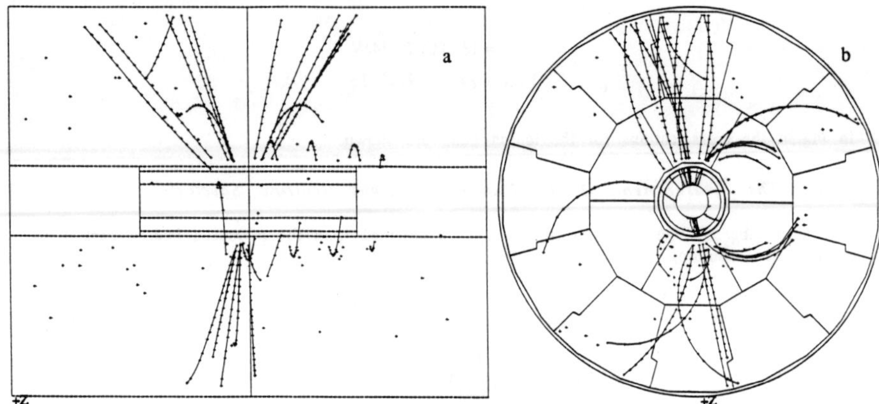

Figure 1: An event displayed by the inner tracking chamber and Time Projection Chamber taken during the pilot run: (a) the R-Z view, and (b) the R-ϕ view.

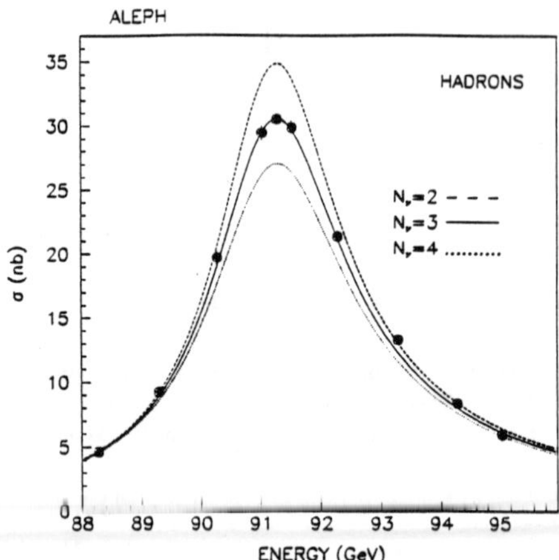

Fig.2 Cross section for $e^+e^- \rightarrow$ hadrons as a function of the centre-of-mass energy. The Standard Model predictions for $N_\nu = 2,3$ and 4 are shown

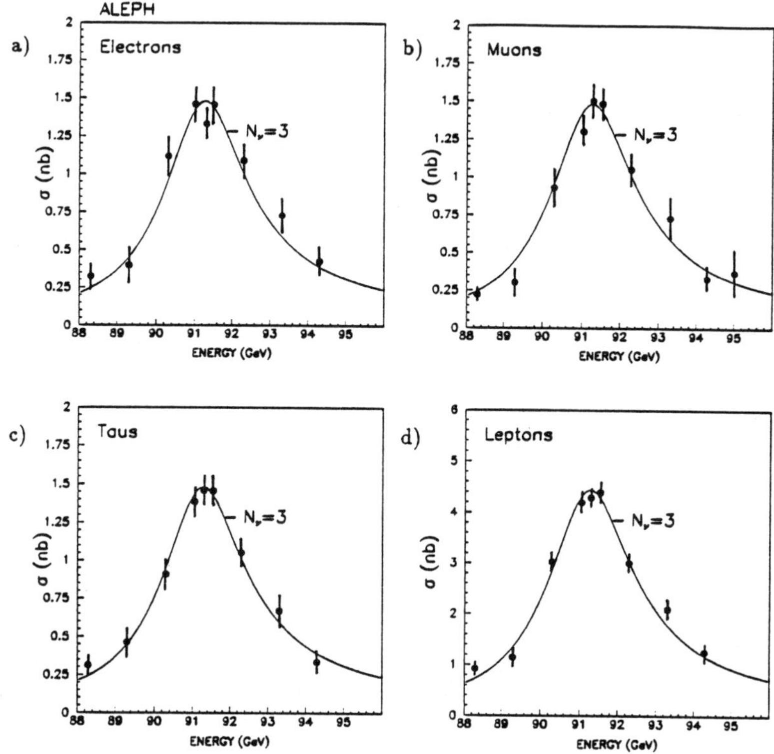

Fig.3 Cross sections for $e^+e^- \rightarrow$ lepton pairs as a function of the centre-of-mass energy. The Standard Model prediction for $N_\nu = 3$ is shown. a) $e^+e^- \rightarrow e^+e^-$, b) $e^+e^- \rightarrow \mu^+\mu^-$, c) $e^+e^- \rightarrow \tau^+\tau^-$ and d) $e^+e^- \rightarrow$ lepton pairs.

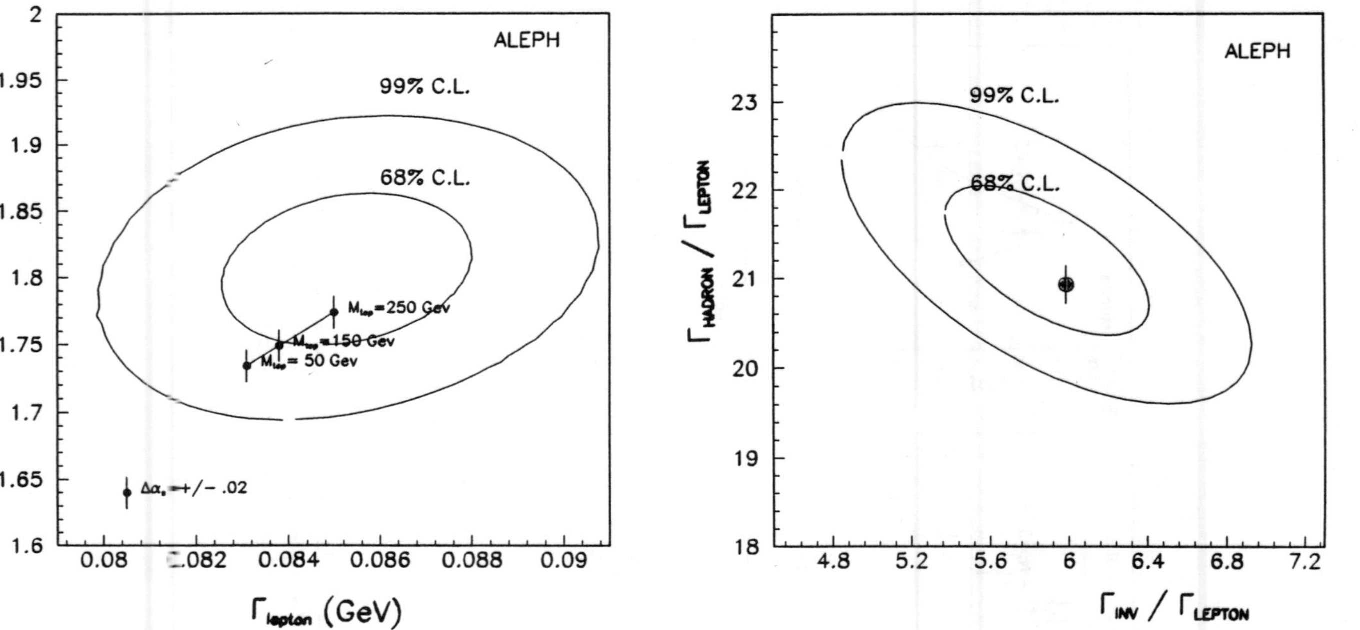

Fig.4 Contours of constant χ^2 for Γ_{had} versus Γ_{ll} together with the Standard Model predictions for various top masses. The uncertainty due to the QCD correction is indicated by the error bars.

Fig.5 Contours of constant χ^2 for Γ_{had}/Γ_{ll} versus Γ_{inv}/Γ_{ll} together with the Standard Model prediction. The uncertainty in the QCD correction is indicated by the error bar.

2.4 The Determination of $\sin^2\theta_w$

The average leptonic partial width can be related to the effective weak mixing angle $\sin^2\theta_w$ in the following manner:

$$\Gamma_{\ell\ell} = (1+\kappa)\frac{\alpha(M_Z)\ M_Z\ (1+(1-4\sin^2\theta_w(M_Z))^2)}{48\sin^2\theta_w(M_Z)\cos^2\theta_w(M_Z)}$$

where $\kappa = (0.2\pm0.3)\%$ represent additional electroweak effects [8] and $\alpha(M_Z)$ is the QCD coupling constant. From our measurement of $\Gamma_{\ell\ell}$ we find:

$$\sin^2\theta_w(M_Z) = 0.227\pm0.005$$

This determination is insensitive to assumptions on the top quark mass, the Higgs boson mass or the Higgs structure of the theory.

Section 3: SEARCH FOR HIGGS BOSONS

3.1 Search for the Neutral Higgs Boson of the Standard Model

The Electroweak Theory of Glashow, Weinberg and Salam [9] has been very successful, giving many theoretical predictions that have subsequently been verified experimentally. The neutral Higgs boson [10], which plays an essential role in the theory, has however not yet been seen. It is fundamentally different from all the other known particles: its observation is essential to our understanding of the "mass" problem.

The ALEPH Collaboration has carried out the search for the neutral Higgs boson twice. The first search [11] is based on the data collected during the early run of LEP, from September 19 to November 7, 1989, which correspond to 11,550 $Z^\circ\rightarrow$hadrons. In the second search [12], we use all the data collected up to the LEP shutdown on December 19, 1989, corresponding to about 25,000 $Z^\circ\rightarrow$hadrons. These data were taken over a range of the center-of-mass energies from 88.3 GeV to 95.0 GeV. In this section 3.1, these two searches are described briefly.

All searches for the neutral Higgs boson at LEP are based on the process

$$Z^\circ\rightarrow H^\circ Z^*,$$

where Z^* is a virtual Z° which decays into a fermion pair. In the first search, we include in the analysis all the decay modes of the Z^*. From the Standard Model, the relative branching ratios of $Z^\circ\rightarrow H^\circ e^+e^-$, $Z^\circ\rightarrow H^\circ\mu^+\mu^-$, $Z^\circ\rightarrow H^\circ\tau^+\tau^-$, $Z^\circ\rightarrow H^\circ\nu\bar{\nu}$ and $Z^\circ\rightarrow H^\circ q\bar{q}(g)$, normalized to the total rate of H° production from Z° decays, are 3.4%, 3.4%, 3.4%, 20.1% and 69.7% respectively.

A detailed account of the analysis for the first search is given in Ref. [11]. For the study below the $H^\circ\rightarrow\mu^+\mu^-$ threshold, i.e. for a Higgs mass below 212 MeV where $H^\circ\rightarrow e^+e^-$ dominates, the Higgs boson is long-lived. For this mass range, we also perform a search with the identification of isolated vertices (V°). Fig.6 gives the number of events expected from the sum, as well as the individual processes, as a function of the Higgs boson mass over the full region of search. While the process $Z^\circ\rightarrow H^\circ\nu\bar{\nu}$ provides the most powerful means for this search, the result of combining all the processes excludes the neutral Higgs boson over the mass range from 32 MeV to 15 GeV at 95% C.L. and 40 MeV to 12 GeV at 99% C.L. This result is obtained within the context of the Standard Model with no further assumption.

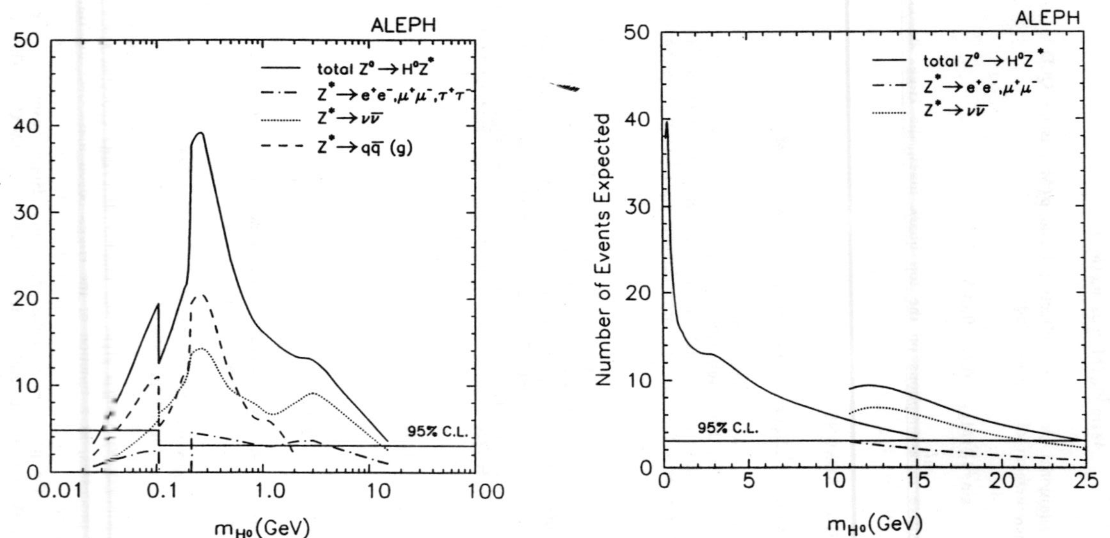

Fig.6   The number of events expected for $Z^\circ \to H^\circ\,\ell^+\ell^-$, $(e^+e^-$, $\mu^+\mu^-$ and $\tau^+\tau^-)$, $Z^\circ \to H^\circ\nu\bar\nu$, $Z^\circ \to H^\circ q\bar{q}(g)$ and the sum of all the above channels as a function of the Higgs mass. The 95% C.L. limit is also indicated corresponding to one observed candidate below 100 MeV and zero above 100 MeV, giving the excluded region of 32 MeV to 15 GeV.

Fig.7   The number of events expected for $Z^\circ \to H^\circ\,\ell^+\ell^-$ (ℓ = e, μ), $Z^\circ \to H^\circ\nu\bar\nu$ and the sum of all the above channels as a function of the Higgs mass for $M_{H^\circ} > 10$ GeV. The 95% C.L. limit at 3 expected events is indicated, corresponding to zero observed candidates. Below 15 GeV the sum of $Z^\circ \to H^\circ\,\ell^+\ell^-$ (ℓ = e, μ or τ), $Z^\circ \to H^\circ\nu\bar\nu$, and $Z^\circ \to H^\circ q\bar{q}$ is shown for the results of the search in Ref. [11].

The second search extends the first one to higher masses of the neutral Higgs boson by using twice the statistics, corresponding to a total integrated luminosity of 1.2pb^{-1}. Only the decay modes $Z^\circ \to H^\circ \nu\bar{\nu}$ and $Z^\circ \to H^\circ \ell\ell$ (ℓ = e and μ) are used. Event selection criteria are similar to those given in Ref. [11] but are modified to optimize the detection efficiency of a Higgs with mass in the range between 11 GeV and 24 GeV. The details of the analysis are given in Ref. [12].

Fig.7 shows the number of events expected from the sum, as well as the individual processes, for $Z^\circ \to H^\circ e^+e^-$, $Z^\circ \to H^\circ \mu^+\mu^-$ and $Z^\circ \to H^\circ \nu\bar{\nu}$ as a function of the Higgs masses. The mass range from 11 GeV to 24 GeV for the neutral Higgs boson is excluded at 95% C.L. from the second analysis. Fig.7 also presents the number of events expected from the sum of all processes as a function of the Higgs boson mass from the results in Ref. [11], giving a combined excluded mass range from 32 MeV to 24 GeV at 95% C.L. limit.

3.2. Limits for the Neutral Higgs Boson from Supersymmetry.

In the minimal supersymmetric extension of the Standard Model, two doublets of Higgs fields are needed, leading to five physical states: two charged scalar bosons heavier than the W; two neutral scalar bosons, one heavier than the Z°, the other, denoted h, lighter than the Z°; one neutral pseudoscalar boson, denoted A, heavier than h. All masses and couplings depend on only two parameters, here chosen as M_h and either M_A or v_2/v_1, the ratio of the vacuum expectation values of the two Higgs doublets [13].

The scalar h behaves very similarly to the standard Higgs boson H°. Limits on the latter can therefore be used to reduce the allowed domain in the space of parameters (M_h, v_2/v_1) or (M_h, M_A), but the sensitivity of this method is reduced when v_2/v_1 is different from unity. However, in this case, the rate of the process $Z^\circ \to hA$ may become substantial and such an analysis has been carried out by the ALEPH Collaboration [14] using all of the data collected in 1989. In our search for $Z^\circ \to hA$, with at least one of h or A decaying into a τ pair, no candidate is found.

The excluded domains, in the (M_h, v_2/v_1) plane, are presented in Fig.8(a): curve (A) limits the domain excluded by our standard Higgs searches; curve (B) limits the domain excluded by the dedicated search for $Z^\circ \to hA$. For $v_2/v_1 > 1$, an alternate presentation of the same results in the (M_h, M_A) plane is given in Fig.8(b). If, as theoretically expected [15], v_2/v_1 is substantially greater than unity, the range $M_h \simeq M_A < 39.2$ GeV is excluded.

3.3 Search for Charged Higgs Boson

The minimal extension of the Standard Model from one having a single complex doublet to one having two such doublets introduces a pair of charged Higgs bosons, H^+ and H^-. A partial width for $H^\pm$ pair production from Z° decay may be obtained without any further specification of the model. This partial width is:

$$\Gamma[Z^\circ \to H^+H^-] = \frac{G_f M_{Z^\circ}^3}{6\sqrt{2}\pi} \left(\frac{1}{2} - \sin^2\theta_w\right)^2 \beta_{H^\pm}^3 \,,$$

where $\beta_{H^\pm} = \sqrt{1 - 4M_{H^\pm}^2/s}$. This gives, for example, a branching ratio of 0.26% for $M_{H^\pm} =$ 35 GeV at $s = M_Z^2$.

We consider the following processes:

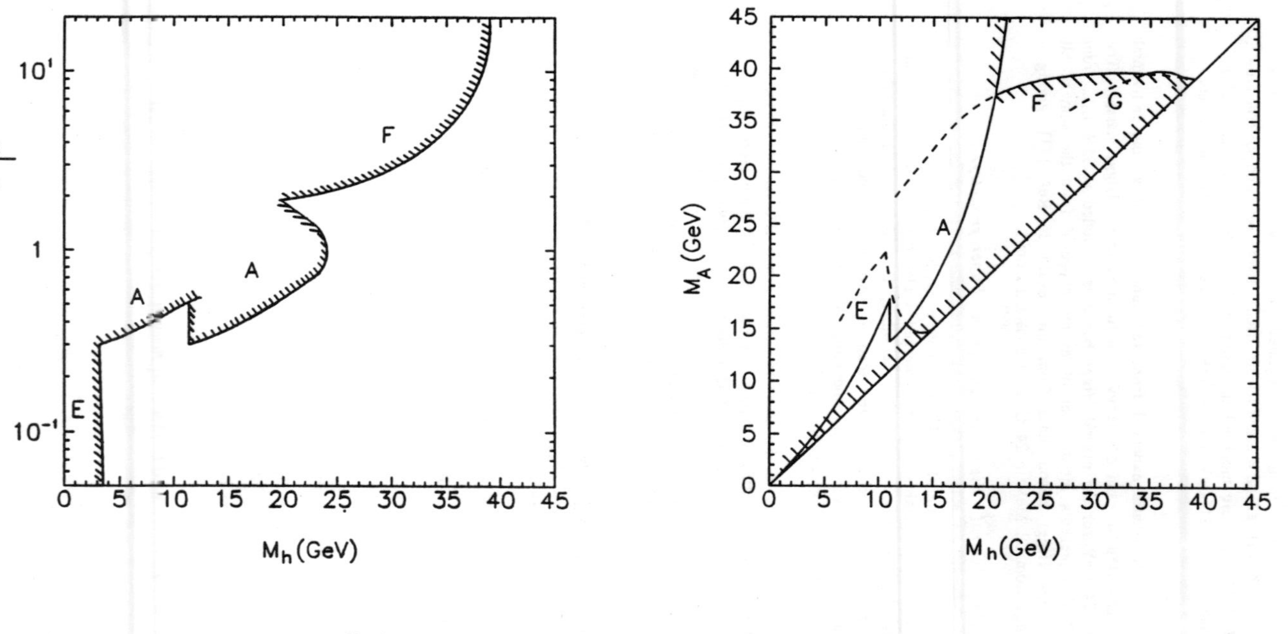

Fig.8   Excluded domains (a) in the $(M_h, v_2/v_1)$ plane, (b) in the (M_h, M_A) plane for $v_2/v_1 > 1$, from:

(A) :   the search for H°
(E) :   low multiplicity final states in $Z^\circ \to hA$

(F) : $\tau^+\tau^-$ jet-jet final states in $Z^\circ \to hA$

(G) :   4 -jet final states in $Z^\circ \to hA$

$$Z^\circ \to \begin{cases} H^+ H^- \to \nu\bar\tau\,\bar\nu\tau \\ H^+ H^- \to \nu\bar\tau\,\bar{c}s \text{ or } H^+ H^- \to \nu\bar\tau\,\bar{c}b \\ H^+ H^- \to c\bar{s}\,\bar{c}s \text{ or } H^+ H^- \to c\bar{b}\,\bar{c}b. \end{cases}$$

Details of searches for these processes are given in Ref. [16]. The two hadronic decay modes $H^\pm \to cs$ and $H^\pm \to cb$ are required to satisfy identical selection criteria, and for each of the processes above which involve hadronic decays two distinct limit contours are derived. The analyses have been performed using all the data obtained by the ALEPH detector at LEP in 1989.

The individual limit contours for the decay channels $H^+H^- \to \nu\bar\tau\,\bar\nu\tau$, $H^+H^- \to \nu\tau\,cs$ (or $\nu\tau\,cb$) and $H^+H^- \to c\bar{s}\,\bar{c}s$ (or $c\bar{b}\,\bar{c}b$) are shown in Fig.9 (or Fig.10). All limits are obtained without the subtraction of expected background. Previous work has excluded the charged Higgs below 19 GeV [17]. In the analyses presented here, the lower limits on $M_{H^\pm}$ are due to the following tendencies in the charged Higgs signatures which reduce the signal detection efficiencies: the increasing collinearity of the final state particles for $H^+H^- \to \nu\bar\tau\,\bar\nu\tau$, the decreasing acoplanarity of the isolated charged particle for $H^+H^- \to \nu\tau\,cs$ ($\nu\tau\,cb$), and the increasing two-jet-like behavior for $H^+H^- \to c\bar{s}\,\bar{c}s$ ($c\bar{b}\,\bar{c}b$). The upper limits are due primarily to the decrease in production rate. The requirement that the angle between the jets $\theta_{jets} <$ 135° for $H^+H^- \to \nu\tau\,cs$ ($\nu\tau\,cb$) and the difficulty in accurately reconstructing four jets for $H^+H^- \to c\bar{s}\,\bar{c}s$ ($c\bar{b}\,\bar{c}b$) also reduce the signal detection efficiency for these decay channels at the larger values of $M_{H^\pm}$.

To extend the limit outside the bounds described by the individual contours, the limits obtainable with any two or all three of the individual contours are calculated. This extended limit is found by using the sum of the number of observed events, N_o^{sum}, and the sum of the expected signals, μ_s^{sum}, from the individual contours. The combination of N_o^{sum} and μ_s^{sum} giving the best limit on the branching ratio is used to make the combined contours shown by the heavier lines in Figs.9 and 10.

To summarize, Fig.9 indicates that at $BR[H^\pm \to \nu\tau] = 100\%$ the charged Higgs is excluded at 95% C.L. in the mass range 7.6 to 43.0 GeV, at $BR[H^\pm \to \nu\tau] = BR[H^\pm \to cs] = 50\%$ in the range 8.3 to 40.6 GeV and at $BR[H^\pm \to cs] = 100\%$ in the range 14.4 to 35.4 GeV. With cs replaced by cb, Fig.10 shows that at $BR[H^\pm \to \nu\tau] = BR[H^\pm \to cb] = 50\%$ the charged Higgs is excluded at 95% C.L. in the range 12.0 to 40.7 GeV, and at $BR[H^\pm \to cb] = 100\%$ in the range 16.2 to 35.7 GeV.

Section 4: SEARCH FOR NEW QUARKS AND LEPTONS

Using data recorded during the first period of LEP operation from September 19 to November 7, 1989, we have performed searches for a top quark (t), a fourth-generation charge $-\frac{1}{3}$ bottom type quark (b'), a sequential heavy charged lepton ($L^\pm$), an unstable neutral heavy lepton (L°) and a stable neutral heavy lepton (ν_L) from Z° decay. In all these cases, a Z° is assumed to decay into a pair of these new quarks or new leptons with a decay rate given by the Standard Model. The processes studied are:

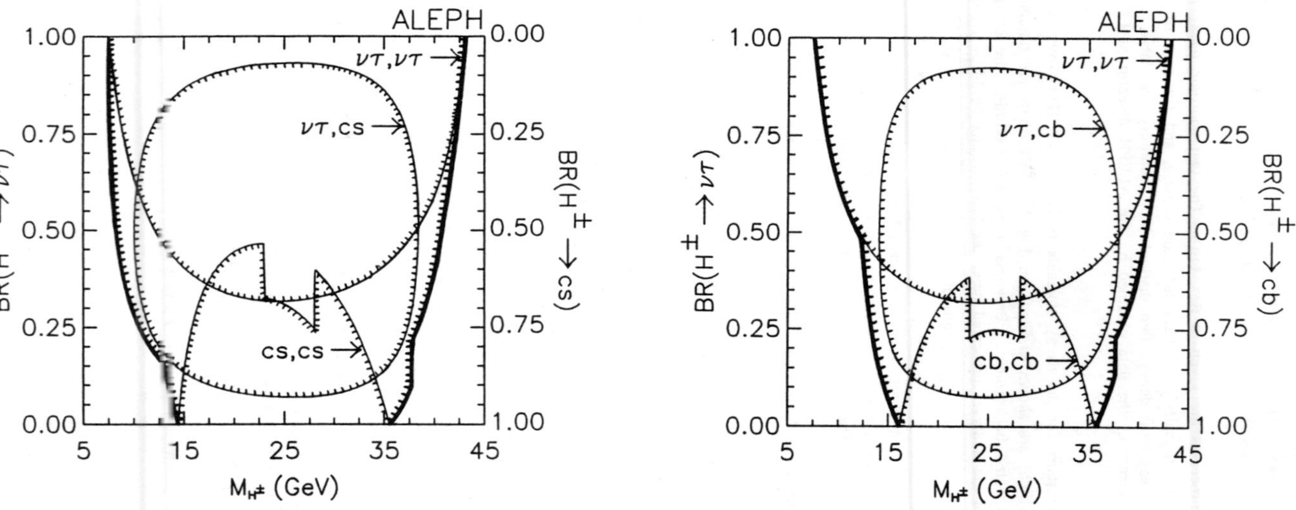

Fig.9   The 95% C.L. limit on the mass of the charged Higgs as a function of $BR[H^{\pm} \to \nu\tau]$ and $BR[H^{\pm} \to cs]$.   The region on the hatched side of the contour is excluded. The combined limit for all channels is indicated by the heavier lines.

Fig.10   The 95% C.L. limit on the mass of the charged Higgs as a function of $BR[H^{\pm} \to \nu\tau]$ and $BR[H^{\pm} \to cb]$.   The region on the hatched side of the contour is excluded. The combined limit for all channels is indicated by the heavier lines.

$$Z^\circ \to t \, \bar{t} \qquad (1)$$

$$Z^\circ \to b' \, \bar{b}' \qquad (2)$$

$$Z^\circ \to L^+ L^- \qquad (3)$$

$$Z^\circ \to L^\circ \bar{L}^\circ \qquad (4)$$

$$Z^\circ \to \nu_L \bar{\nu}_L \qquad (5).$$

The decays of t, b', $L^\pm$, L° are via a charged-current process into a virtual W boson (W*):

$$
\begin{array}{lll}
t & \to & b\,W^* \\
b' & \to & c\,W^* \\
L^\pm & \to & \nu_L W^* \quad (\nu_L \text{ stable}) \\
L^\circ & \to & \ell\,W^* \quad (\ell = e, \mu, \tau).
\end{array}
$$

In addition, the decays of b' via flavor-changing neutral-current are also considered:

$$b' \to b\gamma$$

and

$$b' \to bg.$$

As already discussed in Section 2.2, the number of light neutrino species is found to be 2.99 ± 0.14. This result implies that there is no fourth-generation light neutrino. Because of the high accuracy of this result, one can set a limit on the mass of a stable heavy fourth-generation neutrino (or neutral lepton). On the other hand, if the fourth-generation neutral lepton is unstable, then this result gives a rather poor limit on the mass. For this reason, it is desirable to carry out a direct search for unstable neutral leptons (L°). We consider the case where there is mixing between the neutral heavy lepton and the three light neutrinos. If the mixing parameters of L° with ν_e, ν_μ, and ν_τ are U_{eL°, $U_{\mu L^\circ}$ and $U_{\tau L^\circ}$ respectively, then L°, which is the decay product of the Z° as indicated by (4), contains the terms

$$U_{eL^\circ}|\nu_e\rangle + U_{\mu L^\circ}|\nu_\mu\rangle + U_{\tau L^\circ}|\nu_\tau\rangle. \qquad (6)$$

With such mixing, the heavy neutral lepton L° can decay into a virtual W together with an e, a μ or a τ.

The study of processes (1) to (5) is performed using data collected with the ALEPH detector, corresponding to 11,550 $Z^\circ \to$ hadrons events. This section 4 describes the search for processes (1) to (5), with the following topologies or methods:

(i) Search for charged-current decays of t, b' and L° by the topology of an isolated charged particle. The isolated charged particle is the $e^\pm$, $\mu^\pm$ or one prong decays of the $\tau^\pm$, resulting from the leptonic or semi-leptonic decay of the new quark or new lepton.

(ii) Search for the flavor-changing neutral-current decay of b' by a topology of an isolated photon ($b' \to b\gamma$) or of a four-jet final state ($b' \to bg$, $b' \to \bar{b}g$).

(iii) In the case of $Z^\circ \to L^\circ \bar{L}^\circ$, when the lifetime of L° is long, as explained below in Section 4.3, the topology of the events could be one or two displaced vertices.

(iv) Using the total hadronic cross-section measurement at the Z° peak, mass limits can be set for ν_L and $L^\pm$ on the processes $Z^\circ \to \nu_L \bar{\nu}_L$ and $Z^\circ \to L^+ L^-$, where $L^\pm \to \nu_L W^*$. For the long-lived L°, limits can be set on $|U_{\ell L^\circ}|^2$ as a function of M_{L°.

4.1 Search for Charged-Current Decay of t, b′ and L° by a Topology of an Isolated Charged Particle

A search for new quarks (t, b′) and new neutral heavy leptons (L°) is performed by identifying a topology where there is an isolated charged particle in an event from the leptonic or semi-leptonic decay of the t, b′ or L°.

Figs. 11(a) to (c) give the expected number of events after applying the event selection criteria, as given in Ref. [18], for t, b′ and L° as a function of mass with the 95% C.L. limits indicated. In all cases we set the limit using the number of events expected minus 1.64 (one-sided 95% C.L. limit) times the total systematic error. For the charged-current decays of t, b′, and L°, the following regions of mass are excluded at 95% C.L.:

$$26.0 \text{ GeV} < M_t < 45.8 \text{ GeV}$$
$$26.0 \text{ GeV} < M_b' < 46.2 \text{ GeV}$$
$$25.0 \text{ GeV} < M_{L°} < 45.7 \text{ GeV}.$$

These results extend the earlier ones from TRISTAN [19]. Those on the quarks t and b′ overlap with the results from proton-antiproton colliders [20], and are similar to the recent ones from MARK II [21] and OPAL [22]. The result on the neutral heavy lepton extends the limit given by MARK II [21].

4.2 Search for Flavor-Changing Neutral-Current Decay of b′ by a Topology of an Isolated Photon or of a Four-Jet Final State.

If a b′ exists with mass less than half of the Z° mass and with the present limit of t mass being larger than half of the Z° mass, the charged-current decay is via b′→cW*, which is expected to be suppressed by the small coupling V_{cb}'. The result, shown in Fig.11(b), rules out the existence of a b′ for mass below 45.0 GeV if the rate of the charged-current decay is larger than 10%. It has been pointed out that flavor-changing neutral-current b′ decays could occur with a sizeable rate [23]; such decays are expected to be dominated by b′→bγ and b′→bg. A search for b′→bγ and b′→bg is described in Ref. [18].

(i)   b′→bγ

Similar analysis can apply to both cases: (1) Z°→b′$\bar{b}'$ where both b′ decay to bγ, and (2) Z°→b′$\bar{b}'$ where one b′ decays to bγ and the other to bg. In both cases, we search for a high-energy isolated photon in an event. With the event selection criteria given in Ref. [18], 4 events are observed in the data while a background of 2.0 is expected from the background simulation of Z°→ $q\bar{q}(g)$ based on five quarks. Since a 95% C.L. limit corresponds to 7.2 expected events, we can exclude the branching ratio of b′→bγ greater than 5% for a b′ mass between 26.0 GeV and 46.0 GeV at 95% C.L. as shown in Fig.12 (left-hand side coordinate).

(ii)   b′→bg

A Z°→b′$\bar{b}'$ event with both b′ decaying into bg gives a four-jet final state. With the event selection criteria given in Ref. [18], we observe 6 events in the data and expect 6.2 background events from the Monte Carlo simulation of the background process Z°→$q\bar{q}(g)$ for 5 quarks. Since the 95% C.L. limit corresponds to 6.6 expected events, we exclude the branching ratio of b′→bg greater than 65% for b′ mass between 26.0 GeV and 46.0 GeV at 95% C.L. as shown in Fig.12 (right-hand side coordinate).

In conclusion, if the sum of the branching ratios of b′ decays into $c\ell\nu$ (ℓ = e, μ,τ) by charged current and into bγ or bg by flavor- changing neutral current is assumed to be 100%, then the mass range of b′ from 26 GeV to 46 GeV is excluded at 95% C.L. This result agrees well with Refs. [21] and [22].

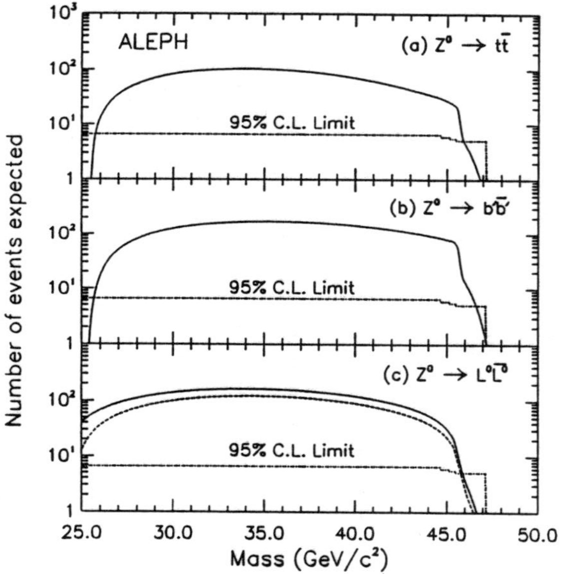

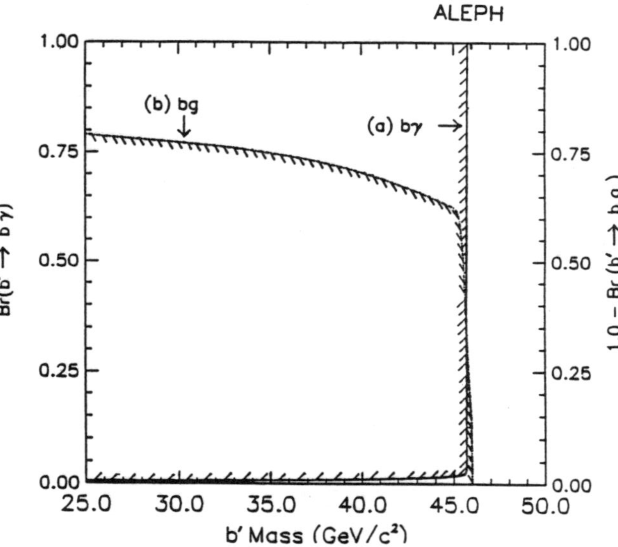

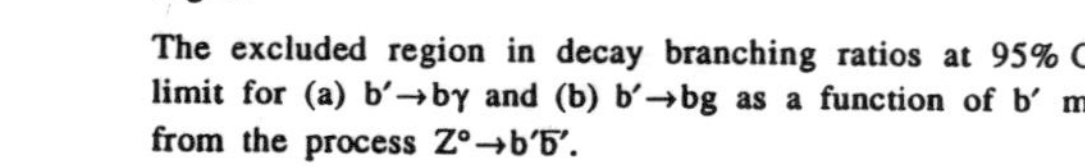

Fig.11 (a) The number of expected events for $Z^\circ \to t\bar{t}$ after applying the event selection criteria as a function of t mass. The 95% C.L. limit is indicated.

(b) The same as (a) but for $Z^\circ \to b'\bar{b}'$.

(c) The same as (a) but for $Z^\circ \to L^\circ \bar{L}^\circ$. The solid curve is for $L^\circ \to eW^*$ or μW^* and the dashed curve for $L^\circ \to \tau W^*$.

Fig.12

The excluded region in decay branching ratios at 95% C.L. limit for (a) $b' \to b\gamma$ and (b) $b' \to bg$ as a function of b' mass from the process $Z^\circ \to b'\bar{b}'$.

4.3 Search for $Z° \to L°\bar{L}°$ from Displaced Vertices

The search for an L° using the isolated particle topology, as described in Section 4.1, requires the mixing parameters $|U_{\ell L°}|^2$ to be large enough that L° decays very close to the e^+e^- interaction point. The lifetime of the L° is inversely proportional to the mixing parameters $|U_{\ell L°}|^2$ when L° couples to a lepton ℓ by:

$$\tau(L° \to \ell^- X^+) = (\frac{m_\mu}{m_{L°}})^5 \frac{\tau_\mu Br(L° \to \ell^- e^+ v_e)}{f(m_{L°}, \ell, X)|U_{\ell L°}|^2} \qquad (7)$$

The factor $f(m_{L°}, \ell, X)$ is a phase-space correction, which is taken to be 1, as the smallest L° mass being considered is 20 GeV. $Br(L° \to \ell^- e^+ v_e)$ is taken to be 11% in this mass range. When the mixing to ℓ is small ($|U_{\ell L°}|^2 \approx 10^{-10}$), the $Z° \to L°\bar{L}°$ signature is one or two vertices, separated from the e^+e^- interaction point, and with no charged particles coming from the interaction point. In the case of two vertices, these are expected to be collinear with the intersection point and on opposite sides of it. A search has been performed to look for such vertices. Details are given in Ref. [18].

The contour (b) in the $|U_{\ell L°}|^2$ versus $M_{L°}$ plane, as shown in Fig.13, corresponds to three expected events (zero observed events) and gives the limit of the excluded region for $Z° \to L°\bar{L}°$ at 95% C.L. This limit applies equally well to L° mixing to e, μ and τ, since the kinematic differences between these decays have a negligible effect upon the vertex finding algorithm. Superimposed in Fig.13 is the 95% C.L. limit contour (a) from the result of the search for prompt L° as described in section 4.1. Hence an L° is excluded up to a mass of 46.0 GeV with mixing parameters $|U_{\ell L°}|^2$ as small as 10^{-13}. This is two orders of magnitude more sensitive than previously reported results [24].

4.4. Mass Limits on New Quarks and New Leptons from the Total Hadronic Cross-Section Measurement at the $Z°$ Peak

From the measurement of the peak hadronic cross-section $\sigma°_{had}$ by the ALEPH Collaboration, the number of light neutrino species is found to be $N_v = 3.01 \pm 0.16$ [1]. This result can be used to set a mass limit on t, b′, $L^\pm$, stable v_L and unstable L°. The updated values of N_v given in Section 2.2 change the result only slightly.

(i) $Z° \to v_L \bar{v}_L$

For $Z° \to v_L \bar{v}_L$, where v_L is the stable fourth-generation neutrino, the result $N_v = 3.01 \pm 0.16$ excludes at 95% C.L. the region where

$$\Gamma_{v_L} > [(3.01-3.00)+1.64 \times 0.16]\ \Gamma_{v_e} = 0.272\ \Gamma_{v_e} \qquad (8).$$

Here Γ_{v_L} and Γ_{v_e} are the partial widths for $Z°$ decays into $v_L \bar{v}_L$ and $v_e \bar{v}_e$ respectively. The factor 1.64 corresponds to the one-sided 95% C.L. limit. The partial width $\Gamma_{1/2}$ for the $Z°$ decaying into a pair of spin one-half particles is a function of the mass of the particle.

$$\Gamma_{1/2} = \frac{N}{24\pi} \frac{G_F M_z^3}{\sqrt{2}} \beta [\frac{1}{2} (3 - \beta^2) v^2 + a^2 \beta^2] \qquad (9)$$

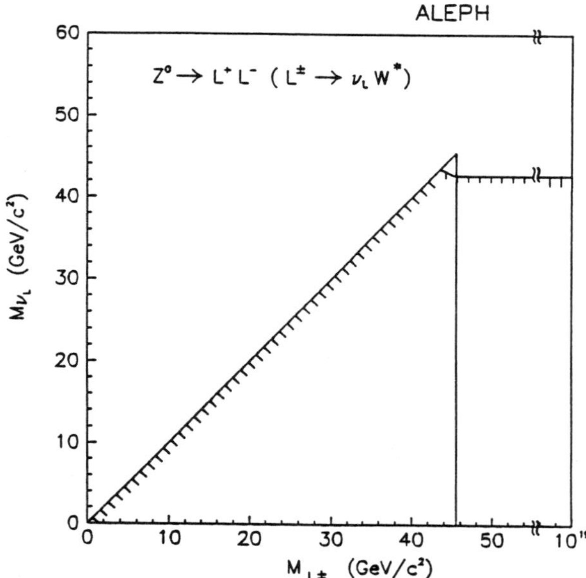

Fig.13 The excluded regions at 95% C.L. for $Z^{\circ} \to L^{\circ}\bar{L}^{\circ}$ in the $|U_{\ell L^{\circ}}|^2$ versus $M_{L^{\circ}}$ plane from (a) the search for prompt L°, (b) the search for long-lived L° from displaced vertices, and (c) the limit for long-lived L° from the total hadronic cross-section measurement at the Z° peak. The bounded regions denote the areas of exclusion.

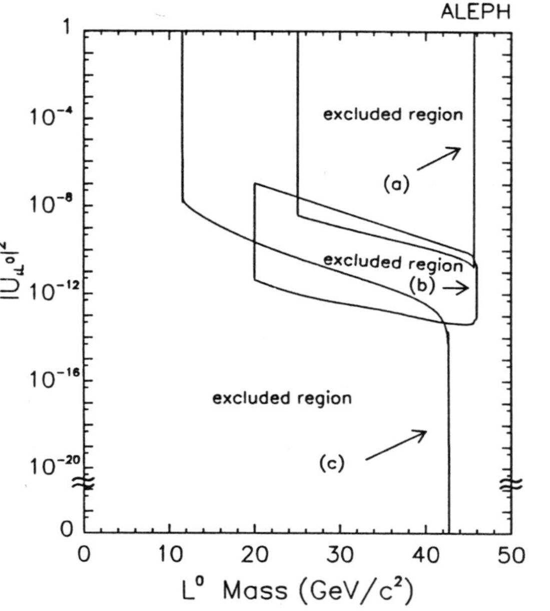

Fig.14

The excluded regions at 95% C.L. for $Z^{\circ} \to L^+L^- \ (L^{\pm} \to \nu_L W^*)$ in the $M\nu_L$ versus $M_{L^{\pm}}$ plane. Note that the region where $M_{L^{\pm}} > M_{Z^{\circ}}/2$ and $M\nu_L < 42.7$ GeV is excluded for all values of $M_{L^{\pm}}$.

246

where N is the color factor, β is the velocity of each of these particles and v and a are the vector and axial vector couplings of the same particle to the Z°. Using eqs (8) and (9) the region where

$$0 \le M_{v_L} < 42.7 \text{ GeV}$$

is excluded at 95% C.L.

(ii) $Z^\circ \to L^+ L^-$

For $Z^\circ \to L^+ L^-$ where $L^\pm \to v_L W^*$, the region where

$$\frac{\Gamma_{had}}{\Gamma_{tot}^2} = \frac{\Gamma_{5 \text{ quarks}} + x\Gamma_{L\pm}}{(\Gamma_{SM} + \Gamma_{v_L} + \Gamma_{L\pm})^2} < \frac{\Gamma_{5 \text{ quarks}}}{(\Gamma_{SM} + 0.272\Gamma_{v_e})^2} \tag{10}$$

is excluded at 95% C.L. The coefficient 0.272 is obtained from eq.(8). Here x is the ratio of the detection efficiency for an $L^\pm$ to be identified as a hadronic event to the detection efficiency for hadronic event using the event selection criteria as described in Ref. [1]. The partial widths $\Gamma_{5 \text{ quarks}}$ and Γ_{SM} take the Standard Model values with three generations: $\Gamma_5 = 1.737$ GeV and $\Gamma_{SM} = 2.487$ GeV.

Using eq. (9) and solving inequality (10) for M_{v_L} in terms of $M_{L\pm}$, we can exclude the full triangle $M_{v_L} < M_{L\pm} < M_Z/2$, as shown in Fig. 14, except for the small area near the top corner.   Using the result given by subsection 4.4(i) above, the region where $M_{L\pm} > M_Z/2$ and $M_{v_L} < 42.7$ GeV is excluded for all values of $M_{L\pm}$ as shown in Fig. 14. This result covers a much larger excluded region in the M_{v_L} versus $M_{L\pm}$ plane than previously reported [25].

(iii) $Z^\circ \to t\bar{t}$, $b'\bar{b}'$ and $L^\circ\bar{L}^\circ$ (L° unstable and short-lived)

For $Z^\circ \to t\bar{t}$, $b'\bar{b}'$ and $L^\circ\bar{L}^\circ$, eq. (10) can be written as

$$\frac{\Gamma_{had}}{\Gamma_{tot}^2} = \frac{\Gamma_{5 \text{ quarks}} + x\Gamma_j}{(\Gamma_{SM} + \Gamma_j)^2} < \frac{\Gamma_{5 \text{ quarks}}}{(\Gamma_{SM} + 0.272\Gamma_{v_e})^2} \tag{11}$$

where Γ_j = partial width of t, b' or L° and x is essentially one.   Applying eq. (9) to eq. (11), the mass regions $M_t < 31.3$ GeV, $M_{b'} < 39.4$ GeV and $M_{L^\circ} < 11.8$ GeV are excluded.   Combining these results with the results from the direct searches described in sections 4.1 and 4.2, the following mass regions are excluded at 95% C.L.:

$$0 \le \quad M_t \quad < \quad 45.8 \text{ GeV}$$
$$0 \le \quad M_{b'} \quad < \quad 46.2 \text{ GeV}$$
$$0 \le M_{L^\circ} < 11.8 \text{ GeV and } 25.0 \text{ GeV} < M_{L^\circ} < 45.7 \text{ GeV}.$$

(iv) $Z^\circ \to L^\circ\bar{L}^\circ$ (L° unstable and long-lived)

We consider an L° which has an average decay length larger than R = 390 cm (the distance between the interaction point and the outer corner of the ECAL). This radius is chosen because it encloses the ITC, TPC and ECAL which are the essential elements for the hadronic event selection used in the measurement of total hadronic cross-section [1]. Eqs. (7) and (9) and inequality (11) are used to determine the excluded

region in the M_{v_L} versus $M_{L\pm}$ plane. The value x in eq. (11) is modified to be the product of the ratio of detection efficiencies as defined previously and the probability that a L° decays inside the radius R. To be conservative the ratio of detection efficiencies is set to one for decay inside R and zero outside. The excluded region in the $|U_{\ell L^\circ}|^2$ versus M_{L° plane at 95% C. L. limit is shown in Fig. 13 (contour c). Combining the results given in Sections 4.1 and 4.3, for
25.0 GeV $< M_{L^\circ} <$ 42.7 GeV all values of $|U_{\ell L^\circ}|^2$ are excluded.

4.5 Search for Excited Leptons

Compositeness [26] has been introduced as a possible solution to Standard Model problems such as the lepton-quark spectrum, the mass generation, the unnaturalness of the Higgs mechanism and the large number of arbitrary parameters. One possible effect of compositeness is the existence of excited states, ℓ^*, of known leptons, ℓ (in this study ℓ = e, μ or τ). If these states carry electroweak charges similar to the ordinary leptons, they will be readily produced at LEP as $\ell^*\ell^*$ pairs if their mass is less than $M_Z/2$. We assume, as is usual, that the branching ratio for ℓ^* decay to $\ell\gamma$ is close to 100% and we search for pair production in the $\ell\ell\gamma\gamma$ final state. Depending on the mass, m*, of the ℓ^* and on the strength of the $\ell^*\ell Z$ coupling, any such new object could also be singly produced in an $\ell^*\ell$ state if m* is less than M_Z and hence observed in the $\ell\ell\gamma$ final state. The details of event selection criteria are given in Ref. [27].

The process $e^+e^- \to \ell^*\ell^*$

Since we have seen no $\ell^*\ell^*$ candidate in any of the three channels we determine the 95% confidence limits for excited lepton masses as 44.6 GeV/c^2, 44.6 GeV/c^2 and 41.2 GeV/c^2 for e*, μ* and τ* respectively. These limits are shown in Fig.15 as vertical lines.

The process $e^+e^- \to \ell^*\ell$

The resulting limits for the three channels are shown in Fig.15 together with previous best limits set for λ^γ/m^* at TRISTAN and at PETRA [28].

Section 5: SEARCH FOR SUPERSYMMETRIC PARTICLES

The search for the neutral Higgs boson from supersymmetry has been described in Section 3.2. In this section, we describe the search for some other supersymmetric particles from Z$^\circ$ decay.

5.1 Search for Supersymmetric Particles Using Acoplanar Charged-Particle Pairs from Z$^\circ$ Decays

We describe here a search for Z$^\circ$ decays to pairs of oppositely-charged particles which are acoplanar (that is, the plane of the charged particles does not include the beam axis) and which have missing energy. Such decays are a signature of the production of pairs of the supersymmetric partners of the known leptons and pairs of charginos, the superpartners of the W$^\pm$ and Higgs boson [29]. These particles are expected to decay immediately to their associated particles and a photino. The photino is assumed here to be the lightest supersymmetric particle, to be stable (as expected from R-parity conservation [30]) and to interact weakly with matter [31]. Thus it escapes detection, carrying away transverse momentum and energy.

The production of a pair of supersymmetric scalar leptons (sleptons, denoted as $\tilde{e}, \tilde{\mu}, \tilde{\tau}$) can occur through the s-channel exchange of a Z$^\circ$ or photon, and also through the t-channel exchange of a photino in the case of scalar electron production. At the Z$^\circ$ peak the contribution from the t-channel is less than 10%, and we include it in the result presented

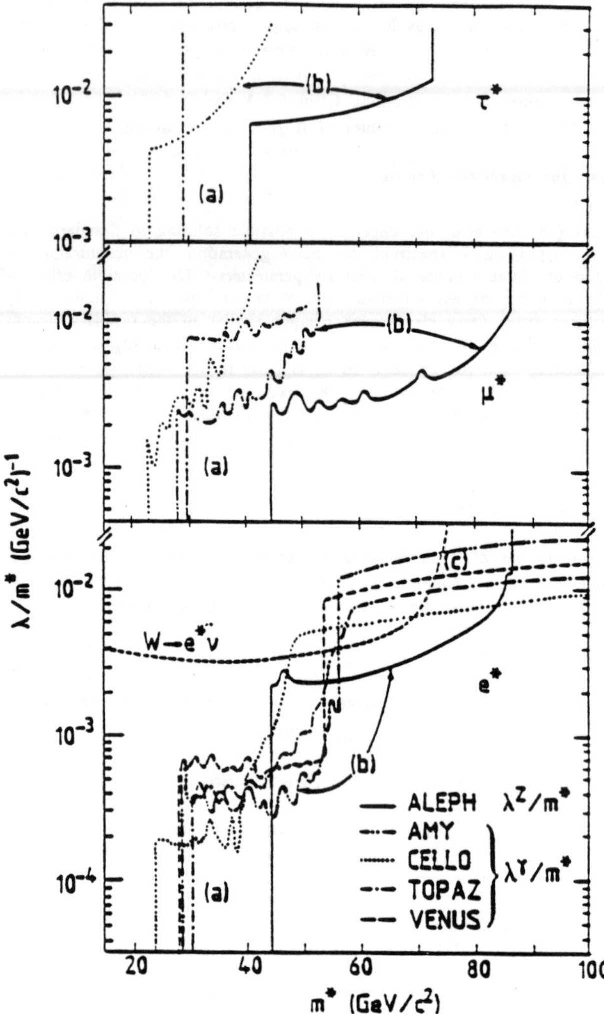

Figure 15: The 95% C.L. upper limits for λ^V/m^* (V = Z for ALEPH, V = γ for previous studies) as a function of ℓ^* mass for e^*, μ^* and τ^*. These limits are derived from the reactions (a) ee→ $\ell\ell\gamma\gamma$, (b) ee→ $\ell\ell\gamma$ (also included ee→eγ(e)) and (c) ee→$\gamma\gamma$. Also shown is the 95% C.L. limit for the coupling We*ν obtained by the UA2 Collaboration from the search of the decay sequence W→e*ν→e$\gamma\nu$. Excluded regions for excited leptons are above and left of the curves.

here. The cross-section depends on the masses of the left- and right-handed sleptons, which may or may not be equal. We present below the limits on these masses for the cases $m\tilde{\chi}_L = m\tilde{\chi}_R$ and $m\tilde{\chi}_L \gg m\tilde{\chi}_R$, which differ by roughly a factor of two in cross-section. The sleptons decay to their associated leptons and a photino.

Pair production of charginos $\tilde{\chi}^{\pm}$, which are the mass eigenstates of admixtures of wino-like and higgsino-like "current" eigenstates, also proceeds through the s-channel. Though no model-independent prediction for the admixture exists, the couplings of the current eigenstates are predicted exactly. We present below our result for the limiting cases of a pure wino and a pure higgsino; the light chargino in any model will lie between these extremes. The coupling of the wino to the $Z°$ is nearly an order of magnitude larger than that of the higgsino [32]. For a pure wino, the cross-section at the $Z°$ pole is on the order of 10 nb, and decreases slowly with increasing wino mass. In this search we are only sensitive to decay modes resulting in one charged particle per chargino. Such final states arise from three-body decays to a lepton, its associated neutrino, and a photino (or, more generally, the lightest neutralino [29]). For simplicity, only decays to electrons and muons in the final state are used to set mass limits; the branching ratios of a chargino to each of these is taken to be 1/9, the minimum of the range given in the model [33].

Details of the data analysis are given in Ref. [34]. Fig.16a shows the region of the $m_{\tilde{e}}$-$m_{\tilde{\gamma}}$ plane excluded by our search to 95% confidence for the two cases $m_{\tilde{e}_L} = m_{\tilde{e}_R}$ and $m_{\tilde{e}_L} \gg m_{\tilde{e}_R}$. In the degenerate-mass (non-degenerate-mass) case we exclude scalar electrons with mass less than 44 (42.5) GeV.

The production of a pair of smuons proceeds through exactly the same mechanism as selectron-pair production, except that there is no contribution from t-channel processes. Fig.16b shows the excluded mass region, which covers nearly the entire kinematically accessible region; for the degenerate-mass (non-degenerate-mass) case, smuons with mass less than 43.5 (41.7) GeV.

For the production of pairs of scalar taus, which also proceeds only through the s-channel, we take advantage of the 85% branching ratio into single charged particles. The efficiency in this channel is lower, however, due to the softer momentum spectrum and the presence of additional photons from $\pi°$'s. Hence we obtain a smaller excluded mass region, shown in Fig.16c. In the degenerate-mass (non-degenerate-mass) case we exclude the region of scalar tau mass below 41.8 (38.3) GeV.

For chargino pairs we arrive at the limits shown in Fig.17, which cover the mass region up to 45.9 GeV in the pure wino case, and to 45 GeV in the limit of a pure higgsino, for photino masses of up to 35 and 30 GeV respectively.

5.2 Mass Limit for Scalar Neutrinos

The partial width for the process $Z° \rightarrow \tilde{\nu}\bar{\tilde{\nu}}$ is related to the partial width of $Z° \rightarrow \nu\bar{\nu}$ by

$$\Gamma(Z° \rightarrow \tilde{\nu}\bar{\tilde{\nu}}) = 0.5 \; \Gamma(Z° \rightarrow \nu\bar{\nu}) \; (1 - \frac{4M_{\tilde{\nu}}^2}{M_Z^2})^{3/2} \; .$$

From equation (8), we have

$$\Gamma^{\text{invisible}}_{\text{new}} < 45.3 \text{ MeV}.$$

In particular, this implies that, if the scalar neutrino is stable and hence invisible,

$$\Gamma(Z° \rightarrow \tilde{\nu}\tilde{\nu}) < 45.3 \text{ MeV}.$$

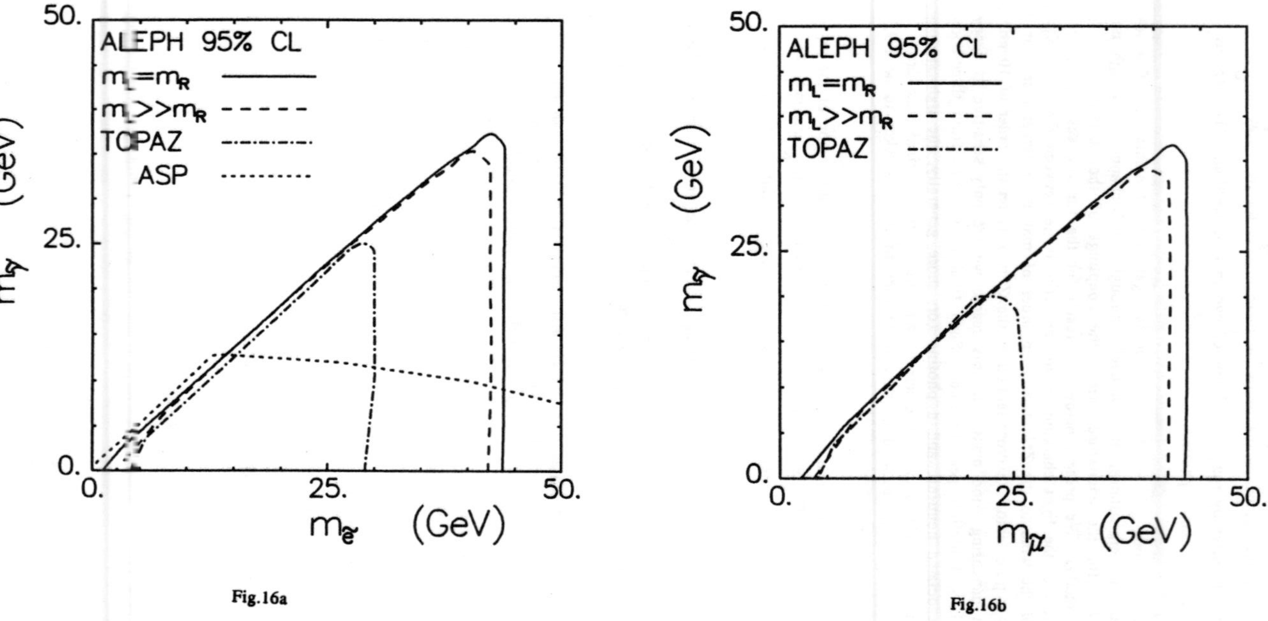

Fig.16 . Region in combined slepton and photino masses excluded to 95% confidence by this search: a, result for scalar electron pair production with previous results from TOPAZ and ASP (the result from ASP is to 90% confidence); b, result for scalar muon pair production with previous result from TOPAZ; c, result for scalar tau pair production with previous result from TOPAZ.

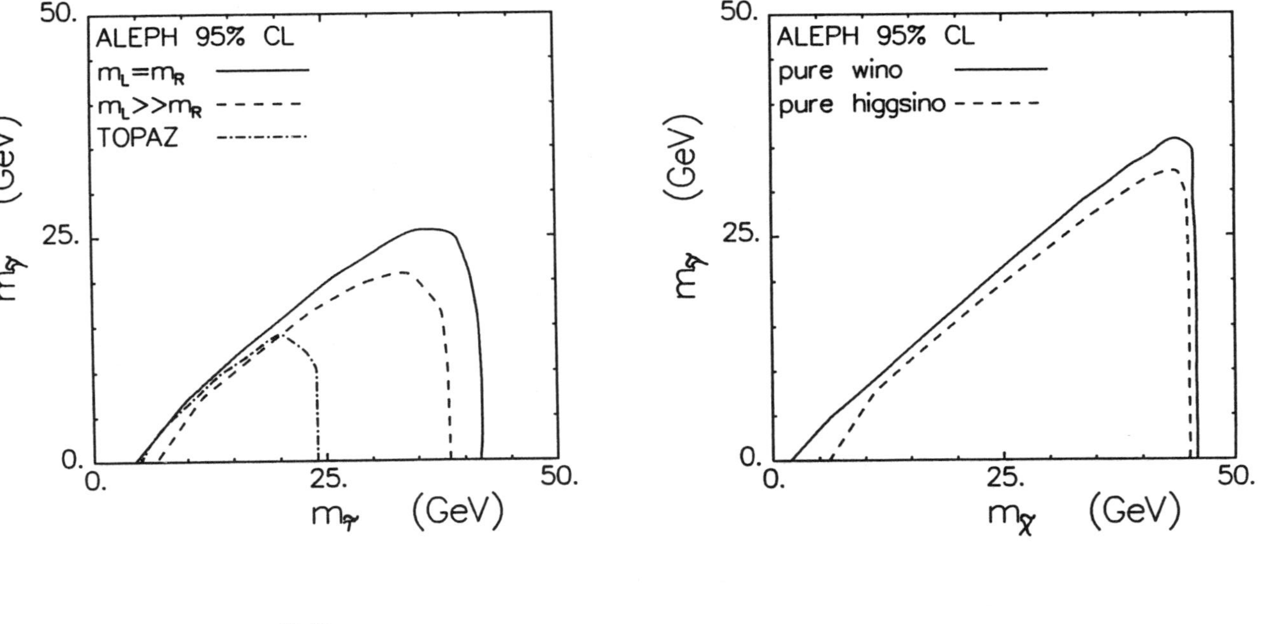

Fig.16c

Fig.17   Region in combined chargino and photino masses excluded to 95% confidence by this search for the extremes of a pure wino and a pure higgsino.

252

From equation (9), we have

$$M_{\tilde{\nu}} > 26.3 \text{ GeV at 95\% C.L.}$$

If we assume that all three types of $\tilde{\nu}$ have the same mass, then

$$M_{\tilde{\nu}} > 37.5 \text{ GeV at 95\% C.L.}$$

ACKNOWLEDGEMENTS

I would like to thank the DPF conference organizers, Prof. B.E. Bonner and Prof. H.E. Miettinen for their hospitality and for a wonderful conference.

REFERENCES

[1]   D. Decamp et al. (ALEPH Collaboration), Phys. Lett. **B231** (1989) 519,
D. Decamp et al., Phys. Lett. **B234** (1990) 399 and
D. Decamp et al., Phys. Lett. **B235** (1990) 339.

[2]   G.S. Abrams et al. (MARK II Collaboration), Phys. Rev. Lett. **63** (1989) 724,
G.S. Abrams et al., Phys. Rev. Lett. **63** (1989) 2173,
B. Adeva et al. (L3 Collaboration), Phys. Lett. **B231** (1989) 509,
M.Z. Akrawy et al. (OPAL Collaboration), Phys. Lett. **B231** (1989) 530,
P. Aarnio et al. (DELPHI Collaboration) Phys. Lett. **B231** (1989) 539,
B. Adeva et al. (L3 Collaboration), Phys. Lett. **B237** (1990) 136,
M.Z. Akrawy et al. (OPAL Collaboration), Phys. Lett. **B240** (1990) xxx,
B. Adeva et al. (L3 Collaboration), L3 preprint #005, submitted to Phys.Lett.B,
P. Abreu et al., (DELPHI Collaboration), CERN-EP/90-31 and
P. Abreu et al., (DELPHI Collaboration), CERN-EP/90-32.

[3]   "ALEPH - a Detector for electron-positron Annihilation at LEP", submitted to Nucl. Instr. Meth.

[4]   F. Bird, J. Harton and R. Johnson, this proceeding (parallel section).

[5]   A. Borelli, M. Consoli, L. Maiani, R. Sisto, CERN-TH-5441 (1989).

[6]   F.A. Berends et al., Z Line Shape group in "Proceedings of the Workshop of Z Physics at LEP", CERN Report 89-08 Vol.I, 89.

[7]   F.A. Berends, G. Burgers and W.L. van Neerven, Nucl. Phys. **B297** (1988) 429 and Nucl. Phys. **B304** (1988) 921.
Computer program ZAPPH, courtesy of G. Burgers.

[8]   D.C. Kennedy, B.W. Lynn, J.C. Im and R.G. Stuart, Nucl. Phys. **B321** (1989) 83.

[9]   S.L. Glashow, Nucl. Phys. **22** (1961) 579,
S. Weinberg, Phys. Rev. Lett. **19** (1967) 1264,
A. Salam, Proc. Eighth Nobel Symp., ed. N. Svartholm (Almqvist and Wiksell, Stockholm, 1968) p.367.

[10]   P.W. Higgs, Phys. Lett. **12** (1964) 132,
Phys. Rev. Lett. **13** (1964) 508,
Phys. Rev. **145** (1966) 1156,
F. Englert and R. Brout, Phys. Rev. Lett. **13** (1964) 321,
G.S. Guralnik, C.R. Hagen and T.W.B. Kibble, Phys. Rev. Lett. **13** (1964) 585,
T.W.B. Kibble, Phys. Rev. **155** (1967) 1554.

[11]   D. Decamp et al. (ALEPH Collaboration), Phys. Lett. **B236** (1990) 233,
and Yi-Bin Pan, this proceeding (parallel section).

[12]   D. Decamp et al. (ALEPH Collaboration), Phys. Lett. **B241** (1990) 141.

[13]   J.F. Gunion and H.E. Haber, Nucl. Phys. **B272** (1986) 1,
B278 (1986) 449 and earlier references therein.

[14]   D. Decamp et al. (ALEPH Collaboration), Phys. Lett. **B237** (1990) 291.

[15]   G.F. Giudice and G. Ridolfi, Z. Phys. **41C** (1988) 447,
M. Olechowski and S. Pokorski, Phys. Lett. **214B** (1988) 393.

[16]   D. Decamp et al. (ALEPH Collaboration), Phys. Lett. **B241** (1990) 623.

[17]   A. Chen *et al.* (CLEO Collaboration), Phys. Lett. **B122** (1983), 317,
W. Bartel *et al.* (JADE Collaboration), Z. Phys. **C31** (1986) 359,
H.-J. Behrend *et al.* (CELLO Collaboration), Phys. Lett. **B193** (1987) 376,
S.M. Ritz, thesis, University of Wisconsin-Madison, Madison, Wisconsin, U.S.A., 1988.

[18]   D. Decamp *et al.* (ALEPH Collaboration), Phys. Lett. **B236** (1990) 511,
and V. Sharma and F. Weber, this proceeding (parallel section).

[19]   I. Adachi *et al.*, TOPAZ Collab., Phys. Lett. **229B** (1989) 427,
S. Eno *et al.*, AMY Collab., Phys. Rev. Lett. **63** (1989) 1910,
S. Eno *et al.*, AMY Collab., KEK Report No. 88-47 (1988), to be published,
K. Abe *et al.*, VENUS Collab., Phys. Rev. **D39** (1989) 3524,
T. Mori *et al.*, AMY Collab., KEK Report No. 88-43 (1988) to be published,
I. Adachi *et al.*, TOPAZ Collab., Phys. Rev. Lett. **60**   (1988) 97.

[20]   C. Albajar *et al.*, UA1 Collab., Z. Phys. **C37** (1988) 505,
G. Altarelli *et al.*, Nucl. Phys. **B308** (1988) 724,
F. Abe *et al.*, CDF Collab., to be published in Phys. Rev. Lett,
T. Akesson *et al.*, UA2 Collab., CERN-EP/89-152, to be published in Z. Phys. C.

[21]   G.S. Abrams *et al.*, MARK II Collab., SLAC PUB-5106 (1989), to be published.

[22]   M.Z. Akrawy *et al.*, OPAL Collab., CERN-EP/89-154, to be published.

[23]   V. Barger *et al.*, Phys. Rev. **D30** (1984) 947,
V. Barger, R.J.N. Phillips and A. Soni, Phys. Rev. Lett. **57** (1986) 1516,
W.S. Hou and R.G. Stuart, Nucl. Phys. **B320** (1989) 227,
W.S. Hou and R. G. Stuart, Phys. Rev. Lett. **62** (1989) 617.

[24]   N.M. Shaw *et al.*, AMY Collab., Phys. Rev. Lett. **63** (1989) 1342,
G.S. Abrams *et. al.*, MARK II Collab., SLAC-PUB-5136 (1989), to be published.

[25]   C. Albajar *et al.*, UA1 Collab., Phys. Lett. **B185** (1987) 241,
G.N. Kim *et al.*, AMY Collab., Phys. Rev. Lett. **61** (1988) 911,
K. Abe *et al.*, VENUS Collab., Phys. Rev. Lett. **61** (1988) 915,
I. Adachi *et al.*, TOPAZ Collab., Phys. Rev. **D37** (1988) 1339.

[26]   See for example H. Harari, Phys. Rep. **104** (1984) 159, Proc. of the 28th Solvay Conf. on
Physics, Austin Texas, U.S.A., Nov. 1982, ed. L. Van Hove and Proc. of the 23rd Int.
Conf. on High Energy Physics, Berkeley Calif., U.S.A., July 16-23, 1986;
L. Lyons, Progress in Particle and Nuclear Physics **10** (1983) 227;
R.D. Peccei, invited talk at the 1985 Int. Symp. on Composite Models of Quarks and
Leptons, Tokyo, Japan, Aug. 13-15, 1985 and lectures given at Lake Louise Winter
Inst., Selected Topics in Electroweak Interactions, Lake Louise, Canada, Feb. 15-21,
1987;
M.E. Peskin, Proc. Int. Symp. on Lepton and Photon Interactions at High Energies, Bonn
(1981), ed. W. Pfeil, p.880, and Proc. Int. Symp. on Lepton and Photon Interactions at
High Energies, Kyoto (1985), ed. M. Konuma and K. Takahashi, p.850;
F.M. Renard, Z. Phys. **C24** (1984) 385.

[27]   D. Decamp *et al.* (ALEPH Collaboration), Phys. Lett. **B236** (1990) 501.

[28]   S.K. Kim *et al.*, AMY Collaboration, Phys. Lett. **223B** (1989) 476,
I. Adachi *et al.*, TOPAZ Collaboration, Phys. Lett. **228B** (1989) 553
K. Abe *et al.*, VENUS Collaboration, Phys. Lett. **213B** (1988) 400,
H.-J. Behrend *et al.*, CELLO Collaboration, Phys. Lett. **168B** (1986) 420,
CELLO Collaboration, EPS Conf., Uppsala, June 25-July 1, 1987 and Int. Symp. on
Lepton and Photon Interactions, Hamburg, July 27-31, 1987,
P. Janot, Thesis, Orsay LAL 87-31 (1987), (unpublished).

[29]   H.E. Haber and G.L. Kane, Phys. Rep. **117** (1985) 75-263, and references therein.

[30]   P. Fayet, Unification of the fundamental particle interactions, Proc. Europhys. Study
Conf. (Erice) (Plenum, New York, 1980) p.587.

[31]   P. Fayet, Phys. Lett. **B86** (1979) 272.

[32]   P. Fayet, Phys. Lett. **B133** (1983) 363.

[33]   T. Schimert, C. Burgess and X. Tata, Phys. Rev. **D32** (1985) 707,
T. Schimert and X. Tata, Phys. Rev. **D32** (1985) 721.

[34]   D. Decamp *et al.* (ALEPH Collaboration), Phys. Lett. **B236** (1990) 86,
and J. Conway, this proceeding, (parallel section).

Z^0 Highlights – First Results from OPAL

The OPAL Collaboration [1]

Presented by

J. William Gary

Physikalisches Institut der Univerisität

Heidelberg, Federal Republic of Germany

Abstract

Results are presented for the mass of the Z^0, its total width, its partial decay widths into hadrons and leptons and for the number of light neutrino generations. Mass limits for a standard model Higgs boson, a charged sequential heavy lepton, supersymmetric particles and for the top and b' quarks are presented, as are studies of global distributions, of jet production rates and of isolated, high energy photons in hadronic Z^0 decays.

1 Introduction

The OPAL detector at LEP had its first physics run from October to December 1989, collecting a data sample of about 1.3 pb^{-1} in integrated luminosity in an energy scan around the mass of the Z^0 boson. On the basis of this data sample, analyses have been performed to measure the Z^0 line shape and number of light neutrino generations, to test other aspects of electro-weak theory, to test QCD and to search for new particles. In this article we present the results of these analyses: these are essentially the same as those presented at the Z^0 highlights session of the DPF-90 Conference at Rice University but have been updated where pertinent.

2 The OPAL Detector

OPAL is a general purpose e^+e^- collider detector. The tracking of charged particles is performed with a jet chamber, a large volume drift chamber which covers a polar angle range of $|cos\theta| < 0.92$, recording up to 159 space hits for charged tracks which traverse its volume. The jet chamber, together with a vertex detector and a z-chamber, is enclosed by a solenoidal magnet which provides an axial field for the momentum measurement. The magnet coil is surrounded by a time-of-flight (TOF) counter array, a lead glass electromagnetic calorimeter with a presampler, an instrumented magnet return yoke serving as a hadron calorimeter and four layers of muon chambers. The TOF system covers the region $|cos\theta| < 0.82$ and is composed of 160 scintillator bars. The electromagnetic calorimeter is divided into a barrel part ($|cos\theta| < 0.82$) and two endcap parts ($0.81 < |cos\theta| < 0.98$): its total solid angle coverage is 98% of 4π. A forward detector, composed of two identical modules placed around the beam pipe at either end of the jet chamber, serves as a luminosity monitor. Each forward detector module contains a lead-scintillator calorimeter and a proportional chamber array. The forward detector covers the polar angle range from 40 to 150 mrad.

[1] Birmingham, Bologna, Bonn, U.C.Riverside, Cambridge, Carleton, CERN, Chicago, Freiburg, Heidelberg, Queen Mary College, Birkbeck College, University College London, Manchester, Maryland, Montréal, NRCC-Ottawa, Rutherford, Saclay, Technion-Israel, Tel Aviv, Tokyo, Kobe, Brunel, Weizmann.

3 The Z^0 line shape from hadronic decays

The Z^0 line shape is determined from its hadronic decays by measuring the hadronic cross section $\sigma(e^+e^- \rightarrow hadrons)$ in an energy scan around the resonance. The elements of this measurement are the c.m. energy $E_{c.m.}$ of the machine and, for each energy, the number $N_{had.}$ of hadronic event candidates, the trigger and selection efficiencies $\epsilon_{trig.}$ and $\epsilon_{sel.}$, the number $N_{bkg.}$ of background events and the integrated luminosity $\int \mathcal{L}dt$:

$$\sigma(e^+e^- \rightarrow hadrons) = \frac{N_{had.} - N_{bkg.}}{\epsilon_{trig.} \cdot \epsilon_{sel.} \cdot \int \mathcal{L}dt} \qquad \text{for each value } E_{c.m.} \qquad (1)$$

The uncertainty in the LEP c.m. energy value has been determined to be 30 MeV. The measurement of other quantities in relation (1) is discussed in [1]. Here will be mentioned only the luminosity measurement, which is determined from the number of bhabha events which appear in a well defined fiducial region of the forward detector between 58 and 124 mrad in polar angle. The identification of the bhabha events relies on energy deposits in the forward detector calorimeters and on the radial and azimuthal positions of the bhabha induced showers measured by the forward detector proportional chamber arrays. The error in the absolute luminosity determination is 2.2%.

Figure 1 (a) shows the measured cross sections for $e^+e^- \rightarrow hadrons$ using 25,801 hadronic decay events distributed over 11 c.m. energy values between 88.28 and 95.04 GeV. A model independent fit in which the mass M_Z, the total width Γ_Z and the pole cross section $\sigma_{had.}^{pole}$ (without QED radiative corrections) are treated as free parameters yields

$$
\begin{aligned}
M_Z &= 91.145 \pm 0.022 \,(\text{exp.}) \pm 0.030 \,(\text{LEP}) \,\text{GeV} \\
\Gamma_Z &= 2.526 \pm 0.047 \,\text{GeV} \\
\sigma_{had.}^{pole} &= 41.2 \pm 1.1 \,\text{nb}
\end{aligned}
\qquad (2)
$$

The number of light neutrinos N_ν can be determined from a Standard Model fit for which $\sigma_{had.}^{pole}$ and the partial decay widths of the Z^0 into hadrons and leptons are taken from theory and N_ν and M_Z are the free parameters. Such a fit yields

$$
\begin{aligned}
M_Z &= 91.141 \pm 0.022 \,(\text{exp.}) \pm 0.030 \,(\text{LEP}) \,\text{GeV} \\
N_\nu &= 3.09 \pm 0.19 \,(\text{exp.}) \,{}^{+0.06}_{-0.12} \,(\text{theor.})
\end{aligned}
\qquad (3)
$$

The theoretical error on N_ν accounts for uncertainties in the values of the top quark mass, the Higgs boson mass and the strong coupling constant. The dashed curve in figure 1 (a) is the line shape predicted by the standard model assuming the measured M_Z and 4 neutrino generations. It is in clear disagreement with the measurements.

4 The Z^0 line shape from a combined analysis of hadronic and leptonic decays

The OPAL cross section measurements for $e^+e^- \rightarrow e^+e^-$, $e^+e^- \rightarrow \mu^+\mu^-$ and $e^+e^- \rightarrow \tau^+\tau^-$ are discussed in detail in [1]. The experimental analysis is restricted to the region $|cos\theta| < 0.70$ for all three processes to ensure good trigger and reconstruction efficiencies. The measured cross sections are shown in figures 1 (b), (c) and (d). The e^+e^- cross section

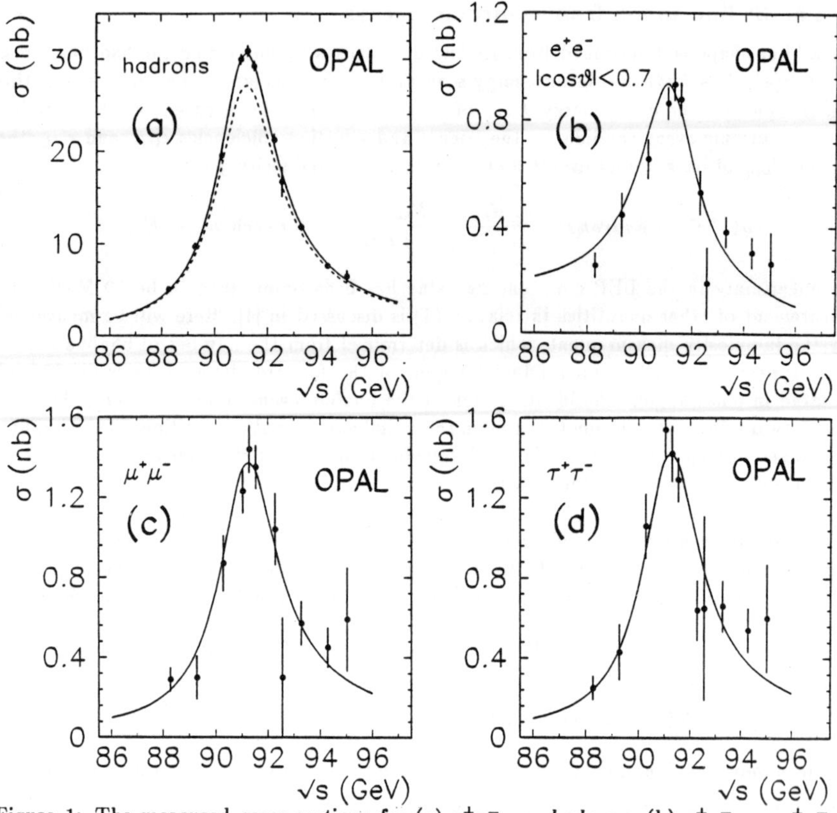

Figure 1: The measured cross sections for (a) $e^+e^- \rightarrow hadrons$, (b) $e^+e^- \rightarrow e^+e^-$ for $|cos\theta| < 0.70$, (c) $e^+e^- \rightarrow \mu^+\mu^-$ and (d) $e^+e^- \rightarrow \tau^+\tau^-$.

is not corrected for the geometrical acceptance because of the singularites associated with the t channel.

OPAL has performed a simultaneous fit to the hadronic and leptonic cross section measurements shown in figures 1 (a)-(d) to measure M_Z, Γ_Z and the partial decay widths $\Gamma_{had.}$, Γ_e, Γ_μ and Γ_τ. This combined fit allows the correlations between the different measurements to be properly included in the errors which are assigned to these parameters. The correlations include the point-to-point uncertainty in the LEP c.m. energy value; the ratio $\Gamma_e/(M_Z^2 \cdot \Gamma_Z^2)$ is common to the theoretical expressions for all four processes [1] and so introduces additional correlations. The line shapes obtained from the combined fit are shown in figures 1 (a)-(d) as the solid curves: the parameter values which are extracted are given in table 4. The values of M_Z, Γ_Z, and $\sigma_{had.}^{pole}$ obtained from the combined fit agree well with those derived using the hadronic data alone. The leptonic widths are in good agreement with leptonic universality. A repetition of the combined fit with all lepton

$$\begin{aligned}
M_Z &= 91.154 \pm 0.021 \text{ (exp.) GeV} & \Gamma_e &= 81.2 \pm 2.6 \text{ MeV} \\
\Gamma_Z &= 2.536 \pm 0.045 \text{ GeV} & \Gamma_\mu &= 82.6 \pm 5.8 \text{ MeV} \\
\sigma_{had.}^{pole} &= 41.4 \pm 1.1 \text{ nb} & \Gamma_\tau &= 85.7 \pm 7.1 \text{ MeV} \\
\Gamma_{had.} &= 1.854 \pm 0.058 \text{ GeV}
\end{aligned}$$

Table 1: Electro-weak parameters from a combined fit to the data of figure 1.

widths constrained to be equal yields $\Gamma_l = 81.9$ MeV and $\Gamma_{had.} = 1.838 \pm 0.046$ GeV for the leptonic and hadronic partial widths.

These results can be combined to provide a model independent measurement of the invisible partial decay width $\Gamma_{inv.}$ of the Z^0:

$$\Gamma_{inv.} = \Gamma_Z - (3 \cdot \Gamma_l) - \Gamma_{had.} = 453 \pm 44 \text{ MeV} \tag{5}$$

Combining $\Gamma_{inv.}$ with the Standard Model value $\Gamma_\nu^{S.M.} = 166.2$ MeV for the partial decay width of the Z^0 to a neutrino pair leads to

$$N_\nu = \frac{\Gamma_{inv.}}{\Gamma_\nu^{S.M.}} = 2.73 \pm 0.26 \text{ (exp.) } {}^{+0.02}_{-0.04} \text{ (theor.)} \tag{6}$$

which excludes four generations of light neutrinos having standard model couplings by more than 4 standard deviations.

5 Mass limits for a standard model Higgs boson

A search for a Standard Model Higgs boson H^0 with a mass above 3 GeV/c^2 has been performed, produced through the reactions

$$e^+e^- \to Z^{*0}H^0 \to (\nu\bar{\nu} \text{ or } e^+e^- \text{ or } \mu^+\mu^-)\,H^0 \quad ; \quad H^0 \to q\bar{q} \text{ or } \tau^+\tau^- \tag{7}$$

The $\nu\bar{\nu}H^0$ channel is especially sensitive because of its asymmetric topology and because of the large rate for $Z^{*0} \to \nu\bar{\nu}$. In figure 2 (a) is shown P_T, which is the magnitude of the missing momentum vector in an event in the direction perpendicular to the beam axis, after application of cuts to eliminate multi-hadron, di-lepton and two-photon events and to require that a large magnitude of missing momentum be present which points into the central region of the detector [2]. For example, an anti-tagging requirement is imposed to discriminate against two-photon events: that less than 2 GeV of energy be observed in the forward detector calorimeters. The P_T value for figure 2 (a) is calculated using both charged tracks and electromagnetic calorimeter clusters. Also shown in figure 2 (a) as the histogram is the expectation for a 24 GeV/c^2 Higgs boson, normalized to the 1.3 pb^{-1} data sample. The experimental data points, shown with error bars, appear at small P_T values and fall off sharply for $P_T \approx 4$ GeV/c. This is the expected signal for two-photon events as it is only for P_T values above about 4 GeV/c that the beam particles in two-photon events are scattered into the forward detector calorimeters and become subject to the anti-tagging restriction. We apply a last cut which eliminates events with P_T values less than 6 GeV/c. This leaves zero Higgs candidates.

Figure 2 (b) shows the expected number of Higgs events as a function of the H^0 mass for the $\nu\bar{\nu}$ channel. Also shown are the results for the e^+e^- and $\mu^+\mu^-$ channels and the sum

258

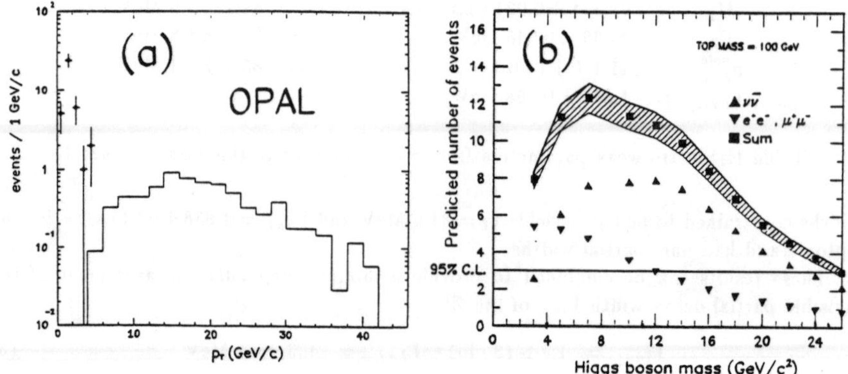

Figure 2: (a) P_T distribution (see text) for the 1.3 pb^{-1} data sample (points with error bars); the histogram shows the expectation for a 24 GeV/c^2 Higgs boson. (b) Expected number of Higgs events vs. the Higgs mass for the different search channels and for their sum.

of the result for all search channels. The e^+e^- and $\mu^+\mu^-$ analyses, for which also no Higgs candidates are observed, are discussed in [2]. The intersection of the curve representing the sum of the search channels with the line corresponding to the 95% c.l. upper limit for zero observed events allows us to place the limits

$$M_{H^0} < 3.0\,\mathrm{GeV} \quad ; \quad M_{H^0} > 25.3\,\mathrm{GeV} \quad \text{at } 95\%\text{ c.l.} \tag{8}$$

for the mass M_{H^0} of the H^0.

6   Global distributions in multi-hadronic Z^0 decays

OPAL has performed a detailed study of the structure of multi-hadronic events in Z^0 decay. Figures 3 (a)-(d) show the OPAL measurements of the T_{major}, Aplanarity, Thrust and charged multiplicity $n_{ch.}$ distributions using a sample of over 20,000 events. The first three of these distributions have been unfolded for detector resolution and acceptance and for initial-state photon radiation to the level which encompasses all stable charged and neutral particles. Also shown in figures 3 (a)-(c) are the predictions of four Monte Carlo programs which simulate perturbative QCD and the hadronization of partons: Jetset version 7.2, Herwig version 4.3, Ariadne version 3.1 and a complete second order matrix element model with a modified perturbation scale (ERT-E0). The $n_{ch.}$ distribution is not unfolded: the Monte Carlo predictions in this case include initial-state radiation and simulation of the OPAL detector. We obtain an unfolded value for the mean charged multiplicity of 21.28 $\pm$ 0.04$\pm$0.84, where the first error is statistical and the second is systematic [4]. The parameter values of the four Monte Carlos have been adjusted to describe the T_{major} distribution but not the other three distributions. Jetset, Herwig and Ariadne, all of which are based on a parton shower, are in good agreement with the distributions, which demonstrates their consistency for describing our 91 GeV hadronic data sample (we do not show the Ariadne prediction for $n_{ch.}$). The ERT-E0 model provides a reasonably good description of Thrust but its predictions are low relative to data for the central part of the Aplanarity distribution.

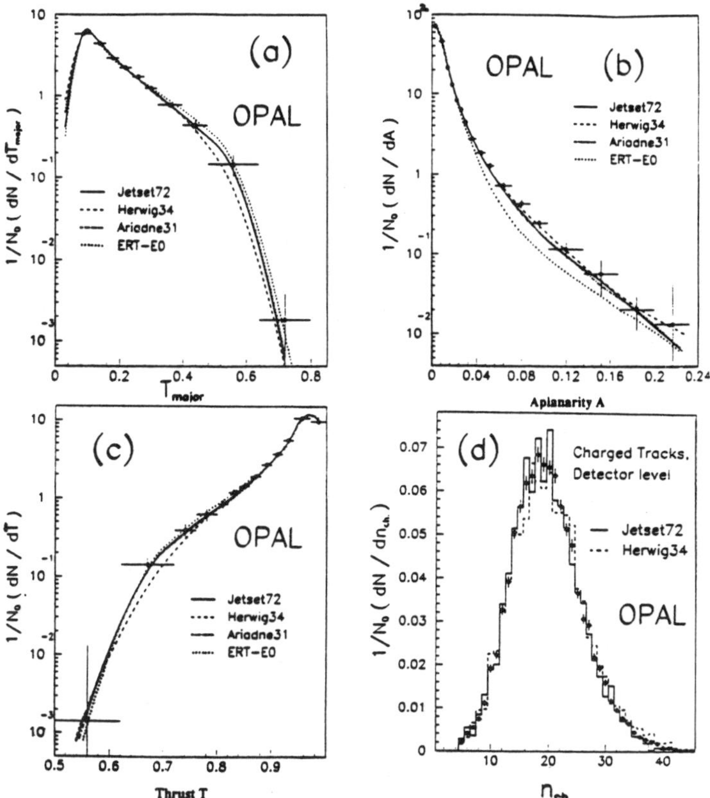

Figure 3: The unfolded (a) T_{major}, (b) Aplanarity and (c) Thrust distributions; (d) the detector level $n_{ch.}$ distribution.

7 Jet rates and the running of α_S

The production rates of multi-jet final-states in Z^0 decay have been measured by OPAL. For this the jet finding algorithm introduced by the JADE Collaboration [3] is employed. An important feature of this algorithm is the close agreement between the number of jets reconstructed at the hadron level and at the parton level of an event over a large range of c.m. energies, as has been determined through studies with QCD simulation programs. The jet finding algorithm depends on a single resolution parameter y_{cut} which specifies the minimum jet-jet invariant mass to be allowed. In figure 4 (a) is shown the relative production rates of 2-, 3-, 4- and 5-jet events measured by OPAL, as a function of y_{cut}. The measurements have been unfolded for detector resolution and acceptance and for initial-state photon radiation. As y_{cut} increases, the 3-, 4- and 5-jet rates decrease since the minimum invariant mass value between jets is larger: the 2-jet rate shows a corresponding

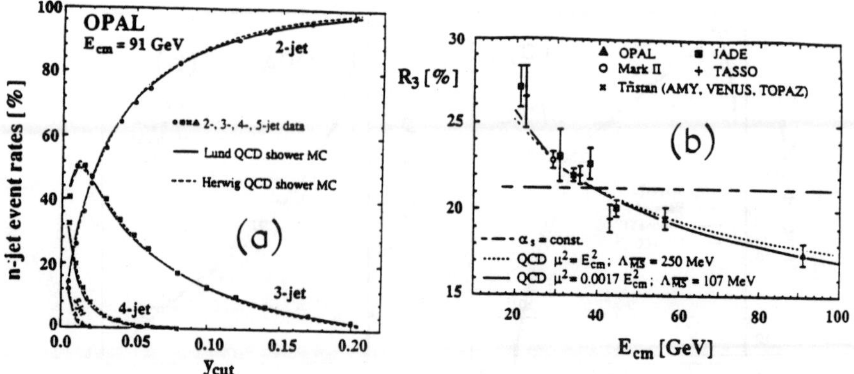

Figure 4: (a) The jet rates at $E_{c.m.}$=91 GeV. (b) The 3-jet rate R_3 vs. $E_{c.m.}$.

increase. Figure 4 (a) also shows the multi-jet rates predicted by Jetset and Herwig, which are in good agreement with the data.

From figure 4 (a) we observe that the 3-jet rate R_3 at 91 GeV is about 17.5% for a y_{cut} value of 0.08. This same OPAL measurement is shown in figure 4 (b) as the data point at $E_{c.m.}$ =91 GeV, in comparison with measurements of R_3 by e^+e^- annihilation experiments at lower c.m. energies using the same jet finding algorithm and value for y_{cut}. A clear scaling violation is observed: the 3-jet rate decreases as the c.m. energy increases. This scaling violation can be interpreted in the context of QCD as being entirely due to the running of the strong coupling constant α_S. The prediction of the 2nd order QCD calculation of Kramer and Lampe [5] is also shown in figure 4 (b), for two choices of the renormalization scale μ. The similarity in the two curves demonstrates the insensitivity of R_3 to the choice of this scale. The QCD predictions describe the measurements well. In contrast, the hypothesis that α_S does not depend on energy results in a constant rate for 3-jet events, as shown in figure 4 (b), and leads to a χ^2 value with the data of 45.8 (8 degrees of freedom) for c.m. energies above about 25 GeV. The hypothesis that α_S is constant is thus ruled out with a significance of 5.7 standard deviations.

The measured jet rates at 91 GeV have been used by OPAL to adjust the values of $\Lambda_{\overline{MS}}$ and μ^2 in order $O(\alpha_S^3)$ QCD [6]. We obtain $\mu^2/E_{c.m.}^2 = 0.001 - 0.003$ and $\Lambda_{\overline{MS}} = 80 - 180$ MeV or if $\mu^2 = E_{c.m.}^2$ then $\Lambda_{\overline{MS}} = 200 - 450$ MeV. This leads to a measured value for the strong coupling constant α_S of

$$\alpha_S\,(91\text{ GeV}) \;=\; 0.117 \pm 0.015 \quad \text{for} \quad (\mu^2/E_{c.m.}^2) = 0.001 - 1.0 \tag{9}$$

where the error includes the uncertainties of hadronization and of the renormalization scale.

8 Searches for new particles which are pair produced in Z^0 decay

OPAL has performed direct searches and set mass limits for the following objects which might be pair produced in Z^0 decays [7]:

$$top \text{ quark} \quad : \quad M_{top} > 45.1\ GeV/c^2 \tag{10}$$

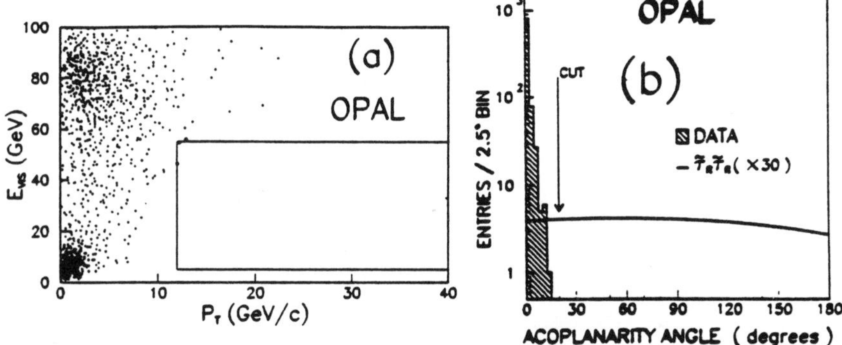

Figure 5: (a) P_T vs. E_{vis}. for a 752 nb^{-1} data sample: 7.6 L^+L^- events with M_{L^+}=44 GeV/c^2 are expected in the region P_T >12 GeV/c and 5 < E_{vis}. < 15 GeV. (b) Acoplanarity angle distribution for low multiplicity 2-jet events in a 550 nb^{-1} sample: the solid curve shows the expectation for right handed supersymmetric $\tilde{\tau}\tilde{\tau}$ events with $M_{\tilde{\tau}}$=40 GeV/c^2 and $M_{\tilde{\gamma}}$=10 GeV/c^2.

$$b' \text{ quark} \quad : \quad M_{b'} \; > \; 45.4 \; GeV/c^2$$

$$\text{Charged heavy lepton } L^+ \quad : \quad M_{L^+} \; > \; 44.3 \; GeV/c^2$$

$$\text{Supersymmetric particles} \quad : \quad M_{\tilde{e}} \; > \; 43.4 \; GeV/c^2 \quad ; \quad M_{\tilde{\mu}} \; > \; 43.0 \; GeV/c^2$$

$$M_{\tilde{\tau}} \; > \; 43.0 \; GeV/c^2 \quad ; \quad M_{\tilde{W}} \; > \; 45.0 \; GeV/c^2$$

where the mass limits are at the 95% c.l. in all cases. The L^+ mass limit is valid for $M_{\nu_L} < 20$ GeV/c^2; for the supersymmetric channels we assume $M_{\tilde{\gamma}} < 20$ GeV/c^2 and mass degeneracy between the right- and left-handed supersymmetric particles.

As an example, figure 5 (a) shows P_T, the magnitude of the missing momentum in the direction perpendicular to the beam axis, vs. the visible energy E_{vis}. for the events in a 752 nb^{-1} sample. The events with large E_{vis}. values are multi-hadronic Z^0 decays; those with small E_{vis}. and small P_T values are two-photon events. A heavy charged lepton L^+ with a mass of 44 GeV/c^2 is expected to contribute 7.6 events to the region P_T >12 GeV/c and 5 < E_{vis}. < 15 GeV which is delineated by the solid line in figure 5 (a) (this example assumes $M_{\nu_L} = 0$). The lack of this signal demonstrates the absence of such an L^+ population in our data. As a second example, figure 5 (b) shows the distribution of acoplanarity angle for low multiplicity 2-jet events in a 550 pb^{-1} sample, where the acoplanarity angle is the angle between the jets in the direction perpendicular to the beam axis. The data peak sharply at low values of acoplanarity angle (this is primarily a $\tau^+\tau^-$ signal): in contrast a supersymmetric $\tilde{\tau}\tilde{\tau}$ signal would exhibit a flat distribution as shown.

9   Isolated photons in hadronic Z^0 decays

Isolated, high energy photons are rare in e^+e^- multi-hadronic events and may arise from initial- or final-state photon radiation or from new phenomena. The evidence for final-state

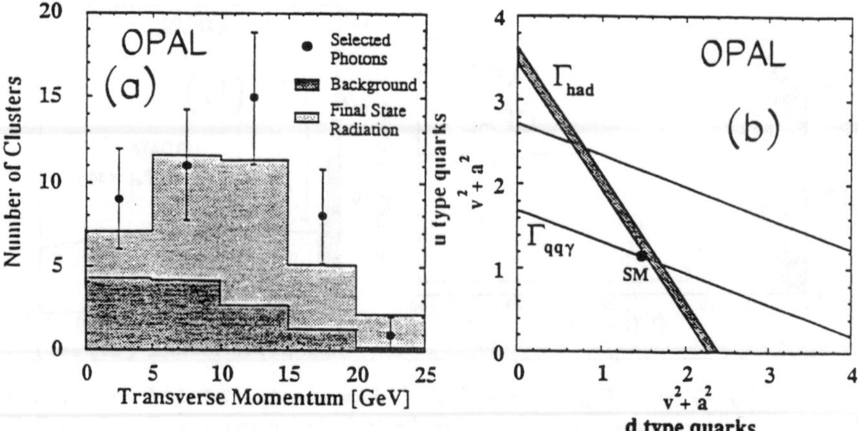

Figure 6: (a) The transverse momentum spectrum of high energy, isolated photons with respect to the thrust axis. (b) Illustration of the extraction of the vector and axial vector coupling constants of *up* and *down*-like quarks.

photon radiation from quarks has been indirect to date, relying on interference between inital- and final-state emission. At LEP, initial-state radiation is suppressed because of the Z^0 pole thus motivating a direct search for final-state photon radiation from quarks.

OPAL has selected multi-hadronic events having a signal consistent with an isolated high energy photon. The isolated photon candidates are required to have at least 10 GeV of energy, to lie at least 20 degrees in angle from any reconstructed charged track or other neutral particle and to exhibit signal characteristics in the electromagnetic and hadron calorimeters which are consistent with those of a photon. In a sample of over 27,000 multi-hadron events, 35 are found to possess an isolated photon which satisifies these requirements. The transverse momentum spectrum of the photon with respect to the thrust axis is shown in figure 6 (a) for these 35 events. Also shown are the predictions of the Jetset Monte Carlo program, with and without final-state photon radiation included in the simulation. The Monte Carlo prediction without final-state radiation is shown as the background curve in figure 6 (a), which may attributed to initial-state radiation and fluctuations in the hadronization process. This background curve is unable to reproduce the measured yield of isolated photons. In contrast, the Monte Carlo prediction with final-state photon radiation is in good agreement with the observed rate and transverse momentum spectrum shown in figure 6 (a). From this and other studies we conclude that we observe direct evidence for final-state photon radiation from quarks: that the measured isolated photon yield can be attributed to standard model sources.

The absence of an anamolous isolated photon signal allows an extraction of the vector and axial vector couplings of up-type and down-type quarks by a new technique, through comparison of the rate for the isolated photon events with the total hadronic event rate in Z^0 decays. The extraction of the coupling constants by this technique is illustrated in figure 6 (b) (see also [8]). Briefly, events with a hard isolated photon preferentially contain an *up*-like rather than a *down*-like quark at the primary vertex because of their larger electric charge (it is expected that hard, isolated photons from final-state radiation are primarily emitted by the primary quark in an event). The rates $\Gamma_{q\bar{q}\gamma}$ and $\Gamma_{had.}$ for multi-hadronic events with a final-state photon and for the total hadronic rate have a dependence on the vector and axial vector quark coupling constants v_i and a_i which are

$$\Gamma_{q\bar{q}\gamma} \sim (3 \cdot c_{1/3} + 8 \cdot c_{2/3}) \quad ; \quad \Gamma_{had.} \sim (3 \cdot c_{1/3} + 2 \cdot c_{2/3}) \tag{11}$$

where $c_i = v_i^2 + a_i^2$ with $i = 2/3$ for *up*-like quarks and $i = 1/3$ for *down*-like quarks. The OPAL measured values for $\Gamma_{q\bar{q}\gamma}$ and $\Gamma_{had.}$ then provide [8]:

$$v_{1/3}^2 + a_{1/3}^2 = 1.24 \pm 0.44 \quad \text{and} \quad v_{2/3}^2 + a_{2/3}^2 = 1.72 \pm 0.64 \tag{12}$$

OPAL has also used the measured yield of high energy, isolated photons to exclude the production of excited quark states over most of the kinematically allowed range and to exclude the possibility of a composite Z^0 for a wide range of parameters [8].

10   Summary

In summary, the OPAL experiment has collected a data sample of 1.3 pb^{-1} during the first 3 months of LEP operation and has performed analyses in electro-weak physics, in QCD and hadronic physics and in searches for new particles, as have been outlined in this article.

11   Acknowledgements

It is a pleasure to thank the LEP Division for the efficient operation of the machine and their continuing cooperation with our experimental group.

References

[1] OPAL Collaboration, M.Z. Akrawy *et al.*, to appear in Phys. Lett. **240** (1990); see also the articles by G. Van Dalen and by D. Karlen, these proceedings.

[2] OPAL Collaboration, M.Z. Akrawy *et al.*, Phys. Lett. **B236** (1990) 224; see also E. Duchovni, these proceedings.

[3] JADE Collaboration, W. Bartel *et al.* Z.Phys. **C33** (1986) 23.

[4] OPAL Collaboration, M.Z. Akrawy *et al.*, CERN-EP/90-48, to be published in Z.für Physik **C**.

[5] G. Kramer and B. Lampe, Z.Phys. **C39** (1988) 101.

[6] OPAL Collaboration, M.Z. Akrawy *et al.*, Phys. Lett. **B235** (1990) 389.

[7] OPAL Collaboration, M.Z. Akrawy *et al.*, Phys. Lett. **B236** (1990) 364; Phys. Lett. **B240** (1990) 261; Phys. Lett. **B240** (1990) 250.

[8] OPAL Collaboration, in preparation; see also P. Mättig, these proceedings.

Z PARAMETERS: DELPHI RESULTS

by The DELPHI Collaboration

presented by Alexander Firestone
Iowa State University and Ames Laboratory, Ames, Iowa 50011

ABSTRACT

We present results of studies of the Z Boson produced in e^+e^- annihilations in the DELPHI detector at the CERN LEP collider. Approximately 600 nb.$^{-1}$ of useful data were collected in the first LEP running period in the last half of 1989. Results are presented on the mass and width of the Z Boson, on the partial widths of the Z decays into hadrons and into lepton pairs, on comparisons with the predictions of QCD, on the number of light neutrino generations, and on the searches for new particles. We find that $M_Z = 91.171\pm0.030$ (stat.)±0.030 (beam) GeV, $\Gamma_Z = 2.511\pm0.065$ GeV, $\Gamma_{had.} = 1741\pm61$ MeV, $\Gamma_{lept.} = 85.1\pm2.9$ MeV, $\Gamma_{inv.} = 515\pm54$ MeV, and $N_\nu = 2.97\pm0.26$ generations. We have set limits on the mass of the charged Higgs particle, and we find no evidence for new quarks, new heavy leptons, or any supersymmetric partners to the known particles. All our results are consistent with the Standard Model of electroweak interactions plus QCD with exactly three generations of elementary fermions.

1. Introduction

We present a short overview of the DELPHI detector in section 2, a discussion of the trigger in section 3, a discussion of the luminosity in section 4, and a discussion of the hadronic event selection in section 5. In the remaining sections we present our analysis and results: $Z^0 \to$ hadrons in section 6, $Z^0 \to e^+e^-$ in section 7, $Z^0 \to \mu^+\mu^-$ in section 8, and $Z^0 \to \tau^+\tau^-$ in section 9. Finally, we present the results of our searches for new particles; Higgs particles in section 10, and other new particles, e.g. supersymmetric particles and heavy quarks or leptons in section 11.

2. Detector Overview

The DELPHI detector is in the standard cylindrical geometry with two large end caps at either end. It consists of 13 sub-systems and the world's largest superconducting solenoid. Those features of the detector relevant for the present analysis are listed below:

• The INNER DETECTOR is a cylindrical drift chamber (inner radius = 12 cm, outer radius = 22 cm) covering polar angles between 29° and 151°. A jet-chamber section providing 24 rϕ coordinates is surrounded by 5 layers providing rϕ and longitudinal coordinates.

• The TIME PROJECTION CHAMBER is a cylinder with 28 cm inner and 122 cm outer radii and a length of 2.7 m. For polar angles between 22° and 158° at least 4 space points are available for track reconstruction, while for angles between 39° and 141° up to 16 space points can be used.

• The OUTER DETECTOR has 5 layers of drift cells at radii between 192 and 208 cm and covers polar angles from 50° to 130°. All layers provide precise rϕ coordinates, and 3 of them also provide longitudinal information.

• The HIGH DENSITY PROJECTION CHAMBER (HPC) measures electromagnetic energy with high granularity over polar angles from 40° to 140°. For fast triggering, a a scintillator layer is inserted behind the first five radiation lengths. The light signals are carried by optical fibers to the outside of the iron yoke.

• The SUPERCONDUCTING SOLENOID provides 1.2 T of axial magnetic field. For the 1989 run part of the data were taken with a reduced field of 0.7 T.

- The TIME OF FLIGHT (TOF) system is composed of a single layer of 172 counters surrounding the solenoid, and covering $|\cos\vartheta| \leq 0.75$.

- The FORWARD ELECTROMAGNETIC CALORIMETER (FEMC) in the endcaps consists of 2 x 4500 lead glass blocks (granularity = $1° \times 1°$) with phototriode readout and covering polar angles from $10°$ to $35.5°$ and from $144.5°$ to $170°$.

- The SMALL ANGLE TAGGER (SAT) calorimeters are used for monitoring the luminosity. They are discussed in detail in section 4.

3. Trigger

The hadronic trigger was based on the fast signals from the HPC and TOF scintillation counters. Individual counters were arranged in two groups of four quadrants each placed symmetrically on both sides of the beam crossing point. The HPC scintillation counters were sensitive to electromagnetic showers with an energy ≥ 2 GeV, while the TOF counters were sensitive to minimum ionizing particles which penetrated the HPC and the coil. The single particle "scintillator" efficiency was measured with cosmic rays to be 90 ± 3 %. The following subtriggers were formed:
1. coincidences of back-to-back TOF sectors,
2. majority ≥ 3 TOF sectors,
3. majority ≥ 2 HPC sectors, and
4. coincidence of any TOF with any HPC sector.

The final "scintillator trigger" was the OR of these four subtriggers. In addition, a "chamber trigger" formed by the back-to-back coincidences of inner and outer tracking detectors was used to measure the efficiency of the four subtriggers from the data by recording the trigger pattern event by event. To enhance the number of recorded Z events a "forward trigger" based on the FEMC with a 3 GeV threshold was added and it was also used to check the systematic error on the total trigger efficiency. For hadronic Z events with sphericity axis between $50°$ and $130°$ the trigger efficiencies were:

TOF triggers alone (1+2)	95.5%
HPC trigger alone (3)	81.4%
TOF $*$ HPC trigger (4)	96.4%
chamber trigger	**97.5%**
forward trigger	90.0%
Overall trigger	99.5%

4. Luminosity Measurement

The trigger for luminosity events is based on analog sums of 24 channels in 24 overlapping sectors of $30°$ per endcap. The trigger requires back-to-back coincidence of of energy depositions above 10 GeV. Due to the overlapping geometry the trigger efficiency was calculated to be $(98\pm2)\%$ away from the dead regions and in the specified energy range.

The SAT calorimeters cover polar angles from 43 to 125 mrad. They are composed of alternating layers of lead sheets - concentric with the beam axis - and scintillating fibers running parallel to the beam. The light is collected behind the calorimeters and is measured by photodiodes. There are 288 towers in each SAT calorimeter.

The inner four rings of read-out elements have an azimuthal segmentation of $15°$, the outer four rings $7.5°$. A small dead region, 2 cm wide, is at the vertical junction of the two half cylinders. A lead (Pb) mask was added to the electron arm to improve the precision of the acceptance cut. The mask was 10 radiation lengths thick. Two set of measurements were performed with two different masks. The first had a radius of 12.0 cm and covered about 2/3 of the innermost SAT ring, while the second had a radius of 13.0 cm and covered it totally.

To reduce backgrounds to $< 1\%$ we required reconstructed showers, defined by clusters of $\geq$ 4 neighboring read-out elements with $E > 0.5$ GeV, to be coplanar to within $8°$, and to have an RMS azimuthal width of $< 11.5°$.

Figure 1 shows a schematic of the electron arm SAT calorimeter. To use the lead mask to define precisely the inner radius of the acceptance region, we require that the energy fraction in ring 1 be less than 0.90 for the 12 cm ring and 0.60 for the 13 cm ring. This guarantees that the electron has entered the calorimeter above the mask and not from below.

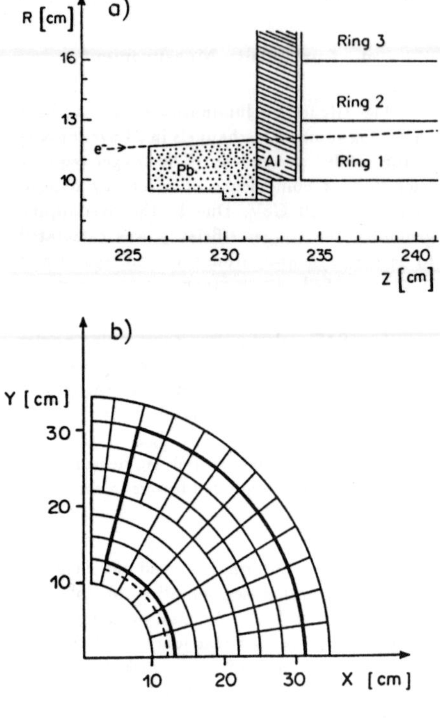

Figure 1. Schematic of the SAT electron calorimeter.

In figure 2 we show plots of the energy ratio E_2/E_{beam} in the positron calorimeter (without a lead ring) versus the fraction of the total electron shower energy deposited in ring 1 of the electron calorimeter. The energy loss due to the lead masks is visible in the reduced values

of R.

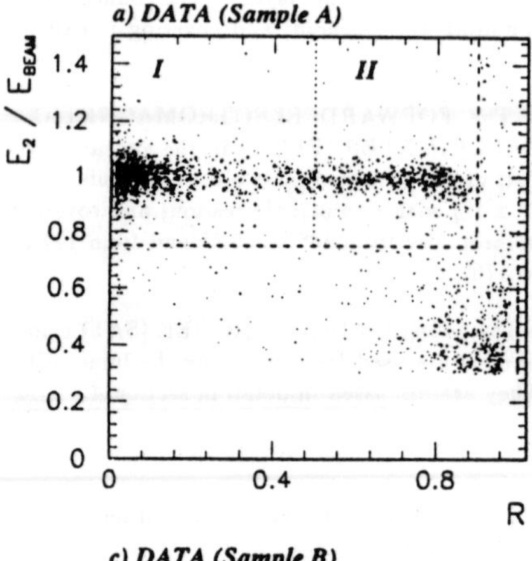

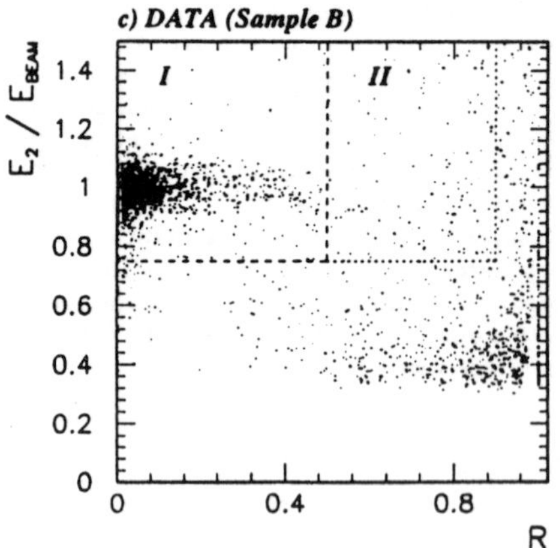

Figure 2. Energy ratio E_2/E_{beam} versus energy fraction R deposited in ring 1 of the electron arm for (a) the 12 cm Pb mask and (b) the 13 cm Pb mask.

In figure 3 we show the energy deposited in one SAT calorimeter arm versus that deposited in the other arm for events triggered with a single arm trigger. An enhancement for Bhabha scatters is clear, and shows that the data are well separated from a background of random coincidences of off-momentum electrons. The last cut made is to demand $E > 0.75 \, E_{beam}$ in each arm, as shown by the dashed lines in figure three.

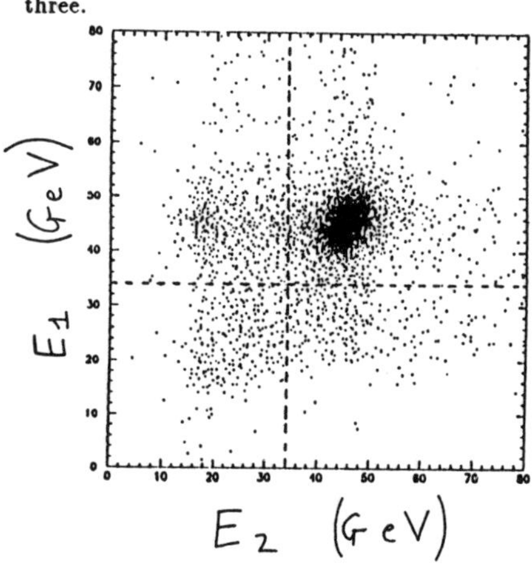

Figure 3. Energy deposited in the SAT e^+ calorimeter versus energy deposited in the SAT e^- calorimeter for events triggered with a single arm trigger.

The contributions to the uncertainty in the luminosity measurement are listed in Table I, in which A and B refer to the 12 cm and 13 cm Pb masks respectively.

Table I

Contribution		%
Trigger efficiency		0.6
ϕ-cut		1.0
Energy cut		1.0
Interaction point position		0.5
Background subtraction (A) B	(1.9)	0.5
MC model		1.0
MC statistics		0.6
Theory		1.0
Total, sample (A) B	(2.9)	2.3

5. Hadronic Event Selection

Two methods were used to select hadronic events: a "soft cut" method which used only charged tracks in the TPC, and a "hard cut" method which used both charged tracks in the TPC and electromagnetic energy in the HPC. In both methods the selection criteria on charged tracks in the TPC were:

$20° < \vartheta < 180°$,
0.1 GeV $< p < 50$ GeV,
track length > 30 cm,
$\Delta p/p < 1.0$,
xy projection of impact parameter < 4 cm, and
z coordinate of track origin $< \pm 10$ cm.

For method A we required ≥ 3 charged tracks in one hemisphere, and the sum over all charged tracks of p_T^2 be $> 9\ GeV^2$. The multiplicity cut eliminates cosmic ray events and leptonic events (except for some $\tau^+\tau^-$ events), and the Σp_T^2 cut eliminates beam-gas events and two-photon events. We estimate residual beam-gas and two-photon contamination to be less than 0.1% each, and the residual $\tau^+\tau^-$ contamination to be (1.3 ± 0.3)%. The efficiency in the barrel region (Sphericity axis $|\ cos\vartheta\ |<$ 0.65) is > 99.9%. Losses in the region $|\ cos\vartheta\ |>$ 0.65 are calculated by extrapolation according to:

$$(1 + (1 - 8/3\alpha_S/\pi)cos^2\vartheta_S).$$

The total efficiency was (93.5 ± 1.0)%.

For method B we used the same requirements on charged tracks as in method A, but added to them energy clusters in the HPC if those clusters were not associated with charged tracks and if $0.1 < E < 50$ GeV. For the events we required ≥ 5 charged tracks and either the invariant mass of the charged particles be > 12 GeV or the total energy be > 16 GeV. As with the first method the multiplicity cut eliminates cosmic ray events and most leptonic events, but in this case the $\tau^+\tau^-$ contamination is reduced to (0.3 ± 0.1)%. The charged mass cut eliminates both beam-gas and two-photon events. The total energy cut improved the full TPC data selection by 1% and helped to recover the runs in which a TPC sector was missing. The total efficiency with this method was (92.1 ± 1.1)%.

6. $Z^0 \rightarrow$ Hadrons

In figures 4, 5, 6, and 7 we show the distributions of the hadronic events in $\sqrt{\Sigma p_T^2}$, the angular distribution of the sphericity axis, the invariant mass of the charged particles, and the charge multiplicity respectively. In each case the predictions of the Monte Carlo calculation are in excellent agreement with the data.

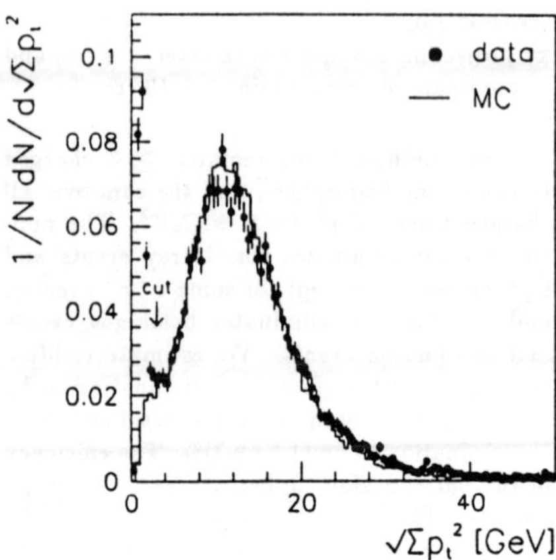

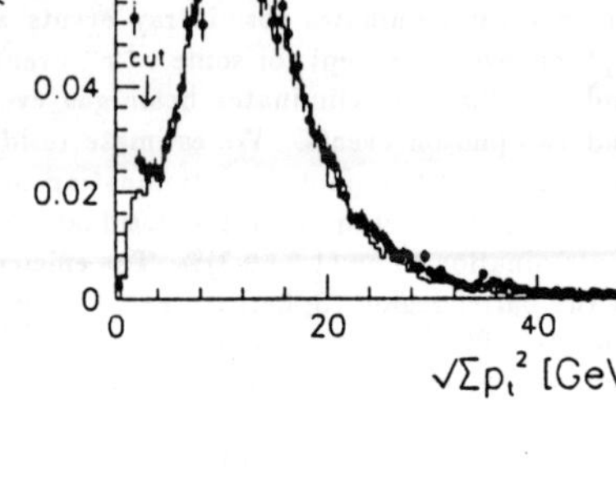

Figure 4. $\sqrt{\Sigma p_T^2}$.

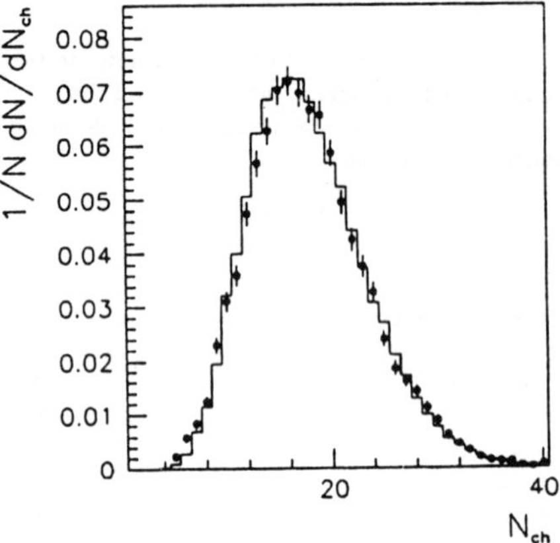

Figure 6. Invariant mass of the charged particles.

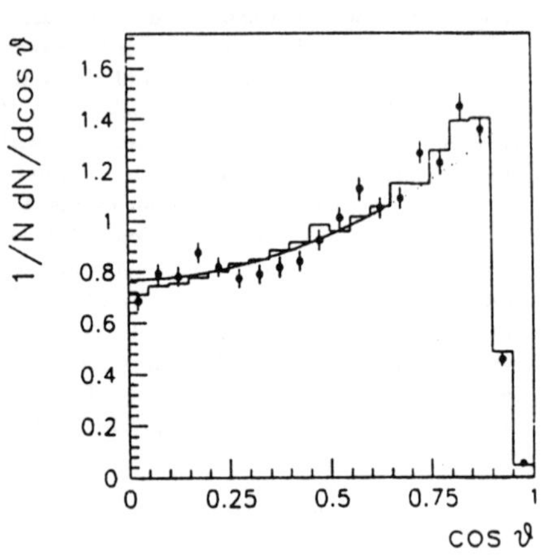

Figure 5. Angular distribution of sphericity axis.

Figure 7. Charge multiplicity.

In Table II we show the numbers of hadronic events obtained with methods A and B, the integrated luminosity, and the resultant cross section at each of the ten energies around the Z peak.

Table II. Z → hadrons

$\sqrt{s}$ (GeV)	No. Events A	No. Events B	$\int L$ (nb.$^{-1}$)	σ (nb.)
88.284	241	236	54.4	4.74 ± 0.32
89.284	427	416	49.8	9.42 ± 0.50
90.283	1094	1060	61.8	19.51 ± 0.73
91.036	1987	1930	73.3	29.15 ± 0.89
91.283	2392	2321	81.9	31.02 ± 0.89
91.536	2984	2918	106.3	29.97 ± 0.76
92.286	785	768	39.8	20.92 ± 0.96
93.284	587	575	54.2	11.57 ± 0.55
94.284	280	270	35.0	8.54 ± 0.57
95.042	95	93	16.3	6.19 ± 0.69
Total	10872	10587	572.8	

The quoted errors do not include the systematic uncertainty of 2.6% on the overall normalization.

At each energy the cross section for hadronin events was computed from the relation

$$\sigma_h = (N_h - N_b)/L\epsilon,$$

where N_h is the number of observed hadronic events listed in Table II, N_b is the number of background events (mostly $\tau^+\tau^-$ described earlier), L is the integrated luminosity and ϵ is the total efficiency.

We fit the cross section data listed in Table II to the sum of the continuum $\sigma_\gamma(s)$ including its leading radiative correction and $\sigma_Z(s)$ for the Z resonance and the interference terms.

$$\sigma_Z(s) = 12\pi \frac{\Gamma_e \Gamma_h}{\left(s - M_Z^2\right)^2 + \frac{s^2}{M_Z^2}\Gamma_Z^2} \times$$

$$[(\frac{s}{M_Z^2} + R\frac{s - M_Z^2}{M_Z^2})F - (2 + R)G]$$

The two-loop self-energy corrections to the vector boson propagator are given by the s^2/M_Z^2 and s/M_Z^2 terms. The factor R includes the real part of the γZ interference terms to lowest order plus the leading log and leading top quark mass corrections. With current limits on the top quark and Higgs masses R could vary from 0.07 to 0.12, and any deviation from the standard

model, e.g. the presence of a Z' pole, would result in an unusual value for R. The factors F and G include mainly the radiative corrections. The soft part of the initial state radiation is handled by the exponentiation formalism, and the hard part is calculated to second order in α. The final state radiation and its interference with the initial state are calculated to lowest order. The QCD corrections to the quark partial widths are calculated to second order in α_S using $\alpha_S = 0.12$, resulting in a 4% effect. We have used a top quark mass of 130 GeV and a Higgs mass of 100 GeV for these fits.

The fits performed vary from the least model dependent to those most restructive and most dependent on the predictions of the standard model. In the first fit we let $\Gamma_e\Gamma_h$ be free (in addition to Γ_Z and M_Z), and this allows a determination of the total width of the Z independent of the standard model constraint on the normalization. In this fit R was fixed at 0.095, which corresponds to the top and Higgs masses listed above. We obtain:
$\Gamma_Z = 2.511\pm0.065$ GeV,
$M_Z = 91.171\pm0.030$ (stat)±0.030 (beam) GeV,
$\Gamma_e\Gamma_h = 0.148\pm0.006$ (stat)±0.004 (syst) GeV^2,
$\chi^2/\text{DoF} = 4.0/6$, and
$\sigma_{Born}^{pole} = 12\pi\Gamma_e\Gamma_h/M_Z^2\Gamma_Z^2$
$= 41.6\pm0.7$ (stat)±1.1 (syst) nb.

The beam error in M_Z is due to the 30 MeV uncertainty in the machine energy, common to all LEP experiments. The correlation between Γ_Z and $\sigma_0 = \sigma_{Born}^{pole}$ is shown clearly in figure 8. We have also repeated the first fit also leaving R free. The best values of M_Z, Γ_Z and $\Gamma_e\Gamma_h$ remain unchanged, and we obtain a best value for R of -0.16$\pm$1.02, in good agreement with the standard model. Clearly, it will require significantly larger statistics to observe a deviation from the standard model by an unusual value for R.

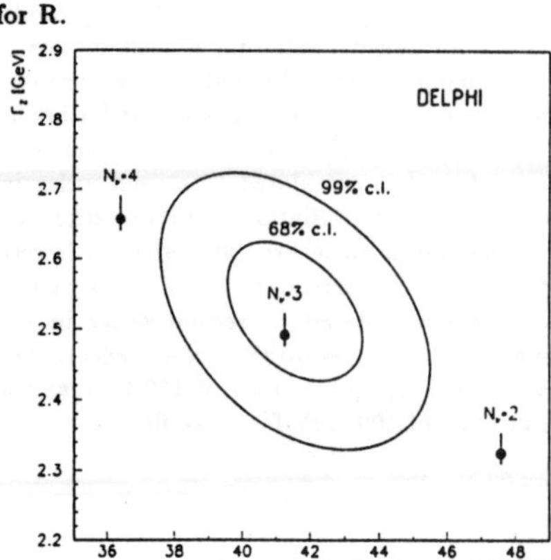

Figure 8. Correlation in the fitted values of Γ_Z and σ_o. The data points are for 2,3, and 4 light neutrino generations.

In the second fit we fix $\Gamma_e\Gamma_h$ at its standard model value. We obtain:
$\Gamma_Z = 2.494\pm0.020$ (stat)±0.039 (syst) GeV,
$M_Z = 91.170\pm0.030$ (stat)±0.030 (beam) GeV,
$\chi^2/\mathrm{DoF} = 4.0/7$.

The systematic error in Γ_Z is due to the normalization error discussed earlier.

If we use the standard model partial widths we obtain:
$\Gamma_{invis} = 495\pm20$ (stat)±39 (syst) MeV and
$N_\nu = 2.97\pm0.12$ (stat)±0.23 (syst).

Thus, a fourth generation massless neutrino is excluded at the 99.9% confidence level, and equivalently a fourth neutrino with $M_\nu < 36$ GeV is excluded at the 95% confidence level.

In the third fit only M_Z and an overall normalization factor, K, are free. All other quantities are fixed at their standard model values with $N_\nu = 3$. We obtain:
$M_Z = 91.171\pm0.030$ (stat)±0.030 (beam) GeV,
K $= 1.005\pm0.013$, and
$\chi^2/\mathrm{DoF} = 4.0/7$.

All these results are in good agreement with the predictions of the standard model with exactly three generations of light neutrino. In figure 9 we show the data listed in Table II with the results of the third fit plotted. The curves are for 2, 3, and 4 light neutrino generations.

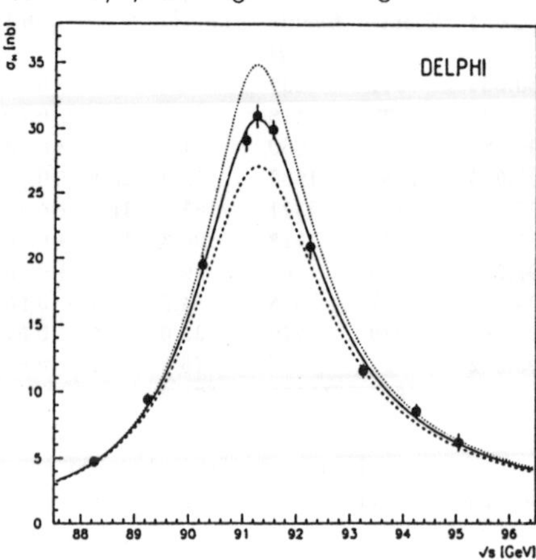

Figure 9. Hadronic cross section as a function of $\sqrt{s}$. The curves show the results of the third fit with 2, 3, and 4 generations of light neutrinos.

If we combine the results of the first fit with the ratio of leptonic to hadronic partial widths (see next sections), i.e. $\Gamma_\ell/\Gamma_h = 0.0489\pm0.023$, we obtain:
$\Gamma_\ell = 85.1\pm2.9$ MeV,
$\Gamma_h = 1741\pm61$ MeV,
$\Gamma_{invis} = 515\pm54$ MeV, and therefore
$N_\nu = 3.05\pm0.28$ generations.

Since the Z peak is 100 MeV above M_Z because of the radiative corrections, the cross section at $\sqrt{s} = 91.28$ GeV measures the peak cross section, which is 31.02 ± 0.89 (stat) $\pm$ 0.81 (syst), where the systematic error on the cross section is from the normalization uncertainty.

7. $Z^0 \rightarrow e^+e^-$

For the general leptonic trigger we required the OR of:
coincidence of back-to-back TOF sectors,
coincidence of ID/OD for back-to-back tracks,
or energy deposited in the HPC > 2 GeV.

The efficiency of this trigger was $> 99.7\%$ for e^+e^- events and $(97\pm2)\%$ for $\mu^+\mu^-$ and $\tau^+\tau^-$ events.

To select e^+e^- events we required:
$\geq$ 1 energy cluster > 25 GeV,
$\geq$ 1 HPC cluster > 10 GeV in opp. hemisphere,
these clusters compatible with TPC/ID tracks,
Acolinearity $< 10°$,
$50° < \vartheta_{E_{max}} < 130°$, and
a fiducial cut on the cluster of maximum energy.

The fiducial cut was imposed to avoid HPC boundaries.

A total of 263 events survived these cuts. Corrections required to obtain cross sections were:

0.94±0.02	$\tau^+\tau^-$ background
1.03±0.01	Acolinearity cut
1.02±0.01	Energy deposition losses
1.27	Geometry (fiducial cut)

The e^+e^- data are shown in Table III. These data correspond to 510.0 nb.$^{-1}$ integrated luminosity.

Table III. Z $\rightarrow e^+e^-$

$\sqrt{s}$ (GeV)	$N(e^+e^-)$	Cross Section (nb.)
88.28	15	0.36 ± 0.10
89.28	16	0.44 ± 0.11
90.28	37	0.83 ± 0.15
91.04	47	0.92 ± 0.13
91.28	43	0.84 ± 0.13
91.54	66	0.87 ± 0.11
92.29	13	0.53 ± 0.15
93.28	12	0.32 ± 0.09
94.28	10	0.37 ± 0.11
95.04	4	0.37 ± 0.18

These data have been fit to a Breit-Wigner function of the form:

$$\sigma(e^+e^- \rightarrow f\bar{f}) =$$

$$\frac{s}{M_Z^2} \times \frac{12\pi\Gamma_e\Gamma_f}{\left(s - M_Z^2\right)^2 + s^2\frac{\Gamma_Z^2}{M_Z^2}} \times [1 + \delta(s)]$$

The radiative corrections are contained in the factor δ. For each lepton species we define the ratio R_ℓ as:

$$R_\ell = \frac{\sigma(e^+e^- \rightarrow \ell^+\ell^-)}{\sigma(e^+e^- \rightarrow hadrons)}$$

The data listed in Table III and the results of the fit are shown in figure 10. After correcting for the t-channel contributions we obtain:
$R_e = (5.09\pm0.32(\text{stat})\pm0.18(\text{syst}))\%$, and
$\Gamma_e = 83.2\pm3.0(\text{stat})\pm2.4(\text{syst})$ MeV.

In the fit we have fixed $M_Z = 91.171$ GeV and $\Gamma_Z = 2.511$ GeV, i.e. at the values obtained from the hadronic fit.

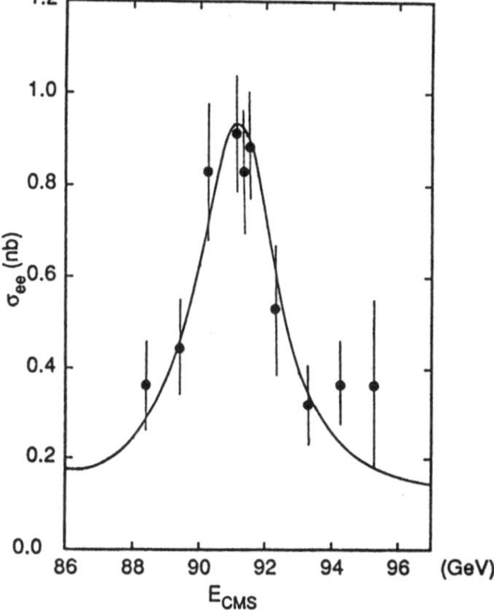

Figure 10. Z $\rightarrow e^+e^-$ cross section as a function of $\sqrt{s}$.

8. $Z^0 \rightarrow \mu^+\mu^-$

To select dimuon events we require events with exactly two charged tracks with:
p > 15 GeV,
$\geq$ 1 track with $50° < \vartheta < 130°$,
Acolinearity $< 10°$.

The cosmic ray events are cut by OD/TOF timing.

We use two methods of muon identification. In method A we require $\geq$ 1 hit in the muon chambers associated with an extrapolated track, and in method B we require a non-showering track to pass through the HPC associated with an extrapolated track. The efficiency for method A was $(95\pm2)\%$ due largely to the dead space in the muon counters, and for method B it was $(98\pm1)\%$. A total of 195 events survived the cuts, and these corresponded to an integrated luminosity of 390.0 nb.$^{-1}$. The overall efficiency was $(99.8\pm0.2)\%$. Corrections applied to the data to obtain the cross sections were:

0.96 ± 0.01	$\tau^+\tau^-$ background
1.07 ± 0.02	TPC dead space
1.03 ± 0.02	Trigger efficiency
1.03 ± 0.01	Acolinearity and p cuts.

The dimuon data are shown in Table IV. These data correspond to 390.0 nb.$^{-1}$ integrated luminosity.

Table IV. $Z^0 \to \mu^+\mu^-$

$\sqrt{s}$ (GeV)	$N(\mu^+\mu^-)$	Cross Section (nb.)
88.28	5	0.13 ± 0.06
89.28	7	0.24 ± 0.09
90.28	14	0.35 ± 0.09
91.04	30	0.72 ± 0.13
91.28	33	0.80 ± 0.14
91.54	70	0.90 ± 0.11
92.29	13	0.50 ± 0.14
93.28	17	0.39 ± 0.09
94.28	3	0.37 ± 0.21
95.04	3	0.25 ± 0.15

We have fit these data to the modified Breit-Wigner form as for the electron data, and as in the electron case we have set M_Z and Γ_Z to their values from the hadronic fit. The data and the results of the fit are shown in figure 11. We obtain:

$R_\mu = (4.96\pm0.35\pm0.17)\%$, and

$$\sqrt{\Gamma_e\Gamma_\mu} = 84.6\pm3.0\pm2.4 \text{ MeV.}$$

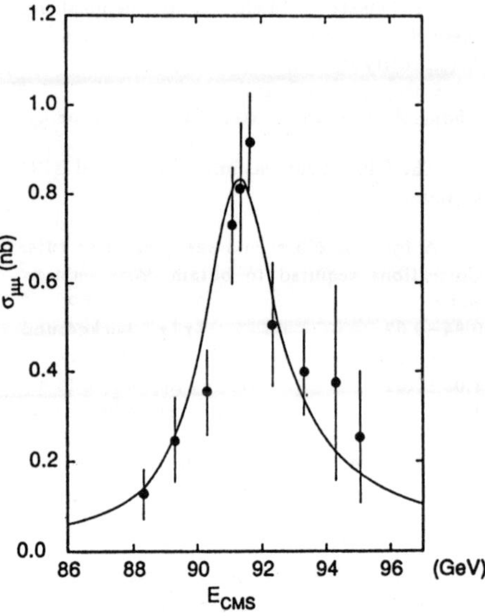

Figure 11. $Z \to \mu^+\mu^-$ cross section as a function of $\sqrt{s}$.

9. $Z^0 \to \tau^+\tau^-$

In order to select τ events from the hadronic events we require one τ decay into a single charged particle which is very isolated, and the other τ to decay into 1-5 charged particles. We also require:

$E_{vis} > 8$ GeV,

$p > 1$ GeV for all tracks,

Angle(isolated track, nearest track) $> 150°$,

$50° < \vartheta_{isolated\ track} < 130°$,

$50° < \vartheta_{\geq 1\ other\ track} < 130°$.

For the 1-1 topology we also require zero net charge, electromagnetic energy < 30 GeV on each side to reduce e^+e^- background, and either > 3 GeV electromagnetic energy or no hits in the muon chambers associated with extrapolated tracks in order to reduce $\mu^+\mu^-$ background. Even with these cuts the electron and muon background is about 30%. However, the kinematics of τ decay allow a clean separation, i.e. we also require:

Acoplanarity $> 1°$ or Acolinearity $> 3°$ and

$\Sigma_i \mid p_i \mid < 60$ GeV.

The systematic error in this selection procedure is 6%.

A total of 158 events survive these cuts, of which 84 are 1-1 topology, 28 are 1-2 topology, and 46 are 1-n topology for n > 2. The efficiency is $(64.5\pm3.6)\%$. The $\tau^+\tau^-$ data are shown in Table V, and these data correspond to 457.0 nb.$^{-1}$ integrated luminosity.

Table V. $Z^0 \to \tau^+\tau^-$

$\sqrt{s}$ (GeV)	$N(\tau^+\tau^-)$	Cross Section (nb.)
88.28	7	0.27 ± 0.10
89.28	5	0.36 ± 0.16
90.28	12	0.41 ± 0.12
91.04	34	0.85 ± 0.15
91.28	34	0.62 ± 0.11
91.54	47	0.93 ± 0.14
92.29	12	0.47 ± 0.14
93.28	3	0.12 ± 0.07
94.28	2	0.25 ± 0.18
95.04	2	0.18 ± 0.13

As for the electrons and muons, these data are fitted to the modified Breit-Wigner form with M_Z and Γ_Z fixed at their hadronic values. We obtain:
$R_\tau = (4.72\pm0.38\pm0.29)\%$ and
$\sqrt{\Gamma_e\Gamma_\tau} = 82.6\pm3.3\pm3.2$ MeV.

These data and the results of the fit are shown in figure 12.

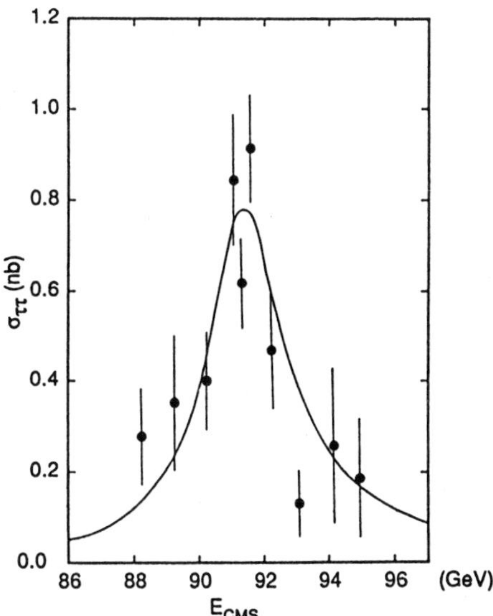

Figure 12. $Z \to \tau^+\tau^-$ cross section as a function of $\sqrt{s}$.

If we assume lepton universality, we obtain the combined result:
$R_\ell = (4.96\pm0.20\pm0.12)\%$.

Subtracting the s-channel photon contribution gives:
$\Gamma_\ell/\Gamma_h = (4.89\pm0.20\pm0.12)10^{-2}$

and a combined leptonic width of
$\Gamma_\ell = 83.6\pm1.8\pm2.2$ MeV.

We note that the standard model predicts 83.4 MeV for $\alpha_s = 0.12$ and Higgs and top masses of 100 GeV and 90 GeV respectively. However, the prediction increases to only 84.6 MeV for a top mass of 220 GeV.

If we assume lepton universality, we may write:

$$\Gamma_Z = \Gamma_h + 3\Gamma_\ell + \Gamma_{invis}$$

or

$$\Gamma_{invis}/\Gamma_\ell = 1/B_\ell - \Gamma_h/\Gamma_\ell - 3$$

where $B_\ell = \Gamma_\ell/\Gamma_Z$ and $B_\ell^2 \sim \sigma_\ell^{pole}$.

We use the muon data to obtain:
$\Gamma_{invis}/\Gamma_\ell = 6.22 \pm 0.47 \pm 0.50$.

We note that the strong anti-correlation of R_ℓ and B_ℓ was first noticed by G. Feldman [1]. Interpreting the invisible width of the Z in terms of the standard model width for a light neutrino generation, we obtain:
$N_\nu = 3.12 \pm 0.24 \pm 0.25$,

where the second error is due to the uncertainty in the luminosity.

10. Higgs Search

We search for the production of a pair of charged Higgs particles which decay either by (a) $H^\pm \to \tau^\pm \nu$ followed by $\tau^\pm \to 1$ prong, or by (b) $H \to \tau\nu$ and $H \to c\bar{s} \to 2$ jets. For decay modes (a) we require 2 charged prongs, $p_{tracks} > 3$ GeV, both prongs in the barrel region, Acolinearity $> 15°$ in the xy plane, and nothing in the HPC or FEMC. Monte Carlo simulations indicate that the product of the trigger and reconstruction efficiencies is then $(17\pm1)\%$.

For decay modes (b) we use two methods: hemispheric isolation and an isolated track. For hemispheric isolation we require one charged track, isolated in one hemisphere, with $p > 3$ GeV. Background from $q\bar{q}$ and $\tau^+\tau^-$ decays of the Z is eliminated by requiring ≥ 4 charged tracks in the opposite hemisphere. Finally, we require that the angle between the isolated track and the thrust axis in the opposite hemisphere be $> 20°$. This technique is inefficient for large Higgs masses as the hadrons are then produced more spherically.

For the isolated track method, efficient at larger Higgs masses, we require a charged track with $p > 5$ GeV be isolated in a jet and p_T to the jet axis be > 2.5 GeV. To eliminate QCD background we also require 2 spectator jets be no closer than $50°$ and no farther than $140°$ and Thrust < 0.9.

Figure 13 shows the efficiency of the two methods as functions of the charged Higgs mass. The sum of the efficiencies is thus approximately constant at about 20% from 10 to 40 GeV Higgs mass. Since no candidate events survived,

we place limits on the Higgs mass at the 95% confidence level as shown in figure 14. The variables are the mass of the Higgs and the branching fraction of the Higgs into hadrons.

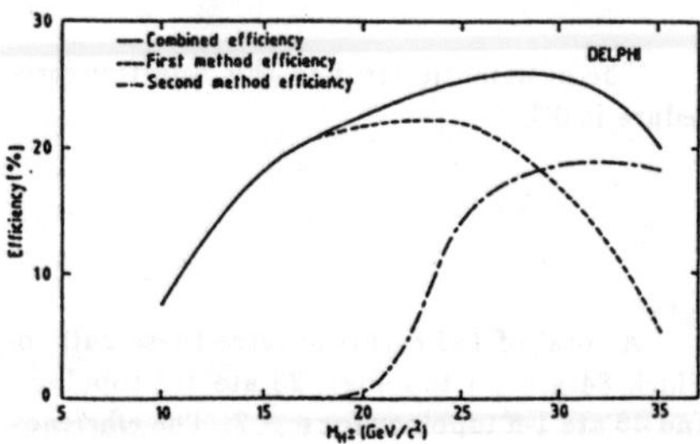

Figure 13. Efficiency of charged Higgs selection as a function of M_H.

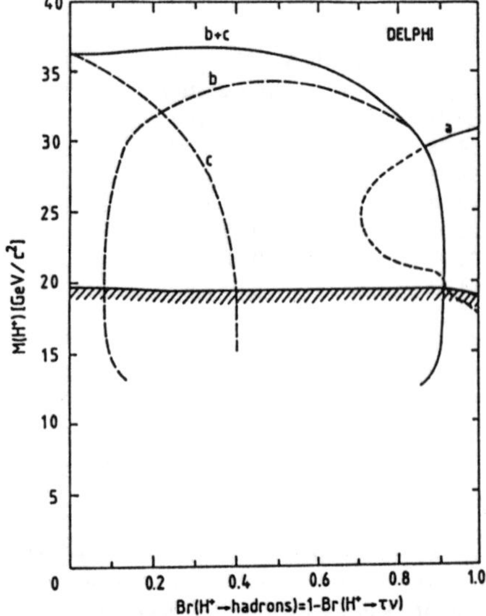

Figure 14. Confidence limits (95%) as functions of M_H and Higgs decay branching fractions.

11. New Particle Searches

We have conducted a series of new particle searches essentially through an isolated track study. The results may be summarized in the following listing:

$t\bar{t}'$ For 45 GeV we find R = 0.3
Suppress threshold behavior.
$$M_t > 44 \text{ GeV}$$

$b'\bar{b}'$ 45 GeV excluded. σ twice as big.

$L^0\bar{L}^0$ $L^0 \to \ell w^*$, $a^2 = v^2 = 1$
factor $(\beta/2)(3 + \beta^2)$ $|U|^2$ is big enough.
$\epsilon = 70\%$ as ≥ 2 leptons/event.
$$M(L^0) > 45 \text{ GeV (kinematic limit)}$$

$\chi^+\chi^-$ Mixing: pure wino to pure Higgsino
$$\chi^+ \to \chi^0 w^*$$
Worst case: $\beta(3 - \beta^2)/2$
Branching ratio 1/9 per lepton. $\epsilon = 20\%$
$$M(\chi) > 45.4 \text{ GeV}.$$
But, if M(χ^+) $\approx$ M(χ^0), no $q\bar{q}$ produced.
Look for misaligned $\ell^+\ell^-$. $\epsilon = 40\%$
$$M(\chi) > 44.6 \text{ GeV}.$$

t$\to$H b, and H$\to \tau\nu$ BR 30%
$$M(H) > 40 \text{ GeV}.$$

Selectron   Look for acoplanar 2-prongs.
Degenerate case : $M_R = M_L$
$$M(\tilde{e}) > 41.85 \text{ GeV}.$$

Conclusion and Acknowledgements

Using data collected in the first physics running period of the new LEP collider we have measured the mass, total width, and partial widths into hadrons and leptons of the Z boson. These measurements are used to compute the number of light neutrino generations.

The essential conclusion from this first half-year of LEP running is that there is no evidence for any deviations from the predictions of the standard model with exactly three generations of light neutrino. There is also no evidence for any Higgs particle, any supersymmetric partners of the known particles, or the top quark.

Please note that although the subjects covered here are identical to what was presented at the DPF meeting, some of the numbers presented there were preliminary. These numbers have been updated for this report.

The DELPHI collaboration wishes to thank the CERN accelerator groups for their outstanding achievement in creating a reliable and stable LEP collider in an astonishingly short time. The collaboration wishes to thank the funding agencies for all their support during the construction and running of the DELPHI detector. The Ames Laboratory group is supported by the U. S. Department of Energy under contract number W-7405-ENG-82-KA-01-01.

References

[1] Gary Feldman, "On the Possibility of Measuring the Number of Neutrino Species to a Precision of 1/2 Species with Only 2000 Z Events", MarkII/SLC-Physics Working Group Note 2-24, (1987).

PHYSICS AT THE Z

Conveners: J. Dorfan & D. Treille

SEARCHES FOR NEW PARTICLES
PRODUCED IN Z BOSON DECAY[*][†]

R. VAN KOOTEN, C.K. JUNG, and S. KOMAMIYA
Representing the Mark II Collaboration

*Stanford Linear Accelerator Center, Stanford University,
Stanford, California 94309*

ABSTRACT

Searches for events with new particle topologies in 455 hadronic Z decays with the Mark II detector at SLC are presented. 95% confidence level lower limits of 40.7 GeV/c^2 for the top quark mass and 42.0 GeV/c^2 for the mass of a fourth generation charge $-1/3$ quark, regardless of decay mode, are obtained. For a fourth generation sequential Dirac neutrino ν_4, a significant range of mixing matrix elements of ν_4 to other generation neutrinos is excluded for a ν_4 mass up to 43 GeV/c^2. Decays of the Z boson to a pair of non-minimal Higgs bosons ($Z \rightarrow H_s^0 H_p^0$), where one of them is relatively light ($\lesssim 10$ GeV/c^2), are also considered. Limits are obtained on the $ZH_s^0 H_p^0$ coupling as a function of the Higgs boson masses.

INTRODUCTION

New Quarks[‡]

We specifically search for top (t) quarks, and fourth-generation charge $-1/3$ (b') quarks. We assume pair-production of the new particles through Z decay with couplings and decay widths given by the Standard Model.

One expects t quarks to decay via the virtual-W (W^*) charged current (CC) process $t \rightarrow bW^*$. A b' quark may not decay 100% of the time via the CC decay $b' \rightarrow cW^*$ ($M_{b'} < M_t$) because of increased suppression of transitions which cross two generations. Consequently, the flavor-changing neutral-current (FCNC) loop decays[2] of $b' \rightarrow bg$ and $b' \rightarrow b\gamma$ must also be considered. Furthermore, in extensions of the Standard Model with two Higgs doublets, t and b' would dominantly decay into charged Higgs particles (H^+) by $t \rightarrow H^+b$ or $b' \rightarrow H^-c$ if $M_{H^\pm} < M_t, M_{b'}$.

---

[*] This work was supported in part by Department of Energy contract DE-AC03-76SF00515 (SLAC).

[†] This report is partly based on previous Mark II publications[1] where more details can be found.

[‡] Presented by R. Van Kooten.

Three types of event topologies are investigated. Type 1 is an event with a high-momentum isolated charged track. The semi-leptonic decays of t and b' or the decays of ν_4 will produce isolated leptons. To keep detection efficiencies high, lepton identification is not used. The type 2 topology is an event with an isolated photon. The decay $b' \rightarrow b\gamma$ motivates the search for this topology. Type 3 is the topology produced by a pair of heavy objects each decaying hadronically into two or more jets. Massive particles decaying into jets (*e.g.* $t, b \rightarrow qW^*$; $t, b' \rightarrow qH$; or $b' \rightarrow bg$) tend to produce spherical events which can be characterized by large momentum sums out of the event plane.

<u>Massive Neutrinos</u>[§]

Recently, measurements have been made at SLC[3] and LEP[4] to determine the number of neutrino species. The results rule out at greater than 95% confidence level (CL) the possibility of a fourth generation neutrino. These measurements, however, assume the neutrino to be massless and stable. Thus, we explore the possibility of a fourth generation massive neutrino ν_4.

We restrict our ν_4 search to a sequential fourth generation Dirac neutrino. We assume that $M_{\nu_4} < M_{L^-}$ in the new lepton doublet (ν_4, L^-), and that the weak eigenstates ν_ℓ and mass eigenstates ν_i of the four generations of neutrinos are mixed in analogy with the quark sector: $\nu_\ell = \sum_{i=1}^{4} U_{\ell i} \nu_i$. Through this mixing, ν_4 could decay by the weak charged current ($\nu_4 \rightarrow \ell W^*$; $\ell = e, \mu, \tau$).

Assuming ν_4 mixes with only one other generation of ℓ, the lifetime of the ν_4 can be expressed in terms of the muon lifetime as

$$\tau(\nu_4 \rightarrow \ell^- X^+) = \left[\frac{m_\mu}{m_4} \right]^5 \frac{\tau(\mu \rightarrow e\nu\bar{\nu}) Br(\nu_4 \rightarrow \ell^- e^+ \nu)}{|U_{\ell 4}|^2 f} \quad ,$$

where m_4 is the mass of the neutrino, and f is a phase-space suppression factor[5] for massive final state particles.

The expected characteristics of the topology of ν_4 events is different depending on the values of the mass m_4 and the mixing matrix element $U_{\ell 4}$. We investigate four types of events topologies each of which corresponds to a different region in the $m_4 - |U_{\ell 4}|^2$ plane. For heavy mass and large mixing (short-lived), the topology is similar to the type 1 isolated charged track topology described above. An invisible or semi-invisible event corresponds to a light mass and small mixing (long lived) region. The existence of these events would increase the invisible width of the Z boson resonance. We reinterpret the Mark II invisible width measurement (neutrino counting) analysis to obtain an exclusion region. An event with detached

---

§ Presented by C.K. Jung.

vertices corresponds to an intermediate mass and mixing region. These events can be characterized by a large fraction of charged tracks with large impact parameters with respect to the primary vertex. Finally, an event with two charged tracks recoiling against many tracks (2 vs. N) corresponds to a light mass and large mixing region.

<u>Non-Minimal Higgs Bosons</u> [¶]

For simplicity the model considered is restricted to two Higgs doublet (not necessarily the minimal supersymmetric model). In two doublet Higgs models there are two physical neutral scalar (CP even) Higgs bosons H_1^0 and H_2^0, one pseudoscalar (CP odd) H_p^0, and two charged Higgs bosons H^+ and H^-.

In the following, H_s^0 denotes either H_1^0 or H_2^0. We consider the decay $Z \rightarrow H_s^0 H_p^0$. The decay width for two doublet models is given by:[6]

$$\Gamma(Z \rightarrow H_s^0 H_p^0) = 0.5 \cdot \Gamma_{\nu\bar{\nu}} \bar{\beta}^3 \cos^2(a - b) \tag{1}$$

where $\bar{\beta} = \{[s - (M_s + M_p)^2][s - (M_s - M_p)^2]\}^{\frac{1}{2}}/s$, $s = E_{\text{cm}}^2$, and a and b are mixing angles of the Higgs doublets. Note that processes like $e^+ e^- \rightarrow Z \rightarrow Z^* H_p^0 \rightarrow f\bar{f} H_p^0$ or $e^+ e^- \rightarrow Z^* \rightarrow Z H_p^0$ are not allowed, since $Z H_p^0 Z$ coupling is forbidden at the tree level. These processes are allowed for H_s^0 but the rate is smaller than for the minimal Higgs boson by a factor[7] of $\sin^2(a - b)$. Therefore, search for the decay $Z \rightarrow H_s^0 H_p^0$ is complimentary to that for the decay $Z \rightarrow Z^* H_s^0$, since $\Gamma(Z \rightarrow H_s^0 H_p^0)$ is proportional to $\cos^2(a - b)$.

Higgs bosons are expected to decay dominantly into the heaviest available fermion pair: $H_i^0 \rightarrow f\bar{f}$, $(i = 1, 2, p)$. If the scalar mass is more than two times the pseudoscalar mass, $H_s^0 \rightarrow H_p^0 H_p^0$ is the dominant decay mode unless it is suppressed by the Higgs mixing.

APPARATUS

Details of the Mark II detector can be found elsewhere.[8] A cylindrical drift chamber in a 4.75 kG axial magnetic field measures charged particle momenta. Photons are detected in electromagnetic calorimeters covering the angular region $|\cos\theta| < 0.96$, where θ is the angle with respect to the beam axis. Barrel lead-liquid-argon sampling calorimeters cover the central region $|\cos\theta| < 0.72$ and the remaining solid angle is covered by end-cap lead-proportional-tube calorimeters. The detector is triggered by two or more charged tracks within $|\cos\theta| < 0.76$ or by neutral-energy requirements of a single shower depositing at least 3.3 GeV in the barrel calorimeter or 2.2 GeV in an end-cap calorimeter. This combination results in an estimated trigger efficiency of greater than 99% for hadronic Z decays.

[¶] Presented by S. Komamiya.

ANALYSIS AND RESULTS

Except for the long-lived ν_4 searches, charged tracks are required to project into a cylindrical volume of radius 1 cm and half-length of 3 cm around the nominal collision point parallel to the beam axis. Tracks are required to be within the angular region $|\cos\theta| < 0.82$, and to have transverse momenta with respect to the beam axis of at least 150 MeV/c. An electromagnetic shower is required to have shower energy greater than 1 GeV and $|\cos\theta| < 0.68$ for the central calorimeter and $0.68 < |\cos\theta| < 0.95$ for the endcap calorimeter.

The expected number of produced exotic events before cuts is normalized to the total number of hadronic events (N_{had}) that fulfill hadronic event selection criteria.[3] The expected number of produced exotic events N_x, $x = t\bar{t}, b'\overline{b'}, \nu_4\bar{\nu}_4$, or $H_s^0 H_p^0$ is given by

$$N_x = \frac{N_{\text{had}}\Gamma_x}{\epsilon_{q\bar{q}}\Gamma_{q\bar{q}} + \epsilon_x\Gamma_x} \quad ,$$

where $\Gamma_{q\bar{q}}$ is the partial width of the Z to u, d, s, c, and b ($udscb$) quarks, $\epsilon_{q\bar{q}}$ is the efficiency for $udscb$ quarks to pass the hadronic event criteria, Γ_x is the partial width of the Z to the exotic particle in question, and ϵ_x is the efficiency for the exotic particle events to pass the hadronic event criteria. First order QCD corrections[9] are used when calculating Γ_q and Γ_x. The data sample corresponds to $N_{\text{had}} = 455$ events and to an integrated luminosity of 19.7 ± 0.8 nb^{-1}.

New Quarks

All events are required to contain at least six charged tracks and the sum of charged particle energy and shower energy (E_{vis}) must be greater than $0.1\, E_{\text{cm}}$. To ensure that the events are well contained within the detector, the polar angle of the thrust axis (θ_{thr}) of each event must satisfy the condition $|\cos\theta_{\text{thr}}| < 0.8$.

Type 1 events must have event thrust less than 0.9 and must contain at least one isolated charged track. An isolated track is one with isolation parameter $\rho_i > 1.8$ where ρ_i is defined as follows: The Lund jet-finding algorithm is applied[10] to the charged and neutral tracks excluding the candidate track i. We then define

$$\rho_i \equiv \min_{\text{jets } j}[(2E_i(1 - \cos\theta_{ij}))^{1/2}],$$

where E_i is the track energy in GeV and θ_{ij} is the angle between the track and each jet axis. The ρ value of an event is defined as the maximum value of ρ_i of all charged tracks in an event.

The detection efficiencies (ϵ_D) for t, b', and ν_4 are calculated with a modified Lund 6.3 parton shower Monte Carlo program with Lund symmetric fragmentation.[11] Uncertainties in detection efficiency ($\Delta\epsilon_D/\epsilon_D$) from Monte Carlo statistics ($\approx 3\%$), detector simulation and beam backgrounds ($\approx 1\%$), theoretical uncertainties in semi-leptonic branching ratios ($\approx 2\%$), and fragmentation models are calculated. The last error is estimated using different Monte Carlo generators and fragmentation schemes.[12] For masses in the range 25 – 30 GeV/c^2 the error can be as large as 12%, and we choose to use the value 12% for all quark masses.

The number of produced events N_x has both a statistical uncertainty from N_h, and a substantial systematic error (as large as 25% depending on the exotic particle mass) due to uncertainties in higher order QCD corrections[9] in the calculation of Γ_x if x is a t or b' quark. The total error on the expected number of events is calculated by summing the individual statistical and maximum systematic errors in quadrature. Our best estimate of the expected number of exotic events minus the total error is used for setting mass limits.

There is one event satisfying the type 1 event selection criteria in our data sample of 455 hadronic Z decays, while 0.9 events (Lund Shower with Peterson fragmentation[12]) to 1.8 events (Webber 4.1[12]) are expected from QCD five-flavor processes. To be conservative, background subtraction is performed using the smallest value (0.9 events) expected. Using a standard approach,[13] the limits shown in Table 1 are obtained.

For the type 2 event topology we require that the event thrust not exceed 0.9 and that there be at least one isolated photon. An isolated photon is defined as a neutral shower with $\rho_i > 3.0$ where ρ_i is defined as for a charged track. No events were found satisfying the type 2 event selection criteria. From this observation, we obtain $M_{b'} > 45.4$ GeV/c^2 (95% C.L.) if B($b' \to b\gamma$) $\geq 25\%$.

The type 3 event topology requires $M_{\mathrm{out}} > 18$ GeV/c^2 where

$$M_{\mathrm{out}} \equiv \frac{E_{\mathrm{cm}}}{E_{\mathrm{vis}}} \frac{1}{c} \sum |p_T^{\mathrm{out}}|.$$

p_T^{out} is the momentum component of a charged track or neutral shower out of the event plane defined by the sphericity tensor, and the sum is over all charged tracks and neutral showers. Six events are observed in the data with $M_{\mathrm{out}} > 18$ GeV/c^2 , while 4.8 events (Lund Matrix Element[12]) to 11.7 events (Webber 4.1[12]) are expected from QCD five-flavor processes. To be conservative, background subtraction is performed using the smallest value (4.8 events) expected.

The above observation allows us to set the limits included in Table 1. The case of the H^- decaying partially into $\tau\bar{\nu}$ is found to weaken the listed limits for $t \to bH$ and $b' \to cH$, but if B($H^- \to \tau\bar{\nu}$) $< 70\%$ both limits remain over 40 GeV/c^2 .

Finally, the analyses of the three event topologies are combined to give mass limits on b' as a function of branching ratio into the CC process, assuming that the remaining decays are only through the FCNC decays bg and $b\gamma$ (we assume a Higgs particle is not kinematically accessible). Detection efficiencies are found for the possible combinations of the above decays, and combined to give the result that $M_{b'} > 42.0$ GeV (95% C.L.) for all possible values of the branching ratio into the CC process and all possible mixtures of bg and $b\gamma$ in the FCNC part.

Table 1. Summary of the mass limits set by searching for the three event topologies described in the text. For $t \to bH^+$ and $b' \to cH^-$ limits, $M_{H^\pm} \geq 25$ GeV/c^2, and $H \to c\bar{s}$. The ν_4 limits are for decays with vertices within the described cylindrical fiducial region (i.e. decay length $\lesssim 1$ cm).

Particle	Decay Products (Br = 100%)	Topology	Mass Limit (95% C.L.) (GeV/c^2)
top	bW^*	Isolated Track	40.0
	bW^*	M_{out}	40.7
	bH^+	M_{out}	42.5
b'	cW^*	Isolated Track	44.7
	cW^*	M_{out}	44.2
	cH^-	M_{out}	45.2
	bg	M_{out}	42.7
	$b\gamma$, Br $\geq 25\%$	Isolated Photon	45.4
ν_4	eW^*	Isolated Track	43.7
	μW^*	Isolated Track	44.0
	τW^*	Isolated Track	41.3

Massive Neutrinos

Mass limits from the isolated track analysis for a short-lived ν_4 ($\beta\gamma c\tau \lesssim 1$ cm) with different generation mixings are given in Table 1 of the previous section, and are shown as exclusion regions in Fig. 1. In Fig. 1, the lower bound of sensitivity along the $|U_{\ell 4}|^2$ axis is due to massive neutrinos decaying at larger radii, and tracks failing to project into the fiducial cylinder centered at the IP described earlier.

For longer ν_4 lifetimes, detached vertices occur, resulting in a large fraction of tracks with large impact parameters. The dominant background of beam-gas and beam-beampipe interaction events are usually forward-scattered, and have low multiplicity and low total energy. To eliminate them, we require an event to have at least eight good charged tracks and to have total visible energy greater than 35% of E_{cm}. In addition, the minimum of energy visible in the forward and backward hemispheres with respect to the electron beam direction must be greater than 7% of E_{cm}. We conservatively estimate that there are less than 0.01 beam-gas and beam-beampipe interaction events in the final 350 event data sample.

The impact parameter b in the plane perpendicular to the beam axis is defined as the distance of closest approach to the average beam position. An event search parameter χ_{imp} is then defined as the fraction of charged tracks with significance

b/σ_b greater than 5.0, where σ_b is the sum in quadrature of the track position error to the track trajectory (≈ 300 μm) and the error in the average beam position (200 μm).

Hadronic background events containing charm, bottom, or strange quark decays rarely yield χ_{imp} greater than 0.5, since there are many other tracks in the event which project to the primary vertex[*]; however, many $\nu_4\bar{\nu}_4$ events with a reasonable lifetime would yield χ_{imp} greater than 0.5. We require $\chi_{\text{imp}} > 0.6$ for an event for it to be tagged as a $\nu\bar{\nu}_4$ event. No events in the data pass the selection criteria, and Monte Carlo simulations predict less than 0.2 events. Detection efficiencies including full trigger emulation are determined for simulated[14] $\nu_4\bar{\nu}_4$ events of different masses and lifetimes. Uncertainties in detection efficiency ($\Delta\epsilon_D/\epsilon_D$) from Monte Carlo statistics ($\approx 2\%$), detector simulation and beam backgrounds ($\approx 4\%$), tracking efficiencies for tracks with large impact parameters ($\approx 10\%$), and different fragmentation models are estimated. Following the procedure of the previous section, the 95% CL limit contour is determined and illustrated in Fig. 1.

From fits to the Mark II Z resonance data where M_Z and the number of *massless* neutrino generations N_ν in the total width $\Gamma_Z^0 = \Gamma_{udscb} + \Gamma_e + \Gamma_\mu + \Gamma_\tau + N_\nu\Gamma_\nu$ are allowed to vary, we find $N_\nu < 3.86$ at 95% CL. A massive neutrino would contribute a fractional neutrino generation due to the mass threshold factor. Assuming that the neutrinos of the first three generations are massless, that ν_4 is a Dirac-type neutrino, and that no other physics intervenes, we obtain a limit of $m_4 > 19.6$ GeV/c^2 for a *stable* ν_4.

In the fits, σ_{peak} is given by $\sigma_{\text{peak}} = 12\pi\Gamma_e\Gamma_{\text{vis}}^0/M_Z^2\Gamma_Z^2$, where Γ_{vis}^0 is constrained to be the Standard Model prediction (3 generations, 5 quarks) for decays visible in the detector. For an unstable ν_4, the region where the relation

$$\frac{\Gamma_{\text{vis}}}{\Gamma_Z^2} = \frac{\Gamma_{\text{vis}}^0 + \epsilon\Gamma_{\nu_4}}{(\Gamma_Z^0 + \Gamma_{\nu_4})^2} < \frac{\Gamma_{\text{vis}}^0}{(\Gamma_Z^0 + (3.86 - 3)\Gamma_\nu)^2} \tag{1}$$

is true is excluded at 95% CL. ϵ is the efficiency for $\nu_4\bar{\nu}_4$ events to satisfy the visible event selection criteria, and Γ_{ν_4} is the partial width (a function of m_4) for ν_4. The values of ϵ for different m_4 and lifetimes are determined and the region where eqn. (1) is satisfied is shown in Fig. 1.

We now consider the region $2.5 \lesssim m_4 \lesssim 20$ GeV/c^2 and $\tau(\nu_4) \approx 10$ psec which is not excluded by any prior direct searches by other collaborations for the case of ν_4 mixing to ν_τ. The search topology is two charged tracks recoiling against many (2 versus N, $N \geq 2$). For the mass range considered, $Br(\nu_4 \to 2\text{ prong}) \gtrsim \frac{1}{3}$, resulting in high detection efficiencies.

---

[*] The same is true for a hadronic event with one or more particles undergoing a nuclear interaction in the beampipe or detector materials.

286

Events are divided into two hemispheres with respect to the thrust axis, and the two versus N charged track topology selected. Additional cuts are $|\cos\theta_{\text{thr}}| < 0.8$; $E_{\min}^{\text{ch}} > 6$ GeV, where $E_{\min}^{\text{ch}}$ is the total charged energy in the hemisphere with the minimum charged energy; and sphericity $S > 0.4$ for $m_4 \geq 10$ GeV/c^2. Four masses (2.5, 5.0, 10.0, and 22.0 GeV/c^2) of ν_4 events are simulated at values of $|U_{\tau 4}|^2$ resulting in ν_4 lifetimes long enough to lie in regions excluded by other experiments. These are the most conservative cases in the search region since detection efficiencies are always higher for larger values of $|U_{\tau 4}|^2$ (*i.e.* shorter lifetimes). Results are shown in Table 2 and as an exclusion region in Fig. 1.

Table 2. Results of applying the selection criteria for the 2 versus N topology for data and simulated $\nu_4\bar{\nu}_4$ events and the resultant exclusion levels (*no* background subtraction).

m_4 (GeV/c^2)	No. of Expected Events	No. of Data Events	Confidence Level for Exclusion
2.5	21.0	4	>99.99%
5.0	15.5	4	>99.94%
10.0	4.9	1	95.6%
22.0	5.8	1	97.9%

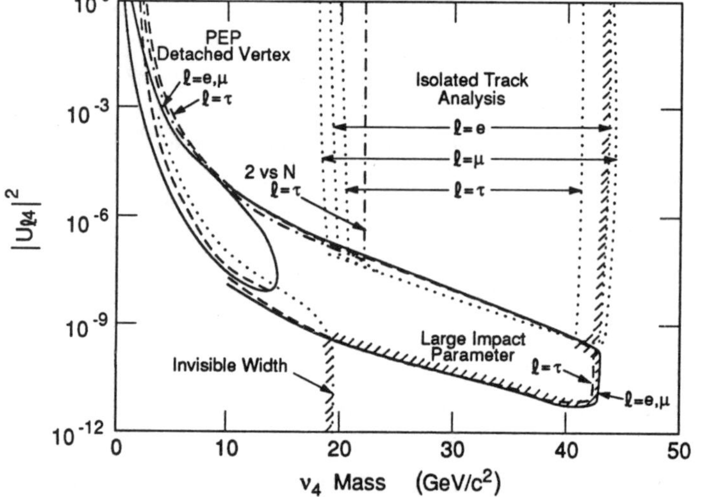

Figure 1. Summary of the 95% CL exclusion regions for a sequential fourth generation Dirac neutrino ν_4 as a function of mass and mixing matrix element (which is related to lifetime). The hatched region is the combined exclusion region for mixing to ν_e only.

Non-Minimal Neutral Higgs Bosons

We concentrate on the case in which one of the produced Higgs bosons $(H_l^0)^\star$ is relatively light (less than $2M_b$). We study four typical cases: [A] $M_{H_l^0} < 2M_\mu$, [B] $2M_\mu < M_{H_l^0} < 2M_\tau$ and H_h^0 decays into $f\bar{f}$, and [C] $2M_\mu < M_{H_l^0} < 2M_\tau$ and H_h^0 decays into $H_l^0 H_l^0$. We also investigate the case in which [D] $2M_\tau < M_{H_l^0} < 2M_b$ and H_l^0 decays into $\tau^+\tau^-$.

In case [A] ($Z \to H_l^0 H_h^0$, $H_l^0 \to e^+e^-$ or $\gamma\gamma$, $H_h^0 \to b\bar{b}, c\bar{c}$ or $\tau^+\tau^-$), H_l^0 is sufficiently long lived to escape detection. If the heavier Higgs boson (H_h^0) decays into a heavy fermion pair ($b\bar{b}$, $c\bar{c}$ or $\tau^+\tau^-$) and the mass is smaller than about the beam energy, the signature of $Z \to H_s^0 H_p^0$ events is a monojet topology. If the mass of the heavier Higgs boson is about equal to or greater than the beam energy, the momentum of the unseen H_l^0 is small and hence the event topology is two jets with a large angle between their axes (acoplanar two-jet events). The monojet events are selected with the following criteria: (M1) $|\cos\theta_{\text{th}}| < 0.7$ and (M2) the sum of the charged and neutral energy in the lower energy hemisphere (defined by the event thrust axis), E_{back}, is smaller than 3.0 GeV. The acoplanar two-jet events are selected by the following cuts: (P1) $|\cos\theta_{\text{th}}| < 0.7$, (P2) P_T of the event must be larger than 15 GeV and (P3) the acoplanarity angle ϕ_{acop} must be greater than 40 degrees. After applying cuts ((M1-2) or (P1-3)), no events survive. The expected number of background events from ordinary quark ($udscb$) production is estimated to be 0.3 to 0.7 using QCD-based Monte Carlo models.[12] In Fig. 2(a), the 95% C.L. contour for the excluded region is shown in the plane of the suppression factor ($\cos^2(a-b)$) vs. $M_{H_h^0}$, assuming H_l^0 is light ($M_{H_h^0} < 2M_\mu$) and stable for the case that H_h^0 decays into $b\bar{b}$ or $c\bar{c}$ [$\tau^+\tau^-$].

For case [B] ($Z \to H_l^0 H_h^0$, $H_l^0 \to \pi^+\pi^-$ or $\mu^+\mu^-$, $H_h^0 \to b\bar{b}, c\bar{c}$ or $\tau^+\tau^-$), the event topology is an isolated particle pair with opposite charge (for instance, $\mu^+\mu^-$, $\pi^+\pi^-$ or K^+K^-) which recoils against jets. We require that E_{vis} be greater than $0.5\sqrt{s}$ and that there be at least one isolated particle pair with opposite charge. An isolated pair of charged particles (i,j) is defined as two oppositely charged particles with momentum sum ($|\vec{p}_i + \vec{p}_j|$) larger than 20 GeV, individual momenta greater than 2 GeV, and isolation parameter $\rho_{ij} > 4.0$ GeV$^{\frac{1}{2}}$. The isolation parameter ρ_{ij} is defined as follows: The Lund jet-finding algorithm is applied to all charged tracks in the event (except the candidate pair ij) and neutral tracks with energy greater than 1.5 GeV. We then define

$$\rho_{ij} \equiv \min_{\text{jets } J} \sqrt{2E_{ij}(1 - \cos\chi_{ijJ})},$$

$\star$ We also use notations H_l^0 and H_h^0, where H_l^0 is defined to be lighter than H_h^0 and the two have opposite CP eigenvalues.

288

where E_{ij} is the pair energy assuming the pair to be $\pi^+\pi^-$ and χ_{ijJ} is the angle between the pair momentum direction and the jet axis. For $H_s^0 H_p^0$ events, a peak is seen at $|\vec{p}_i + \vec{p}_j| \approx (\sqrt{s}/2)(1 - M_{H_h^0}^2/s)$. Events are selected if $0.75\,(\sqrt{s}/2)(1 - M_{H_h^0}^2/s) < |\vec{p}_i + \vec{p}_j| < 1.25\,(\sqrt{s}/2)(1 - M_{H_h^0}^2/s)$ for an assumed value for $M_{H_h^0}$. No events survive the selection criteria. The number of expected background events increases with $M_{H_h^0}$ from 0.1 ($M_{H_h^0} = 5$ GeV) to 0.5 ($M_{H_h^0} = 60$ GeV), and is estimated using Monte Carlo models.[12] The limits are shown in Fig. 2(b).

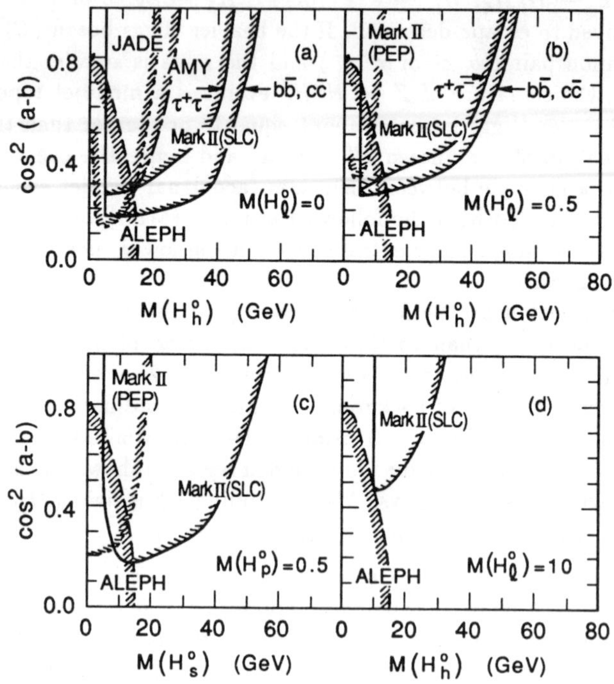

Figure 2. The 95% C.L. contours for the excluded region in the plane of the suppression factor ($\cos^2(a - b)$) vs $M_{H_h^0}$ for: (a) H_ℓ^0 is light ($M_{H_\ell^0} < 2M_\mu$) and stable; (b) H_ℓ^0 decays into a particle pair of opposite charges and H_h^0 decays into $b\bar{b}$, $c\bar{c}$ (solid curve) or $\tau^+\tau^-$ (dashed curve) (assuming that $M_{H_\ell^0} = 0.5$ GeV, but the limit is valid for $M_{H_\ell^0}$ smaller than a few GeV as long as it decays dominantly into a particle pair of opposite charges); (c) $Z \to H_s^0 \to H_p^0 H_p^0 \to 3(\mu^+\mu^-)$ or $3(e^+e^-)$ and assuming that $M_{H_p^0} = 0.5$ GeV, but the limit is valid for $M_{H_p^0}$ smaller than a few GeV as long as it decays dominantly into a particle pair of opposite charges; and (d) H_ℓ^0 decays into $\tau^+\tau^-$ with 100% branching fraction and H_h^0 decays into $b\bar{b}$, $c\bar{c}$ or $\tau^+\tau^-$.

For case [C] ($Z \to H_s^0 H_p^0 \to H_p^0 H_p^0 H_p^0$, $H_p^0 \to \mu^+\mu^-$), the event topology is three pairs of oppositely charged particles. The $H_p^0 \to \mu^+\mu^-$ decay mode is dominant since $\pi\pi$ or $\pi\pi\pi$ modes are suppressed for the H_p^0 decay. We require that the total charged particle energy E_{ch} be greater than $0.5\sqrt{s}$ and that exactly three jets are found using the Lund jet-finding algorithm. We require for each jet that the energy be larger than 4 GeV, the invariant mass be smaller than 4 GeV, and the total charge of each jet be -1, 0 or 1. We further require that the maximum charged multiplicity of the jets be either 2 or 3 and the minimum is either 1 or 2. No events survive the selection criteria. The excluded region is shown in Fig. 2(c).

For case [D] ($Z \rightarrow H_l^0 H_h^0 \rightarrow \tau^+\tau^- + \text{jets}$), the Lund jet-finding algorithm is applied. We select events with only two jets in either of the hemispheres defined by the plane perpendicular to the event thrust axis. Further, we require that the two jets be consistent with a tau pair (the invariant mass of each jet is smaller than 2 GeV, the number of charged particles in each jet is one, and charge of the two jets is opposite). Since a $\tau^\pm$ decay involves missing neutrinos, we cannot look for an invariant mass peak of $\tau^+\tau^-$. We look for the peak in the $\tau^+\tau^-$ opening angle. Events are selected between 75% and 150% of the Jacobian peak of the opening angle (24 degrees at $M_{H_h^0} = 10$ GeV and 31 degrees at $M_{H_h^0} = 45$ GeV). After the cuts no events survive in the angular region and the expected number of background events is 0.3-0.5. The excluded region is shown in Fig.2(d) assuming the H_l^0 decays into $\tau^+\tau^-$ with 100% branching fraction.

SUMMARY

The study of Z decays provides an ideal laboratory to search for new particles. The presented results significantly extend limits from e^+e^- annihilation before the advent of Z physics at SLC and LEP. Due to space considerations, detailed comparisons with new limits from LEP with a larger data sample presented in these proceedings are not made; however, our results are consistent with the LEP limits.

REFERENCES

1. G.S. Abrams *et al.*, Phys. Rev. Lett. **63**, 2447 (1989); C.K. Jung, R. Van Kooten, *et al.*, Phys. Rev. Lett. **64** 1091 (1990); S. Komamiya *et al.*, SLAC–PUB–5175 (1990), submitted to Phys. Rev. Lett.

2. W. Hou and R.G. Stuart, Phys. Rev. Lett. **62**, 617 (1989); V. Barger *et al.*, Phys. Rev. Lett. **57**, 1518 (1986).

3. Mark II Collaboration, G.S. Abrams *et al.*, Phys. Rev. Lett. **63**, 2173 (1989).

4. ALEPH Collaboration, D. Decamp *et al.*, CERN preprint EP/89-132; OPAL Collaboration, M.Z. Akrawy *et al.*, CERN preprint EP/89-133; L3 Collaboration, B. Adeva *et al.*, CERN preprint L3-preprint-001; DELPHI Collaboration, P. Aarnio *et al.*, CERN preprint EP/89-134

5. R.E. Shrock, Phys. Rev. **D24**, 1275 (1981); Y.S. Tsai, Phys. Rev. **D4**, 2821 (1971) and Phys. Rev. **D19**, 2809 (1979).

6. G. Pocsik and G. Zsigmond, Z. Physik, **C10** (1981) 367.

7. J.F. Gunion and H.E. Haber, Nucl. Phys. **B272**, 1 (1986).

8. G.S. Abrams *et al.*, Nucl. Instrum. Methods A **281**, 55 (1989).

9. J.H. Kühn, A. Reiter, and P.M. Zerwas, Nucl. Phys. **B272**, 560 (1986).

10. T. Sjöstrand, Comput. Phys. Comm. **28**, 229 (1983). In the algorithm, the jet-forming cutoff parameter d_{join} is changed from its default value to $d_{join} = 0.5$ GeV.

11. Lund 6.3 parton shower model, T. Sjöstrand, Comput. Phys. Comm. **39**, 347 (1986); M. Bengtsson and T. Sjöstrand, Nucl. Phys. **B289**, 810 (1987). In the hadronic decay of t and b' quarks, gluon emission effects are incorporated by the parton shower model based on a leading-log-approximation.

12. Lund 6.3 parton shower (Ref. 11) and Lund 6.3 matrix element (T.D. Gottschalk and M.P. Shatz, Phys. Lett. B **150**, 451 (1985)), both with Lund symmetric fragmentation (Ref. 11) and Peterson fragmentation (C. Peterson *et al.*, Phys. Rev. D **27**, 105 (1983)); and Webber 4.1 (G. Marchesini and B.R. Webber, Nucl. Phys. **B238**, 1 (1984); B.R. Webber, Nucl. Phys. **B238**, 492 (1984)).

13. Particle Data Group, Phys. Lett. B **204**, 81 (1988); O. Helene, Nucl. Inst. Meth. **212**, 319 (1983).

14. Using a modified form of LULEPT, a heavy lepton Monte Carlo (D.P Stoker *et al.*, Phys. Rev. D **39**, 1811 (1989)), which includes polarization and spin spin correlation effects.

Search for Supersymmetry in ALEPH

John Conway
University of Wisconsin
Madison, Wisconsin

May 8, 1990

Abstract

We have performed a search for supersymmetric particles in decays of the Z^0, using the ALEPH detector at LEP. Using 1.2 pb^{-1} of integrated luminosity near the Z^0 peak, no evidence for signal events is found, allowing limits to be placed on the masses of scalar leptons, charginos, and neutral Higgses from supersymmetry.

1 Introduction

One way to overcome the naturalness and hierarchy problems of the Standard Model [1] is to introduce a symmetry between bosons and fermions, known as supersymmetry [2]. If such a symmetry exists then every fermion (boson) would have a bosonic (fermionic) partner. Additionally, supersymmetry requires the existence of at least two Higgs doublets, giving rise to three neutral and two charged Higgs states. The two lightest neutral Higgs states, a scalar lighter than the Z^0 and a pseudoscalar heavier than the scalar, are potentially accessible at LEP energies.

We describe here searches for scalar leptons, charginos, and neutral Higgses. In the case of the former, we search for decays of the Z^0 to pairs of oppositely-charged particles which are acoplanar and which have missing energy. Such decays are a signature of the production of pairs of the supersymmetric partners of the known leptons and pairs of charginos, the superpartners of the $W^\pm$ and Higgs bosons [3]. These particles are expected to decay immediately to their associated particles and a photino. The photino is assumed here to be the lightest supersymmetric particle, to be stable (as expected from R-parity conservation [4]), and to interact weakly with matter [5]. Thus it escapes detection, carrying away transverse momentum and energy.

To search for the neutral Higgses from supersymmetry, which we denote as h for the scalar and A for the pseudoscalar states, one uses the decays $Z^0 \to hZ^*$ and $Z^0 \to hA$. In the case of the former process, which is enhanced in the case of equal Higgs vacuum expectation values, the final states are essentially the same as those of the standard model Higgs, and we can apply the result [6] obtained there directly. In the case of unequal vacuum expectation values, the process $Z^0 \to hA$ is enhanced, and we search for the individual decay modes of the h and A.

2 The Apparatus

The ALEPH detector is described elsewhere [7], but we briefly describe here the components relevant to this measurement. The central charged particle tracking detectors are an axial-wire drift chamber (ITC) at small radii and a large cylindrical time projection chamber (TPC). The ITC provides trigger information and up to 8 coordinates in azimuth and radius while the TPC provides up to 21 three-dimensional coordinates along the trajectory of a charged particle. The momentum of charged particles is determined from the curvature of the particle trajectory in the 1.5 Tesla magnetic field from the superconducting solenoid. Between the TPC and the coil cryostat lies the electromagnetic calorimeter (ECAL), which is a lead sheet/proportional wire chamber sandwich having 74000 cathode pad towers in a projective geometry. The 45 wire planes in each of the 36 modules are also read out individually and are used in triggering. At small angles is a similarly-constructed electromagnetic calorimeter, the LCAL, used to measure the luminosity. Finally, the iron return yoke of the solenoid houses the limited streamer tube chambers of the hadronic calorimeter (HCAL), used to measure the energy of hadrons and also to trigger on and identify muons.

Data were collected at seven energies around the Z^0 peak, from 88.3 GeV to 94.3 GeV in steps of 1 GeV, during the first main running period of the LEP machine. Half of the total integrated luminosity of 1.2 pb^{-1} was collected at the energy nearest the peak, 91.3 GeV. A total of over 24000 hadronic events were observed in the sample.

3 Search for Scalar Leptons and Charginos

The production of a pair of supersymmetric scalar leptons (sleptons, denoted as $\tilde{e}$, $\tilde{\mu}$, $\tilde{\tau}$) can occur through the s-channel exchange of a Z^0 or photon, and also through the t-channel exchange of a photino in the case of scalar electron production. The cross section depends on the masses of the left- and right-handed sleptons, which may or may not be equal. We present below the limits on these masses for the cases $m_{\tilde{l}_L} = m_{\tilde{l}_R}$ and $m_{\tilde{l}_L} \gg m_{\tilde{l}_R}$, which differ by roughly a factor of two in cross section. The signal for this process is a pair of acoplanar charged particles carrying some fraction of the center-of-mass energy, the balance being carried away by the photinos.

Pair production of charginos $\tilde{\chi}^{\pm}$, which are the mass eigenstates of admixtures of wino-like and higgsino-like "current" eigenstates, also proceeds through the s-channel. The couplings of the current eigenstates, and hence the cross sections, are predicted exactly if one knows the mixing. Hence we present below our result for the limiting cases of a pure wino and a pure higgsino; the light chargino in any model will lie between these extremes. We employ the same search for acoplanar charged particle pairs; thus in this search we are only sensitive to decay modes resulting in one charged particle per chargino. Such final states arise from from three-body decays to a lepton, its associated neutrino, and a photino. We do not consider here the case where the scalar neutrino is lighter than the chargino. For simplicity only decays to electrons and muons in the final state are used to set mass limits; the branching ratios of a chargino to each of these is taken to be 1/9, the minimum of the range given in the model [8].

Our most basic criterion is that there be two and only two tracks from oppositely-charged particles in the tracking detectors. Tracks are required to originate from a point less than 7 cm from the interaction point along the beam axis and less than 3 cm in radius from the beam axis, have an angle of greater than $18.2°$ with the beam axis, have at least 5 TPC coordinates, and have momentum greater than 100 MeV. To eliminate events which are acoplanar due to accompanying high-energy photons, events are rejected if there are any ECAL energy deposits over 5 GeV which are more than $25.4°$ away from the tracks' initial directions. Also, to eliminate events with large initial-state radiation at small angles, or two-photon events with high-energy small-angle beam particles, we require that there be no energy deposit of 5 GeV or more in any of the four modules of the luminosity calorimeter.

The signal we seek is characterized by two high-energy charged particles which are acoplanar, and missing energy. To eliminate the Standard Model background we require

- momentum of each track > 2 GeV

- $18.2° <$ acoplanarity angle $< 161.8°$

- missing energy < 80 GeV

After these cuts, 7 events remain; the simulation predicts 9.3 ± 0.3 background events; our observation is consistent with this. Thus we can exclude at the 95% confidence level any signal process for which we would have seen more than 6.04 signal events. The regions in the plane of sparticle mass and photino mass where more than this number of events would have been observed is shown in Figure 1 for the four signal processes. Limits from previous colliders are also shown. [10] We exclude scalar electrons below a mass of 44 (42.5) GeV, scalar muons below 43.5 (41.7) GeV, and scalar taus below 41.8 (38.3) GeV in the degenerate (non-degenerate) case. For charginos, pure higgsinos are excluded below 45 GeV, and a pure wino is excluded below 45.9 GeV. Similar results have been obtained by other LEP collaborations. [11]

4 Search for Higgs in Supersymmetry

In the context of the minimal supersymmetric standard model [9], all masses and couplings of the Higgs states depend on two parameters, which we can take as the scalar and pseudoscalar masses. The light Higgs scalar h behaves essentially as a Standard Model Higgs, and we use the result [6] to exclude a region in the $m_h - m_A$ plane, as shown as A in figure 2 for the ratio of the vacuum expectaion values $v_2/v_1 > 1$. The excluded region can be extended by searching for the process $Z^0 \to hA$. In this process, the h and A both decay preferentially to the heaviest fermion pair available; however even above the threshold for b-quark production there is a significant rate for τ-pair production. We search, therefore, for events with two isolated high-energy tracks and two jets, using the criteria outlined in ref. [12]. No events satisfying the criteria are observed, allowing one to exclude the region denoted F in Figure 2.

5   Acknowledgements

We would like to thank our colleagues of the LEP division for their outstanding performance in bringing the LEP machine into operation. Thanks are also due to the many engineering and technical personnel at CERN and at the home institutes for their contribution toward ALEPH's success. Those of us not from member states wish to thank CERN for its hospitality.

References

[1] C.H. Llewellyn Smith and G.G. Ross, Phys. Lett. **105B** (1981) 38.

[2] J. Wess and B. Zumino, Phys. Lett. **49B** (1974) 52; for a review see P. Fayet and S. Ferrara, Phys. Rep. **32C** (1977) 249.

[3] H.E. Haber and G.L. Kane, Phys. Rep. **117**, Nos. 2-4 (1985) 75-263 and references therein.

[4] P. Fayet, "Unification of the Fundamental Particle Interactions", Proc. Europhys. Study Conf., Erice (Plenum, 1980) p. 587.

[5] P. Fayet, Phys. Lett. **86B** (1979) 272; P. Fayet, Phys. Lett. **133B** (1983) 363.

[6] D. Decamp *et al.*, *Search for Neutral Higgs Boson from Z Decay in the Higgs Mass Range Between 11 and 24 GeV*, CERN-EP/90-16.

[7] D. Decamp *et al.*, *ALEPH: A Detector for Electron- Positron Annihilations at LEP*, CERN-EP/90-25.

[8] T. Schimert, C. Burgess, and X. Tata, Phys. Rev. **D32** (1985) 707; T. Schimert and X. Tata, Phys. Rev. **D32** (1985) 721.

[9] H. Nilles, Phys. Rep. **C10** (1984) 1, and references therein; R. Barbieri, Riv. Nuovo. Cim. **11** (1988); M. Chen, C. Dionisi, M. Martinez, and X. Tata, Phys. Rep. **159** (1988) 201.

[10] I. Adachi *et al.*, Phys. Lett. **218B** (1989) 105; C. Hearty *et al.*, Phys. Rev. **D39** (1989) 3207.

[11] "Mass limits for scalar smuons, scalar electrons, and winos from e^+e^- collisions near $\sqrt{s} = 91$ GeV," L3 Collaboration, submitted to Phys. Lett. B. (L3 Preprint 2); M.Z. Akrawy *et al.*, A Search for Acoplanar Pairs of Leptons or Jets in Z^0 Decays: Mass Limits on Supersymmetric Particles, CERN-EP/89-170.

[12] D. Decamp *et al.*, *Search for Neutral Higgs from Supersymmetry in Z Decays*, CERN-EP/89-168.

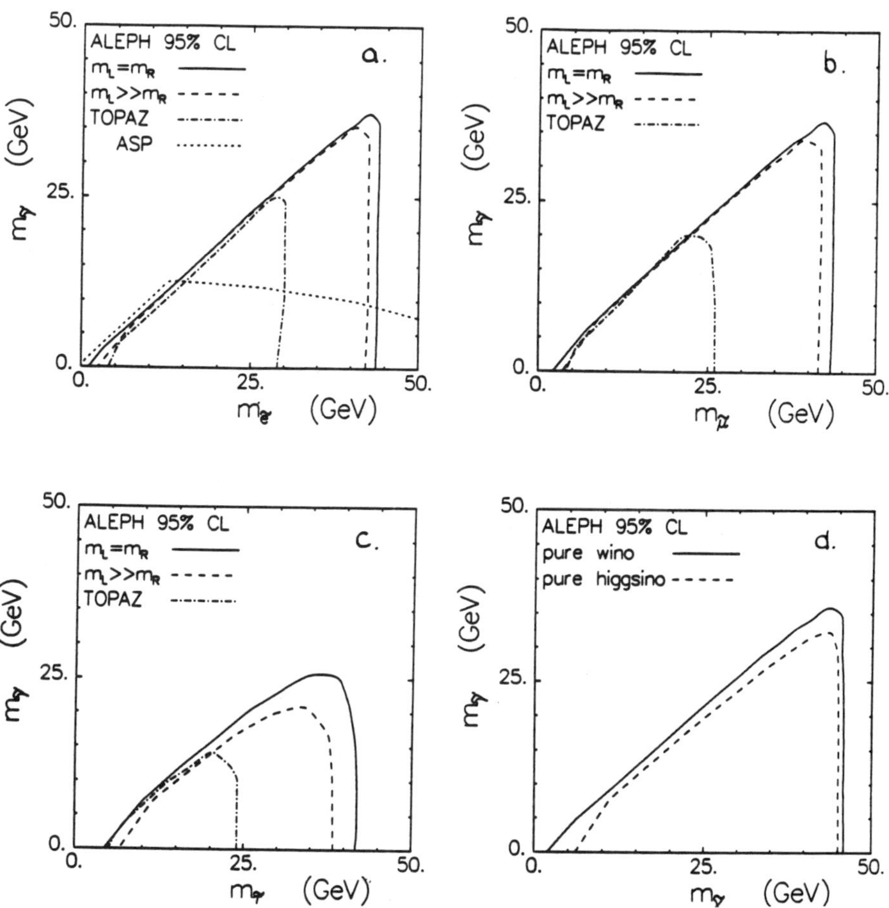

Figure 1: Excluded mass regions for a) scalar electrons, b) scalar muons, c) scalar taus, d) charginos.

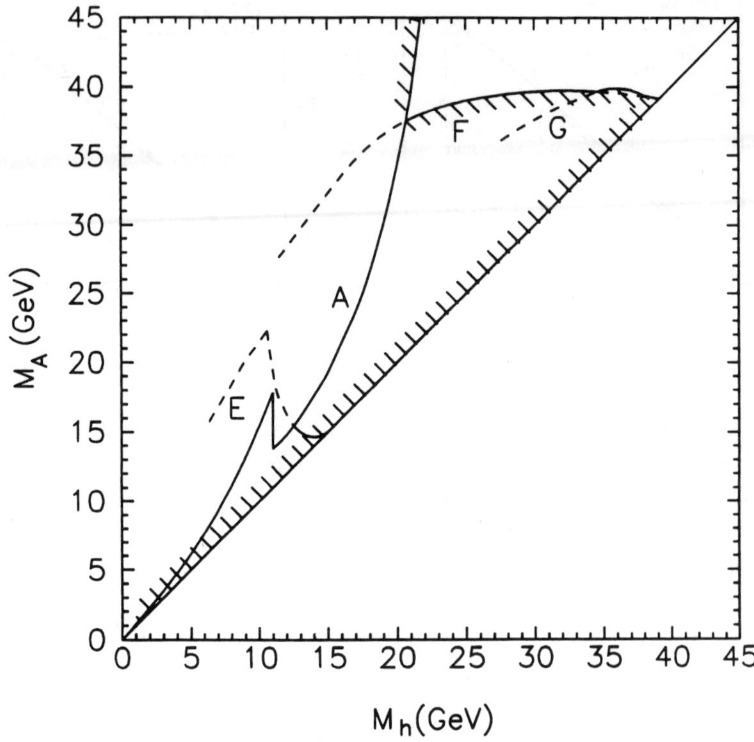

Figure 2: Excluded mass region for light Higgs scalar and pseudoscalar. See text for explanation of regions.

Search for the Neutral Higgs Boson from Z° Decay in ALEPH

Yi-Bin Pan

Department of Physics, University of Wisconsin
Madison, Wisconsin 53706
Representing the **ALEPH** Collaboration

ABSTRACT

A search for the neutral Higgs boson, using the processes $Z° \rightarrow H°e^+e^-$, $Z° \rightarrow H°\mu^+\mu^-$, $Z° \rightarrow H°\tau^+\tau^-$, $Z° \rightarrow H°\nu\bar{\nu}$ and $Z° \rightarrow H°q\bar{q}$, is performed on data collected by the ALEPH detector. Combining all these processes, the mass range excluded is 32 MeV to 24 GeV at 95% C.L. The result from this experiment is unambiguous in the context of the Standard Model.

INTRODUCTION

In this paper we search for the neutral Higgs boson of the Standard Model. The search is carried out for the process

$$Z° \rightarrow H°Z^*,$$

where Z^* is a virtual $Z°$ which decays into a fermion pair. We include in the analysis all the decay modes of the Z^*. From the Standard Model, the relative branching ratios of $Z° \rightarrow H°e^+e^-$, $Z° \rightarrow H°\mu^+\mu^-$, $Z° \rightarrow H°\tau^+\tau^-$, $Z° \rightarrow H°\nu\bar{\nu}$ and $Z° \rightarrow H°q\bar{q}(g)$, normalized to the total rate of $H°$ production from $Z°$ decays, are 3.4%, 3.4%, 3.4%, 20.1% and 69.7% respectively.

SECTION 1: 32 MeV $< M_H <$ 15 GeV

In this search, we are able to cover the mass range of the Higgs boson from a few tens of MeV to about 15 GeV. We use the data collected during the early running of LEP, from September 19 to November 7, 1989, which correspond to 11,550 $Z° \rightarrow$ hadrons. These data were taken over a range of the center-of-mass energies from 88.3 GeV to 94.3 GeV, corresponding to a total integrated luminosity of 542 nb^{-1}. For the study below the $H° \rightarrow \mu^+\mu^-$ threshold, i.e. for a Higgs mass below 212 MeV where $H° \rightarrow e^+e^-$ dominates, the Higgs boson is long-lived. For this mass range, we also perform a search with the identification of isolated vertices ($V°$).

The analysis for this Higgs mass range is described in Ref. [1]. Since the decay channel $Z° \rightarrow H°\nu\bar{\nu}$ provides the best means to search for a Higgs over a wide range of Higgs masses, we give some details here. For a light Higgs, the signal is a monojet with large missing energy and momentum. The event selection criteria are given here:

(1) The event is required to have a minimum of two oppositely charged particles and the total observed charge of the event should not exceed ± 4.

(2) The angle between the beam axis and the sphericity axis of the event, calculated with charged particles only, must be greater than 40°.

(3) The magnitude of the vector sum of the transverse momenta of all charged particles with respect to the beam axis must be greater than 2 GeV/c.

(4) $\cos\theta_{max} > -0.1$.
θ_{max} is the maximum angle between any charged particle with momentum greater than 0.3 GeV/c and the direction of the vector sum of the momenta of all charged particles.

(5) $\cos \theta_{\text{jet-jet}} > -0.1$.

The event is divided into two hemispheres, by using the plane perpendicular to its sphericity axis, in the rest system of all charged particles. Charged particles in each hemisphere then define a jet. In the laboratory system, the vector sum of the momenta of all the charged particles in each jet gives the jet axis and $\theta_{\text{jet-jet}}$ is the angle between these two jet axes.

(6)   The events which satisfy the selection criteria (1) to (5) are monojet-like. Consider all the electromagnetic calorimeter modules in the hemisphere opposite to the direction of the monojet (defined by the vector sum of the momenta of the charged particles). The event is rejected if the energy deposition of any module exceeds 2 GeV. Similarly, events are rejected if the total energy in the Luminosity Calorimeter (LCAL) exceeds 5 GeV. From a study of the randomly triggered events, this requirement on the LCAL energy introduces a 1.5% inefficiency to the acceptance of the Higgs events.

Selection criteria (2) and (3) remove background from two-photon and beam gas interactions, while (4), (5) and (6) primarily remove the background from $Z^\circ \rightarrow q\bar{q}(g)$. Selection criterion (4) defines the monojet and is the most powerful cut. Fig. 1(a) shows the $\cos\theta_{\text{max}}$ distribution for the data and Fig. 1 (b) for the Monte Carlo simulation for Higgs mass of 5 GeV after the selection criteria (1), (2), (3) and (6) have been applied. The cut is at $\cos\theta_{\text{max}} = -0.1$. Applying the additional selection criterion (5) to the data, no event survives.

The result of the search for the neutral Higgs boson, based on the processes $Z^\circ \rightarrow H^\circ e^+e^-$, $H^\circ \mu^+\mu^-$, $H^\circ \tau^+\tau^-$, $H^\circ \nu\bar{\nu}$ and $H^\circ q\bar{q}(g)$ using 542 nb^{-1} of data collected by the ALEPH detector, is shown in Fig.2. Fig.2 gives the number of events expected from the sum, as well as the individual processes, as a function of the Higgs boson mass over the full region of search. While the process $Z^\circ \rightarrow H^\circ \nu\bar{\nu}$ provides the most powerful means for this search, the result of combining all the processes excludes the neutral Higgs boson over the mass range from 32 MeV to 15 GeV at 95% C.L. and 40 MeV to 12 GeV at 99% C.L. This result is obtained within the context of the Standard Model with no further assumption.

SECTION 2: 11 GeV $< M_H < 24$ GeV

This analysis extends the neutral Higgs boson search to a mass range up to about 24 GeV for the neutral Higgs boson resulting from doubling the event statistics and a modified analysis procedure. We use all the data collected in 1989, which correspond to about 25,000 $Z^\circ \rightarrow$ hadrons. These data were taken over a range of the center-of-mass energies from 88.3 GeV to 95.0 GeV, corresponding to a total integrated luminosity of 1.2 pb^{-1}. The decay modes of $Z^\circ \rightarrow H^\circ \nu\bar{\nu}$ and $Z^\circ \rightarrow H^\circ \ell\ell$ (ℓ = e and μ) are used. Event selection criteria are similar to those given in Ref. [1] but are modified to optimize the detection efficiency of a Higgs with mass in the range between 11 GeV and 24 GeV. The details of this analysis are given in Ref. [2].

After applying the selection criteria given in Ref. [2], we find that no event survives in our data sample in the range of this search. Fig.3 shows the number of events expected from the sum, as well as the individual processes, for $Z^\circ \rightarrow H^\circ e^+e^-$, $Z^\circ \rightarrow H^\circ \mu^+\mu^-$ and $Z^\circ \rightarrow H^\circ \nu\bar{\nu}$ as a function of the Higgs masses. To set a mass limit conservatively, we reduce the number of expected events by 4% to take into account the systematic error. The mass range from 11 GeV to 24 GeV for the neutral Higgs boson is excluded at 95% C.L. from this analysis. Fig.3 also presents the number of events expected from the sum of all processes as a function of the Higgs boson mass from the results in Ref. [1], giving a combined excluded mass range from 32 MeV to 24 GeV at 95% C.L. limit.

REFERENCES

[1]   D. Decamp *et al.* (ALEPH Collaboration), Phys. Lett. **B236** (1990) 233.
[2]   D. Decamp *et al.* (ALEPH Collaboration), Phys. Lett. **B241** (1990) 141.

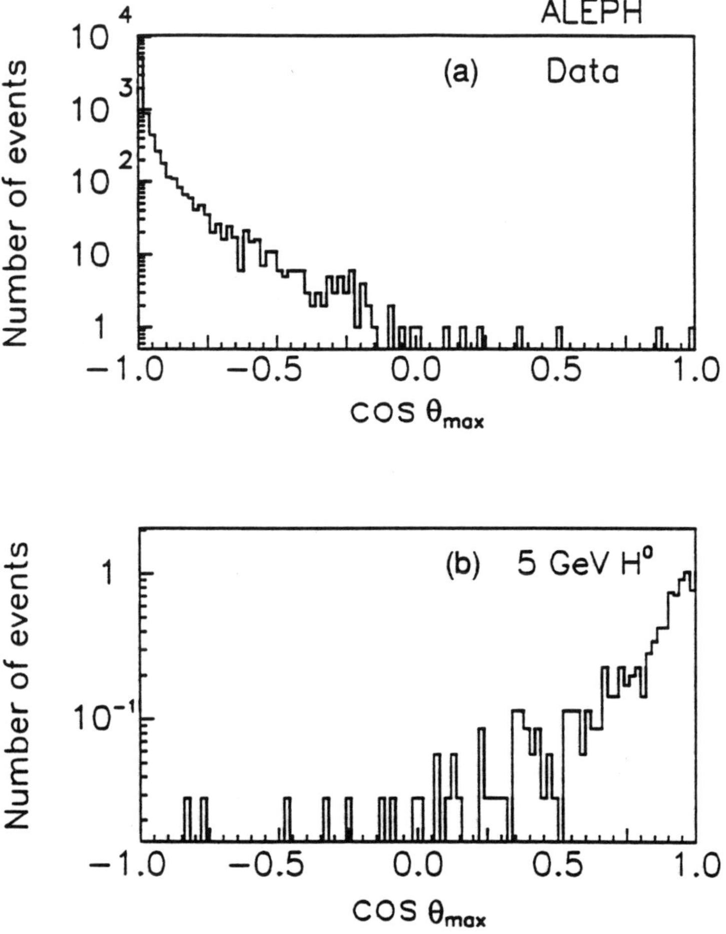

Fig.1 $\cos\theta_{max}$ distribution for (a) data and (b) Monte Carlo simulation of $Z^\circ \to H^\circ \upsilon\bar{\upsilon}$ for 5 GeV Higgs mass, both after applying the selection criteria (1), (2), (3) and (6) of section 1. (See section 1 for definition of $\cos\theta_{max}$). The cut is at $\cos\theta_{max} = -0.1$. The distribution in (b) is normalized to the data.

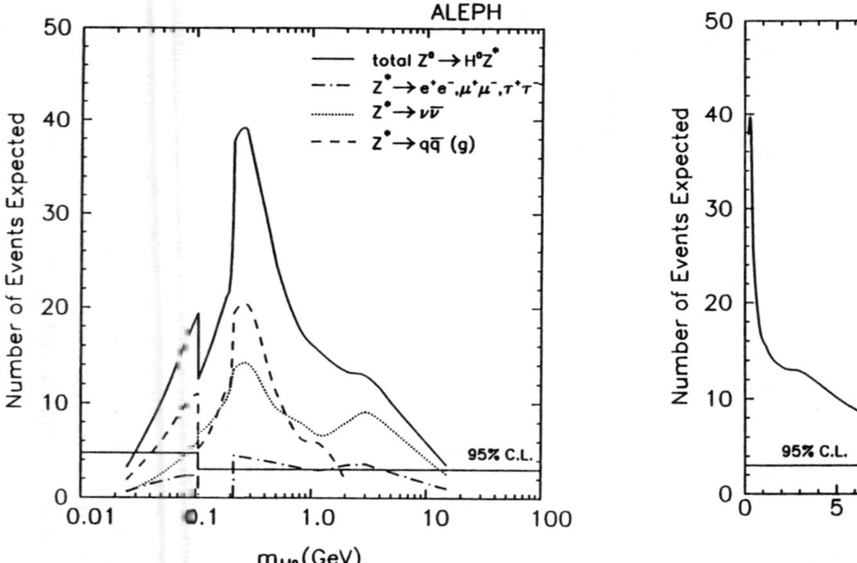

Fig.2 The number of events expected for $Z^\circ \to H^\circ \ell^+\ell^-$, $(e^+e^-, \mu^+\mu^-$ and $\tau^+\tau^-)$, $Z^\circ \to H^\circ \nu\bar{\nu}$, $Z^\circ \to H^\circ q\bar{q}(g)$ and the sum of all the above channels as a function of the Higgs mass. The 95% C.L. limit is also indicated corresponding to one observed candidate below 100 MeV and zero above 100 MeV, giving the excluded region of 32 MeV to 15 GeV.

Fig.3 The number of events expected for $Z^\circ \to H^\circ \ell^+\ell^-$ ($\ell =$ e, μ), $Z^\circ \to H^\circ \nu\bar{\nu}$ and the sum of all the above channels as a function of the Higgs mass for $M_{H^\circ} > 10$ GeV. The 95% C.L. limit at 3 expected events is indicated, corresponding to zero observed candidates. Below 15 GeV the sum of $Z^\circ \to H^\circ \ell^+\ell^-$ ($\ell =$ e, μ or τ), $Z^\circ \to H^\circ \nu\bar{\nu}$, and $Z^\circ \to H^\circ q\bar{q}$ is shown for the results of the search in Ref. [1].

Search for New Quarks and Leptons from Z° Decay at LEP

Vivek Sharma and Frederick Weber
Department of Physics, University of Wisconsin
Madison, Wisconsin 53706
Representing the **ALEPH** Collaboration

ABSTRACT

A search for Z° decays into pairs of new heavy quarks (t and b′), new heavy charged leptons ($L^{\pm}$), stable heavy neutral leptons (ν_L) and unstable heavy neutral leptons (L°) is performed on data collected by the ALEPH detector corresponding to 11,550 events of Z°→hadrons. The limits on the masses of the heavy quarks are $M_t > 45.8$ GeV and $M_{b'} > 46.0$ GeV, allowing for both charged-current and flavor-changing neutral-current decays of the b′. If an $L^{\pm}$ decays into a stable ν_L, then for $M_{\nu_L} < 42.7$ GeV, the mass of $L^{\pm}$ is excluded for all values of $M_{L\pm} > M_{\nu_L}$. Finally, while the mass of the stable ν_L is excluded up to 42.7 GeV, the mass of the unstable L° is excluded up to 45.7 GeV with the mixing parameters $|U_{\ell L°}|^2$ down to 10^{-13} at this mass. For 25.0 GeV $< M_{L°} < 42.7$ GeV, all values of $|U_{\ell L°}|^2$ are excluded. All limits are given at 95% C.L.

1. INTRODUCTION

Using data recorded during the first period of LEP operation, searches have been performed for a top quark (t), fourth-generation bottom type quark (b′), sequential heavy charged lepton ($L^{\pm}$), unstable neutral heavy lepton (L°) and stable neutral heavy lepton (ν_L) from Z° decay. In all these cases, a Z° is assumed to decay into a pair of these new quarks or leptons with a decay rate given by the Standard Model. The decays of t, b′, $L^{\pm}$, L° are via a charged-current process into a virtual W boson (W*). The decays of b′ via flavor-changing neutral-current are also considered.

Recently, the ALEPH Collaboration has determined, on the basis of the cross section for e^+e^-→hadrons at the Z° peak, the number of light neutrino species to be 3.01 ± 0.16 [1, 2]. This can be used set a limit on the mass of a stable ν_L. However, this approach gives a poor limit for an unstable L°, so that it is desirable to also perform a direct search for an L°. We consider the case where there is mixing between the L° and the three light neutrinos. If the mixing parameters with ν_e, ν_μ, and ν_τ are $U_{eL°}$, $U_{\mu L°}$ and $U_{\tau L°}$ respectively, then the L° wavefunction contains the terms

$$U_{eL°}|\nu_e\rangle + U_{\mu L°}|\nu_\mu\rangle + U_{\tau L°}|\nu_\tau\rangle.$$

With such mixing, the L° can decay into a virtual W together with an e, μ or τ.

The study of these processes is performed using data collected with the ALEPH detector with an integrated luminosity of 542 nb^{-1} at center-of-mass energies from 88.3 GeV to 94.3 GeV. This corresponds to 11,550 Z°→hadrons events. The following studies were performed:

(i)   Search for charged-current decays of t, b′ and L° by the topology of an isolated charged particle. The isolated charged particle is the $e^{\pm}, \mu^{\pm}$ or one prong decays of the $\tau^{\pm}$, resulting from the leptonic or semi-leptonic decay of the new quark or new lepton.

(ii)   Search for the flavor-changing neutral-current decay of b′ by a topology of an isolated photon (b′→bγ) or of a four-jet final state (b′→bg, $\bar{b}′→\bar{b}g$).

(iii)   In the case of Z°→L°$\bar{L}$°, when the lifetime of L° is long, as explained below, the topology of the events could be one or two displaced vertices.

(iv)   Using the total hadronic cross-section measurement at the Z° peak, mass limits can be set for v_L and $L^{\pm}$ on the processes Z°→$v_L\bar{v}_L$ and Z°→L^+L^-, where $L^{\pm}$→v_LW^*. For the long-lived L°, limits can be set on $|U_{\ell L°}|^2$ as a function of $M_{L}°$.

2.   THE ALEPH DETECTOR

The ALEPH detector is described in detail elsewhere [3]. The parts of the detector relevant to this analysis are the inner tracking chamber (ITC), the large time projection chamber (TPC), the electromagnetic calorimeter (ECAL) and the hadronic calorimeter (HCAL). A 1.5 Tesla magnetic field is provided by the superconducting solenoid surrounding the TPC and ECAL. The luminosity calorimeter (LCAL) provides energy and position measurements of the showers produced by the small-angle Bhabha scattering $e^+e^-→e^+e^-$.

3.   SEARCH FOR CHARGED-CURRENT DECAY OF t, b′ AND L°

A search for new quarks (t, b′) and new neutral heavy leptons (L°) is performed by identifying a topology where there is an isolated charged particle in an event from the leptonic or semi-leptonic decay of the t, b′ or L°. In this analysis, as well as the analysis described in Section 4, both charged and neutral particles are used. The ALEPH hadronic event selection criteria were used [1].

Selection criteria for the events are: (i) the vector sum of the transverse momenta $|\Sigma\vec{P}_T|$ of all particles with respect to the beam axis must be greater than 8 GeV/c; (ii) the aplanarity (A) of the event is greater than 0.03.

Selection criteria for an isolated charged particle are: (i) the charged particle is required to have $P_T > 3$ GeV/c where P_T is defined as the transverse momentum of the charged particle with respect to the thrust axis of the event; (ii) for each charged particle i, the isolation parameter ρ_i is calculated by performing a jet analysis using the LUND cluster algorithm [4] on all the charged particles of the event, except particle i. Here ρ_i is defined as [5]

$$\rho_i = \min_{\text{jet } j} \sqrt{2E_i(1-\cos\theta_{ij})}$$

where E_i is the energy in GeV of the ith charged particle and θ_{ij} is the angle between the ith particle and the axis of the jth jet. Events are accepted if $\rho > 1.8$, where ρ = maximum of (ρ_i). Fig.1 shows the distributions of the isolation parameter ρ for the data and for the simulated events, normalized to the same integrated luminosity as the data, of b′ at 40 GeV mass after applying all other selection criteria.

Using the event and isolated charged particle selection criteria, 6 events are observed in the data and 6.5 events are expected from the Monte Carlo simulation of the background from $Z^\circ \to q\bar{q}(g)$. Hence we can exclude at 95% C.L. mass ranges of t, b' and L$^\circ$ for which we would have observed more than 6.2 events.

First-order QCD corrections [6] are included to calculate the cross-section and hence the expected number of events for t and b'. The simulated events for $Z^\circ \to t\bar{t}$, $b'\bar{b}'$ and $L^\circ \bar{L}^\circ$ are processed by the same reconstruction and analysis programs as those used for the real data. To implement the jet fragmentation, the LUND parton shower model version 6.3 is used, which is known to describe the data well [7,8,9]. The detection efficiencies after all event selection criteria depend on the quark and lepton masses and are about 14% for t and b', 28% for $L^\circ \to eW^*$ or μW^* and 22% for $L^\circ \to \tau W^*$ at a mass of 40 GeV.

The systematic errors on the expected number of detected events is dominated by an uncertainty of 25% in the production cross-section due to higher order QCD corrections [6]. Figs. 2(a) to (c) give the expected number of events after applying the event selection criteria for t, b' and L$^\circ$ as a function of mass with the 95% C.L. limits indicated. For the charged-current decays of t, b', and L$^\circ$, the following regions of mass are excluded at 95% C.L.:

$$26.0 \text{ GeV} < M_t < 45.8 \text{ GeV}$$
$$26.0 \text{ GeV} < M_{b'} < 46.2 \text{ GeV}$$
$$25.0 \text{ GeV} < M_{L^\circ} < 45.7 \text{ GeV}.$$

4. SEARCH FOR FLAVOR-CHANGING NEUTRAL-CURRENT DECAY OF b'.

If a b' exists with mass less than half of the Z° mass and with the present limit of t mass being larger than half of the Z° mass, the rate of the charged-current decay via $b' \to cW^*$ is expected to be suppressed by the small coupling $V_{cb'}$ [10]. The result, shown in Fig.2(b), rules out the existence of a b' for mass below 45.0 GeV if the rate of the charged-current decay is larger than 10%. It has been pointed out that flavor-changing neutral-current b' decays could occur with a sizeable rate [10,11]; such decays are expected to be dominated by $b' \to b\gamma$ and $b' \to bg$. A search for such processes is described below. The selection of charged and neutral particles, and the simulation of the $Z^\circ \to b'\bar{b}'$ process [12] as well as the background process of $Z^\circ \to q\bar{q}(g)$ follow the description given in Section 3.

(i) $b' \to b\gamma$: The following description of the analysis applies to both cases: (1) $Z^\circ \to b'\bar{b}'$ where both b' decay to $b\gamma$, and (2) $Z^\circ \to b'\bar{b}'$ where one b' decays to $b\gamma$ and the other to bg. In both cases, a search is performed for a high-energy isolated photon. Using the same definition of variables as given in Section 3, the event selection criteria are: thrust (T) < 0.8, aplanarity (A) > 0.03, photon energy and transverse momentum with respect to the thrust axis greater than 10 GeV and 3 GeV/c respectively. Finally, the isolation parameter ρ for the photon is required to be greater than 2.5. The detection efficiency for $b' \to b\gamma$ is 30% at 40 GeV mass. With these event selection criteria, 4 events are observed in the data while a background of 2.0 is expected from the simulation of $Z^\circ \to q\bar{q}(g)$ based on five quarks. Since a 95% C.L. limit corresponds to 7.2 expected events, we can exclude the branching ratio of $b' \to b\gamma$ greater than 5% for a b' mass between 26.0 GeV and 46.0 GeV at 95% C.L. as shown in Fig.3.

(ii) $b' \to bg$: A $Z^\circ \to b'\bar{b}'$ event with both b' decaying into bg gives a four-jet final state. In each event with $T < 0.8$ and $A > 0.2$, particles are required to group into four jets using the LUND cluster algorithm [4]. The energy of each jet (E_j) is then determined from the solution of the four equations of energy-momentum conservation:

$$\sum_{j=1}^{4} E_j \, \vec{\beta}_j = 0 \quad \text{and} \quad \sum_{j=1}^{4} E_j = E_{cm} \qquad (1)$$

where the velocity $\vec{\beta}_j$ is the vector sum of the momenta of all the particles in jet j divided by the sum of their energies. The direction of this vector sum defines the direction of the jet. The jet-jet invariant mass is then computed for all pairs of the four jets. Each pair of two jets is assigned to a b' for that pairing which gives the smallest difference in invariant masses ($\triangle M$) of the two pairs. Events are accepted if $|\triangle M| \leq 2\sigma_M$ where σ_M, the invariant mass resolution of two jets, is 5 GeV. The detection efficiency is 9%. With these event selection criteria, we observe 6 events in the data and expect 6.2 background events from the Monte Carlo simulation of the background process $Z^\circ \to q\bar{q}(g)$ for 5 quarks. Since the 95% C.L. limit corresponds to 6.6 expected events, we exclude the branching ratio of $b' \to bg$ greater than 65% for b' mass between 26.0 GeV and 46.0 GeV at 95% C.L. as shown in Fig. 3.

In conclusion, if the sum of the branching ratios of b' decays into $c\ell v$ ($\ell = e, \mu, \tau$) by charged current and into $b\gamma$ or bg by flavor-changing neutral current is assumed to be 100%, then the mass range of b' from 26 GeV to 46 GeV is excluded at 95% C.L.

5. SEARCH FOR $Z^\circ \to L^\circ \bar{L}^\circ$ FROM DISPLACED VERTICES

The search for an L° using the isolated particle topology, as described in Section 3, requires the mixing parameters $|U_{\ell L^\circ}|^2$ to be large enough that L° decays very close to the e^+e^- interaction point. This can, of course, be related to the lifetime of the L°:

$$\tau(L^\circ \to \ell^- X^+) = \left(\frac{m_\mu}{m_{L^\circ}} \right)^5 \frac{\tau_\mu \mathrm{Br}(L^\circ \to \ell^- e^+ v_e)}{f(m_{L^\circ}, \ell, X)|U_{\ell L^\circ}|^2} \qquad (2).$$

The factor $f(m_{L^\circ}, \ell, X)$ is a phase-space correction, which is taken to be 1, as the smallest L° mass being considered is 20 GeV. $\mathrm{Br}(L^\circ \to \ell^- e^+ v_e)$ is taken to be 11%. When the mixing to ℓ is small ($|U_{\ell L^\circ}|^2 \approx 10^{-10}$), the $Z^\circ \to L^\circ \bar{L}^\circ$ signature is one or two vertices, separated from the e^+e^- interaction point, and with no charged particles coming from the interaction point. A search has been performed for such vertices.

Only charged particles having at least six TPC coordinates are considered for this analysis. A vertex finding algorithm is used to find all vertices of these charged particles in the event with χ^2 probability at least 1%. A possible $L^\circ \bar{L}^\circ$ pair is defined as any pair of vertices where the cosine of the angle θ in the $r\varphi$ plane between the two vectors joining the vertices to the interaction point is less than -0.4. (i.e. $\theta > 114°$). If no such pair exists, the single vertex with the largest number of charged particles and lowest χ^2 is used. If there is more than one possible topology, the one using the most tracks (and widest opening angle, in two-vertex events) is chosen. A vertex is now required to have a minimum of three charged particles, to be well contained within the TPC volume, and to have the vector sum of the momenta of all the charged particles in it pointing away from the interaction point. Events with a vertex having a radius in the x-y plane within 0.5 cm of the beam pipe, the ITC outer wall or the TPC inner wall are also rejected.

To eliminate the background process $Z^\circ \to q\bar{q}(g)$, the vertices found are required to have $R_{r\phi} > 1$ cm for a two-vertex topology and $R_{r\phi} > 7.8$ cm for a single-vertex topology. Beam-gas and beam-pipe interactions are removed by requiring the total charged energy of the vertex or vertices in an event to be greater than 10 GeV. To remove $Z^\circ \to \tau^+\tau^-$ events, which are highly collimated and hard to vertex, the cosine of the angle θ_{12} between the two fastest charged particles from the same vertex is required to be greater than -0.98 (i.e. $\theta_{12} < 169°$). The majority of the cosmic ray background events are rejected by requiring a difference between the time of beam crossing and the time of energy deposition in the ECAL modules to be less than 200ns. This timing selection introduces less than 2.5 % inefficiency in the $Z^\circ \to q\bar{q}(g)$ data. Cosmic ray events are further reduced by rejecting events where high-momentum tracks have several TPC coordinates between their vertex point and the interaction point. Applying the above event selection criteria to the data of $540\ nb^{-1}$, we find no candidate with a displaced vertex.

Monte Carlo simulated events for $Z^\circ \to L^\circ \overline{L^\circ}$ are generated using the TIPTOP program [13] for different values of the mass and of the mixing parameter $|U_{\ell L^\circ}|^2$, taking into account the tracking systematics at their most pessimistic levels. The selection algorithm is applied to obtain the expected number of events at each point and a bicubic spline fit is then performed to obtain the contour (b) in the $|U_{\ell L^\circ}|^2$ versus M_{L° plane as shown in Fig.4. This corresponds to three expected events and gives the limit of the excluded region for $Z^\circ \to L^\circ \overline{L^\circ}$ at 95% C.L. This limit applies equally well to L° mixing to e, μ and τ, since the kinematic differences between these decays have a negligible effect upon the vertex finding algorithm. Shown in Fig.4 is the 95% C.L. limit contour (a) from the result of the search for prompt L° as described in section 3. Hence an L° is excluded up to a mass of 46.0 GeV with mixing parameters $|U_{\ell L^\circ}|^2$ as small as 10^{-13}.

6. MASS LIMITS ON NEW QUARKS AND NEW LEPTONS FROM THE TOTAL HADRONIC CROSS-SECTION MEASUREMENT AT THE Z° PEAK

From the measurement of the peak hadronic cross-section (σ°_{had}) by the ALEPH Collaboration, the number of light neutrino species is found to be $N_\nu = 3.01 \pm 0.16$ [2]. This result can be used to set a mass limit on t, b′, $L^\pm$, stable ν_L and unstable L°.

(i) $Z^\circ \to \nu_L \bar{\nu}_L$: For $Z^\circ \to \nu_L \bar{\nu}_L$, this result $N_\nu = 3.01 \pm 0.16$ excludes at 95% C.L. the region where

$$\Gamma_{\nu_L} > [(3.01-3.00)+1.64\times0.16]\ \Gamma_{\nu_e} = 0.272\ \Gamma_{\nu_e} \qquad (3)$$

Here Γ_{ν_L} and Γ_{ν_e} are the partial widths for Z° decays into $\nu_L \bar{\nu}_L$ and $\nu_e \bar{\nu}_e$ respectively. The factor 1.64 corresponds to the one-sided 95% C.L. limit. The partial width $\Gamma_{1/2}$ for the Z° decaying into a pair of spin one-half particles is a function of the mass of the particles,

$$\Gamma_{1/2} = \frac{N}{24\pi}\ \frac{G_F M_z^3}{\sqrt{2}}\ \beta\ [\frac{1}{2}\ (3 - \beta^2)\ v^2 + a^2 \beta^2] \qquad (4)$$

where N is the color factor, β is the velocity of each of these particles and v and a are the vector and axial vector couplings of the same particle to the Z°. Using eqs (3) and (4) the region where $0 \leq M_{v_L} < 42.7$ GeV is excluded at 95% C.L.

(ii) $\quad Z^\circ \to L^+ L^-$: For $Z^\circ \to L^+ L^-$ where $L^\pm \to v_L W^*$, the region where

$$\frac{\Gamma_{had}}{\Gamma_{tot}^2} = \frac{\Gamma_{5\ quarks} + x\Gamma_{L\pm}}{(\Gamma_{SM} + \Gamma_{v_L} + \Gamma_{L\pm})^2} < \frac{\Gamma_{5\ quarks}}{(\Gamma_{SM} + 0.272\Gamma_{v_e})^2} \qquad (5)$$

is excluded at 95% C.L. Here x is the ratio of the detection efficiency for an $L^\pm$ to be identified as a hadronic event to the detection efficiency for hadronic event using the event selection criteria as described in [1]. The partial widths $\Gamma_{5\ quarks}$ and Γ_{SM} take the Standard Model values with three generations: $\Gamma_5 = 1.737$ GeV and $\Gamma_{SM} = 2.487$ GeV.

Using eq. (4) and solving inequality (5) for M_{v_L} in terms of $M_{L\pm}$, we can exclude the full triangle $M_{v_L} < M_{L\pm} < M_Z/2$, as shown in Fig. 5, except for the small area near the top corner. In obtaining this excluded region the values of x are taken from Monte Carlo computations. The result is insensitive to x near the area where $M_{L\pm}$ and M_{v_L} are close in mass (x is near zero). Using the result given by subsection 6(i) above, the region where $M_{L\pm} > M_Z/2$ and $M_{v_L} < 42.7$ GeV is excluded for all values of $M_{L\pm}$ as shown in Fig. 5.

For $Z^\circ \to L^+ L^-$ where the $L^\pm$ are stable, eq. (5) can be used to set limits on $M_{L\pm}$ if we assume that $M_{v_L} > M_Z/2$ and hence it is not produced in Z° decay. In this case, in addition to Γ_{v_L} being zero, x is also zero due to the fact that a two charged particle final state does not pass the hadronic event selection criteria. It follows from eq.(5) that the mass region $0 \leq M_{L\pm} < 26.5$ GeV for stable $L^\pm$ is excluded at 95% C. L.

(iii) $\quad Z^\circ \to t\,\bar{t},\ b'\bar{b}'$ and $L^\circ \bar{L}^\circ$: For $Z^\circ \to t\,\bar{t},\ b'\bar{b}'$ and $L^\circ \bar{L}^\circ$, eq. (5) can be written as

$$\frac{\Gamma_{had}}{\Gamma_{tot}^2} = \frac{\Gamma_{5\ quarks} + x\Gamma_j}{(\Gamma_{SM} + \Gamma_j)^2} < \frac{\Gamma_{5\ quarks}}{(\Gamma_{SM} + 0.272\Gamma_{v_e})^2} \qquad (6)$$

where Γ_j = partial width of t, b' or L° and x is essentially one. Applying eq. (4) to eq. (6), the mass regions $M_t < 31.3$ GeV, $M_{b'} < 39.4$ GeV and $M_{L^\circ} < 11.8$ GeV are excluded. Combining these results with the results from the direct searches described in sections 3 and 4, the following mass regions are excluded at 95% C.L.:

$$0 \leq M_t < 45.8 \text{ GeV}$$
$$0 \leq M_{b'} < 46.0 \text{ GeV}$$
$$0 \leq M_{L^\circ} < 11.8 \text{ GeV and } 25.0 \text{ GeV} < M_{L^\circ} < 45.7 \text{ GeV}.$$

(iv) $\quad Z^\circ \to L^\circ \bar{L}^\circ$: We consider an L° which has an average decay length larger than R = 390 cm (the distance between the interaction point and the outer corner of the ECAL). This radius is chosen because it encloses the ITC, TPC and ECAL which are the essential elements for the hadronic event selection. Eqs. (2) and (4) and inequality (6) are used to determine the excluded region in the M_{v_L} versus $M_{L\pm}$ plane. The value x in eq. (6) is modified to be the product of the ratio of detection efficiencies as defined previously and the probability that a L° decays inside the radius R. To be conservative the ratio of detection efficiencies is set to one for decay inside R and zero outside. The excluded region in the $|U_{\ell L^\circ}|^2$ versus M_{L° plane at 95% C. L. limit is shown in Fig. 4 (contour c). Combining the results given in Sections 3 and 5, for $25.0\,\text{GeV} < M_{L^\circ} < 42.7\,\text{GeV}$ all values of $|U_{\ell L^\circ}|^2$ are excluded.

7. CONCLUSION

A search for new heavy quarks (t and b'), new heavy charged ($L^\pm$) and unstable neutral leptons (L°), has been performed using 542 nb^{-1} of data collected by the ALEPH detector. For $Z^\circ \to t\bar{t}$, the t mass is excluded up to 45.8 GeV. For $Z^\circ \to b'\bar{b}'$, the b' mass is excluded up to 46.0 GeV taking into account both charged-current and flavor-changing neutral-current decays of the b'. For $Z^\circ \to \nu_L \bar{\nu}_L$ and $Z^\circ \to L^+ L^-$ ($L^\pm \to \nu_L W^*$) where ν_L is a stable fourth-generation neutrino, results are deduced from the total hadronic cross-section measurement at the Z° peak. The mass of ν_L is excluded up to 42.7 GeV. For $M\nu_L < 42.7$ GeV, the mass of $L^\pm$ is excluded for all values of $M_{L\pm} > M\nu_L$. Finally, for $Z^\circ \to L^\circ \bar{L}^\circ$, the L° is excluded up to $M_{L^\circ} = 45.7$ GeV with the mixing parameters $|U_{\ell L^\circ}|^2$, where $\ell = $ e, μ or τ, down to 10^{-13} at this mass. For $25.0\,\text{GeV} < M_{L^\circ} < 42.7\,\text{GeV}$, all values of $|U_{\ell L^\circ}|^2$ are excluded. All limits are given at 95% C.L, and are in good agreement with published results from TRISTAN, SLC, and LEP.

9. REFERENCES

[1]   D. Decamp *et al.*, ALEPH Collab., Phys. Lett. **231B** (1989) 519.

[2]   D. Decamp *et al.*, ALEPH Collab., Phys. Lett. **235B** (1990) 399.

[3]   D. Decamp *et al.*, ALEPH Collab., 'ALEPH: A Detector for Electron-Positron Annihilations at LEP', submitted to Nucl. Inst. Meth.

[4]   T. Sjöstrand, Comput. Phys. Comm. **28** (1983) 229. In the algorithm, the jet-forming cutoff parameter d_{join} is changed from its default value to $d_{join} = 2.0$ GeV.

[5]   T. Barklow, SLAC Report-315 (1987) 210.

[6]   J.H. Kühn, A. Reiter and P.M. Zerwas, Nuclear Physics **B272** (1986) 560.

[7]   G.S. Abrams *et al.*, MARK II Collab., Phys. Rev. Lett 63 (1989) 1558.

[8]   D. Decamp *et al.*, ALEPH Collab., Phys. Lett. **234B** (1989) 209.

[9]   M.Z. Akrawy *et al.*, OPAL Collab., Phys. Lett. **235B** (1990) 389.

[10]   V. Barger *et al.*, Phys. Rev. **D30** (1984) 947.

[11]   V. Barger, R.J.N. Phillips and A. Soni, Phys. Rev. Lett. **57** (1986) 1516;
        W.S. Hou and R.G. Stuart, Nucl. Phys. **B320** (1989) 227;
        W.S. Hou and R. G. Stuart, Phys. Rev. Lett. **62** (1989) 617.

[12]   The simulation for b'→bγ and bg is implemented by P. Mättig and T. Sjöstrand in JETSET, version 7.2.

[13]   S. Jadach, J.H. Kühn, 'TIPTOP, a Monte Carlo for Heavy Fermion Production and Decay', MPI PAE Pth 86-64 (1986);
        T. Sjöstrand *et al.* in *Z Physics at LEP 1*, CERN 89-08, Vol.3, p.293, eds. G. Altarelli, R. Kleiss and C. Verzegnassi.

308

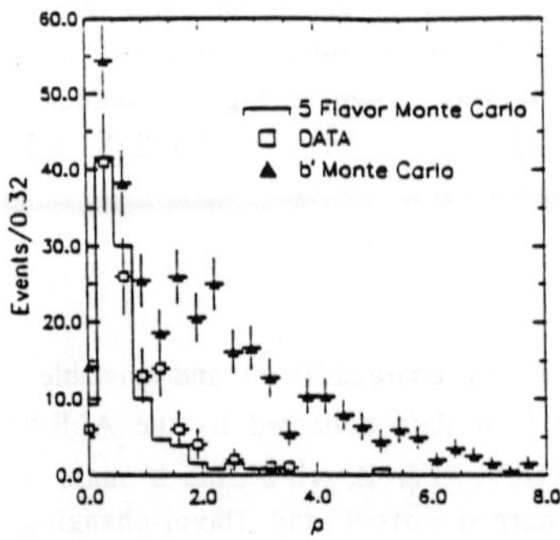

Fig. 1. Distributions of the isolation parameter ρ, after applying all other selection criteria, for the data and the Monte Carlo simulated events, normalized to the same integrated luminosity as the data, for $Z^0 \to b'\bar{b}'$ at 40 GeV mass. The cut is at $\rho = 1.8$.

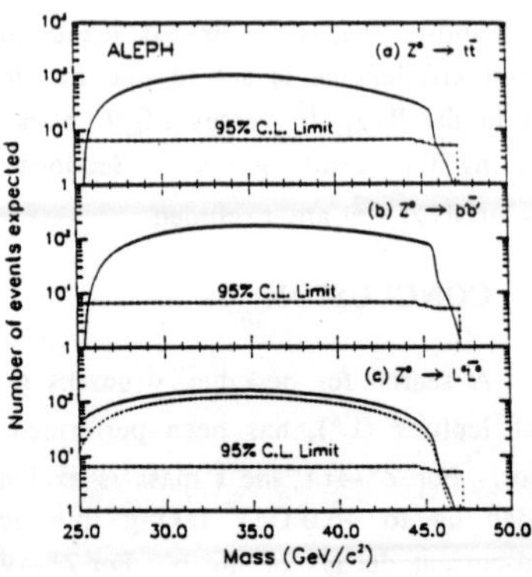

Fig. 2. (a) The number of expected events for $Z^0 \to t\bar{t}$ after applying the event selection criteria as a function of t mass. The 95% CL limit is indicated. (b) The same as (a) but for $Z^0 \to b'\bar{b}'$. (c) The same as (a) but for $Z^0 \to L^0\bar{L}^0$. The solid curve is for $L^0 \to eW^*$ or μW^* and the dashed curve for $L^0 \to \tau W^*$.

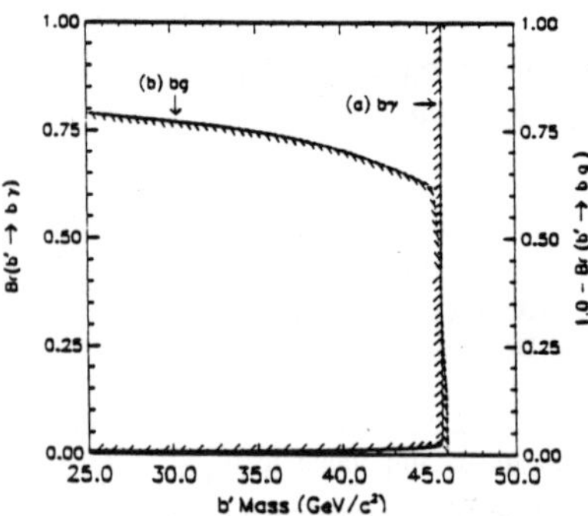

Fig. 3. The excluded region in decay branching ratios at 95% CL limit for (a) $b' \to b\gamma$ and (b) $b' \to bg$ as a function of b' mass from the process $Z^0 \to b'\bar{b}'$.

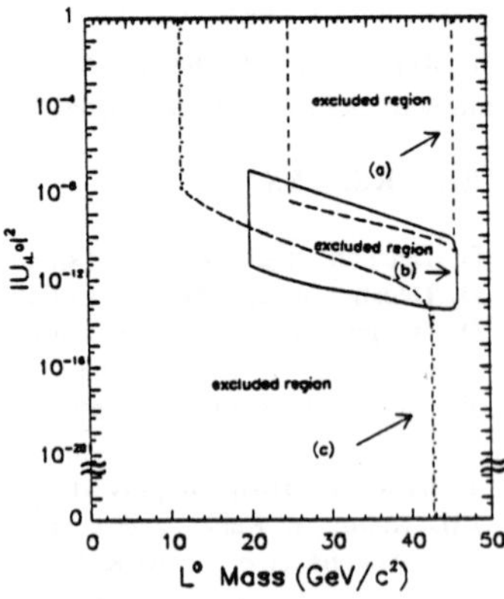

Fig. 4. The excluded regions at 95% CL for $Z^0 \to L^0\bar{L}^0$ in the $|U_{\ell\ell^0}|^2$ versus M_{L^0} plane from (a) the search for prompt L^0, (b) the search for long-lived L^0 from displaced vertices, and (c) the limit for long-lived L^0 from the total hadronic cross section measurement at the Z^0 peak. The bounded regions denote the areas of exclusion.

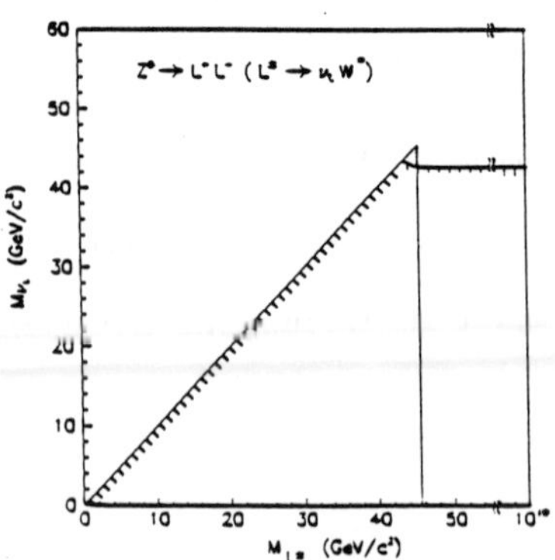

Fig. 5. The excluded regions at 95% CL for $Z^0 \to L^+L^-$ ($L^\pm \to \nu_L W^*$) in the M_{ν_L} versus $M_{L^\pm}$ plane. Note that the region where $M_{L^\pm} > M_{Z^0}/2$ and $M_{\nu_L} < 42.7$ GeV is excluded for all values of $M_{L^\pm}$.

A Search for Acoplanar Pairs of Leptons or Jets in Z^0 Decays: Mass Limits on Supersymmetric Particles

J. Raab

CERN EP Division

CH-1211 Geneva 23, Switzerland

Representing the OPAL Collaboration

Abstract

We report on a search for Z^0 decays into acoplanar pairs of leptons or jets using data collected with the OPAL detector during the 1989 LEP run. We present model independent limits on branching ratios for the Z^0 to decay into pairs of heavy particles with subsequent decays leading to the above topologies. We place lower bounds on the masses of the scalar leptons and the chargino.

Introduction

We describe a general search for Z^0 decays into pairs of massive particles that decay promptly to conventional quarks or leptons and undetected neutrino-like particles. Such events would be characterised by large missing energy and momentum. In the background to this signature from conventional sources the missing momentum vector points either along the e^+e^- beam direction (for initial state bremsstrahlung or two-photon interactions) or along the flight direction of the detected particles (e.g. in τ decays). We therefore searched for events in which the observed particles or jets are acollinear in the plane perpendicular to the beam direction (acoplanar).

Although our search for events containing acoplanar leptons (e, μ, τ) or jets was motivated primarily by supersymmetric models [1,2], similar signatures can be produced by other sources of new physics [3]. The absence of a signal allows us to set model independent limits on the rate of Z^0 decays of this kind and, within the supersymmetric framework, to place bounds on the masses of supersymmetric particles [4].

Detector and Trigger

The data were collected during a scan over the Z^0 resonance; the centre of mass energies varied between 88.3 and 94.3 GeV. The data correspond to a measured integrated luminosity of about $550\ nb^{-1}$ and approximately 12000 observed hadronic Z^0 decays. The pertinent components of the OPAL detector for this analysis are a large volume tracking chamber, providing up to 159 space-point measurements per track, surrounded by a solenoidal coil, a time-of-flight (TOF) scintillator-counter array, an electromagnetic calorimeter with over 10000 lead glass blocks, an iron return yoke for the magnetic field that is instrumented with 9 layers of streamer tubes, and 4 layers of muon chambers. The luminosity is measured with 2 lead-scintillator calorimeters surrounding the beam pipe in the forward and backward directions. The detector covers essentially the full 4π solid angle.

The trigger conditions relevant to this search were (1) an energy deposit of more than 6 GeV in the barrel electromagnetic calorimeter (2) at least 2 tracks with $|\cos\theta| < 0.7$ (3) hits in at least 3 of 4 layers of muon chambers which were associated in azimuthal angle with either a track or a TOF counter hit (4) an energy deposit greater than 4 GeV in the barrel electromagnetic calorimeter in coincidence with at least one track or one TOF-counter hit.

Analysis

We describe four separate searches for events containing acoplanar pairs of (a) electromagnetic clusters (b) muons (c) low-multiplicity jets and (d) hadronic jets. For the searches (a) (b) and (c) the selection criteria followed the same pattern, namely first to select events containing two energetic particles or jets, then to demand that there was no other significant activity in the event and finally to require that the particles or jets were acoplanar. These criteria were chosen to give negligible background from standard processes. The trigger efficiencies for events that would have passed the selection criteria (described below) were above 99% for (a) and (d), $96 \pm 2\%$ for (b), and $93 \pm 1\%$ for (c).

(a) In the *search for events with acoplanar electromagnetic clusters* the following selection criteria were applied: (1) at least two clusters with energy greater than 10% of the beam energy and with $|\cos\theta| < 0.7$ (2) the number of clusters had to be less than 15 and the total electromagnetic energy outside cones with a half-width of $10°$ around the two most energetic clusters had to be less than 2 GeV (3) no tracks that were separated by more than $20°$ in azimuth from the two most energetic clusters (4) at least one of the two most energetic clusters was matched to a TOF hit within $5°$ in azimuth, and the measured time had to be within 8 ns of the expected time-of-flight from the interaction point (5) excluding the first layer, the number of hits in the remaining 8 layers of the barrel hadron calorimeter was required to be less than 6 (6) the two most energetic clusters had to have an acollinearity angle greater than $10°$. (7) the acoplanarity angle between the two most energetic clusters had to exceed $20°$.

Cut 1 favoured events containing two energetic electrons (or photons). Cuts 2 and 3 removed hadronic events, cuts 4 and 5 removed cosmic rays, while most e^+e^- and $\tau^+\tau^-$ events were rejected by cut 6. After the first six cuts 12 events remained. The acoplanarity distribution for these events is shown in Fig. 1a. After the final cut on the acoplanarity angle no candidates survived. The curve in the figure represents the distribution expected from supersymmetric electron pairs and will be discussed later.

(b) In our *search for events with acoplanar muons* we applied the following selection criteria: (1) the events had to contain exactly two tracks, each with momentum greater than 4 GeV/c and $|\cos\theta| < 0.7$ (2) at least one track had to be found in the muon chambers, with hits in 3 out of 4 layers (3) the energy in electromagnetic clusters outside cones with $20°$ half-width around the two tracks was required to be less than 2 GeV (4) the acoplanarity angle between the two tracks had to exceed $20°$.

After requiring cuts 1 to 3, we were left with 383 events for which the distribution of the acoplanarity angle is shown in Fig. 1b. The observed number of events is consistent with the expectation from Z^0 decays to $\mu^+\mu^-$ and $\tau^+\tau^-$ pairs. After the acoplanarity cut no candidates survived.

(c) We now describe the *search for events with acoplanar low-multiplicity jets* (e.g. τ decays). Most multihadronic and cosmic ray events were removed by requiring that the track multiplicity be less than 9 and that the tracks emanate from the interaction region. The remaining events were analysed in terms of low-multiplicity jets. In this context a "jet" could be a single charged or neutral particle or a narrow jet of a few charged or neutral particles, as produced by τ decays. To estimate the energy of a jet, the momenta of the tracks and the clusters were added separately. To avoid double counting, the larger of the two sums was taken as the energy of the jet. With the jets defined in this way the following selection criteria were applied: (1) exactly 2 jets were required, each with at least one track (2) at least one jet had to be within $|\cos\theta| < 0.8$ (3) the energy of each jet was required to be greater than 3 GeV (4) the missing transverse momentum calculated from the two jets had to be greater than 3 GeV/c (5) the energy deposited in the forward detector was required to be less than 10 GeV (6) the acoplanarity angle between the two jets had to exceed $20°$.

Cuts 1 through 3 selected pairs of jets with little other activity in the event. Cut 4 eliminated two-photon events which would not be tagged by the forward detector due to the acceptance cutoff at 40 mrad in polar angle. Cut 5 discriminated against two-photon interactions that deposited energy in the forward detectors. After cuts 1 to 5, we were left with 1000 events consistent with the expectation for lepton pair production. In Fig. 1c we show the acoplanarity distribution for these events. No events remained after the last cut.

d) In the *search for acoplanar hadron jets* the events had to satisfy the following requirements: (1) the number of tracks had to exceed five (2) in both hemispheres defined by the plane normal to the beam axis the electromagnetic energy had to be greater than 2 GeV (3) the total electromagnetic energy had to exceed 7.5 GeV (4) in order to ensure good containment of the events, the thrust axis (calculated from the electromagnetic clusters) had to satisfy $|cos\theta| < 0.86$ (5) the event was divided into two hemispheres using the plane normal to the thrust axis. In each hemisphere, the vector sum of the momenta of the clusters was formed. The acoplanarity angle between these two vector sums was required to be greater than 50°.

After applying cuts 1 to 4, a sample of 9837 events remained. The distribution of the acoplanarity angle for these events is shown in Fig. 1d. The distribution computed for hadronic decays of the Z^0 (obtained using the Jetset Monte Carlo [5]) is also shown in the figure and agrees well with our data. Only 11 events have an acoplanarity angle larger than 50°. This number is in agreement with the simulation which predicted 12 ± 4 events.

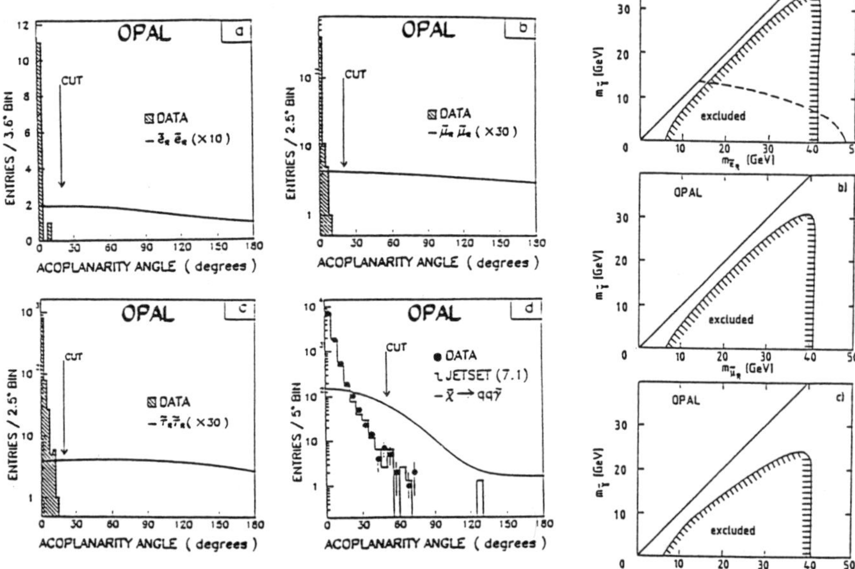

Fig. 1 Acoplanarity distributions for the 4 searches.

Fig. 2 The 95% CL lower mass limit contour plot for $\tilde{l}_R$. The dashed curve in (a) is the 90% CL ASP result [6].

Limits on Supersymmetric Particle Masses and on Topological Z^0 Branching Ratios

We now proceed to place limits on branching ratios for the Z^0 to decay into a pair of massive particles that subsequently decay into leptons or quarks and undetected neutral particles. Such events are expected to be a signature for the production of supersymmetric (SUSY) particles. We assume that the photino ($\tilde{\gamma}$), the SUSY partner of the photon, is the lightest stable SUSY particle. With this assumption, the scalar lepton ($\tilde{l}$) would decay promptly to the corresponding lepton and a $\tilde{\gamma}$. In most SUSY models the right-handed scalar leptons ($\tilde{l}_R$) are lighter than the left-handed ($\tilde{l}_L$). The partners of the charged bosons ($W^\pm, H^\pm$) are called charginos ($\tilde{\chi}^\pm$). The unknown mixing between these states leads to some model dependence in the production cross section and decay branching ratio of the $\tilde{\chi}^{\pm}$. The $\tilde{\chi}^{\pm}$ decay is expected to have leptonic modes $\tilde{\chi}^{\pm} \to l\nu\tilde{\gamma}$, and hadronic modes $\tilde{\chi}^{\pm} \to q\bar{q}\tilde{\gamma}$. Masses below 25 GeV/$c^2$ have already been excluded for most of the supersymmetric unstable charged particles [6].

The production cross sections of SUSY particles were obtained in the Born approximation from refs. [7,8]. Initial state bremsstrahlung was taken into account according to the formalism of ref. [9] which reduced the cross section by about 30% relative to the Born approximation. Using this framework we generated Monte Carlo events to determine the selection efficiencies for the various SUSY processes. The fragmentation of the quarks and the decays of the $\tau^{\pm}$ were included using the Jetset Monte Carlo. The events were then passed through a full simulation of the OPAL detector. The cross sections and selection efficiencies were calculated as a function of the masses of the various supersymmetric particles. The total detection efficiencies for the $\tilde{l}$, using $m_{\tilde{\gamma}} = 20$ GeV/c^2 and $m_{\tilde{l}} = 40$ GeV/c^2, were 44±2%. For a $\tilde{\chi}^{\pm}$ with a mass of 40 GeV/c^2 and $m_{\tilde{\gamma}} = 0$ GeV/c^2 an efficiency of $15 \pm 1\%$ was obtained. Of course, these efficiencies dropped off rapidly when the mass difference between the decaying SUSY particle and the $\tilde{\gamma}$ became small. In general, for $\tilde{\chi}^{\pm}$ and $\tilde{l}$ masses greater than 40 GeV/c^2 the detection efficiencies of these events are independent of spin and determined by the decay kinematics. The expected distributions for the relevant SUSY particles, after the corresponding cuts, are shown as smooth curves in figs. 1a-d.

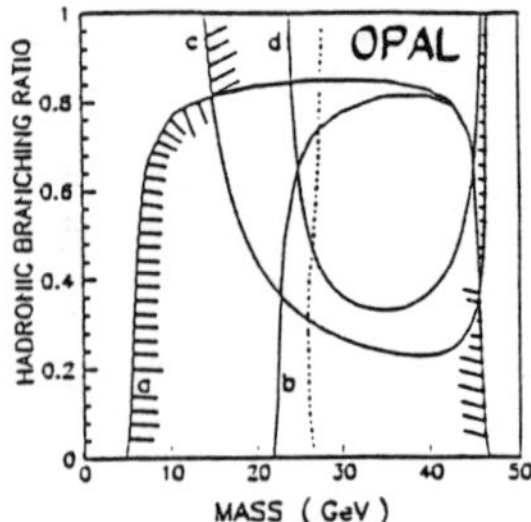

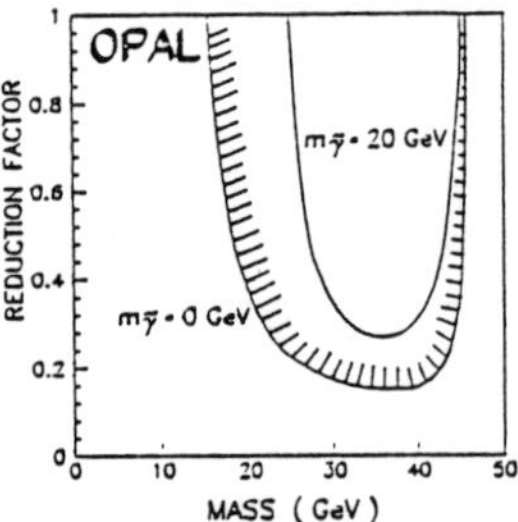

Fig. 3 Curves (a) and (c) correspond to $m_{\tilde{\gamma}} = 0$ GeV/c^2, (b) and (d) to $m_{\tilde{\gamma}} = 20$ GeV/c^2. The interior of (a) and (b) are from search (a), while the interior of (c) and (d) are from analysis (d). The dash-dotted line represents the limit from TOPAZ [6].

Fig. 4 The $\tilde{\chi}^{\pm}$ mass limit as a function of $m_{\tilde{\gamma}}$. The curves are obtained from the combined limit of analysis (a) and (d).

For obtaining lower mass limits on the $\tilde{l}$, we considered the cases where the $\tilde{l}_L$ and $\tilde{l}_R$ have different masses and also the case where they have equal mass. For the $\tilde{e}$ the limits were derived from search (a), for $\tilde{\mu}$ from search (b), and for $\tilde{\tau}$ from analysis (c). For $m_{\tilde{\gamma}} < 30$ GeV/c^2 the 95% CL lower mass limits for the $\tilde{e}$ are 43.4, 42.3, and 41.5 GeV/c^2 if $m_R = m_L$, $m_L \gg m_R$, and $m_R \gg m_L$ respectively. Similarly, for both $\tilde{\mu}$ and $\tilde{\tau}$, with $m_{\tilde{\gamma}} < 30$ GeV/c^2 and $m_{\tilde{\gamma}} < 23$ GeV/c^2 respectively, we obtain at 95% CL the lower mass bounds of 43.0, 41.9, and 41.0 GeV. The excluded mass range for each $\tilde{l}_R$ as a function of $m_{\tilde{\gamma}}$ is shown in fig. 2.

For the chargino ($\tilde{\chi}^{\pm}$), we calculated mass limits as a function of the hadronic branching ratio assuming equal branching ratios into each leptonic channel. We further assumed that the $\tilde{l}$ and the associated scalar neutrinos are heavy. Otherwise, the $\tilde{\chi}^{\pm}$ would decay predominately as $\tilde{\chi}^{\pm} \to l\nu_l$ or $\tilde{\chi}^{\pm} \to \tilde{l}\nu_l$, a case that is covered by our search for acoplanar leptons. On the Z^0 resonance and for a $\tilde{\chi}^{\pm}$ mass of 40 GeV/c^2 the model leads to a Born cross section of about 8.5 nb. Treating the hadronic branching ratio as a free parameter, Fig. 3 shows the region of the $\tilde{\chi}^{\pm}$ mass that is excluded by searches (a) and (d). The regions are given for two choices of the $\tilde{\gamma}$ mass. Combined limits from the two searches are obtained from the crossing points of the curves in Fig. 3. For photino masses of 0 (20) GeV/c^2, we exclude a $\tilde{\chi}^\pm$ with a mass less than 45.7 (45.0) GeV/c^2 irrespective of the hadronic branching ratio. In general, different SUSY models and different parametrisations can give other $\tilde{\chi}^\pm$ production cross sections than the one used above. Therefore, we show in Fig. 4 the variation in the mass limits (independent of the hadronic branching ratio) if the cross section is scaled by an overall reduction factor.

The selection efficiencies, typically 40% for heavy particles (mass greater than 40 GeV/c^2) decaying into leptons and 15% for decays into jets, are largely model independent. Therefore, we calculated limits on topological Z^0 decays, without reference to any particular model. The limits were obtained using the relation $Br < N_0/(\epsilon \cdot N_Z)$, where N_0 is the expected number of events at 95% CL, ϵ the detection efficiency, and N_Z the total number of produced Z^0's ($\approx$ 17000 for analysis (a) and $\approx$ 15000 for the others). The 95% CL upper limits for Z^0 decays into pairs of massive particles that decay promptly to a lepton or photon plus undetected neutrino-like particles are: 4.4×10^{-4} for electron or photon pairs and 5.0×10^{-4} for μ or τ pairs. We observed 11 events with a pair of acoplanar hadronic jets. The corresponding upper limit on the branching ratio for Z^0 to acoplanar hadronic jets is then 8.0×10^{-3}, independent of the fragmentation models.

Summary

We found no evidence for the decay of the Z^0 into acoplanar pairs of leptons or hadronic jets above background. We set upper limits on Z^0 decays into pairs of massive particles that decay promptly to conventional quarks, leptons, or photons and undetected neutrino-like particles. For the leptonic final states we obtained upper limits on the branching ratios of 5×10^{-4} and for the hadronic final states 8×10^{-3} . These limits are essentially model independent. For mass degenerate scalar leptons the lower mass limits of 43 GeV/c^2 were obtained [10]. For the chargino we excluded the mass range below 45.0 GeV/c^2. All the above numbers are given at 95% confidence level. Recently similar limits were obtained by other experiments at LEP [11].

References

[1] Y. Gol'fand and E. Rikhtman, JETP Lett. 13 (1970) 323; D. Volkov and V. Akulov, JETP Lett. 16 (1972) 438; J. Wess and B. Zumino, Nucl. Phys. B70 (1974) 39.

[2] P. Fayet and S. Ferrara, Phys. Rep. 32C (1977) 249; H. E. Haber and G. L. Kane, Phys. Rep. 117C (1985) 75; H. P. Nilles, Phys. Rep. 110C (1984) 1.

[3] Physics at LEP, CERN-86-02, editors: J. Ellis and R. Peccei, vol. 1, (1986) 297.

[4] To be published in Phys. Lett. B240, April 19 1990.

[5] T. Sjöstrand, Comp. Phys. Comm. 39 (1986) 347; JETSET, Version 7.1. For the quark fragmentation we used the QCD shower formalism.

[6] ASP Collaboration, C. Hearty et al., Phys. Rev. Lett. 58 (1987) 1711; Phys. Rev. D39 (1989) 3207; TOPAZ Collaboration, I. Adachi et al., Phys. Lett. B218 (1989) 105; For a review on recent mass limits and further references: T. Kamae, Proceedings of the 24th International Conference in High Energy Physics, editors: R. Kotthaus and J. H. Kühn, (1988) 156.

[7] M. Chen, C. Dionisi, M. Martinez, and X. Tata, Phys. Rep. 159 (1988) 201.

[8] A. Bartel, H. Fraas, and W. Majerotto, Z. Phys. C 30 (1986) 441; Z. Phys. C 41 (1988) 475.

[9] F. A. Berends, R. Kleiss, and S. Jadach, Nucl. Phys. B202 (1982) 63; Comp. Phys. Comm. 29 (1983) 185; Z-Physics at LEP 1, Vol. 3, CERN 89-08, editors: G. Altarelli, R. Kleiss, and C. Verzegnassi.

[10] Using the full 1989 data sample ($\approx 1.1pb^{-1}$ the limits are expected to shift upward by about 3% for the degenerate case, and 1–2% for the others.

[11] L3 Collaboration, B. Adeva et al., L3 preprint 2 (1989); ALEPH Collaboration, D. Decamp et al., CERN-EP 89-158 (1989).

FIRST MEASUREMENTS OF HADRONIC
DECAYS OF THE Z BOSON[*]

Eric Wicklund

California Institute of Technology
Pasadena, Ca 91125
representing the Mark II Collaboration

ABSTRACT

We have observed hadronic final states produced in the decays of Z bosons. In order to study the parton structure of these events, we compare the distributions in sphericity, thrust, aplanarity and number of jets to the predictions of several QCD-based models and to data from lower energies. The data and models agree within the present statistical precision.

1. INTRODUCTION

Studies of the decays of Z bosons to hadronic final states are presented in this talk.[1] The experiment was performed with the Mark II detector at the SLAC e^+e^- Linear Collider (SLC). We focus on measurables which relate to testing the partonic structure that underlies the observed hadrons, namely sphericity, aplanarity, thrust and jet multiplicity.

2. EVENT DETECTION AND EVENT SELECTION

The Mark II detector has been described in detail elsewhere.[2] The momenta of charged tracks are measured with a 72 layer drift chamber in a 4.75 kG solenoidal magnetic field. The energy and direction of photons are measured in two electromagnetic calorimeter systems of strip readout geometry which cover the angular region $|\cos\theta| < 0.96$, where θ is measured relative to the incident beam direction. The trigger for hadronic Z decays is a combination of charged particle and neutral energy triggers and is described in detail in Ref. 2. Two charged tracks in the angular region $|\cos\theta| < 0.76$ are sufficient for an event trigger. The system has considerable redundancy and has an estimated efficiency of greater than 0.99 for hadronic events from Z decay.

Events are selected by requiring at least 7 well-reconstructed charged tracks and that the sum of charged particle energies and shower energies (E_{vis}) be greater than $0.5\,E_{cm}$. This selection ensures that the events are well contained within the sensitive regions of the detector and eliminates lepton pairs from the data sample.

---

[*] Work supported in part by the US Department of Energy, contract DE-AC03-81ER40050

Charged tracks must originate from the region of the e^+e^- collision point defined by a cylinder around the nominal collision point of radius 1 cm and half-length 3 cm along the beam direction. This requirement reduces the number of background tracks due to secondary interactions in the material of the detector and reduces the number of tracks from beam-gas events. In order to ensure well-measured momenta and a high tracking efficiency, the tracks must be within the angular region $|\cos\theta| < 0.82$ and must have transverse momenta with respect to the beam axis of at least 150 MeV/c. In the central calorimeters, showers are selected within a fiducial volume bounded by $|\cos\theta| < 0.68$ and within the active volume of the calorimeter in the azimuthal angle ϕ. The total solid angle coverage of the central calorimeters is 63.5%. In the forward and backward calorimeters, showers are selected within a fiducial volume of $0.74 < |\cos\theta| < 0.95$. All showers in the fiducial volume of the calorimeters are retained if their energy exceeds 0.5 GeV, irrespective of any association with a charged track.

The data used in this analysis have been collected during a scan of center-of-mass energy (E_{cm}) ranging from 89.2 to 93.0 GeV, corresponding to an integrated luminosity of 19.7 nb^{-1}. The criteria given above select 394 events. The selection efficiency is estimated using QCD-based Monte Carlo generators[3-5] (described in detail later) to be 0.80 ± 0.02, where the error is the systematic uncertainty from different Monte Carlo models. The events removed by this selection tend to have large undetected energy close to the beam line, and studies with the Monte Carlo models demonstrate their removal does not bias the Z decay characteristics presented in this paper.

Event backgrounds resulting from beam-gas scattering, lepton pairs, and two-photon production are small. The number of beam-gas events has been estimated to be less than 0.4 by searching for events with vertices outside the required region. The numbers of background events from lepton pairs and from two photon production have been estimated from Monte Carlo simulations to be less than 0.1 events and 0.01 events, respectively.

3. PROPERTIES OF HADRONIC DECAYS OF THE Z BOSON

The characteristics of the selected events are described in terms of the event shape parameters sphericity (S), aplanarity (A) and thrust (T), and these quantities are compared to the expectations of QCD models. These shape parameters are commonly used in the literature and their full description can be found in Ref. 6. In addition, a cluster algorithm is used to estimate the number of jets (N_{jet}) observed in each event. The analysis method for calculating N_{jet} is described in detail elsewhere[7,8].

The quantities have been corrected for detector acceptance inefficiencies and for machine-related backgrounds using the following procedure. Simulated events have been generated using 5-flavor QCD fragmentation models which were adjusted to fit the Mark II data at a center-of-mass energy of 29 GeV[6]. The models are the Lund parton shower model (Lund 6.3 shower)[3], the Webber-Marchesini parton shower model (Webber 4.1)[4], and the parton shower model of Gottschalk and Morris (Caltech II 86)[5]. The generated particle four-momenta are then used in a simulation of the responses of the Mark II detector components.

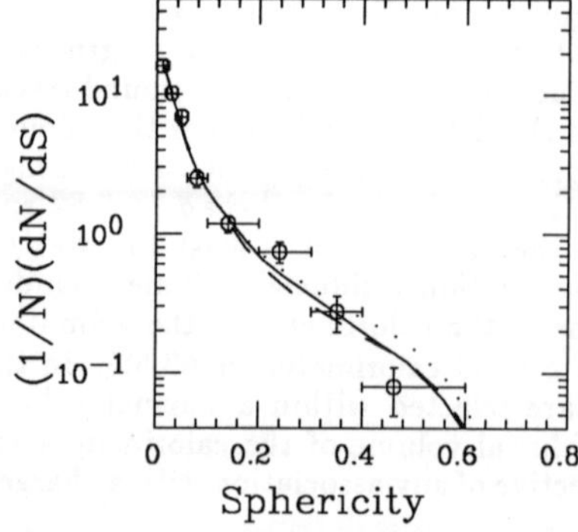

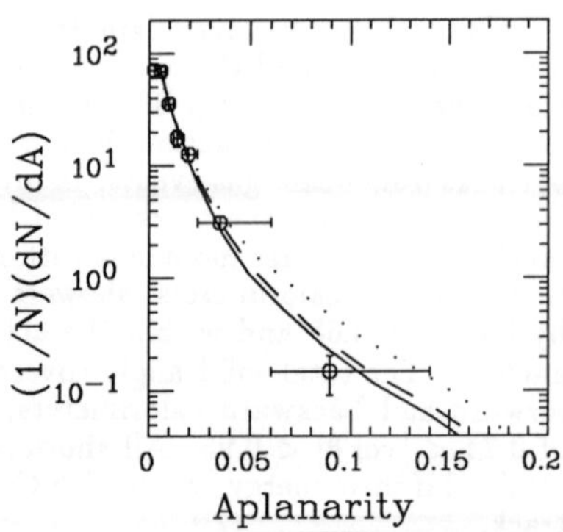

Fig. 1. Event sphericity, for data (circles, with statistical errors) and three QCD Monte Carlo models (curves). Data points are plotted at the mean data value within each range and the horizontal bars show the extent of the range.

Fig. 2. Event aplanarity, for data (circles, with statistical errors) and three QCD Monte Carlo models (curves).

In addition, the contamination of the data arising from machine-related backgrounds is included. These backgrounds include soft photons which are absorbed in the drift chamber gas and result in spurious, saturated drift chamber signals and muons produced near the final focus of the SLC which deposit energy in the calorimeters. These backgrounds have been taken into account with a technique in which signals recorded in the detector during random beam crossings have been mixed with Monte Carlo simulated events. These mixed, simulated events have then been subjected to the same reconstruction and analysis procedures as the data.

The corrected distribution in a shape parameter is obtained from the measured distribution using a bin-by-bin correction function[6]. The correction factors varied within the range of 0.7 to 1.2.

The corrected distributions for the shape parameters sphericity, aplanarity and thrust are shown in Figs. 1–3. Also shown are the predictions from the Lund 6.3 shower model (solid lines), the Webber 4.1 model (dashed lines) and the Caltech II 86 model (dotted lines) for these quantities. These QCD model predictions are in good agreement with the data. The mean values of the shape quantities have been measured to be $< S >= 0.075 \pm 0.005$, $< A >= 0.0117 \pm 0.0007$ and $< T >= 0.931 \pm 0.003$ where the corrections for detector acceptance have been applied and the errors are statistical only.

The corrected fractions of 2, 3, 4 and 5-jet events are shown in Fig. 4 as a function of y_{cut}. In Fig. 3 each event contributes at all values of y_{cut} and hence the statistical errors for different values of y_{cut} are not independent. As illustrated in Ref. 8, the corrected jet multiplicity is expected to reproduce rather closely the produced jet (parton) multiplicity. At a standard value of $y_{cut} = 0.08$, the fraction of three jet events in hadronic events is 0.22 ± 0.02.

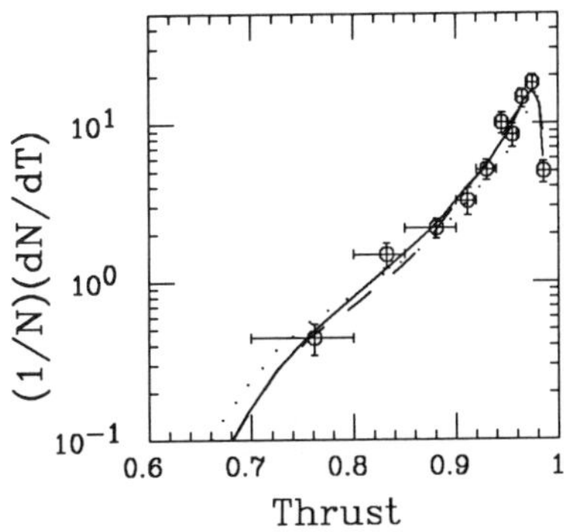

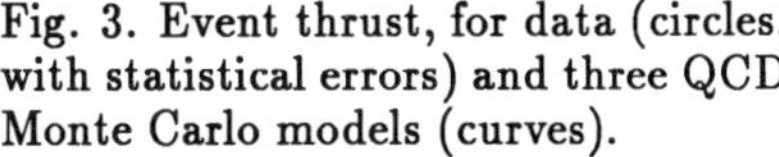

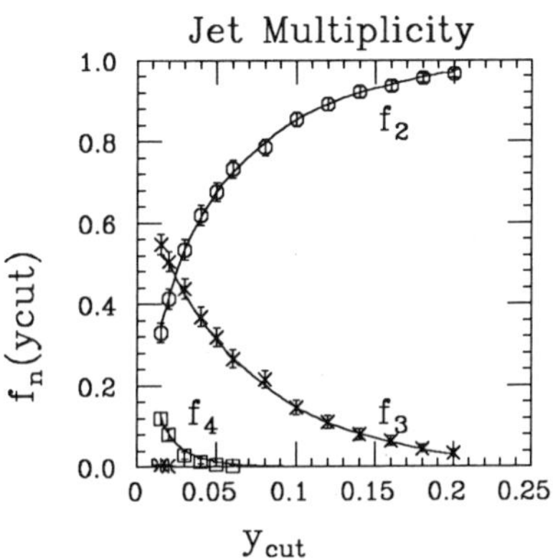

Fig. 3. Event thrust, for data (circles, with statistical errors) and three QCD Monte Carlo models (curves).

Fig 4. The observed 2, 3, 4 and 5-jet fractions for different values of y_{cut} (symbols, with statistical errors) and two QCD Monte Carlo model predictions (curves). Note that each event contributes at all values of y_{cut} and hence the statistical errors for different values of y_{cut} are not independent.

Statistical errors are the dominant uncertainties in these studies. One source of systematic error is the correction for machine-related backgrounds determined using the mixing technique described above. The effects of these backgrounds are increases in mean sphericity of 7% and mean aplanarity of 12% and a decrease in mean thrust of 0.6%. The fraction of three jet events is increased by less than 4% for the range of y_{cut} values used. The uncertainties in these corrections are small compared to the statistical errors.

The mean values of the corrected quantities are compared in Fig. 5 to mean values from other experiments[9] at different center-of-mass energies and to the values from this experiment at 29 GeV [6,8]. As the energy increases, the mean values of sphericity and aplanarity decrease and the mean value of thrust increases. These trends can be interpreted as evidence that the underlying parton kinematics are more distinct at higher center-of-mass energies due to the lessened importance of soft fragmentation effects. For comparison, the solid curves show the expectations from the Lund parton shower model which follows the data over a wide range of energies.

With the current statistical precision, all three QCD-based models account well for the data. No significant anomalies are seen and the global properties of the Z hadronic decays appear to be correctly predicted by QCD-based fragmentation models. Until higher statistics tests are available at these energies, it is reasonable to use the standard Monte Carlo models as a guide to the partonic properties of Z decays.

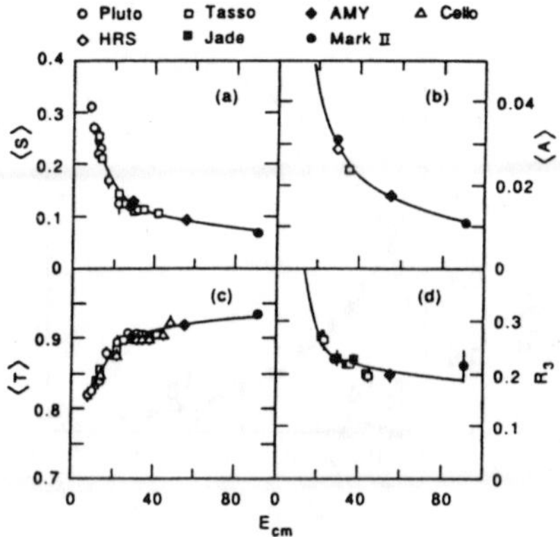

Fig 5. The mean values of (a) sphericity (b) aplanarity and (c) thrust and (d) the 3-jet fraction for a y_{cut} of 0.08 compared to data from several center-of-mass energies. The errors are statistical only. The curves are the predictions from the Lund parton shower model.

In conclusion, we have presented the first study of the global properties of the decays of Z bosons to hadrons. The data are in agreement with the predictions of the QCD-based fragmentation models optimized at lower energies at PEP using the same detector.

ACKNOWLEDGMENTS

We are indebted to the SLAC staff and to the contributing universities whose efforts culminated in the successful operation of the SLC which produced the events for this talk.

REFERENCES

1. G. Abrams et al., Phys. Rev. Lett. **63**, 1558 (1989).

2. G. Abrams et al., Nucl. Instrum. Methods. A **281**, 55 (1989).

3. T. Sjöstrand, Comp. Phys. Comm. **39**, 347 (1986); T. Sjöstrand and M. Bengtsson, Comp. Phys. Comm. **43**, 367 (1987); M. Bengtsson and T. Sjöstrand, Nucl. Phys. B **289**, 810 (1987).

4. G. Marchesini and B.R. Webber, Nucl. Phys. B **238**, 1 (1984); B. R. Webber, Nucl. Phys. B **238**, 492 (1984).

5. T.D. Gottschalk and D. Morris, Nucl. Phys. B **288**, 729 (1987).

6. A. Petersen et al., Phys. Rev. D **37**, 1 (1988).

7. W. Bartel et al., Z. Phys C **33**, 23 (1986).

8. S. Bethke et al., Z. Phys C **43**, 325 (1989).

9. Ch. Berger et al., Z. Phys C **12**, 297 (1982); M. Althoff et al., Z. Phys C **22**, 307 (1984); D. Bender et al., Phys. Rev. D **31**, 31 (1985); S. Bethke et al., Phys. Lett. B **213**, 235 (1988); W. Braunschweig et al., Z. Phys. C **41**, 359 (1988); W. Braunschweig et al., Phys. Lett. B **214**, 286 (1988); I.H. Park et al., Phys. Rev. Lett. **62**, 1713 (1989); H.J. Behrend et al., DESY Preprint 89-019, to be published in Z. Phys. C; Y.K. Li et al., KEK Preprint 89-34 (1989), Y.K. Li,S. Olsen (private communication).

CHARGED-PARTICLE INCLUSIVE DISTRIBUTIONS FROM HADRONIC Z^0 DECAYS

K. O'SHAUGHNESSY

representing the Mark II Collaboration*

Stanford Linear Accelerator Center
Stanford University, Stanford, California, 94305

ABSTRACT

We have measured inclusive distributions for charged particles in hadronic decays of the Z boson. The variables chosen for study were the mean charged-particle multiplicity ($< n_{ch} >$), scaled momentum (x), and momenta transverse to the sphericity axes ($p_{\perp in}$ and $p_{\perp out}$). The distributions have been corrected for detector effects and are compared with data from $e^+ e^-$ annihilation at lower energies and with the predictions of several QCD-based models. The data are in reasonable agreement with expectations.

*This work was supported in part by Department of Energy contracts DE-AC03-81ER40050 (CIT), DE-AM03-76SF00010 (UCSC), DE-AC02-86ER40253 (Colorado), DE-AC03-83ER40103 (Hawaii), DE-AC02-84ER40125 (Indiana), DE-AC03-76SF00098 (LBL), DE-AC02-76ER01112 (Michigan), and DE-AC03-76SF00515 (SLAC), and by the National Science Foundation (Johns Hopkins).

We present measurements of charged-particle inclusive distributions in hadronic decays of the Z^0 boson. The data were taken with the Mark II detector at the SLAC e^+e^- Linear Collider (SLC) running on and near the Z boson resonance peak at 91.1 GeV.[1] These data correspond to a total integrated luminosity of 19.7 nb^{-1}. Comparisons with lower energy data are presented, in particular with measurements at 29 GeV taken with the same detector.

The Mark II detector has been described in detail elsewhere.[2] Charged particles are measured with a 72-layer cylindrical drift chamber in a 4.75 kG solenoidal magnetic field. The momentum (p) resolution was determined from Bhabha scattering events at 29 GeV to be $\sigma(p)/p = 0.0046p$ (p in GeV/c). When the charged tracks are constrained to originate at the e^+e^- interaction point (IP), the momentum resolution improves to $\sigma(p)/p = 0.0031p$. The trigger includes charged particle and neutral energy components and has an estimated efficiency of greater than 99% for hadronic Z decays.[1]

Events were selected based on the reconstructed charged tracks and electromagnetic showers. The charged tracks were required to pass through a cylinder around the measured IP of radius 0.01 m and half-length 0.03 m along the beam direction. The polar angles had to satisfy $|\cos\theta| < 0.82$, where θ is the polar angle relative to the beam direction. The momenta transverse to the beam direction were required to exceed 0.3 GeV/c.

Electromagnetic showers were measured in two systems. In the central calorimeter (lead–liquid argon) they were required to satisfy $|\cos\theta| < 0.68$ and be away from cracks between modules. The fiducial volume for the endcap calorimeter (lead–proportional tube) was defined to be $0.74 < |\cos\theta| < 0.95$. An energy greater than 0.5 GeV was required. Showers were not retained if associated with a charged track.

Events were required to have at least 5 charged tracks passing these cuts and an additional cut on the invariant hemisphere mass was used to eliminate τ decays with a 3–3 topology. The visible energy (calculated from showers and charged tracks assuming pion masses) was required to be greater than $0.4E_{cm}$. The number of events passing all selection criteria was 398. Backgrounds from beam-gas scattering, Z decays into lepton pairs and two photon scattering were estimated to be less than 0.8 event.[3] Contamination from accelerator-related backgrounds was included by superimposing data from random beam crossings onto Monte Carlo (MC) events with detector simulation.

The data are compared with events simulated by three QCD-based MC event generators. The models used are the Lund parton shower model with string fragmentation (JETSET 6.3 shower),[4] the Webber-Marchesini parton shower model with cluster fragmentation (BIGWIG 4.1),[5] and the parton shower model of Gottschalk and Morris (Caltech-II 86) with a combined fragmentation method.[6] The parameters of these models were tuned to fit Mark II data at 29 GeV.[7] The Lund model based on second-order QCD matrix element calculations, again with string fragmentation, was not used because an extrapolation to 91 GeV is not possible without changing parameters which should be kept constant.[8]

The data were corrected for detector inefficiencies, resolutions and machine backgrounds using bin-by-bin correction factors derived from the JETSET 6.3 shower MC with full detector simulation. Charged particles from all K_S^0 and Λ decays were included in the corrected distributions. Typical correction factors were ~ 1.2, with a spread of $\sim 30\%$ for the different bins in each distribution. Systematic errors included differences between the QCD models. Corrections for QED radiative effects were included but were less than 2% for these data. All errors shown for these data have statistical and systematic uncertainties added in quadrature.

The charged-particle multiplicity distribution was not corrected using this bin-by-bin method because the correlations between bins are large. An unfold procedure[9] was used to measure the mean corrected charged-particle multiplicity to be $<n_{ch}> = 20.1 \pm 1.0 \pm 0.9$. This number is consistent with the extrapolation from the lower energy data. For comparison, the JETSET 6.3 shower model gives a multiplicity of 21.4, BIGWIG 4.1 predicts 20.1 and the Caltech-II model gives 22.5.

Figure 1(a) shows the corrected inclusive distribution $1/\sigma_{had} \, d\sigma_{trk}/dx$, where $x = 2p/E_{cm}$, compared with the predictions of the models. The quantities σ_{had} and σ_{trk} are the total hadronic and charged-particle inclusive cross sections, respectively. All of the models predict a spectrum consistent with the observed distribution. Figure 1(b) compares the results of this analysis with data from other e^+e^- experiments.[7,9,10] The solid line is the prediction of the JETSET 6.3 shower model. The higher x bins show small scaling violations, in agreement with this model and qualitatively expected from QCD. The increase in the lower x bins are due to the increase in available phase space for particle production.

The transverse momenta defined by the sphericity axes[11] in the event

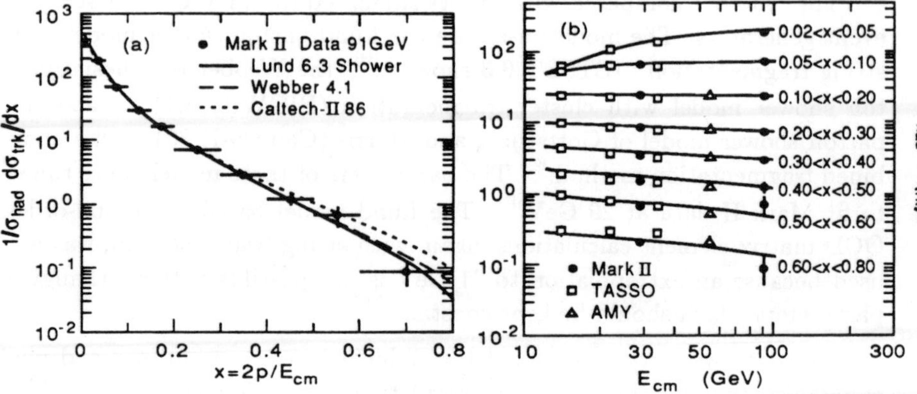

Fig. 1. (a) Corrected charged-particle inclusive distribution $1/\sigma_{had}\, d\sigma_{trk}/dx$, where $x = 2p/E_{cm}$, compared with several models. (b) Comparison between charged-particle inclusive distribution in x for hadronic Z^0 decays and e^+e^- experiments at lower E_{cm}.

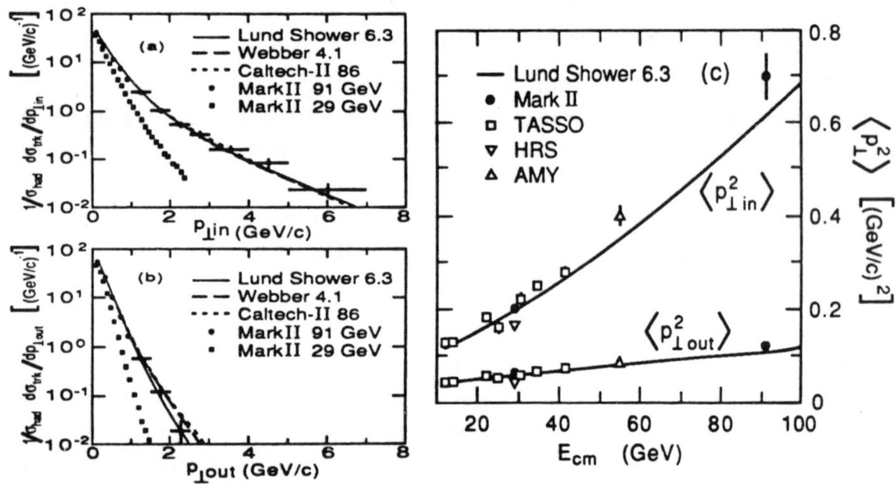

Fig. 2. Corrected charged-particle distributions (a) $1/\sigma_{had}\, d\sigma_{trk}/dp_{\perp in}$ and (b) $1/\sigma_{had}\, d\sigma_{trk}/dp_{\perp out}$ compared with the predictions of several models and with Mark II data at 29 GeV. (c) Comparison between means of $p^2_{\perp out}$ and $p^2_{\perp in}$ distributions for e^+e^- experiments at various E_{cm}.

plane ($p_{\perp in}$) and out ($p_{\perp out}$) are shown in figures 2(a) and 2(b). The $p_{\perp in}$ distribution is sensitive to 3–jet events whereas the $p_{\perp out}$ distribution sees effects from events with 4 or more jets. The distributions compare well with the model predictions. Mark II data taken at 29 GeV[7] is also shown for comparison. The corrected mean square values were measured to be $<p^2_{\perp in}>= 0.70 \pm 0.05$ (GeV/c)^2 and $<p^2_{\perp out}>= 0.121 \pm 0.005$ (GeV/c)^2, and these are compared with the results from other experiments in Figure 2(c).[7,9,10,12] The solid lines show the JETSET 6.3 shower model predictions, which are slightly below our measured values for both $<p^2_{\perp in}>$ and $<p^2_{\perp out}>$. The differences arise mainly from the tails of the data distributions which are broader than the MC predictions.

The charged-particle inclusive distributions presented here for hadronic decays of Z bosons are consistent with our extrapolations of the three models and lower energy data. These models also described the detected event shapes, such as sphericity, thrust, aplanarity and number of jets.[3] The small differences observed when compared with the data, e.g. in the momenta transverse to the sphericity axes, are not indicative of significant inadequacies in the models.

REFERENCES

1. G. S. Abrams *et al.*, Phys. Rev. Lett. **63**, 724 (1989); G. S. Abrams *et al.*, Phys. Rev. Lett. **63**, 2173 (1989).

2. G. Abrams *et al.*, Nucl. Instrum. Methods. **A281**, 55 (1989).

3. G. S. Abrams *et al.*, Phys. Rev. Lett. **63**, 1558 (1989).

4. T. Sjöstrand, Comp. Phys. Comm. **39**, 347 (1986); T. Sjöstrand and M. Bengtsson, Comp. Phys. Comm. **43**, 367 (1987); M. Bengtsson and T. Sjöstrand, Nucl. Phys. **B289**, 810 (1987).

5. G. Marchesini and B.R. Webber, Nucl. Phys. **B238**, 1 (1984); B. R. Webber, Nucl. Phys. **B238**, 492 (1984).

6. T.D. Gottschalk and D. Morris, Nucl. Phys. **B288**, 729 (1987).

7. A. Petersen *et al.*, Phys. Rev. **D37**, 1 (1988).

8. T. Sjöstrand, Int. J. Mod. Phys. **A3**, 751 (1988), see page 764. When changing energies in the JETSET second-order matrix element model, the parameter specifying the minimum invariant mass, M_{min}, between two partons should be kept constant. However, the value used in reference 7 at 29 GeV, $M_{min} = 3.5$ GeV, gives a negative two parton cross-section at 91 GeV.

9. M. Althoff *et al.*, Z. Phys. **C22**, 307 (1984).

10. Y. Li (AMY), private communication; KEK preprint 89-149.

11. J. D. Bjorken and S. J. Brodsky, Phys. Rev. **D1**, 1416 (1970); G. Hanson *et al.*, Phys. Rev. Lett. **35**, 1609 (1975); R. Brandelik *et al.*, Phys. Lett. **B86**, 243 (1979).

12. D. Bender *et al.*, Phys. Rev. **D31**, 1 (1985).

DETERMINATION OF α_s FROM A DIFFERENTIAL JET MULTIPLICITY DISTRIBUTION AT SLC AND PEP[*†]

SACHIO KOMAMIYA

Representing the Mark II Collaboration

Stanford Linear Accelerator Center, Stanford University,
Stanford, California 94309

ABSTRACT

We measure the differential jet multiplicity distribution in e^+e^- annihilation with the Mark II detector. This distribution is compared with the second order QCD prediction and α_s is determined to be $0.123 \pm 0.009 \pm 0.005$ at $\sqrt{s} \approx M_Z$ (at SLC) and $0.149 \pm 0.002 \pm 0.007$ at $\sqrt{s} = 29$ GeV (at PEP). The running of α_s between these two center of mass energies is consistent with the QCD prediction. The Q^2 dependence of the $\Lambda_{\overline{MS}}$ determination is also discussed.

1. INTRODUCTION

In determining $\Lambda_{\overline{MS}}$ (or α_s), it is better to use observables which are insensitive to fragmentation and higher order QCD effects. The three-jet event fraction appears relatively insensitive to fragmentation effects, if one chooses a reasonable jet algorithm and if one deals only with hard three-jet events.[3] However, the actual dependence of the three-jet event fraction on the jet resolution parameter (y_{cut}) used to select hard three-jet events is not statistically easy to handle. This problem can be solved by using a differential jet multiplicity as described below.

2. DIFFERENTIAL JET MULTIPLICITY

To define the number of jets (jet multiplicity) in an event we use the algorithm proposed by the JADE collaboration.[4] The scaled invariant mass cut-off (y_{cut}) is used for the jet resolution in the algorithm. The three-jet fraction $f_3(y_{cut})$ is defined to be the number of three-jet events obtained with the algorithm, divided by the total number of hadronic events. The two-jet fraction $f_2(y_{cut})$ and the four jet fraction $f_4(y_{cut})$ are similarly defined.

---

[*] This work was supported in part by Department of Energy contract DE-AC03-76SF00515 (SLAC).

[†] This report is based on the paper by the Mark II Collaboration.[1] The experimental method using a differential jet multiplicity was presented at the International Europhysics Conference (EPS meeting) in Madrid, September 1989.[2]

This jet algorithm has the important feature that mapping from parton jets to hadron jets in Monte Carlo hadronic events is close to one-to-one for reasonably large y_{cut} (≥ 0.04) values.[3] However, it is not easy to extract α_s by fitting the $f_3(y_{cut})$ (or $f_2(y_{cut})$) distribution because the same events contribute at different y_{cut} values and one must take into account all the correlations in this distribution.

To overcome this difficulty, a differential jet multiplicity is defined in the following way. The clustering is terminated when the number of jets has reached a pre-selected value n. For each event, particles are assigned to n-jets using this method and y_n is defined to be the minimum value of the scaled invariant mass $y_{ij} = M_{ij}^2/E_{\mathrm{vis}}^2$ ($i \neq j$, $i,j = 1,2,...,n$). In other words, y_n is the y_{cut} value corresponding to the transition from n-jet to $(n-1)$-jet for a given event. The distribution function of y_n is denoted $g_n(y_n)$. Integrating $g_3(y_3)$ over y_3 from 0 to y_{cut}, one recovers $f_2(y_{cut})$ because all the events with $y_3 < y_{cut}$ are categorized as two-jet events for the given jet resolution y_{cut}.

$$\text{Hence,} \qquad g_3(y_3)|_{y_3=y_{cut}} = \frac{\partial}{\partial y_{cut}} f_2(y_{cut}).$$

$$\text{Similarly,} \qquad g_4(y_4)|_{y_4=y_{cut}} = \frac{\partial}{\partial y_{cut}}[f_2(y_{cut}) + f_3(y_{cut})].$$

Note that only the leading term ($\propto \alpha_s^2$) is available for g_4 in second order QCD calculations. Similarly, $g_5(y_5)|_{y_5=y_{cut}} \equiv 0$ in second order. Therefore we restrict our analysis to the differential jet fraction $g_3(y_3)$ to determine α_s.

3. EXPERIMENTAL METHODS

Multihadron events are selected by requiring that the number of charged tracks is at least seven at SLC [at least five at PEP] and that the sum of charged and neutral particle energies (E_{vis}) is greater than 0.50 $\sqrt{s}$ at SLC [0.55 $\sqrt{s}$ at PEP].[‡] The detection efficiency for multihadron events is estimated using QCD-based Monte Carlo generators [8-10] to be 0.80 ± 0.02 at SLC [0.51 ± 0.02 at PEP]. A total of 391 events from the SLC data and 7348 events from the PEP data pass the selection cuts.

Second order perturbative QCD predictions are directly compared with the data for testing the hard QCD processes and for determining α_s. Detector effects, biases due to event selection and initial state radiation effects are corrected with bin-by-bin correction factors. The corrected $g_3^{corr}(y_3)$ distributions for the two data samples are shown in Fig.1.

---

‡ In order to reduce the bias due to initial state radiation and background from two photon processes for the PEP data, events with large missing energy or with a large energy photon are eliminated by applying additional cuts described in Ref.6. For the Z-resonance data such effects are small, hence we do not apply any cuts other than those mentioned above.[7]

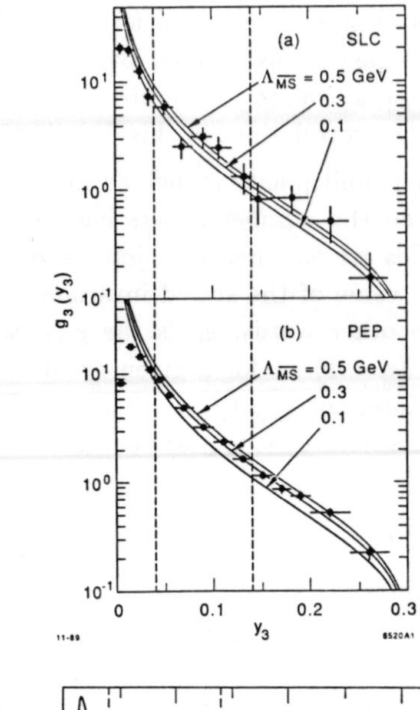

Fig.1:
The experimental distributions of y_3 at (a) $\sqrt{s} = 91$ GeV, and (b) $\sqrt{s} = 29$ GeV. Only the statistical errors are indicated in the figures. The curves below $y_3 = 0.14$ indicate the QCD predictions with $\Lambda_{\overline{MS}} = 0.1$ GeV, 0.3 GeV and 0.5 GeV for $Q^2 = s$. The y_3 range used in the fit for the determination of α_s is defined by the two dashed lines. The curves above $y_3 = 0.14$ are extrapolated from the QCD predictions in the low y_3 range.

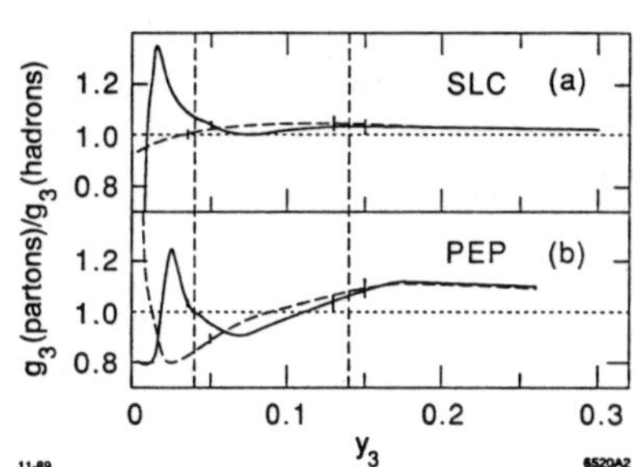

Fig.2:
The ratio $g_3^{partons}/g_3^{hadrons}$ as a function of y_3 for partons and for hadrons (after fragmentation and decay of unstable particles) at (a) $\sqrt{s} = 91$ GeV, and (b) $\sqrt{s} = 29$ GeV. The solid curve corresponds to the Lund model based on $\mathcal{O}(\alpha_s^2)$ matrix element and the dashed curve to the Lund parton shower model. The error bars indicate the Monte Carlo statistical errors.

Corrections are not applied for fragmentation effects. Rather, they are accounted for as systematic errors. In Fig.2, the ratio $g_3^{partons}/g_3^{hadrons}$ is shown as a function of y_3 for two models.[8,10] In the range $0.04 \leq y_3 \leq 0.14$, the bin-to-bin systematic errors associated with fragmentation effects are 3-5% at SLC [5-10% at PEP]. The normalization uncertainty is estimated to be 2% at SLC [4% at PEP].

4. RESULTS (RUNNING α_s)

The α_s value is obtained from a fit of the corrected $g_3(y_3)$ distribution to the $\mathcal{O}(\alpha_s^2)$ QCD prediction.[11] The fit is performed within the range of $0.04 \leq y_3 \leq 0.14$ using a likelihood method which accounts for the statistical errors and the various systematic errors. The lower y_3 limit of the fitted range is chosen in order to limit the fragmentation effects, while the upper limit arises only because the QCD prediction for $y_3 > 0.14$ is not available in Ref.11. Choosing the renormalization point Q^2 to be s, we obtain

$$\alpha_s = 0.123 \pm 0.009 \pm 0.005 \quad \text{at SLC,}$$
$$\alpha_s = 0.149 \pm 0.002 \pm 0.007 \quad \text{at PEP.}$$

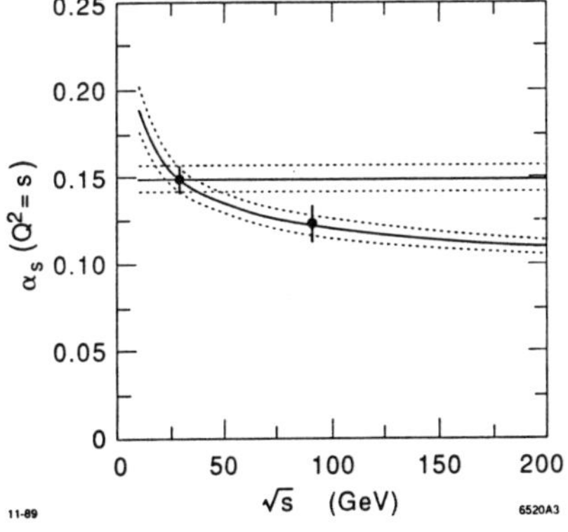

Fig.3:
The strong coupling $\alpha_s(Q^2 = s)$ as a function of $\sqrt{s}$. The errors include statistical and systematic uncertainties added in quadrature. Also shown are the extrapolations of the α_s measurement at $\sqrt{s} = 29$ GeV to higher energies using the formula of Ref.5, or assuming a constant α_s. The dotted lines indicate the extrapolation of the measured $\alpha_s \pm 1\sigma$ from 29 GeV.

The running of α_s from 29 GeV to 91 GeV is consistent with the QCD prediction, as shown in Fig.3. The running of α_s with Q^2 is governed by the Renormalization Group Equation (RGE) which, to second order in α_s, is given by

$$\frac{\partial}{\partial \ln Q^2} \frac{\alpha_s}{2\pi} = -b_0 \left(\frac{\alpha_s}{2\pi}\right)^2 \left(1 + b_1 \frac{\alpha_s}{2\pi}\right) \quad .$$

Denoting by n_f the effective number of flavors at a given Q^2, QCD predicts $b_0 = (33 - 2n_f)/6$ and $b_1 = (153 - 19n_f)/(33 - 2n_f)$. The RGE can be integrated to express b_0 in terms of our two measurements of the coupling constant α_s^{SLC} and α_s^{PEP} and of the $\ln Q^2$ variation $\Delta \ln Q^2 = 2\ln(91/29) = 2.29$. One gets

$$b_0 = \frac{F(\alpha_s^{SLC}) - F(\alpha_s^{PEP})}{\Delta \ln Q^2}, \qquad \text{with} \qquad F(\alpha_s) = \frac{2\pi}{\alpha_s} - b_1 \ln\left(\frac{2\pi}{\alpha_s} + b_1\right).$$

We obtain $b_0 = 3.4^{+2.1}_{-1.4}$ where the errors take into account the partial cancellation of the normalization uncertainties. This value, which is almost independent of b_1, agrees with the QCD prediction of $b_0 = 3.83$ for $n_f = 5$.

To express the α_s measurements in terms of the QCD scale parameter $\Lambda_{\overline{MS}}$, we use the approximate solution of the RGE given in Ref.5. We obtain $\Lambda_{\overline{MS}} = 0.29^{+0.17+0.11}_{-0.12-0.06}$ GeV at SLC, and $\Lambda_{\overline{MS}} = 0.28^{+0.02+0.08}_{-0.02-0.07}$ GeV at PEP, in agreement with the value $0.33 \pm 0.04 \pm 0.07$ GeV previously obtained using the energy-energy-correlation by Mark II at 29 GeV.[12]

5. Q^2 DEPENDENCE OF $\Lambda_{\overline{MS}}$

In finite order perturbative QCD, the predictions depend on the renormalization scheme (RS) and on the renormalization point (Q^2). Therefore the Λ_{QCD} value, which is extracted from the data using finite order QCD predictions, depends on both the RS and the Q^2. Triggered by the work of Kramer and Lampe,[13] several experimental papers were published in an attempt to optimize Q^2 for the determination of $\Lambda_{\overline{MS}}$.[14−17] The simultaneous determination of Q^2 and $\Lambda_{\overline{MS}}$ using jet multiplicity favors very small Q^2 values,[16,17] but the results are very sensitive to perturbative QCD predictions in the very soft region where the size of the second order term is large compared to the first order term, i.e. where the $\mathcal{O}(\alpha_s^2)$ perturbative expansion is not reliable. Therefore the simultaneous determination of Q^2 and $\Lambda_{\overline{MS}}$ is based on the instability of the $\mathcal{O}(\alpha_s^2)$ perturbative expansion in the infrared region, and hence is highly questionable. If we restrict ourselves only in the region where we expect the fragmentation and higher order QCD effects to be small, $\Lambda_{\overline{MS}}$ and Q^2 cannot be determined independently. For example, using $g_3(y_{cut})$ in the range $0.04 \leq y_3 \leq 0.14$, where $\mathcal{O}(\alpha_s^2)$ perturbative QCD works well, the resultant one sigma contour in the $\Lambda_{\overline{MS}}$-Q^2 plane is a band along a curve starting from $\Lambda_{\overline{MS}} = 0.1$ GeV at $Q^2 = (3\text{ GeV})^2$ and extending to $\Lambda_{\overline{MS}} = 2.3$ GeV at $Q^2 = (1000\text{ GeV})^2$.

In the second order calculations, difference of the predictions for different RS's can be absorbed into the 'Q^2 ambiguity'.[18] Therefore 'RS ambiguity' and 'Q^2 ambiguity' are degenerated, to the $\mathcal{O}(\alpha_s^2)$. A reasonable renormalization point Q^2 must be chosen as we choose the $\overline{MS}$ scheme for RS.[*]

Several prescriptions have been proposed to choose a particular value of Q^2.[19−21] For the purpose of illustrating and exploring the effect of the choice of Q^2, we use the Brodsky-Lepage-Mackenzie (BLM) method[21] to eliminate the Q^2 ambiguity for g_3 at each y_3 value. In this picture, the source of the running α_s is the vacuum polarization of gluons (in analogy with QED), hence Q^2 might be the typical momentum scale involved in the vacuum polarization loops; the energy scale is related to the allowable invariant mass (virtuality) of gluons, which can be as small as a few GeV. The choice of Q^2 depends on the kinematical variable y_3 because the gluon virtuality depends on y_3. The Q value prescribed by the BLM method (Q^*) is 4 GeV [1.3 GeV] at $y_3 = 0.05$ and increases to 6 GeV [2.0 GeV] at $y_3 = 0.10$ for $\sqrt{s} = 91$ GeV [$\sqrt{s} = 29$ GeV]. Choosing $Q^2 = (Q^*)^2$ at each value of y_3 and $\sqrt{s}$, and n_f values appropriate to the small Q^* values ($n_f = 4$ for SLC and $n_f = 3$ for PEP), the $\Lambda_{\overline{MS}}$ values obtained using the BLM method are

---

[*] Conventionally, $Q^2 = s = q^2_{\gamma^*,Z^*}$ is chosen for e^+e^- collision and $Q^2 = -q^2_{\gamma^*}$ for deep inelastic lepton-nucleon scattering. However, these choices are not directly connected to the $q\bar{q}g$ vertex where α_s should be determined.

$0.17^{+0.08+0.05}_{-0.06-0.03}$ GeV at SLC and $0.17^{+0.01+0.03}_{-0.01-0.03}$ GeV at PEP. These $\Lambda_{\overline{MS}}$ values are significantly smaller than the values which are determined with $Q^2 = s$.

REFERENCES

1. Mark II Collab., S. Komamiya and F. Le Diberder et al., Phys. Rev. Lett. **64**, 987 (1990).

2. S. Komamiya, SLAC-PUB-5154, to be published in the Proceedings of Int. Europhysics Conf. on High Energy Physics, Madrid Spain, September 1989.

3. JADE Collab., S. Bethke et al., Phys. Lett. **B213**, 235 (1988); Mark II Collab., S. Bethke et al., Z. Physik **C43**, 325 (1989).

4. JADE Collab., W. Bartel et al., Z. Physik, **C33**, 23 (1986).

5. Particle Data Group, M. Aguilar-Benitez et al., Phys. Lett. **B170**, 78 (1986).

6. Mark II Collab., A. Petersen et al., Phys. Rev. **D37**, 1 (1988).

7. Mark II Collab., G.S. Abrams et al., Phys. Rev. Lett. **63**, 1558 (1989)

8. T. Sjöstrand, Comp. Phys. Comm. **39**, 347 (1986); T. Sjöstrand and M. Bengtsson, Comp. Phys. Comm. **43**, 367 (1987).

9. G. Marchesini and B.R. Webber, Nucl. Phys. **B238**, 1 (1984); B. R. Webber, Nucl. Phys. **B238**, 492 (1984).

10. T.D. Gottschalk and M. P. Shatz, Phys. Lett. **B150**, 451 (1985)

11. G. Kramer and B. Lampe, Fortschr. Phys. **37**, 161 (1989).

12. Mark II Collab., D.R. Wood et al., Phys. Rev. **D37**, 3091 (1988).

13. G. Kramer and B. Lampe, Z. Physik, **C39**, 101 (1988).

14. N. Magnussen, PhD thesis, Universität Wuppertal, 1988.

15. CELLO Collab., H.J. Behrend et al., Z.Phys. **C44**, 63 (1989).

16. S. Bethke, Z. Physik, **C43**, 331 (1989); AMY Collab., I.H. Park et al., KEK-89-53, submitted to the Lepton Photon Symposium (Stanford, 1989).

17. OPAL Collab., M.Z. Akrawy et al., Phys. Lett. **B235**, 389 (1990)

18. for example, K. Hagiwara, Suppl. Progr. of Theor. Phys. **77**, 100 (1983).

19. G. Grunberg, Phys. Lett. **B95**, 70 (1980).

20. P.M. Stevenson, Phys. Rev. **D23**, 2916 (1981).

21. S. Brodsky, G.P. Lepage and P.B. Mackenzie, Phys. Rev. **D28**, 228 (1983).

A Search for *top* and b' Quarks and an Examination of Global Event Shape Distributions in Hadronic Z^0 Decays

The OPAL Collaboration

Presented by

J.William Gary

Physikalisches Institut der Univerisität
Heidelberg, Federal Republic of Germany

Abstract

Results of a search for the *top* and b' quarks are presented, using the OPAL detector at LEP. The following limits are set at the 95% c.l.: $M_{top} > 45.1$ GeV/c^2 and $M_{b'} > 45.4$ GeV/c^2. A detailed study of global event structure in hadronic Z^0 decays is also presented. This study demonstrates the consistency of the 5 flavor QCD models to describe the Z^0 hadronic data.

1 Introduction

In this article is described the results of two analyses based on event shape variables in hadronic Z^0 decays: a search for the *top* and b' quarks and a study of global event structure. The data have been collected with the OPAL detector at the CERN e^+e^- collider LEP. Event shape variables such as Acoplanarity and Thrust [1] are sensitive measures for the presence of new, slowly moving heavy quarks because the final states they produce are rather spherical, in contrast to the case of the known light quarks. A search using event shape variables is very general, allowing the possibilities (1) charged current, (2) flavor changing neutral current and (3) decay through a charged Higgs boson to be considered (amongst others) for the heavy quark decay. The study of global event shape distributions differs from the heavy quark search in that final states from light quarks provide the signal rather than the background. These studies establish the consistency of QCD based models to describe the experimental data.

2 The OPAL Detector

The main features of the OPAL detector are given in [2]. The two analyses presented here are based primarily on the central tracking chamber and the electromagnetic calorimeter. The tracking of charged particles is performed with a large volume drift chamber with jet chamber geometry, covering the polar angle range $|cos\theta| < 0.92$ and positioned inside a solenoidal magnet which provides an axial magnetic field for the momentum measurement. The electromagnetic calorimeter consists of a cylindrical array of 9,440 lead glass blocks, 25 radiation lengths in depth, covering the angular range $|cos\theta| < 0.82$, and of two endcap arrays of 2,264 lead glass blocks each, 20 radiation lengths, covering the range $0.81 < |cos\theta| < 0.98$. The total solid angle coverage of the calorimeter is 98% of 4π.

The OPAL trigger and initial event selection for multi-hadronic events are described in [3]. The combined trigger and selection efficiency is estimated to be $97.0 \pm 0.7\%$ while

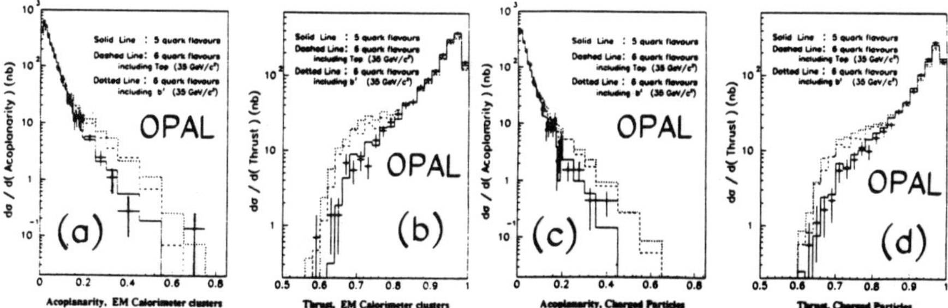

Figure 1: Acoplanarity and Thrust as measured by the electromagnetic calorimeter [(a) and (b)] and by the jet chamber [(c) and (d)]. The solid curve shows the Monte Carlo prediction for 5 quark flavors; the dashed and dotted curves show the expectation should a 6th quark, either *top* or b', be present with a mass of 35 GeV/c^2.

the residual background from tau pairs, two-photon processes, beam-pipe interactions and cosmic radiation is less than 1%.

3 *top* and b' quark search

The *top* and b' quark search is performed using a sample of about 15,000 events collected near the peak of the Z^0 resonance, at center-of-mass values between 91.28 and 91.54 GeV. This corresponds to an integrated luminosity of about 0.5 pb^{-1}.

For this analysis, at least 8 deposits of energy ("clusters") are required to be present in the electromagnetic calorimeter. Clusters in the barrel region are used if they have an energy larger than 200 MeV; those in the endcap are used if they have an energy larger than 500 MeV. The total energy deposit in the calorimeter must exceed 10 GeV. The polar angle θ of the thrust axis determined from the clusters must satisfy $|cos\theta| < 0.96$. A total of 14,491 events is obtained for subsequent analysis. In figure 1 (a) and (b) are shown the Acoplanarity and Thrust distributions measured by the calorimeter, using a smaller sample of 2,185 events for this illustration. The solid curves show the predictions of the Jetset parton shower Monte Carlo with 5 quark flavors, including simulation of the OPAL detector and initial-state photon radiation. They agree well with the data. Also shown are the Monte Carlo predictions should a *top* or a b' quark of mass 35 GeV/c^2 be present which decays by the standard model charged current. These curves demonstrate the sensitivity of the Acoplanarity and Thrust distributions to the presence of a new, heavy quark flavor and that there is no compelling evidence for such a particle in our data. Figures 1 (c) and (d) show the same distributions as measured by the jet chamber. The agreement with the 5 flavor model is again good. The similarity in the Monte Carlo description relative to data for the electromagnetic and jet chamber measurements allows us to establish that our results are not biased by imperfections in the event simulation or by a misunderstanding of detector performance.

Mass limits are extracted by comparing the measured number of events having Acoplanarity A larger than 0.25 to the number which is expected for the case (1) of five light quark flavors (N_5) and for the case (2) where a 6th quark flavor of mass M is present ($N_6(M)$). We

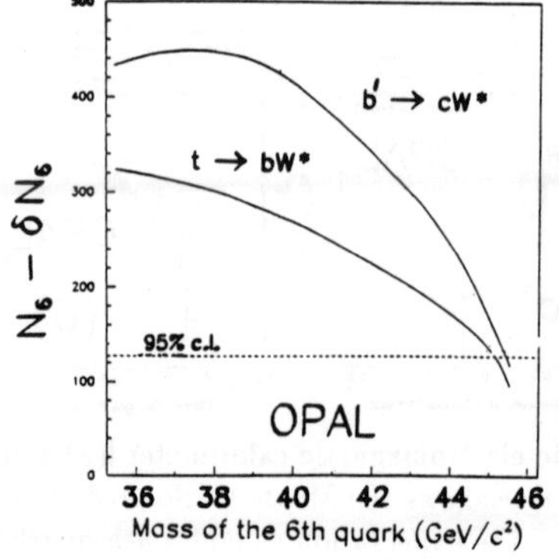

Decay channel		upper mass limit [GeV]	
		95% CL	99% CL
$t \to bW^*$		45.1	45.0
$b' \to cW^*$		45.4	45.3
$b' \to bg$	BR=0.00	45.4	45.3
	BR=0.25	45.4	45.3
	BR=0.50	45.1	45.0
$b' \to b\gamma$	BR=1.00	41.1	40.8
$t \to bH^+$		45.3	45.2
$b' \to cH^-$		45.5	45.5

Figure 2: The extraction of mass limits for charged current decays; a table giving the mass limits which are set for all the considered decay channels.

use the calorimeter measurements to derive the mass limits, which are calculated separately assuming charged current, flavor changing neutral current and charged Higgs decay modes for the heavy quark. Normalizing to N_0, the number of events in the data sample, N_5 and $N_6(M)$ are given by:

$$N_5 = N_0\, \epsilon_5 \qquad ; \qquad N_6(M) = N_0\left[(1 - R_6(M))\,\epsilon_5 + R_6(M)\epsilon_6(M)\right] \qquad (1)$$

with $R_6(M) = \sigma_6(M)/\sigma_{had.}(M)$ the fraction of hadronic Z^0 decays into the 6th quark and ϵ_5 and $\epsilon_6(M)$ the fractions of events which meet the Acoplanarity condition $A > 0.25$ for the case of 5 or of 6 quark flavors, respectively. The branching fraction $R_6(M)$ is calculated from the program ZHADRO [5]. The efficiencies ϵ_5 and $\epsilon_6(M)$ are determined from Monte Carlo for the different decay channels and values of M. The value of $N_6(M)$ has an uncertainty δN_6 which varies from 5% to 20% depending on M, due to uncertainties in the higher order QCD corrections to $R_6(M)$: δN_6 includes an additional error of about 20% because of uncertainties in ϵ_5 and $\epsilon_6(M)$ due to fragmentation.

Figure 2 illustrates the extraction of mass limits for the case of charged current decays. The figure shows $N_6(M)$, after subtraction of its global uncertainty δN_6, vs. M. We observe 107 events with Acoplanarity $A > 0.25$: the 95% c.l. upper limit for this measurement is shown as the dashed line. The intersection of this line with the curves delineates the mass limit which is set (the mass range we exclude extends down to about 23 GeV/c² [4]). The mass limits which are set for all the decay hypotheses are given in the table of figure 2. From this table we extract a mass limit for the *top* quark of: $M_{top} > 45.1$ GeV/c² at the 95% c.l. The weakest limit for the b' quark mass is for the flavor changing neutral current decay for which $BR = \Gamma(b' \to b + \gamma)/[\Gamma(b' \to b + \gamma) + \Gamma(b' \to b + gluon)]=1.0$: this possibiltity is highly improbable given that BR is expected to be about 0.2 [6]. Therefore we set the limit: $M_{b'} > 45.4$ GeV/c² at the 95% c.l.

4 Global event shape distributions

The OPAL study of global event shape variables is based on a data sample of about 1.3 pb⁻¹. The selection of events for this analysis is presented in [7]. The strategy for testing the con-

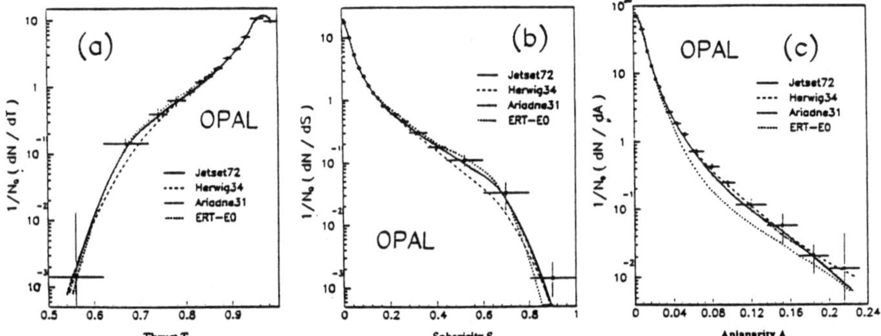

Figure 3: The unfolded (a) Thrust, (b) Sphericity and (c) Aplanarity distributions.

sistency of the QCD model description of our data is the following. (1) A large number of event shape distributions are unfolded for detector resolution and acceptance and for initial-state photon radiation, using electromagnetic calorimeter measurements and a sample of 24,511 events. The unfolding is performed using bin-by-bin corrections to the level which encompasses all charged and neutral particles with lifetimes larger than $3 \cdot 10^{-10}$ s. (2) The parameter values of several Monte Carlo programs which simulate perturbative QCD and the hadronization of partons are adjusted to best describe two distributions: the Thrust major variable T_{major} and the 2nd Fox-Wolfram moment normalized by the 0th moment (H_2/H_0). Three parton shower programs – Jetset version 7.2, Herwig version 3.4 and Ariadne version 3.1 – and a complete 2nd order matrix element program with a modified perturbation scale (ERT-E0) are considered. (3) The predictions of the models with their optimized parameters are compared to the other unfolded distributions. The T_{major} and (H_2/H_0) distributions do not constrain all degrees-of-freedom for multi-hadronic structure [7]: the agreement or disagreement of the models with the 91 GeV data allows a reasonably direct interpretation of their consistency to describe this sample. (4) Monte Carlo events are generated which include detector simulation and initial-state radiation. The model predictions are compared to charged track based distributions at this detector level, using a sample of 21,402 events. This allows a systematic check of the unfolded measurements and of the models' descriptions. (5) The Monte Carlo predictions with optimized parameters are compared to event shape measurements performed at lower c.m. energies to test their consistency at describing multi-hadronic structure over a wide energy range.

The unfolded OPAL measurements of the Thrust, Sphericity and Aplanarity distributions are shown in figures 3 (a)-(c). Jetset, Herwig and Ariadne with their adjusted parameter sets provide a good description of the distributions. The ERT-E0 matrix element model provides a reasonable description of Thrust and Sphericity but is low compared to data for the central region of Aplanarity. In figure 4 (a) is shown the detector level charged track multiplicity distribution n_{ch} in comparsion to Jetset and Herwig – again there is good agreement with data. A correction constant from Jetset provides an unfolded mean charged multiplicity value of $21.28 \pm 0.04 \pm 0.84$, where the first error is statistical and the second is systematic [7]. In figure 4 (b) and (c) the predictions of Jetset and Herwig with their

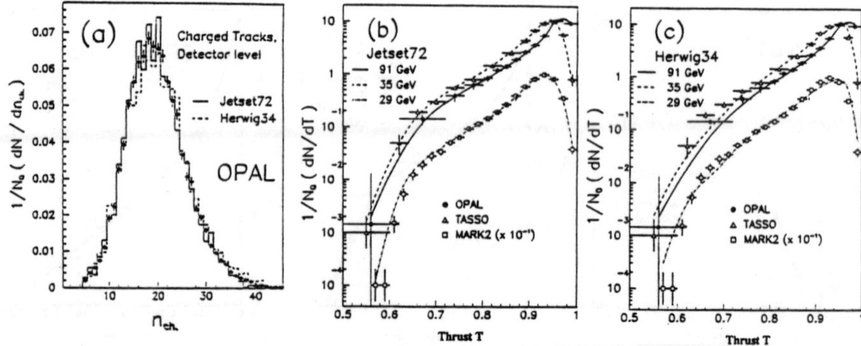

Figure 4: (a) The detector level $n_{ch.}$ distribution at 91 GeV; (b) Jetset and (c) Herwig predictions for Thrust compared to measurements at 91, 35 and 29 GeV.

optimized parameters are compared to the measured Thrust distribution at 91, 35 [8] and 29 GeV [9]. The results for Ariadne resemble those for Jetset. The models provide quite satisfactory descriptions of the measurements for the three energies.

These results demonstrate the consistency of the QCD models to describe the detailed features of global event structure in e^+e^- annihilations over a very large energy range.

5   Acknowledgements

It is a pleasure to thank the LEP Division for the efficient operation of the machine and their continuing cooperation with our experimental group.

References

[1] A. De Rujula *et al.*, Nucl. Phys. **B138** (1978) 387.

[2] OPAL Collaboration, K. Ahmet *et al.*, to be submitted to Nucl.Instr. and Meth.; OPAL Technical Proposal (1983) and CERN/LEPC/83-4.

[3] OPAL Collaboration, M.Z. Akrawy *et al.*, Phys.Lett. **B231** (1989) 530; Phys.Lett. **B235** (1990) 379.

[4] OPAL Collaboration, M.Z. Akrawy *et al.*, Phys.Lett. **B236** (1990) 364.

[5] G. Burgers in Polarization at LEP, CERN Report 88-06 (1988).

[6] W.S. Hou and R.G. Stuart, Nucl. Phys. **B320** (1989) 227.

[7] OPAL Collaboration, CERN-EP/90-48, to be published in Z.für Physik C.

[8] TASSO Collaboration, W. Braunschweig *et al.*, Z.Phys. C41 (1988) 375.

[9] MARK2 Collaboration, A. Petersen *et al.*, Phys.Rev. **D37** (1988) 1.

Studies of Jet Production Rates
and Tests of QCD on the Z^0 Resonance

Siegfried Bethke

Physikalisches Institut der Universität

Heidelberg, Germany

representing

The OPAL Collaboration

Abstract. Production Rates of multijet hadronic final states of Z^0 boson decays are studied. They can be well described by analytic $O(\alpha_s^2)$ QCD calculations with parameters as determined at lower energies. The data cannot be described by an abelian, QED-like vector gluon model. The QCD parameters $\Lambda_{\overline{MS}}$ and μ^2 are adjusted to describe the measured differential 2-jet invariant mass distribution. In second order QCD, including theoretical uncertainties due to the ambiguous choice of the renormalisation scale μ^2, the result is $\Lambda_{\overline{MS}}^{(5)} = (80 - 450)$ MeV, which corresponds to $\alpha_s(M_{Z^0}) = 0.117 \pm 0.015$. Significant scaling violations are observed when the 3-jet fractions are compared to the corresponding results from smaller centre of mass energies, thus increasing, in good agreement with QCD, the evidence for a running coupling constant α_s.

I. Introduction

Hadronic decays of the Z^0 boson are of particular interest for testing the validity of Quantum Chromodynamics (QCD), the nonabelian gauge theory of the strong interactions. Within the framework of this theory, hadronic decays of the Z^0 are associated with the production of quarks and gluons, which subsequently materialise into jets of hadrons. The relative production rates of multijet events are determined by the value of the strong coupling constant, α_s, which - according to the concept of asymptotic freedom - is expected to decrease with increasing energy. Experimental studies of jet production rates are therefore well suited to test the basic ideas of QCD and to determine the free parameters of the theory.

In this letter, the relative production rates of multijet hadronic final states as already published by the OPAL collaboration at LEP [1] will be reviewed. The data, which have been shown to be well described by QCD shower models with parameters adjusted to lower energy data [1], will be compared to analytical $O(\alpha_s^2)$ QCD calculations as well as to the predictions of an abelian, "QED"-like vector gluon model calculated in complete $O(\alpha_A^2)$. Special emphasis will be put on the practical importance of the choice of the renormalisation scale μ^2, on the determination of $\Lambda_{\overline{MS}}$ and $\alpha_s(M_{Z^0})$, and on significant tests of the nonabelian nature of the strong interactions as a proof of the validity of QCD.

II. Theory and Experimental Method

In second order $(O(\alpha_s^2))$ QCD perturbation theory, the relative production rates R_n of 2-, 3- and 4-parton events are quadratic functions of α_s:

$$R_n \equiv \frac{\sigma_n}{\sigma_{tot}} \propto C_{n,1} \cdot \alpha_s + C_{n,2} \cdot \alpha_s^2 \tag{1}$$

where σ_{tot} is the total hadronic cross section, σ_n are the corresponding cross sections of n-parton event production and $C_{n,k}$ are the $k-$th order QCD coefficients for $n-$jet production. The coupling constant $\alpha_s(\mu)$ can be written as a function of $ln(\frac{\mu^2}{\Lambda^2})$ [2], where Λ is the QCD scale parameter which is to be determined by experiment and μ is the QCD renormalisation scale. The number of active quark flavours in the α_s expression is taken to be 5. In e^+e^- annihilation, μ^2 is usually chosen to be the centre of mass energy of the hadronic system ($\mu^2 = E_{cm}^2$). Recent studies have shown, however, that a more general definition like $\mu^2 = f \cdot E_{cm}^2$ with f = 0.001 to 0.01 results in a better overall description of the observed production rates of 2−, 3- and 4-jet events [3,4,5,6,7] in $O(\alpha_s^2)$ QCD.

The "optimisation" of μ, concurrently with the scale parameter Λ, is possible only for observables which are calculated in next-to-leading order, since the corresponding next-to-leading order coefficients like $C_{2,2}$ and $C_{3,2}$ in Eq. 1 exhibit an explicit dependence on the scale factor f [3]. Note that the energy scale μ has no direct physical interpretation because it is a parameter of which the *true* result of a calculation, if performed to *all* orders in perturbation theory, is independent. Finite order calculations, however, may depend on the detailed choice of μ.

In this analysis, hadronic events measured with the OPAL detector at LEP at e^+e^- annihilation energies around the Z_0 mass peak ($E_{cm} = 88 - 94$ GeV) are analysed in terms of the jet finding algorithm introduced by the JADE collaboration [8]. Briefly, resolvable jets of particles are defined by demanding that in each event the scaled jet pair masses y_{kl} exceed a threshold value y_{cut}:

$$y_{kl} = \frac{M_{kl}^2}{E_{vis}^2} > y_{cut}, \tag{2}$$

where M_{kl} is the jet pair mass and E_{vis} is the total visible energy of the event. In [1] and in previous analyses in the centre of mass energy range of 22 to 60 GeV [8,9,5,10] the close agreement between jet rates, reconstructed with the JADE algorithm, and parton rates for identical values of y_{cut} was demonstrated. Hadronisation corrections are small but model dependent; they are therefore not explicitly applied to the data but are usually taken into account as additional systematic uncertainties.

III. Jet Production Rates in Hadronic Decays of the Z^0

In Figs. 1 and 2 the jet production rates as observed by OPAL [1], based on a sample of 10385 well contained hadronic events corresponding to an integrated luminosity of 540 nb^{-1}, are shown as a function of y_{cut}. The data are corrected for the limited acceptance and resolution of the apparatus. Note that the entire data statistics contributes at each value of y_{cut}. The statistical errors of the data are smaller than the size of the plotted symbols.

In Fig. 1, the OPAL data (discrete symbols) are compared with the corresponding results obtained from the Mark-II collaboration at $E_{cm} = 29$ GeV [10]. The difference between the experimental jet rates at these different energies for identical values of y_{cut} constitutes the observation of significant scaling violations. Within the framework of perturbative QCD, these scaling violations are entirely due to the characteristic energy dependence ("running") of α_s, since the QCD parameters $C_{n.k}$ in Eq. 1 are energy independent for fixed y_{cut}. In Fig. 2, the OPAL data are directly compared to different analytical calculations in complete $O(\alpha_s^2)$ QCD [3] and $O(\alpha_A^2)$ "QED". "QED" is an alternative, abelian (toy-)model of the strong interaction which does not contain the process of gluon self-coupling (for details of the "QED"-calculations see [4]). The parameters for calculating the two QCD curves ($\Lambda_{\overline{MS}}^{(5)}$ = 230 MeV for $\mu^2 = E_{cm}^2$ and $\Lambda_{\overline{MS}}^{(5)} = 110$ MeV for $\mu^2 = 0.0017 \cdot E_{cm}^2$) are taken from a fit to the Mark-II at 29 GeV [10,4]. The qualitative description of the Z^0 data is similar as observed at 29 GeV: calculations with energy scale $\mu^2 = E_{cm}^2$ describe the jet rates at large values of $y_{cut} \geq 0.05$, while small renormalisation scales give in addition a better agreement with the jet rates observed at lower values of y_{cut} and of the 4-jet rates in general. The $O(\alpha_A^2)$ abelian vector model does generally not reproduce the dynamic behaviour of the data. As an example, the qualitative disagreement is demonstrated in Fig.2 for the parameters $\alpha_A = 0.22$ and $\mu^2 = 0.023 \cdot E_{cm}^2$, adjusted to describe the data at $y_{cut} = 0.04$.

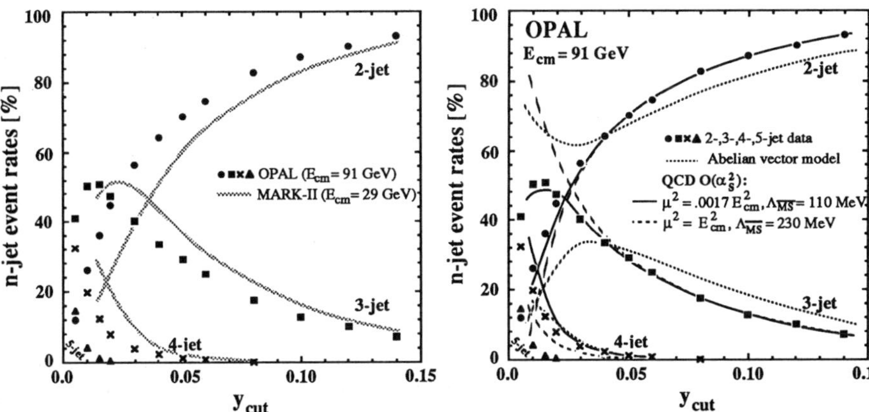

Fig. 1: Relative production rates of 2-, 3-, 4- and 5-jet events as a function of the jet resolution parameter y_{cut}, observed at e^+e^- centre of mass energies around 91 GeV and 29 GeV.

Fig. 2: Jet production rates observed at $E_{cm} = 91$ GeV, compared with analytic calculations in second order QCD and "QED". The QCD parameters come from adjustments to the 29 GeV data shown in Fig. 1.

IV. Differential Jet Rates and a Determination of $\Lambda_{\overline{MS}}$ and $\alpha_s(M_{Z^0})$

In order to minimise the influence of the 4-jet production rates, which are only described in their leading $(O(\alpha_s^2))$ order, and to avoid the correlations between the data at different values of y_{cut} in integral distributions as Figs. 1 and 2, the QCD parameters are determined from the *differential* distribution $D_2(y)$,

$$D_2(y) = \frac{R_2(y) - R_2(y - \Delta y)}{\Delta y} \; ; \; (y \equiv y_{cut}). \tag{3}$$

$D_2(y)$ measures the differential increase of the 2-jet rate R_2 as a function of y_{cut}. In

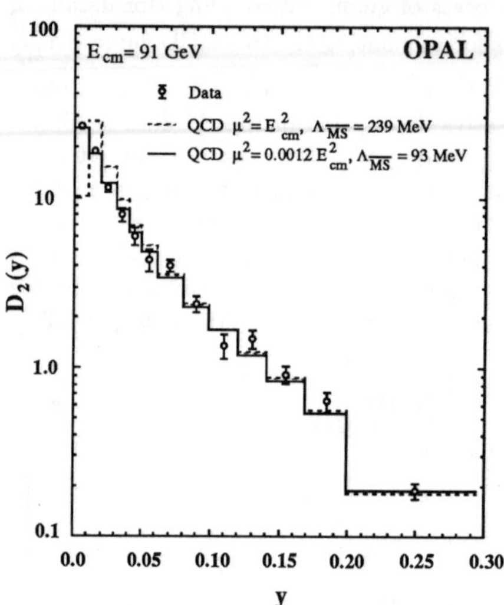

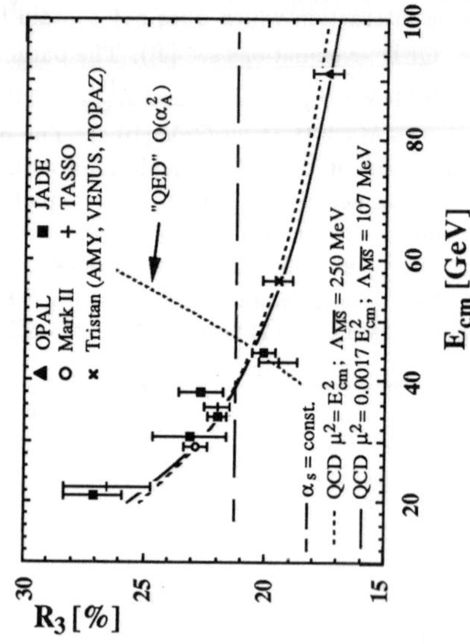

Fig. 3: The measured distribution $D_2(y)$, compared with the $O(\alpha_s^2)$ expextations for optimised QCD parameters $\Lambda_{\overline{MS}}$ and μ^2.

Fig. 4: Energy dependence of 3-jet rates at $y_{cut} = 0.08$, compared to several assumptions about the energy dependence of α_s.

Fig. 3 the D_2 distribution as measured by OPAL around the Z^0 peak is plotted. They are compared to the analytic $O(\alpha_s^2)$ QCD calculations of Kramer and Lampe [3], whereby the QCD parameter $\Lambda_{\overline{MS}}$ is fitted for two different cases: First, the renormalisation scale μ^2 is fixed to E_{cm}^2, resulting - for the region of $y_{cut} > 0.05$ - in $\Lambda_{\overline{MS}}^{(5)} = 239^{+25}_{-24}$ MeV and $\chi^2 = 8.2$ for 8 degrees of freedom. Secondly, both $\Lambda_{\overline{MS}}$ and μ^2 are treated as free parameters, resulting in $\Lambda_{\overline{MS}}^{(5)} = 93^{+12}_{-11}$ MeV and $\mu^2/E_{cm}^2 = 0.0012^{+0.0003}_{-0.0002}$ with $\chi^2 = 12.9$ for 11 degrees of freedom. Since the latter fit results in a better description of the data at low values of y_{cut} (c.f. Fig. 3), the measured jet rates for $y_{cut} \geq 0.01$ are included in this fit. Taking into account the systematic uncertainties on the experimental jet rates, the uncertainties about hadronisation corrections and variations of the results if determined in different intervals of y [1], results in $\Lambda_{\overline{MS}}^{(5)} = (200 - 450)$ MeV for $\mu^2 = E_{cm}^2$ and $\Lambda_{\overline{MS}}^{(5)} = (80 - 180)$ MeV for

$\mu^2/E_{cm}^2 = 0.001 - 0.003$. These results, in particular the dependence of $\Lambda_{\overline{MS}}$ on the choice of μ^2 and the preference of small renormalisation scales by the jet rate data, are in good agreement with previous analyses; see [11] for a recent review. Including the full range of $\mu^2/E_{cm}^2 = 0.001 - 1.0$, the systematic error of $\Lambda_{\overline{MS}}$ finally results in

$$\Lambda_{\overline{MS}}^{(5)} = (80 - 450) \text{ MeV} \quad \text{or} \quad \alpha_s(M_{Z^0}) = 0.117 \pm 0.015.$$

V. Evidence for the Running of α_s

In previuos analyses it was shown that hadronisation corrections to the jet rates are small and, for not too small y_{cut} and $E_{cm} > 25$ GeV, energy independent [8,1]. Since the QCD parameters $C_{n,k}$ do also not depend on E_{cm} for constant values of y_{cut} [3], the energy dependence of experimental jet production rates is a direct measure for the energy dependence of α_s; see Eq. 1. A summary of the 3-jet production rates at $y_{cut} = 0.08$ observed in the c.m. energy range of 22 to 91 GeV is displayed in Fig. 4. Also shown are the fit results for different analytic assumptions of the energy dependence of the coupling strength, namely for QCD in $O(\alpha_s^2)$ with two different choices of μ^2, for "QED" in $O(\alpha_A^2)$ (where α_A is adjusted to describe the 44 GeV data point and extrapolated to other energies according to the 2nd order abelian renormalisation group equation [12]), and for the hypothesis of an energy $independent$ coupling strength. Independent of the choice of μ^2, QCD provides an excellent description of the data, while a constant α_s is now excluded with a significance of 5.7 standard deviations [1]. The abelian "QED", where α_A - due to the absence of the gluon self coupling - $rises$ with increasing energy, is clearly excluded by the measurements.

References

[1] OPAL Collaboration, M. Z. Akrawy et al. , Phys Lett. , **B235** (1990)389 .

[2] Review of Particle Properties , Phys. Lett. , **204B** (1988)96.

[3] G. Kramer, B. Lampe , J. Math. Phys. , **28** (1987)945 ;
DESY 86-103; DESY 86-119 (1986) ;
G. Kramer, B.Lampe , Z. Phys. C , **39** (1988)101 ;
G. Kramer , private communication.

[4] S. Bethke , Z. Phys. C , **43** (1989)331.

[5] AMY collab., I. Park et al. , Phys. Rev. Lett. , **62** (1989)1713 ;
I. Park, PhD Thesis Rutgers University of New Jersey (1989) .

[6] N. Magnussen , DESY Report F22-89-01 (1989) .

[7] L. Smolik , PhD Thesis University of Heidelberg (1989).

[8] JADE collab., W. Bartel et al. , Z. Phys. C , **33** (1986)23;
JADE collab., S. Bethke et al. , Phys Lett , **B213** (1988)235.

[9] TASSO-collab., W. Braunschweig et al. , Phys Lett , **B214** (1988)286.

[10] Mark-II collab., S. Bethke et al. , Z. Phys. C , **43** (1989)325.

[11] S. Bethke , Proc. of the Workshop on the Standard Model at Present and Future Accelerator Energies, Budapest, Hungary (July 1989); LBL-28112 (1989) .

[12] W. Bernreuther , private communication .

Study of Hadronic Decays of the Z⁰ Boson

Ingo Herbst
University of Wuppertal
Gauβstr 20, D5600 Wuppertal, RFA.

DELPHI Collaboration

Abstract

Hadronic decays of Z^0 bosons are studied in the Delphi detector. Global event variables and single particle inclusive distributions are compared with QCD−based predictions. The mean charged multiplicity is found to be 20.6 ± 1.0 (stat + syst). The mean values of the thrust and minor distributions, p_t^{in} and p_t^{out} are compared with values found at lower energy e^+e^- colliders.

1. Introduction

A study of the general properties of charged particles produced in hadronic decays of the Z^0 boson observed in the DELPHI detector during the first months of operation of the new e^+e^- storage ring LEP at centre of mass energies of $91.0 − 91.5$ GeV is presented [1]. Measured distributions of global event shape variables and of inclusive single particle variables as well as the energy dependence of some characteristic mean values are compared with the expectations of four QCD−based fragmentation models pretuned to lower energy data.

2. Detector and event sample

The DELPHI detector is a multipurpose detector with 4π coverage by tracking detectors and electromagnetic and hadronic calorimetry. The essential detector components used for this study are:

- the time projection chamber (TPC) for measuring charged particle tracks. Up to 16 space points are used for reconstructing tracks with a momentum resolution of $\delta p/p^2 = 0.02 \, (GeV/c)^{-1}$ in a magnetic field of 0.7 T,
- the inner detector, a jet chamber providing 24 $r\phi$ coordinates, and the outer tracking detector, both used for a track trigger in the barrel region $40^0 \leq \theta \leq 140^0$,
- the electromagnetic calorimeter with its scintillation counters and the time of flight scintillator system covering an angular region of $40^0 \leq \theta \leq 140^0$, used for the trigger.

For the analysis only charged particle tracks are used assuming always the π mass. The tracks are retained if: (1) they extrapolate back to within 5 cm of the beam axis in the radial distance r and to within 10 cm of the nominal crossing point in z, (2) their momentum p is larger than 0.1 GeV/c, (3) their measured track length is above 50 cm and (4) their polar angle is between 25^0 and 155^0.

Hadronic events are then selected by requiring that (1) the total energy of charged particles seen in the event E_{ch} exceeds 15 GeV, (2) the energy of charged particles seen in each detector hemisphere ($\cos\theta \leq 0$, $\cos\theta \geq 0$) exceeds 3 GeV, (3) there are at least 5 charged particles with momenta above 0.2 GeV/c and (4) the polar angle θ of the sphericity axis is in the range $40^0 \leq \theta \leq 140^0$. After these cuts, 2073 hadronic events remain with a contamination of 0.24% $\tau^+\tau^-$ events (MC − study) and less than 0.1% beam gas (data) and $\gamma\gamma$ events (MC − study).

3. Results

Our data are compared to four QCD − based event generators pretuned to low energy data taken at PETRA and PEP.

- The Lund parton shower Monte Carlo followed by string fragmentation.
- The Marchesini − Webber Monte Carlo followed by cluster fragmentation.
- The second order matrix element program of Gutbrod − Kramer − Schierholz (GKS) followed by string fragmentation of Lund.
- The second order matrix element program of Ellis − Ross − Terano (ERT) as implemented by N.Magnussen followed by string fragmentation of Lund. A scale optimized to jet rates at PETRA is used, i.e. α_S was evaluated using $\Lambda_{\overline{MS}} = 120 \mathrm{MeV}$ at $\mu^2 = 0.005\ Q^2$

Several event shape distributions and single particle inclusive distributions are compared to the model predictions in Figure 1. The distributions are corrected on a bin to bin basis for geometrical acceptance, kinematical cuts, detector resolution and inefficiency, particle interactions in the detector and the effect of radiated photons. The event shape distributions shown are the oblateness O, a variable linear in the particle momenta and the aplanarity A, a variable quadratic in the momenta coming from the momentum tensor. For all variables the standard definitions are taken (see e.g. the Lund program JETSET). The following single particle inclusive distributions are shown: the momentum fraction variable $x_p = 2p/\sqrt{S}$ and the rapidity $y_\pi = 0.5\ln((E+p_\parallel)/(E-p_\parallel))$ where $p_\parallel$ refers to the momentum component parallel to the thrust axis and the π mass is assumed. Integration of the rapidity distribution gives, after applying a small correction due to converted photons, the mean charged multiplicity in hadronic Z^0 events: 20.6±1.0. The error contains the statistic and systematic error added in quadrature.

The dependence of the mean values on the center − of − mass energy is shown in Figure 2 for the event shape variables thrust T and minor m, both variables linear in the momenta and the single particle inclusive distributions p_t^{in} and p_t^{out}. The mean values are corrected for neutral particles with a correction amounting to $\leq 4\%$.

All distributions show a good agreement between data and the Lund parton shower Monte Carlo. The Marchesini − Webber Monte Carlo describes variables measuring the momentum component out of the event plane like minor, aplanarity or p_t^{out} less well. The matrix element Monte Carlos need retuning, as can best be seen in the rapidity distribution. A too low multiplicity of 18.1 is predicted by these models. The effect of a retuning is shown in the rapidity plot. Using Lund Jetset 7.2 with the ERT matrix element and optimized parameters, the distribution coincides nicely with the data. The only significant difference between our data and the different model predictions is found for the p_t^{in} distribution, which shows a smaller mean value than predicted by all models.

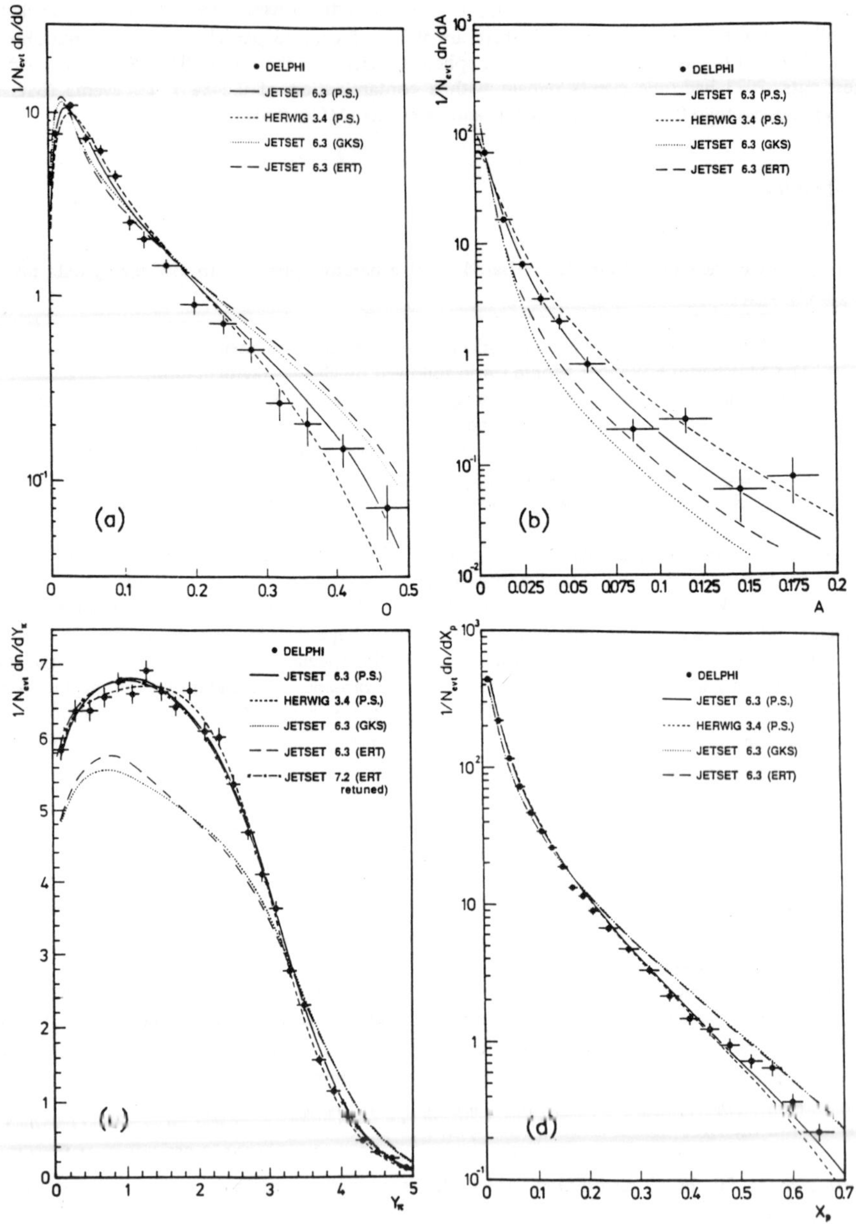

Figure 1: The event shape variables (a) oblateness O and (b) aplanarity A and the single particle inclusive distributions (c) rapidity y_π and (d) momentum fraction x_p.

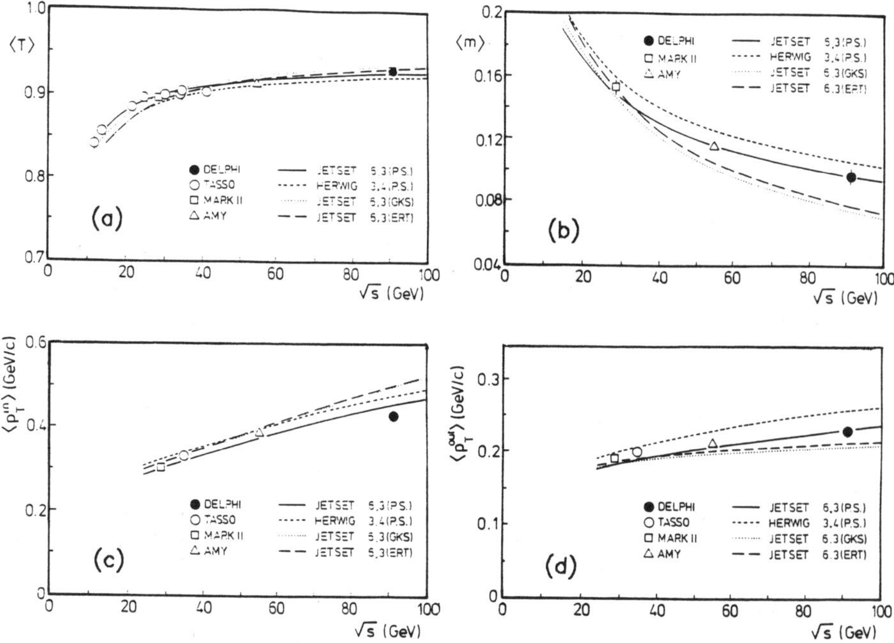

Figure 2: The energy dependence of the event shape variables (a) thrust T, (b) minor m and of the single particle inclusive distributions (c) p_t^{in} and (d) p_t^{out}.

4. Summary

Global event shape variables and single particle inclusive distributions of hadronic decays of Z^0 bosons have been studied with the Delphi detector. The mean charged multiplicity is found to be 20.6 ± 1.0 (stat + syst). The mean values of the thrust and minor distributions, p_t^{in} and p_t^{out} corrected for missing neutral particles are compared with values found at lower energy e^+e^- colliders. The best overall agreement between our data and Monte Carlo predictions based on pertubative QCD is found for the Lund parton shower Monte Carlo with string fragmentation. QCD Monte Carlos based on second order matrix element calculations need retuning of model parameters.

5. References

[1] DELPHI collaboration, P.Aarnio et al., CERN−EP/90−19, submitted to Physics Letters. Study of Hadronic Decays of the Z^0 Boson. This summary is a presentation of this paper, all further references or details may be found there.

Properties of Hadronic Events in e^+e^- Annihilation at $\sqrt{s} = 91\,\mathrm{GeV}$

R.P. Johnson
Department of Physics, University of Wisconsin
Madison, Wisconsin 53706

Abstract

We report on properties of hadronic events from e^+e^- annihilation observed by the ALEPH detector at LEP in the center-of-mass energy range from 91.0 to 91.3 GeV. Measured distributions are presented for the global event-shape variables sphericity, aplanarity, thrust and minor value, and the inclusive variables x_p, $p_\perp^{\mathrm{in}}$, $p_\perp^{\mathrm{out}}$, and y. The mean charged multiplicity is measured to be $\langle N_{\mathrm{ch}} \rangle = 21.3 \pm 0.1\,(\mathrm{stat.}) \pm 0.6\,(\mathrm{syst.})$. The data are in good agreement with QCD-based models which use the leading-log approximation, and are less well described by a model using $O(\alpha_s^2)$ QCD.

1. Introduction

From data taken by the ALEPH detector at energies near the peak of the Z resonance during the first three weeks of running at LEP, properties of hadronic events from e^+e- annihilation have been measured. We present measured distributions of several global event-shape and inclusive variables, based on charged particles. These are compared with the predictions of QCD-based hadron production models which are in reasonable agreement with data at $\sqrt{s} = 29\,\mathrm{GeV}$. Extrapolating these models to a higher energy while retaining the same parameter values provides an important test of the hadron-production mechanism.

2. Apparatus and Event Selection

The ALEPH detector is described in detail in Ref. 1. Only a brief description of those features relevant to this analysis is given here. The momenta of charged particles are measured by two tracking chambers, a large time-projection chamber (TPC) surrounding an inner drift chamber (ITC), situated in a 1.5 T magnetic field. The TPC measures up to 21 three-dimensional coordinates, with a resolution in $r\phi$ of approximately 200 μm and a resolution in z of 1–2 mm. The ITC provides eight additional measurements in $r\phi$, each with a resolution of approximately 150 μm. The momentum resolution achieved for 45 GeV isolated muons is $\delta p/p^2 = 0.0011\,(\mathrm{GeV})^{-1}$. The track detection efficiency is estimated by Monte Carlo studies to be $(97.5 \pm 2.0)\%$ for the momentum and angular range used in this analysis. The triggers for hadronic events are based on the total-energy deposit in the electromagnetic calorimeter surrounding the TPC or on charged particle penetration through the hadron calorimeter, situated in the iron return yoke of the magnet, in coincidence with a track in the ITC. The overall trigger efficiency for hadronic Z decays within our selection is effectively 100%.

Only those charged tracks are considered which have at least 4 coordinates in the TPC, a polar angle within $20° < \theta < 160°$, momentum $p \geq 200\,\mathrm{MeV}$, and pass within 3 cm of the beam axis and within 5 cm of the origin in z. Hadronic events are selected by requiring at least 5 charged tracks and a total charged energy of at least 15 GeV. In addition, the sphericity axis is required to be at least $35°$ away from the beam axis. This results in a sample of 1491 events, with a predicted background of 4 events from τ-pair production.

3. Summary of Results

The measured distributions are corrected using Monte Carlo simulation (based on the program JETSET=6.3 [2]) for effects of geometrical acceptance, detector efficiency and resolution, decays of long-lived particles, particle interactions in the material of the detector, initial state photon radiation and effects of event and track selection. The bin-by-bin correction factors are typically in the range 0.8 to 1.2, and are in all cases between 0.6 and 1.8. Systematic errors in the correction factors were studied by varying some crucial aspects of the detector simulation, such as the momentum resolution and coordinate efficiency. In all cases the systematic errors are smaller than or comparable with the statistical errors. Only statistical errors are shown on the plots.

The measured distributions are compared with the Lund Parton-Shower (PS) model [3], the Lund Matrix-Element (ME) model [2] (both from JETSET-6.3), and the Webber-Marchesini model as implemented in HERWIG-3.4 [4,5]. The parameters used in Lund ME and Lund PS were determined by fitting to data from the Mark II experiment at $\sqrt{s} = 29\,\mathrm{GeV}$ [6]. For HERWIG, the default parameters have been derived from its predecessor, BIGWIG [5], which is in reasonable agreement with data at lower energies [6,7].

We have measured the distributions of the global event shape variables sphericity, aplanarity, thrust and minor value (see Ref. 8 for definitions of these variables). The measured distributions of thrust and minor value are shown in Figs. 1(a) and 1(b). Also shown are data from the Mark II experiment at $\sqrt{s} = 29\,\mathrm{GeV}$ [6]. The data show an increased jet-like structure at higher energy (from the thrust distribution) and more aplanar events than predicted by Lund ME (from the distribution of minor value). Both the Lund PS and HERWIG models describe the data well. Similar conclusions follow from the distributions of sphericity and aplanarity, which are not shown here.

We also have measured the single-particle inclusive distributions $p_\perp^{\mathrm{in}}$, $p_\perp^{\mathrm{out}}$, which are the momentum components in and out of the event plane defined by the momentum tensor, the scaled momentum $x_p = p/p_{\mathrm{beam}}$ and the rapidity, $y = \frac{1}{2}\ln\left[(E + p_\parallel)/(E - p_\parallel)\right]$, where $p_\parallel$ is the momentum component parallel to the thrust axis. The x_p and rapidity distributions are shown in Figs. 1(c) and 1(d). The predictions of Lund PS and HERWIG are in good agreement with the data while Lund ME predicts far too few charged particles for $y < 3.5$ and does not predict properly the scaling violation in x_p. By integrating the y distribution we obtain a measurement of the mean charged multiplicity: $N_{\mathrm{ch}} = 21.3 \pm 0.1\,(\mathrm{stat.}) \pm 0.6\,(\mathrm{syst.})$. Due to lack of space, the other distribution are not shown here but may be found in Ref. 8.

In general, the Lund PS and HERWIG models, both based on leading-logarithm approximation of QCD implemented as a parton shower, describe the data well both at $\sqrt{s} = 29\,\mathrm{GeV}$ and $\sqrt{s} = 91\,\mathrm{GeV}$ with a single set of parameters. The $O(\alpha_s^2)$ matrix element formulation of QCD, as implemented in Lund ME, fails to describe properly the data at $\sqrt{s} = 91\,\mathrm{GeV}$ when tuned to fit the 29 GeV data. In particular, it produces too low a charged multiplicity, too few aplanar events, too little momentum out of the event plane, and fails to predict the scaling violation of x_p between the two energies.

Acknowledgements

The results presented here are some of the fruits of a cooperative effort of about 360 physicists from 30 institutions, forming the ALEPH collaboration. We wish to express our appreciation to the technical staffs of CERN and the outside institutions for their support in constructing ALEPH and to the LEP division for the construction and rapid commissioning of the accelerator.

References

[1] D. Decamp et al., ALEPH Collab., CERN-EP/90-25, submitted to Nucl. Inst. Meth.

[2] T. Sjöstrand and M. Bengtsson, Comp. Phys. Com. **43** (1987) 367.

[3] M. Bengtsson and T. Sjöstrand, Phys. Lett. B **185** (1987) 435.

[4] G. Marchesini and B. Webber, Nucl. Phys. **B310** (1988) 461;

[5] G. Marchesini and B. Webber, Nucl. Phys. **B238** (1984) 1; **B238** (1984) 492.

[6] A. Petersen et al. (Mark II), Phys. Rev. D **37** (1988) 1.

[7] W. Braunschweig et al. (TASSO), Z. Phys. C **41** (1988) 359.

[8] D. Decamp et al. (ALEPH), Phys. Lett. B **234** (1990) 209.

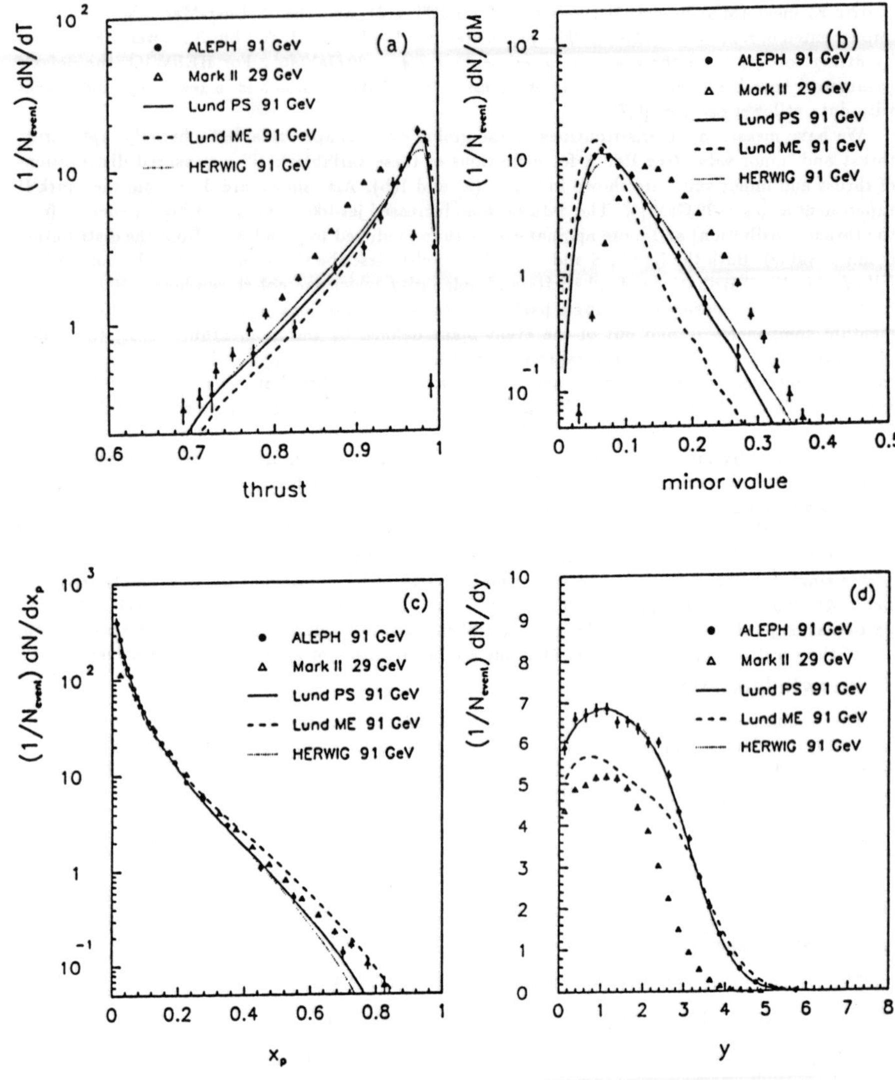

Figure 1: The measured distributions of event shape variables thrust (a) and minor value (b) and of inclusive variables $x_p = p/p_{beam}$ (c) and rapidity (d), compared with data at $\sqrt{s} = 29\,\mathrm{GeV}$ and three QCD hadron production models.

MEASUREMENTS OF Z BOSON
RESONANCE PARAMETERS IN e^+e^- ANNIHILATION[*]

BRIAN D. HARRAL

Representing the Mark II Collaboration

Department of Physics, The Johns Hopkins University
Baltimore, Maryland 21218

ABSTRACT

We present results of the measurement of the mass, width, and cross section of the Z boson. We give results of fits to the data for which the resonance width is unconstrained, and fits for which the visible width is constrained to its Standard Model value, allowing us to derive limits on the number of light neutrino species.

1. Introduction

The mass and decay width of the Z boson have become, with the advent of SLC and LEP, experimental quantities which can now be measured with great precision. Knowledge of these quantities is crucial to extending understanding of the Standard Model of elementary particle physics. In the following are detailed results of analysis of data from the Mark II detector at SLC, among which results were one of the first high-precision measurements of the Z mass, and one of the first significant measurements to use the Z width to infer the number of light neutrino species.

The methods used in this analysis are the same as those used to study the production of meson and baryon resonances: the program is to measure a cross section and fit a Breit-Wigner resonance shape to the data, from which the quantities of interest are extracted. The cross section is computed using the standard formula

$$N_X = \mathcal{L}\, \sigma_X\, \epsilon_X, \tag{1}$$

[*] This work was supported in part by Department of Energy contracts DE-AC03-81ER40050 (CIT), DE-AM03-76SF00010 (UCSC), DE-AC02-86ER40253 (Colorado), DE-AC03-83ER40103 (Hawaii), DE-AC02-84ER40125 (Indiana), DE-AC03-76SF00098 (LBL), DE-AC02-84ER40125 (Michigan), and DE-AC03-76SF00515 (SLAC), and by the National Science Foundation (Johns Hopkins).

where

$$N_X = \text{number of events with final state } X,$$
$$\mathcal{L} = \text{luminosity},$$
$$\sigma_X = \text{cross section to final state } X, \text{ and}$$
$$\epsilon_X = \text{detection efficiency for final state } X. \tag{2}$$

(All the above quantities are in general functions of the center-of-mass energy E.) Assuming that the luminosity is defined by a similar equation where a different final state L is considered, and assuming that the luminosity cross section can be calculated, then

$$\sigma_X(E) = \frac{N_X(E)}{N_L(E)} \frac{\epsilon_L(E)}{\epsilon_X(E)} \, \sigma_L(E). \tag{3}$$

Therefore to measure the desired cross section we must (1) measure the center-of-mass energy E, (2) count N_X and N_L, (3) estimate ϵ_X and ϵ_L by Monte Carlo methods, and (4) calculate the luminosity cross section σ_L. In the following will be discussed how these things were done, and how the Breit-Wigner formula was used to extract the Z mass and width.

2. Apparatus

Since the SLC is not a storage ring, the center-of-mass energy must be measured for every beam crossing. This was accomplished using an energy spectrometer system which has been described in detail elsewhere.[1] It consisted of precisely measured dipole magnets through which the beams passed on their way to the beam dumps. The beam particles emitted synchrotron radiation as they passed through the magnetic fields of horizontal dipoles preceding and following the main (vertical) spectrometer magnet. This radiation then fell on phosphorescent screens viewed by cameras. The character of these synchrotron radiation stripes gave both the mean energy and the spread of particle energies within the pulse. The energy spread (~ 200 MeV) and the "jitter" around the nominal collision energy (~ 70 MeV) was accounted for in the fit procedure by using a weighted average of the Z cross section over a suitable range of energies, rather than simply the value of σ_Z at the nominal collision energy. The absolute error on the energy measurement (due to survey uncertainty, detector resolution, beams not colliding head-on, finite dispersion, and the like) was about 40 MeV; this was the largest systematic error of the Z mass.

The Mark II detector has also been described in detail elsewhere.[2] The detector elements used in this analysis for detecting the products of Z decays (process X mentioned above) were (1) the central drift chamber (CDC) to measure the momenta of charged particles, (2) the lead/liquid argon electromagnetic calorimeter (LAC) to measure the energy of electrons and photons emitted in the angular range $|\cos\theta| \lesssim 0.65$ and to differentiate them from muons and hadrons, and (3) the lead/proportional tube "end cap" electromagnetic calorimeters (ECC) to perform the same function as the LAC in the angular range $0.71 \lesssim |\cos\theta| \lesssim 0.97$. The uses to which each detector was put will be apparent from the discussion of the analysis below.

There were two detector elements used to find luminosity events (process L mentioned above), both of which are described in detail in Ref. 2. The Small Angle Monitor (SAM) was a combination tracking chamber (drift tubes) and electromagnetic calorimeter (lead/proportional tubes) with very good position resolution ($\sim 1/4$ mm for tracking, ~ 1 mm for calorimetry alone), which covered the angular range 50 mrad $\lesssim |\cos\theta| \lesssim$ 160 mrad. The Mini Small Angle Monitor (Mini-SAM) was a tungsten/scintillator electromagnetic calorimeter covering the angular range 15 mrad $\lesssim |\cos\theta| \lesssim$ 25 mrad with four quadrants, each covering $90°$ in azimuth. Each quadrant was read out by a photomultiplier tube. The use of these detectors will be discussed in the following section.

3. Luminosity Analysis

The process used to measure luminosity at e^+e^- colliders is $e^+e^- \to e^+e^-$, known as Bhabha scattering. This process can be mediated both by the photon and the Z; in order to minimize the effect of the Z on this cross section, this process should be measured at small polar angles near the pole in the photon t-channel process. To lowest order, this t-channel diagram has a cross section

$$\sigma(\theta_{\min}, \theta_{\max}) \propto \frac{1}{\theta_{\min}^2} - \frac{1}{\theta_{\max}^2} \tag{4}$$

between the (small) angles $\theta_{\min}$ and $\theta_{\max}$. We have used various Monte Carlo programs to calculate the cross sections subtended by the SAM and Mini-SAM;[3] although the various methods differ by less than 1%, we have been conservative and attributed a 2% uncertainty to higher-order radiative corrections.

The effects of beam pipe intrusions into the SAM acceptance window and possible misalignments have led us to a procedure for calculating the luminosity which minimizes certain systematic errors at the expense of introducing the statistical uncertainty in this scale factor. We begin with a well-defined fiducial volume of the SAM and use the Monte Carlo programs to calculate the cross section using the "gross/precise" counting scheme, as follows: for an event having more than one SAM shower with $E_{\mathrm{shower}} > 0.2\ E_{\mathrm{cm}}$, (1) if all showers are within 65 mrad $< |\cos\theta| <$ 160 mrad, count the event with weight 1 ("precise"); (2) if one shower is in the above range, and a shower on the other side of the IP is in the range 60 mrad $< |\cos\theta| <$ 65 mrad, count the event with weight $\frac{1}{2}$ ("gross"). This counting procedure minimizes systematic errors due to motion of the interaction point (IP) and misalignment of the detectors. Systematic errors due to detector resolution and reconstruction effects have been estimated as 2%, giving (in combination with higher-order radiative corrections) a total systematic error of 2.8%.

In order to use the statistics from the full SAM acceptance, we define a set of events with the simple criterion that each SAM module must have showers with $E_{\text{shower}} > 0.2\,E_{\text{cm}}$, without regard to angle. This "inclusive" cross section is computed from the above "gross/precise" cross section by

$$\sigma_S(E) = \frac{\sum N_S}{\sum N_{GP}}\,\sigma_{GP}(E), \tag{5}$$

where the sums are taken over the entire run. This scale factor has a statistical error of 2.9%; in combination with the systematic errors listed above, the overall luminosity-scale error is then 4.0%. By viewing SAM energy spectra, we estimate negligible inefficiency in Bhabha finding.

Bhabha finding in the Mini-SAM is complicated by synchrotron radiation accompanying the beam as it passes through the detector (the inner edge of the Mini-SAM is about 3 cm from the beam axis). Therefore we require an excess energy of $E_{\text{shower}} > 0.3\,E_{\text{cm}}$ above the average energy in other quadrants. We also require timing signals compatible with particles striking the detector from the side facing the IP (rather than associated with the incoming beam). Bhabha-finding efficiency is estimated on a run-by-run basis by superimposing Monte-Carlo generated Bhabha events upon randomly-triggered events. This efficiency is rarely less than 90%, and is usually greater than 99%.

We have chosen to scale the Mini-SAM cross section to the SAM inclusive cross section by the same procedure used to scale the SAM inclusive cross section to the SAM gross/precise cross section. One reason this was done was that detailed information from the SAM and information from studies by the Mark II at PEP[4] indicates that the Bhabha kinematics may not be well simulated by the Monte Carlo programs at the level enforced by the small acceptance of the Mini-SAM. Small misalignments of the masks defining the Mini-SAM acceptance, and motions of the IP on the order of 1 mm, both induce changes in the Mini-SAM cross section. Realignment of the beam pipe in mid-run caused us to use two independent scale factors, each of which had a 4.5% statistical error. This error does not affect the overall scale factor, but only appears in the relative luminosity between data taken at different energies.

4. Z Analysis

In order to maximize the statistics of our final event sample, we define "process X" mentioned above as (1) all the $Z \rightarrow$ hadrons events the detector can reasonably accept, and (2) $Z \rightarrow \mu^+\mu^-$ and $Z \rightarrow \tau^+\tau^-$ events within $|\cos\theta| < 0.65$, so that the LAC can be used for particle ID; assuming $\frac{d\sigma}{d\Omega} \propto 1 + \cos^2 \theta$, this allows 55.6% of such events to be identified. Event selection for hadronic events is kept simple to assure high efficiency, but is sufficient to reject background to a high degree. We require ≥ 3 charged tracks with $p_\perp > 0.11$ GeV/c (no looping tracks), $|\cos \theta| < 0.92$ (assures good momentum measurement), and distances of closest approach to the IP bringing the track to within 1 cm of the beam axis and within 3 cm of the IP along the beam axis (*i.e.*, in z). We also require $> 5\%$ of the center-of-mass energy to be projected into each hemisphere in z. The estimated efficiency of these cuts (independent of Monte-Carlo hadronization scheme) is $95.3 \pm 0.6\%$. Background from beam-gas interactions is estimated at < 0.2 events, and background from $e^+e^- \rightarrow e^+e^-q\bar{q}$ is estimated at 0.02 events. Identification of leptonic events is unambiguous and has high efficiency. Mu pairs are identified by high-momentum tracks which deposit little energy in the LAC; tau pairs are identified by low multiplicity, high thrust, and low invariant mass where applicable. Low-angle tau pairs are rejected from the hadronic event sample by these same criteria. A summary of the data taken is given in the following table.

$< E_{cm} >$ (GeV)	N_S	N_M	ϵ_M	$\mathcal{L}$ (nb^{-1})	Z Decays Had.	Lep.	Tot.	σ_Z (nb)
89.24	24	166	0.99	0.68 ± 0.05	3	0	3	$4.5^{+4.5}_{-2.5}$
89.98	36	174	0.99	0.76 ± 0.05	8	2	10	$13.5^{+6.0}_{-4.3}$
90.35	116	617	1.00	2.61 ± 0.10	60	2	62	$24.8^{+3.8}_{-3.3}$
90.74	54	266	0.96	1.21 ± 0.07	33	3	36	$31.7^{+6.8}_{-5.5}$
91.06	170	923	0.99	4.08 ± 0.12	114	6	120	$31.6^{+3.4}_{-3.1}$
91.43	164	879	0.91	4.12 ± 0.13	108	6	114	$29.8^{+3.3}_{-2.9}$
91.50	53	275	0.99	1.23 ± 0.07	33	6	39	$34.3^{+7.0}_{-5.7}$
92.16	31	105	0.97	0.54 ± 0.05	11	0	11	$21.5^{+9.2}_{-6.6}$
92.22	128	680	0.98	3.05 ± 0.11	67	4	71	$24.3^{+3.4}_{-3.0}$
92.96	39	214	0.98	1.00 ± 0.07	13	1	14	$14.6^{+5.4}_{-4.0}$
Totals	815	4299		19.3 ± 0.9	450	30	480	

Table 1: Average energy, number of SAM and Mini-SAM events, Mini-SAM efficiency, luminosity, number of Z events, and Z cross section for each energy scan point.

5. Fit Procedure

Owing to the small number of events at various energies, we were forced to use the maximum-likelihood method to find the best values of the resonance parameters. For a given luminosity, we have two different types of events: luminosity events and Z events. Thus the probability is given by binomial statistics.[5]

The prediction for the Z cross section used in the likelihood function was of the form

$$\sigma_Z(E) = \sigma_0 \, \frac{s\Gamma^2}{(s - m_Z^2)^2 + s^2\Gamma^2/m_Z^2} \, (1 + \delta(E)), \tag{6}$$

where

$$
\begin{aligned}
s &\equiv E^2, \\
\sigma_0 &= \frac{12\pi}{m_Z^2} \, \frac{\Gamma_e\Gamma_X}{\Gamma^2}, \\
\Gamma &= \text{total width of } Z, \\
&= \Gamma_{\text{hadrons}} + \Gamma_{\text{leptons}} + N_\nu \, \Gamma_\nu, \\
m_Z &= Z \text{ mass}, \\
\Gamma_e &= \Gamma(Z \to e^+e^-), \\
\Gamma_X &= \Gamma_{\text{hadrons}} + 0.556(\Gamma_\mu + \Gamma_\tau),
\end{aligned}
\tag{7}
$$

and $(1 + \delta(E))$ represents corrections due to initial-state photon emission.[6] The factor of 0.556 in Γ_X comes from the geometric acceptance for mu pairs and tau pairs mentioned earlier. The luminosity cross section is given by

$$\sigma_L = \sigma_S + \epsilon_M \, \sigma_M, \tag{8}$$

where σ_S and σ_M are the scaled SAM and Mini-SAM cross sections mentioned previously, and ϵ_M is the average Mini-SAM efficiency listed in the above table.

The Breit-Wigner resonance has three parameters: the pole position (m_Z), the width (Γ), and the normalization (σ_0). This allows us to do various fits making different Standard Model assumptions. A one-parameter fit was done allowing m_Z to be a free parameter and calculating all Z partial widths using the Standard Model and this mass assumption. A two-parameter fit was done by also allowing m_Z to be free and calculating all partial widths, but allowing N_ν to be a free parameter when calculating the total width Γ. Then a three-parameter fit was done allowing m_Z, Γ, and σ_0 to be free, which is essentially a model independent fit. The following table lists the fit results; all the errors are statistics limited except for the error on N_ν, which has approximately equal contributions from statistics and from the uncertainty in the luminosity scale. This value of N_ν allows us to set a 95% CL upper limit of $N_\nu < 3.9$ within the Standard Model framework.

Fit	m_Z GeV/c^2	N_ν	Γ GeV	σ_0 nb
1	91.14 ± 0.12	–	–	–
2	91.14 ± 0.12	2.8 ± 0.6	–	–
3	91.14 ± 0.12	–	$2.42^{+0.45}_{-0.35}$	45 ± 4

Table 2: Z resonance parameters from maximum-likelihood fits.

6. Conclusions

Below is displayed a plot of the data and the best-fit resonance shapes. The results of "Fit 1" are the dashed line, and the (indistinguishable) results of fits 2 and 3 are the solid line. The main finding that nothing beyond minimal Standard Model expectations (five quarks, three light neutrinos, three charged leptons) was found has been confirmed by experiments with higher statistics.[7]

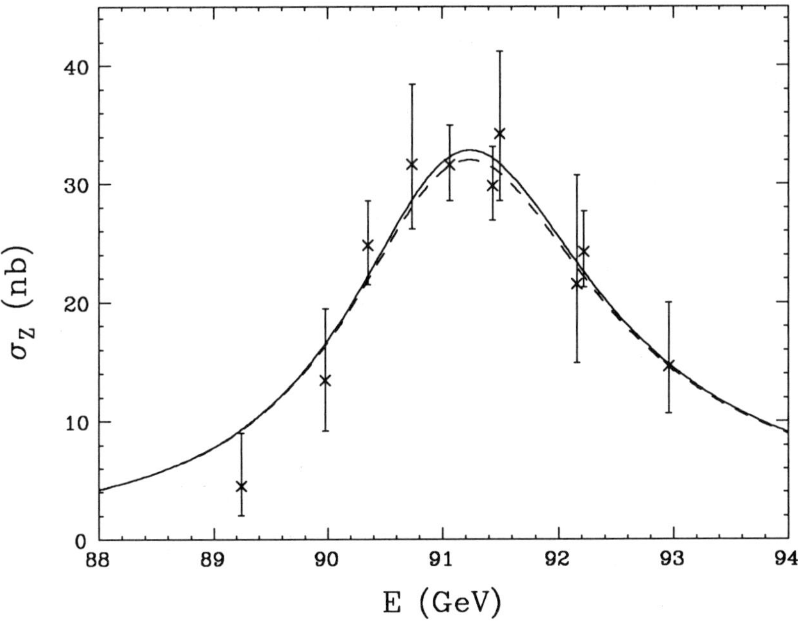

Figure 1: e^+e^- annihilation cross section to all hadronic events plus μ and τ pairs with $|\cos\theta| < 0.65$. Curves explained in text.

REFERENCES

1. C. K. Jung *et al.*, SLAC-PUB-5135 (1990).
   J. Kent *et al.*, SLAC-PUB-4922 (1989).
   M. Levi, J. Nash, and S. Watson, Nucl. Instrum. Meth. **A281** (1989) 265.

2. G. Abrams *et al.*, Nucl. Instrum. Meth. **A281** (1989) 55.

3. F. A. Berends and R. Kleiss, Nucl. Phys. **B228** (1983) 537.
   F. A. Berends, R. Kleiss, and W. Hollik, Nucl. Phys. **B304** (1988) 712.
   S. Jadach and B. F. L. Ward, UTHEP-88-11-01 (1988).

4. C. von Zanthier *et al.*, SLAC-PUB-5213 (1990).

5. A. G. Frodesen, O. Skjeggestad, and H. Tøfte, *Probability and Statistics in Particle Physics* (Universitetsforlaget, Bergen, Norway, 1979), p. 85.

6. R. N. Cahn, Phys. Rev. **D36** (1987) 2666.

7. P. Abreu *et al.*, CERN-EP/90-32 (1990).
   B. Adeva *et al.*, Phys. Lett. **B237** (1990) 136.
   M. Z. Akrawy *et al.*, CERN-EP/90-27 (1990).
   D. Decamp *et al.*, Phys. Lett. **B235** (1990) 399.

Z DECAYS INTO LEPTON PAIRS[*]

Michael Kuhlen[†‡]

California Institute of Technology
Pasadena, Ca 91125

ABSTRACT

The first measurements of leptonic Z decays by the Mark II experiment are presented.

1. INTRODUCTION

Before the advent of the e^+e^- colliders SLC and LEP operating at the Z resonance, information on the neutral current lepton couplings has been obtained indirectly through processes involving virtual Z bosons. The first direct measurements of Z decays into lepton pairs from e^+e^- annihilation at the SLC[1] are presented in this talk.

We measure the ratio of the partial decay width into lepton pairs to the partial decay width into hadrons: $\Gamma_{ll}/\Gamma_{had} = Br(Z \to ll)/Br(Z \to \text{hadrons})$. In the Standard Model (SM) this ratio is expected[2] to be 0.048. For muons and taus at the Z resonance it is simply given by the ratio of produced leptonic to hadronic events, which is independent of the center of mass energy, the luminosity measurement and assumptions about the mass and width of the Z and its line shape. For electrons however QED effects must also be taken into account.

2. EVENT SELECTION

Data were collected with the Mark II detector[3] in the e^+e^- center of mass energy (E_{cm}) range of 89.2 to 93.0 GeV and correspond to an integrated luminosity of 19 nb^{-1} . The central drift chamber allows for charged particle track reconstruction within $|\cos\theta| < 0.92$, where θ is the angle with respect to the beam axis, and the electromagnetic calorimeter covers the angular range $|\cos\theta| < 0.96$. The data acquisition system is triggered by either two or more charged tracks within $|\cos\theta| < 0.76$, or localized energy depositions in the calorimeter. In this analysis both, charged tracks emerging from the interaction region with $|\cos\theta| < 0.82$ with $p_T > 0.15$ GeV/c, and neutral showers with $E > 1$ GeV are used.

Events from Z decays are required to have visible energy larger than 15% of E_{cm} and charged energy larger than 5% of E_{cm} to minimize contributions from two-photon processes and beam-gas interactions. Hadronic and leptonic decays are separated on the basis of the charged multiplicity by requiring at least 7 charged tracks for hadronic events, and fewer than 7 for lepton event candidates.

---

[*] Work supported in part by the US Department of Energy, contract DE-AC03-81ER40050
[†] representing the Mark II Collaboration
[‡] Feodor Lynen fellow, Alexander von Humboldt Foundation

356

The separation of the leptonic Z decays into e, μ and τ pairs requires additional criteria. Electron events are expected to deposit the full available center of mass energy in the calorimeter, and they are separated from μ and τ events by requiring that the total calorimeter energy E_{cal} is larger than 80% of E_{cm}(see Fig. 1). The missing momentum carried away by neutrinos in τ events is used to divide the remaining 2-prong sample into μ and τ candidates. Events are taken as muon candidates if both tracks have momentum larger than 60% of the beam momentum, and as τ events otherwise (see Fig. 2). In addition, muon events are restricted to the central calorimeter region, $|\cos\theta| < 0.68$, and the shower energies associated with the tracks have to be smaller than 15 GeV. Lepton candidates with more than one track in either thrust hemisphere are classified as τ pairs if the effective mass of the charged tracks in that hemisphere is smaller than 2.2 GeV/c^2 and the event thrust is greater than 0.99. This cut suppresses background from hadronic events. In order to reduce the contributions from QED and electroweak interference in the electron channel, the forward region is eliminated by requiring $-0.82 < \cos\theta < 0.68$ for the positron scattering angle.

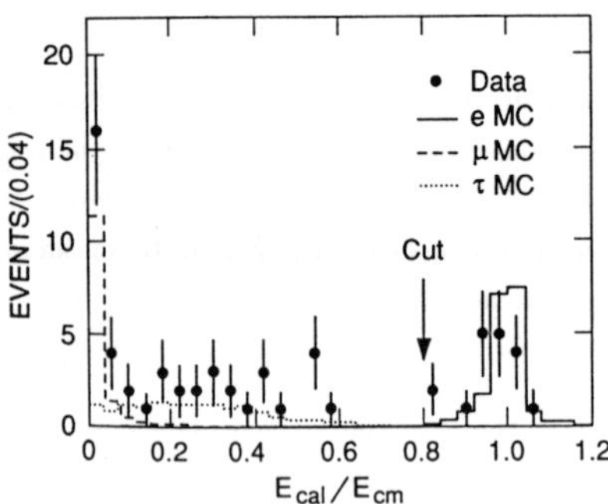

Fig. 1. Calorimeter energy divided by E_{cm} for data and Monte Carlo predictions for e, μ and τ pairs. The arrow indicates the cut at 0.8 to separate the e from the μ and τ events.

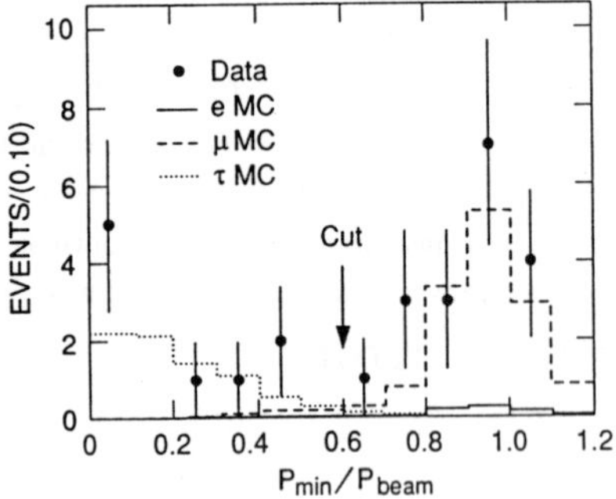

Fig. 2. Ratio of p_{min} to the beam momentum for two-track events after the cut to remove electron events. p_{min} is the smaller of the two momenta in the event. The data are shown with a Monte Carlo simulation for e, μ and τ events. A cut at 0.6 is used to separate μ and τ events.

The numbers of detected events after all cuts, the backgrounds from misclassified Z decays and other processes, and the detection efficiencies are given in the following table:

channel	e	μ	τ	hadrons
detected events	12	13	21	397
background events	0.14 (+1.22 QED)	0.22	1.5	<0.4
acceptance	0.62	0.54	0.66	0.88

3. RESULTS

The resulting ratios of the partial decay widths are $\Gamma_{ee}/\Gamma_{had} = 0.037^{+0.016}_{-0.012}$, $\Gamma_{\mu\mu}/\Gamma_{had} = 0.053^{+0.020}_{-0.015}$ and $\Gamma_{\tau\tau}/\Gamma_{had} = 0.066^{+0.021}_{-0.017}$. The errors are dominated by the statistical uncertainties. These results are consistent with each other confirming lepton universality and agree well with the SM prediction of 0.048. Under the assumption of lepton universality the combined lepton sample yields $\Gamma_{ll}/\Gamma_{had} = 0.053^{+0.010}_{-0.009}$. If we assume a SM leptonic width ($\Gamma_{ll} = 0.083$ GeV) and compare the derived hadronic width $\Gamma_{had} = 1.56^{+0.28}_{-0.24}$ GeV with the SM value of $\Gamma_{had} = 1.73$ GeV, we see no evidence for contributions other than the standard ones to the hadronic Z width.

The number of leptonic events can also be used to provide measurements of the leptonic couplings. At the Z, the cross section is proportional to $(a_e^2 + v_e^2) \cdot (a_l^2 + v_l^2)$, where a and v denote the axial-vector and vector couplings of the incoming electrons and the final state leptons. If we fix the experimentally well constrained quantities a_e, v_e and a_l to their SM values, and use our measured values for the Z mass and width[4], $M_Z = 91.14 \pm 0.12$ GeV/c^2 and $\Gamma = 2.42^{+0.45}_{-0.35}$ GeV, we find $v_\tau^2 = 0.31 \pm 0.31\ ^{+0.43}_{-0.30}$ and $v_\mu^2 = 0.05 \pm 0.30\ ^{+0.34}_{-0.23}$. The first error represents the statistical error, and the second represents the systematic error, which is entirely dominated by our uncertainty in the Z width. These results can be compared with previous measurements, $v_\tau = -1.04 \pm 1.25$ [5], $v_\mu = -0.24 \pm 0.32$ [6] and the SM value, $v_l = -0.07$.

Our data sample provides also a first look at the forward-backward asymmetry A_{FB} in the Z region. For the combined lepton sample of 9 muon and 15 tau events from the region around the Z pole, 90.6 GeV $< E_{cm} <$ 91.5 GeV, we measure $A_{FB} = 0.05 \pm 0.22$. For our data set we expect $A_{FB} = 0.026$ for $M_Z = 91.14$ GeV/c^2, in good agreement with the measurement, but the result is not precise enough to place useful limits on SM parameters.

4. CONCLUSION

In conclusion, our measurements of the leptonic Z decays confirm within their statistical precision the weak couplings of the leptons and quarks to the Z boson as predicted by the Standard Model.

REFERENCES

1. G. S. Abrams et al., Phys. Rev. Lett. 63, 2780 (1989).

2. W. F. L. Hollik, DESY 88-188, December 1988. A top quark mass of 100 GeV/c^2 is assumed, and a QCD correction of 5% has been applied for the hadronic width.

3. G. S. Abrams et al., Nucl. Instr. Meth. A281, 55 (1989).

4. G. S. Abrams et al., Phys. Rev. Lett. 63, 2173 (1989).

5. W. T. Ford et al., Phys. Rev. D36, 1971 (1987).

6. A. Argento et al., Phys. Lett. B120, 245 (1983). This measurement assumes SM couplings for the quarks.

Measurement of the $b\bar{b}$ Fraction in Hadronic Z Decays

J. Frederic Kral

Lawrence Berkeley Laboratory and Department of Physics,

University of California, Berkeley, CA 94720

Using isolated leptons reconstructed in the Mark II detector to tag $b\bar{b}$ events, we measure the fraction of $b\bar{b}$ events in hadronic Z^0 decays to be $0.23^{+0.11}_{-0.09}$, in good agreement with the Standard Model prediction of 0.22.

We measure the fraction of $b\bar{b}$ events in hadronic events produced through e^+e^- annihilation near the Z^0 peak. The Standard Model (SM) predicts that for five kinematically accessible quarks, this fraction $r_b = \frac{\Gamma(Z \rightarrow b\bar{b})}{\Gamma(Z \rightarrow had)}$ is 0.22, considerably larger than 0.09, the predicted fraction of $b\bar{b}$ in hadronic events produced by the single photon exchange process which dominates e^+e^- annihilation at lower energies. Details and references for this analysis can be found in J.F. Kral *et al.*, Phys. Rev. Lett. **64**, 1211 (1990).

The data were taken near the Z^0 pole with the Mark II detector at the SLAC Linear Collider. We select 413 hadronic events with seven or more charged tracks and a visible energy greater than 15% of the center-of-mass energy, E_{cm}. The corresponding efficiencies are estimated by Monte Carlo (MC) simulations based on the Lund parton shower model with string fragmentation (JETSET 6.3 shower) and the Webber-Marchesini parton shower model with cluster fragmentation (BIGWIG 4.1). We use the average of the two models as the prediction to be compared with data, and account for differences between them in the systematic errors. The resulting efficiencies are 0.86 ± 0.02 for detecting produced $udsc$ events and 0.88 ± 0.02 for produced $b\bar{b}$ events. We expect < 0.5 events from other processes.

We tag $b\bar{b}$ event candidates in the hadronic event sample with isolated charged tracks identified as leptons. We define the transverse momentum, p_t, of each track with respect to the nearest cluster formed by the other charged and neutral particles in the event. The p_t spectrum for tracks identified as electrons is shown in Fig. 1(a) together with predictions for the contributions from real electrons and hadrons misidentified as electrons. The p_t spectrum for tracks identified as muons is shown in Fig. 1(b) together with predictions for the contributions from real muons and hadrons misidentified as muons.

To separate $b\bar{b}$ events from $udsc$ events, we assign a p_t value to each event containing an identified lepton. The overall efficiency for tagging produced $b\bar{b}$ events is 0.100 ± 0.012, resulting from the B semi-leptonic branching ratios (here taken to be 0.11 ± 0.01), the fiducial acceptance of the detector, the lepton identification efficiencies and the isolation cut, $p_t > 1.25$ GeV/c. The cuts retain only a small fraction, $0.011^{+0.004}_{-0.003}$, of produced $udsc$ events.

Among the 413 hadronic events in the data, we observe 15 high p_t events, 9 tagged by electrons and 6 by muons. The SM prediction is 14.7 events, with 10.3 events expected from $b\bar{b}$ events containing real leptons. Fig. 1(c) shows the observed p_t spectrum together with the expected quark flavor composition of events with a track identified as a lepton. From the observed numbers of hadronic and tagged events, together with the efficiencies described above, we construct two equations to be solved for the unknown $udsc$ and $b\bar{b}$ populations. The resulting value of r_b $= 0.23^{+0.10}_{-0.08}{}^{+0.05}_{-0.04} \pm 0.02$, where the errors are, in the order quoted, the statistical errors, the systematic errors from uncertainties in the event efficiencies and the systematic error from the uncertainty in the semi-leptonic branching ratio. As shown in Fig. 1, our measurements are in good agreement with the SM predictions.

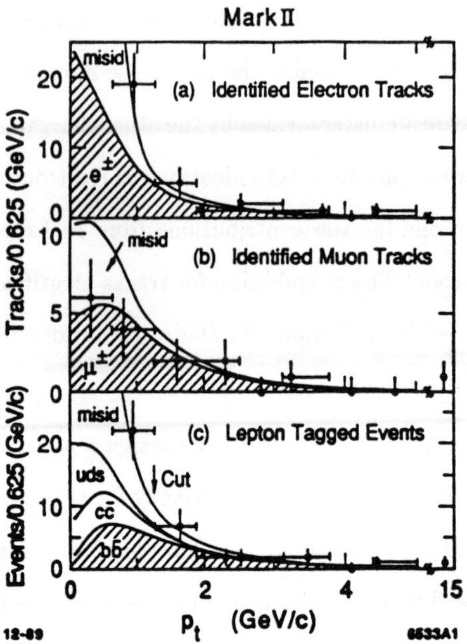

FIG. 1. The p_t spectra for tracks identified as (a) electrons and (b) muons. The shaded and unshaded regions show the expected contributions from real leptons and hadrons misidentified as leptons, respectively. (c) The p_t distribution for leptons ($e^\pm$ or $\mu^\pm$) with one entry per event. The shaded region is the expected contribution from $b\bar{b}$ events with real leptons. Also indicated are the contributions from $c\bar{c}$ and uds events, as well as events tagged by hadrons misidentified as leptons. These predictions come from Monte Carlo simulations normalized to 413 observed events, assuming $r_b = 0.22$.

This work was supported in part by Department of Energy contracts DE–AC03–81ER40050 (CIT), DE–AM03–76SF00010 (UCSC), DE–AC02–86ER40253 (Colorado), DE–AC03–83ER40103 (Hawaii), DE AC02 84ER40125 (Indiana), DE–AC03–76SF00098 (LBL), DE–AC02–76ER01112 (Michigan), and DE–AC03–76SF00515 (SLAC), and by the National Science Foundation (Johns Hopkins).

Measurement of Electroweak Parameters
from Z Decays into Fermion Pairs

F. Bird
European Laboratory for Particle Physics (CERN)
1211 Geneva 23, Switzerland

J.L. Harton and R.P. Johnson
Department of Physics, University of Wisconsin
Madison, Wisconsin   53706

Abstract

We report on the properties of the Z resonance from 31,700 Z decays into fermion pairs collected with the ALEPH detector at LEP. We find $M_Z = (91.197 \pm 0.021_{\text{exp}} \pm 0.030_{\text{LEP}})\,\text{GeV}$, $\Gamma_Z = (2562 \pm 47)\,\text{MeV}$, $\sigma^0_{\text{had}} = (41.30 \pm 0.75)\,\text{nb}$, and for the partial widths, $\Gamma_{\text{inv}} = (503 \pm 33)\,\text{MeV}$, $\Gamma_{\text{had}} = (1803 \pm 37)\,\text{MeV}$, $\Gamma_{e^+e^-} = (85.1 \pm 2.4)\,\text{MeV}$, $\Gamma_{\mu^+\mu^-} = (83.1 \pm 3.6)\,\text{MeV}$ and $\Gamma_{\tau^+\tau^-} = (85.1 \pm 3.7)\,\text{MeV}$, in good agreement with the Standard Model. Assuming lepton universality we obtain $\Gamma_{l^+l^-} = (85.1 \pm 1.7)\,\text{MeV}$. From these measurements the number of light neutrino species is $N_\nu = 2.97 \pm 0.16$, and the electroweak mixing angle is $\sin^2\theta_W(M_Z^2) = 0.227 \pm 0.005$.

1. Introduction

From the total 1989 ALEPH data sample, corresponding to about 25,000 hadronic and 3700 leptonic decays, we present precision measurements of the Z mass and partial widths into hadrons and lepton pairs. From these measurements, the number of light neutrinos and the electroweak mixing angle are determined. These results provide a stringent test of electroweak theory.

2. Apparatus and Trigger

The ALEPH detector is described in detail in Ref. 1. The detector components relevant for these measurements are, going from the collision point outward.

- The Inner Tracking Chamber, ITC, an 8-layer cylindrical drift chamber with sense wires parallel to the beam axis from 13 cm to 29 cm in radius. Tracks with polar angles from $14°$ to $166°$ traverse all 8 layers.

- The large cylindrical Time Projection Chamber, TPC, extending from an inner radius of 31 cm to an outer radius of 180 cm over a length of 4.4 m. Up to 21 space coordinates are recorded for tracks with polar angles from $37°$ to $143°$. Requiring at least 4 coordinates, tracks are reconstructed down to $15°$.

- The Electromagnetic Calorimeter, ECAL, a lead wire-chamber sandwich covering the polar angle range from $11°$ to $169°$. The cathode readout is subdivided into a total of 73,728 projective towers. Each tower of about $1° \times 1°$ solid angle is read out in three stacks of 4, 9 and 9 radiation lengths. The signals from the 45 wire planes of each module are also read out.

- The luminosity calorimeters, LCAL, which is described in detail below.

- The superconducting solenoidal coil, providing a magnetic field of 1.5 T.

- The Hadron Calorimeter, HCAL, consisting of 23 layers of streamer tubes interleaved in the iron of the magnet return yoke, read out in a total of 4608 projective towers. It covers a polar angle range from $6°$ to $174°$. Signals from each of the tubes are also read out. The modules are rotated in azimuth by about $2°$ with respect to the electromagnetic calorimeter so that inactive zones do not align in the two calorimeters.

Two kinds of triggers are used for the Z decays in this analysis:

- A total-energy trigger derived from the electromagnetic calorimeter by summing the signals from the wire planes of the 12 modules separately in the barrel and the two endcaps. An energy of 6.6 GeV in the barrel, or 3.8 GeV in either endcap, or 1.5 GeV in both endcaps is required by this trigger.

- Two charged particle triggers requiring a track candidate in the inner tracking chamber in coincidence with a wire signal from a group of ECAL or HCAL modules which form a trigger segment. A track candidate must have hits in at least five out of eight ITC layers, and the associated energy in an ECAL module must exceed 1.3 GeV, or the particle must penetrate the ECAL and at least 30 cm to 90 cm of iron in the HCAL, depending on the polar angle.

The trigger efficiency depends on the characteristics of the selected event sample and is determined in each case from the data themselves using redundant trigger information. All trigger efficiencies are found to be close to 100%, leaving little room for systematic uncertainties.

3. Selection of Hadronic Events

Events from hadronic Z decays are selected in two independent ways, one based on charged tracks only and the other on calorimetric energy. Both methods have an efficiency very close to unity, and the two event samples overlap by 95%. Only the charged track selection is described here. It requires at least 5 charged tracks measured in the TPC with a total energy larger than 10% of the centre-of-mass energy. The tracks must have a polar angle above 18.2°, which ensures that at least 6 TPC pad rows are traversed, and must have at least 4 reconstructed coordinates. The distance of closest approach of the tracks to the origin must be less than 10 cm along the beam direction and 2 cm transverse to it. After these cuts 25,035 events are retained.

Distributions in global event parameters such as multiplicity, sphericity and charged energy, as well as in track parameters such as momentum and angle, are in good agreement [2] with hadronization models [3]. For the accepted events, the total charged energy, the multiplicity and the sphericity axis are shown in Fig. 1. The efficiency for $q\bar{q}$ events is calculated by Monte Carlo simulation to be $(97.5 \pm 0.6)\%$. The error corresponds to a change of the energy cut by 20% and the charged multiplicity by ± 1. Uncertainties in the hadronization models are reduced to a small level by using the measured sphericity distribution for the acceptance calculation.

The background from $\tau^+\tau^-$ events is estimated to be 42 ± 14 events and is subtracted. Background of beam-gas interactions is estimated from the number of events found passing all the selection cuts except for the cut on the vertex position along the beam and is found to be negligible.

Below the cut of $0.10\sqrt{s}$ in the total charged energy, the background from two-photon processes, $e^+e^- \rightarrow e^+e^- + \text{hadrons}$, is dominant (see Fig. 1a). The background in the hadronic event sample has been estimated by exploiting the different centre-of-mass energy dependence of the resonant and non-resonant contributions. A plot is made of the cross section for charged energies between $0.10\sqrt{s}$ and $0.15\sqrt{s}$ versus the cross section for energies greater than $0.3\sqrt{s}$. A linear extrapolation to zero for the cross section of the large charged energy interval predicts a background of (12 ± 17) pb in the lower energy interval. The extrapolation to the full acceptance has been done with a simulation of the two-photon process and gives (20 ± 35) pb.

4. Measurement of the Luminosity

The luminosity is determined from small angle Bhabha events which shower in a calorimeter (LCAL) consisting of 00 layers of lead sheets and proportional-wire chambers. It is read out in projective towers, of which there are 784 on each side. The tower layout of one half of one module is shown in Fig. 2. Each tower is read out in 3 stories of 4.8, 10.6 and 9.25 radiation lengths in depth. There are in total 4 semicylindrical modules, two each at ± 262 cm along the beamline from the nominal center of ALEPH, covering polar angles from 42 to 160 mrad. Within each module,

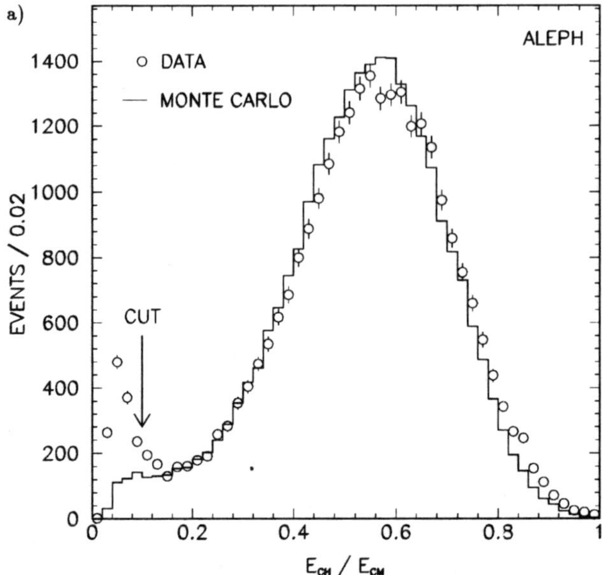
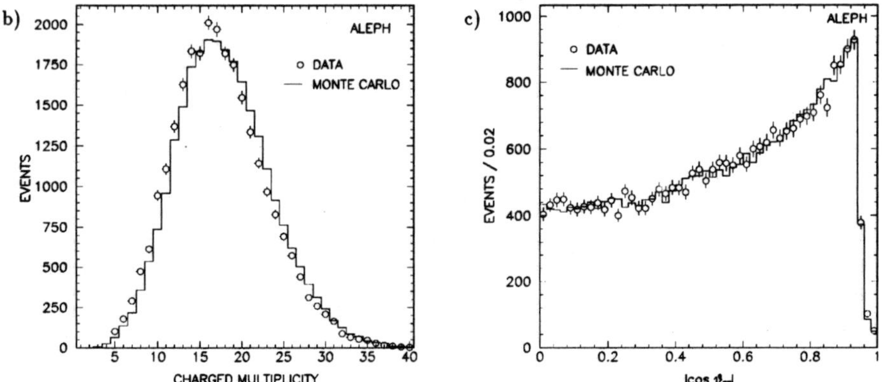

Figure 1: Distributions for track selected events: a) Total charged energy divided by the beam energy. b) Multiplicity of charged tracks. c) Absolute value of the cosine of the polar angle of the sphericity axis.

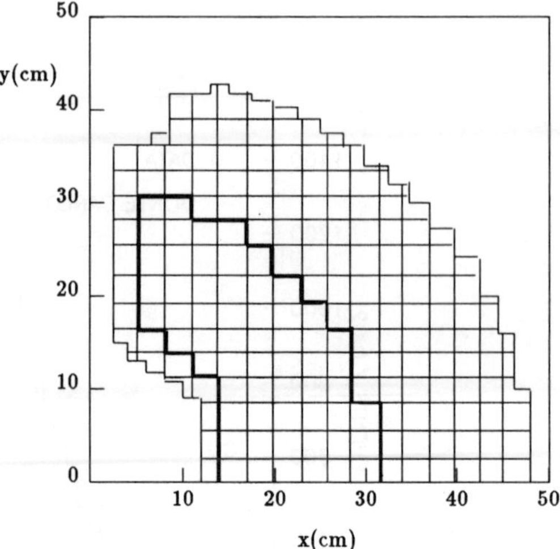

Figure 2: One half of one module of the Luminosity Calorimeter, showing the tower structure and the fiducial region (in bold).

the mechanical precision in the relative position of the towers is $120\,\mu$m. The uncertainty on the inner radius due to the relative positioning of the two modules is $140\,\mu$m. The energy resolution at 45 GeV is 3%, in agreement with test beam measurements. The tower-to-tower response is uniform within $\pm 2\%$, and the spatial resolution of the shower position is 1.5 mm in the horizontal and vertical directions.

Three triggers are used in the luminosity determination, based on the analog sum of pulse heights in the towers:

- A basic luminosity trigger, which requires a coincidence between the two sides with more than 20 GeV on one and 16 GeV on the other side.

- A high energy single arm trigger, requiring more than 31 GeV on either side. This trigger is used to determine the efficiency of the basic trigger. The efficiency varies slightly in time; on average it is 99.7%, while it is always larger than 98%. This trigger was prescaled by a factor of 4 for more than 70% of the data.

- Prescaled single arm triggers of more than 20 GeV and 16 GeV on either side. These triggers allow a determination of the background due to random coincidences of off-momentum electrons in the LEP storage ring.

To minimize the systematic uncertainty, a fiducial region is defined which excludes towers at the inner edges of the calorimeter (see Fig. 2). The large outer region, approximately for angles larger than 110 mrad, is excluded because it is shadowed by dense interior support structures (this results in a loss of only about 15% of the luminosity events, and substantially reduces the tails of the energy distributions, which are otherwise difficult to simulate). Together with the projective nature of the tower geometry, the energy requirement across the fiducial boundary gives a precise definition of the fiducial region: the position uncertainty at the boundary is approximately $150\,\mu$m.

Events are accepted if the electron or the positron deposits more energy inside the fiducial region than outside. The sum of the energies in the first storeys of the three adjacent towers bordering the inside of the fiducial region are compared with the corresponding three towers outside the

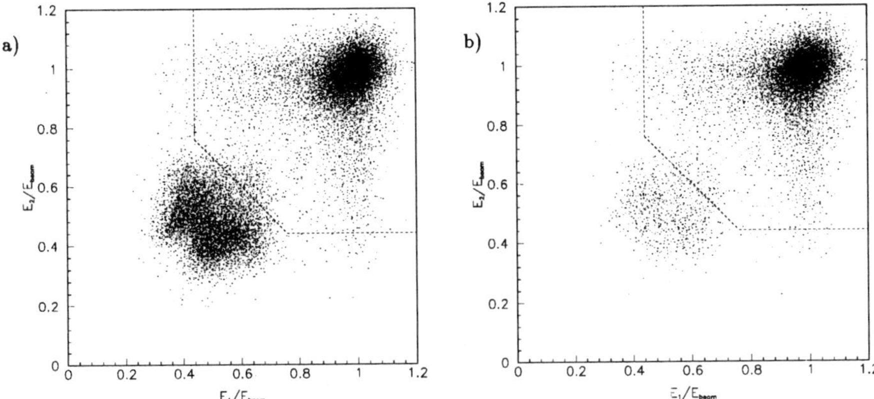

Figure 3: a) Shower energy in the luminosity calorimeter on the fiducial side versus the non-fiducial side. The dashed line shows the event selection cuts. b) The same plot for events with $\Delta\phi > 170°$.

fiducial region. There are no such cuts on the opposite side. This asymmetry in the acceptance condition, together with the requirement of fiducial-side alternation from event to event, ensures that the acceptance is independent, to first order, of transverse and longitudinal displacements of the collision point and of small displacements in the beam direction. The horizontal and vertical vertex distribution at the nominal beam crossing is measured for each LEP fill on the basis of the shower positions on the two sides with a precision of about $200\,\mu$m, resulting in corrections which are negligible.

An event is selected as a Bhabha if the energy on each side is larger than 44% of the beam energy and if the sum of the two energies is larger than 60% of the centre-of-mass energy. Figure 3a shows the correlation of the two energies and the cuts. To further reduce the beam related background, the difference in the azimuth, $\Delta\phi$, between the e^+ and e^- is required to be larger than 170°. Figure 4a shows the $\Delta\phi$ distribution for accepted events, for data and simulation, after correction for the beam position and magnetic deflection. The Bhabha events are near 180°, and a cut at 170° is applied. The effect of this cut in background reduction is clearly seen by comparing the energy correlation plot of Fig. 3b with Fig. 3a. About 31,000 luminosity triggers remain after the $\Delta\phi$ cut, of which 11.6% are removed by the energy cuts shown by the dashed lines in Fig. 3. The $\Delta\phi$ distribution also provides an estimate of the background due to random coincidences in the two sides between single off-momentum particles. This background is distributed much more broadly in $\Delta\phi$ than Bhabha events. The events in the region $0° < \Delta\phi < 10°$ and $160° < \Delta\phi < 170°$ are used to subtract background events at $170° < \Delta\phi < 180°$. The background is very small, at a level of about 0.4%, as can be seen in Fig. 4a and in the Bhabha energy distribution in Fig. 4b. Finally, Fig. 4c shows that the polar-angle distribution of accepted events on the fiducial side is in excellent agreement with the simulation.

The effective cross section is calculated using an event generator that includes first-order radiative corrections [6] and a full simulation of the detector. The events are generated at $91.0\,$GeV centre-

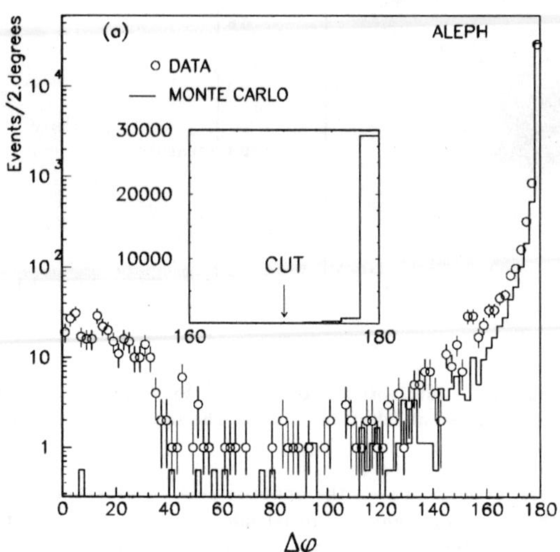

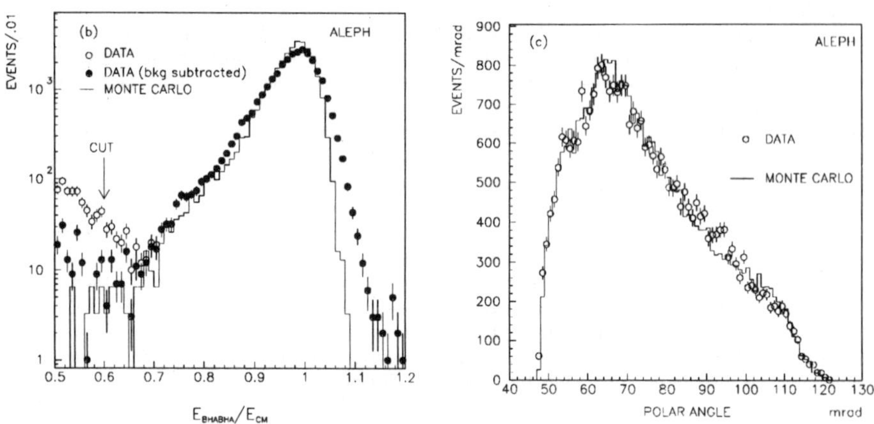

Figure 4: a) Distribution of the difference in the azimuthal angle between the two after correction due to the magnetic deflection. b) Distribution of the total Bhabha energy relative to the centre-of-mass energy before and after background subtraction. The open circles are data without background subtraction; filled circles are with subtraction. c) Distribution of the polar-angle of accepted events on the fiducial side.

Position of pad towers within a module and between two halves	0.003
Energy resolution and cell-to-cell calibration	0.001
Energy scale	0.002
Beam parameters	0.001
Inadequacy of simulation	0.009
Statistics of simulation	0.004
Total experimental uncertainty	0.011
Uncertainty in theory	0.007

Table 1: Summary of systematic errors in the luminosity measurement. Values listed are fractions of the luminosity.

of-mass energy with a Z mass of 91.0 GeV. The corresponding cross section for our acceptance is found to be $(26.53 \pm 0.10_{exp} \pm 0.20_{theory})$ nb. The contribution to this cross section due to radiative corrections is $+2.5\%$. For other energies this cross section is multiplied by a factor of $(91\,\mathrm{GeV})^2/s$ and corrected for small ($< 0.6\%$) electroweak interference effects [6]. The hadronic vacuum polarization has been included [4]. The contribution from $e^+e^- \rightarrow \gamma\gamma$ has been estimated to be 5×10^{-4} of the Bhabha cross section for our acceptance and is therefore believed to be negligible. Our estimate of various systematic errors are given in Table 1.

The first error of the Bhabha cross section represents the effect of uncertainties in the simulation, calibration and positioning of the calorimeter. The second is an estimate of the uncertainty in the radiative corrections [5]. We checked that the cross section within our cuts is not very sensitive to radiative corrections. The difference between lowest-order and first-order calculation is about 2.5%. The systematic error introduced by neglecting higher orders is expected to be smaller than this; we assume an error of 0.7%. The total systematic error attributed to the luminosity measurement is 1.3%.

5. Selection of Leptonic Events

The analysis of the leptonic decays of the Z is made in two different ways. The first method selects charged leptonic decays using only the tracking information, thus including all three leptons types without making any attempt to distinguish between them. The second uses in addition the particle identification capabilities of ALEPH to isolate separate samples of e^+e^-, $\mu^+\mu^-$ and $\tau^+\tau^-$ pairs.

The following requirements are designed to separate leptonic from hadronic decays and are common to the two methods:

1. A track must have at least 4 coordinates measured in the TPC, have a momentum exceeding 0.1 GeV and originate from within 5 cm along the beam direction and within 1.5 cm in the transverse direction.

2. The event is required to have from 2 to 6 tracks in the polar angle range $|\cos\theta| < 0.95$.

3. The event, divided into two hemispheres by a plane perpendicular to the thrust axis, has to have at least one track in each hemisphere (giving two "jets").

4. At least one track must have momentum larger than 3 GeV, and at least one of the jets must have a transverse momentum, with respect to the beam axis, greater than 2.5 GeV. If in a two-track event both tracks have $p < 6$ GeV, the ratio of their transverse momenta must differ from unity by greater than 15%.

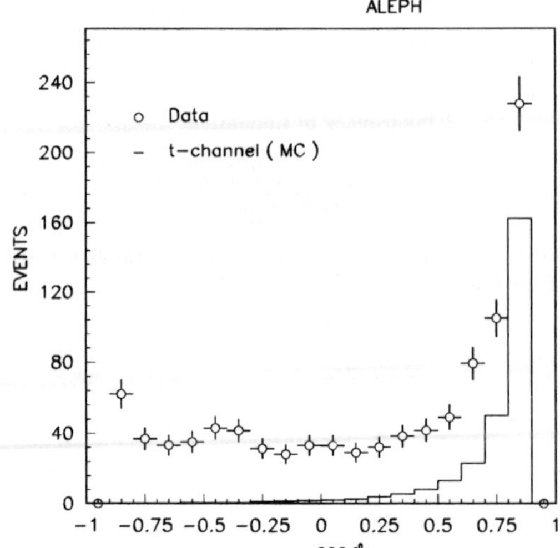

Figure 5: Distribution of the polar angle for $Z \to e^+e^-$ before t-channel subtraction. The histogram shows the predicted t-channel distribution.

5. The track momenta in each hemisphere are summed, giving vectors with polar angles θ_1 and θ_2. The acollinearity η, defined as the angle between the vectors, has to exceed $160°$.

6. Events with more than 4 tracks are rejected if any track makes an angle greater than $31.8°$ with the vector sum of the track momenta in the same hemisphere.

The center-of-mass scattering angle θ^* is defined by

$$\cos\theta^* = \cos\tfrac{1}{2}(\theta_1 + \pi - \theta_2)/\cos\tfrac{1}{2}(\theta_1 + \pi + \theta_2). \tag{1}$$

For analyses which include the e^+e^- final state, the scattering angle is restricted to the range $-0.9 < \cos\theta^* < 0.7$; otherwise the accepted range is $|\cos\theta^*| < 0.9$.

To select samples of $Z \to l^+l^-$ decays with specific final state flavors, additional cuts are necessary. $Z \to e^+e^-$ events are selected by requiring the sum of the momenta of the two most energetic tracks and the sum of the energies of the ECAL clusters associated with the tracks both to exceed 45% of the center-of-mass energy. Furthermore, the acollinearity angle of the two tracks must be larger than $160°$. Fig. 5 shows the corrected angular distribution together with the predicted t-channel contribution.

$Z \to \mu^+\mu^-$ events are selected by requiring, first, that the most energetic track have $p > 35\,\mathrm{GeV}$ and the second most energetic track have $p > 22\,\mathrm{GeV}$. The HCAL track trigger is required to have fired, since it is the only trigger which is efficient for muon pairs. This removes the majority of the $Z \to e^+e^-$ events. Two approaches have been used to eliminate the remainder of the e^+e^- background. The first is based on tracking the muons with the digital readout of the HCAL and is not described in detail here. The second is based on the measured energy deposits in the ECAL and HCAL (the HCAL energy measurement is important only in the cases where one of the electrons goes through a crack in the ECAL). Of the clusters of calorimetric energy associated with the two most energetic tracks, the larger is required to be less than $25\,\mathrm{GeV}$ and the smaller less than $12\,\mathrm{GeV}$. The cross sections derived from these two different selections agree within 0.8%, and an average of the two is used for the final results.

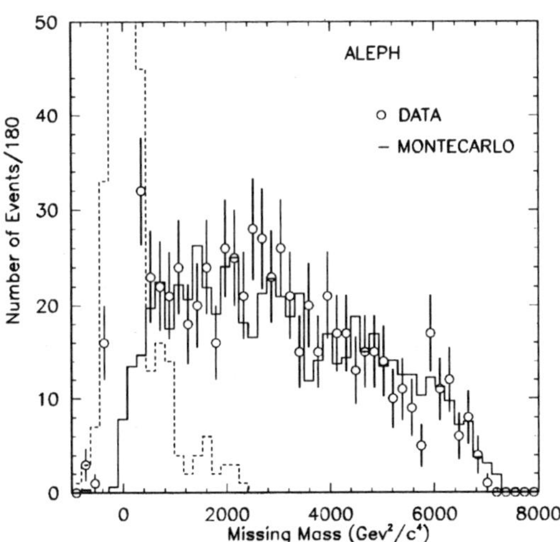

Figure 6: Missing mass squared for the full lepton sample after the cut requiring ECAL wire energy greater than 55 GeV. It is compared with τ Monte Carlo (solid line) and data biased with the requirement of ECAL energy less then 4 GeV or ECAL energy greater than 70 GeV (dotted line) to show the $\mu^+\mu^-$ and e^+e^- contributions.

$Z \to \tau^+\tau^-$ events also are identified by two separate selections, only one of which is described here. This selection is applied only to events with both jets in the angular range $|\cos\theta| < 0.9$. Cosmic ray background is reduced by requiring that in all events with only two charged tracks, at least one of the tracks pass within 0.5 cm of the beam axis. To reject electron pairs and muon pairs, the total ECAL energy is required to be less than 55 GeV, and the square of the missing mass, calculated from the detected charged particles and the known centre-of-mass energy, is required to exceed 400 GeV^2 (see Fig. 6). This last cut is applied only if the multiplicity of charged particles in both jets is below 3. In order to reduce further the background from two photon processes, we require that one track in the event have momentum greater than 5 GeV.

Both for μ and τ pairs, the various selection methods are in all cases in good agreement with each other. The only significant backgrounds in the flavor-specific channels are from the other leptonic channels. Therefore it is advantageous not to separate the channels according to flavor when measuring the inclusive cross section for $Z \to l^+l^-$.

6. The Z Resonance Parameters

The total cross section for fermion pair production in e^+e^- annihilation can be expressed in a model independent formulation [8,9,10] as a function of the physical parameters of the Z resonance. The total cross section contains terms from Z exchange, photon exchange and Z-photon interference. The weak, strong and some electromagnetic corrections naturally separate from the initial state QED corrections. The corrections other than initial-state QED can be calculated first to obtain a kernel cross section. The large effect of the initial state radiation can be included by convoluting the kernel cross section with a suitable radiator function [9,10,11].

The peak cross section due to Z exchange, when unfolded from initial state radiation, is

$$\sigma^0_{f\bar{f}} = \frac{12\pi}{M_Z^2} \frac{\Gamma_{ee}\Gamma_{f\bar{f}}}{\Gamma_Z^2}. \tag{2}$$

Γ_{ee} and $\Gamma_{f\bar{f}}$ are the partial widths of Z decay into e^+e^- and any fermion pair $f\bar{f}$ respectively,

Final State	Partial Width (MeV)	Branching ratio	Peak cross section (nb)
hadrons	1796 ± 50	0.702 ± 0.018	41.23 ± 0.75
e^+e^-	85.1 ± 2.4	0.0333 ± 0.0008	1.95 ± 0.09
$\mu^+\mu^-$	83.1 ± 3.6	0.0325 ± 0.0015	1.91 ± 0.08
$\tau^+\tau^-$	85.1 ± 3.7	0.0333 ± 0.0014	1.95 ± 0.07
invisible	508 ± 50	-	-

Table 2: Z resonance parameters derived from the measured hadron and lepton cross sections.

Γ_Z is the total width and M_Z the mass of the Z boson. The invisible width is defined as $\Gamma_{inv} = \Gamma_Z - \Gamma_{had} - \Gamma_{ee} - \Gamma_{\mu\mu} - \Gamma_{\tau\tau}$.

In the fits that follow, a parameterization of the interference term has been made according to the Standard Model (top-quark mass 130 GeV, Higgs mass 100 GeV). Since this term is very small, this assumption does not bias the results, which have been checked to be stable against large variations of the assumed parameters.

The resonance parameters have been determined by means of three different computer programs. The first two programs, based on formulae by Borelli $et\ al.$ [8] and by Berends $et\ al.$ [9] use analytic expressions for the initial state QED corrections. The third program, following Burgers [10], uses a numerical integration for the initial state QED corrections. Differences in the results from the three programs are less than 10% of the experimental error. Results quoted are from the program using the formula of Ref. 10.

The hadronic and three lepton pair cross sections are fit simultaneously to determine six parameters: the Z mass, the full width, the peak hadronic cross section and the three lepton partial widths in units of the hadronic width. For the electron final state we use only the five cross section points closest to the peak in order to avoid large errors due to the t-channel subtraction. The fit results for the six parameters are

$$
\begin{aligned}
M_Z &= (91.203 \pm 0.021_{\text{exp.}} \pm 0.030_{\text{LEP}})\ \text{GeV} \\
\Gamma_Z &= (2.558 \pm 0.046)\ \text{GeV} \\
\sigma^0_{\text{had}} &= (41.23 \pm 0.75)\ \text{nb} \\
\Gamma_{ee}/\Gamma_{\text{had}} &= 0.0474 \pm 0.0021 \\
\Gamma_{\mu\mu}/\Gamma_{\text{had}} &= 0.0463 \pm 0.0018 \\
\Gamma_{\tau\tau}/\Gamma_{\text{had}} &= 0.0474 \pm 0.0017
\end{aligned}
\tag{3}
$$

with $\chi^2 = 19.5$ for 28 degrees of freedom. These parameters are generally uncorrelated; the only important correlation is between the hadronic peak cross section and the total width (-46%). The second error on the Z mass is due to the uncertainty in the LEP centre-of-mass energy [12]. The possible changes in the beam energy of a few times 10^{-5} [12] from one run period to another are negligible.

The hadronic cross section and the Standard-Model expectations for two, three and four neutrinos are shown in Fig. 7. The cross sections for the three lepton-pair channels are shown in Figs. 8a–c.

Other parameters, including the partial widths and branching ratios, can be derived from these results (see Table 2). The three leptonic widths agree well with each other, as expected from lepton universality.

Using the hadron sample and the lepton sample selected without distinguishing the final state flavour, we fit to determine four parameters: M_Z, σ^0_{had}, Γ_Z and $\Gamma_{ll}/\Gamma_{\text{had}}$. Again, these are mainly uncorrelated except for the peak cross hadronic section and full width. The result of the fit is

$$
M_Z = (91.197 \pm 0.021_{\text{exp}})\ \text{GeV}
$$

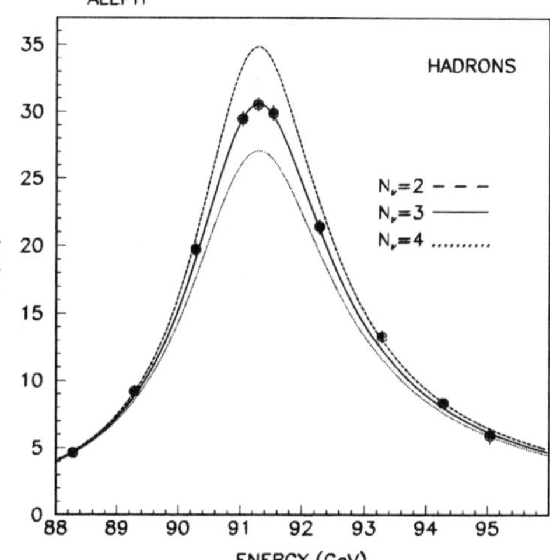

Figure 7: Cross section for $e^+e^- \to$ hadrons as a function of the centre-of-mass energy. The Standard Model predictions for $N_\nu = 2, 3$ and 4 are shown.

Final State	Partial Width (MeV)	Branching ratio	Peak cross section (nb)
hadrons	1803 ± 37	0.704 ± 0.012	41.30 ± 0.75
lepton	85.1 ± 1.7	0.0332 ± 0.0005	1.95 ± 0.06
invisible	503 ± 33	-	-

Table 3: Z resonance parameters derived from the measured hadron and the lepton combined cross sections.

$$\sigma^0_{\text{had}} = (41.30 \pm 0.75)\ \text{nb} \tag{4}$$
$$\Gamma_Z = (2.562 \pm 0.047)\ \text{GeV}$$
$$\Gamma_{ll}/\Gamma_{\text{had}} = 0.0473 \pm 0.0012$$

with χ^2=8.8 for 12 degrees of freedom. Equivalently, the results can be expressed as Γ_{ll} and $\Gamma_{\text{inv}}/\Gamma_{ll}$, rather than as σ^0_{had} and Γ_Z:

$$\Gamma_{ll} = (85.1 \pm 1.7)\ \text{MeV} \tag{5}$$
$$\Gamma_{\text{inv}}/\Gamma_{ll} = 5.91 \pm 0.33 .$$

Other parameters are given in Table 3. In Fig. 8d the cross sections for the lepton pairs are shown. If we use the cross sections from the three identified lepton pair samples instead of the this lepton sample, Γ_{ll} is slightly smaller though statistically consistent.

The number of light neutrino species can be obtained from the measured hadron peak cross section assuming $\Gamma_{\text{inv}} = N_\nu\Gamma_\nu$, with

$$\Gamma_z = \Gamma_{\text{had}} + 3\Gamma_{ll} + N_\nu\Gamma_\nu \tag{6}$$

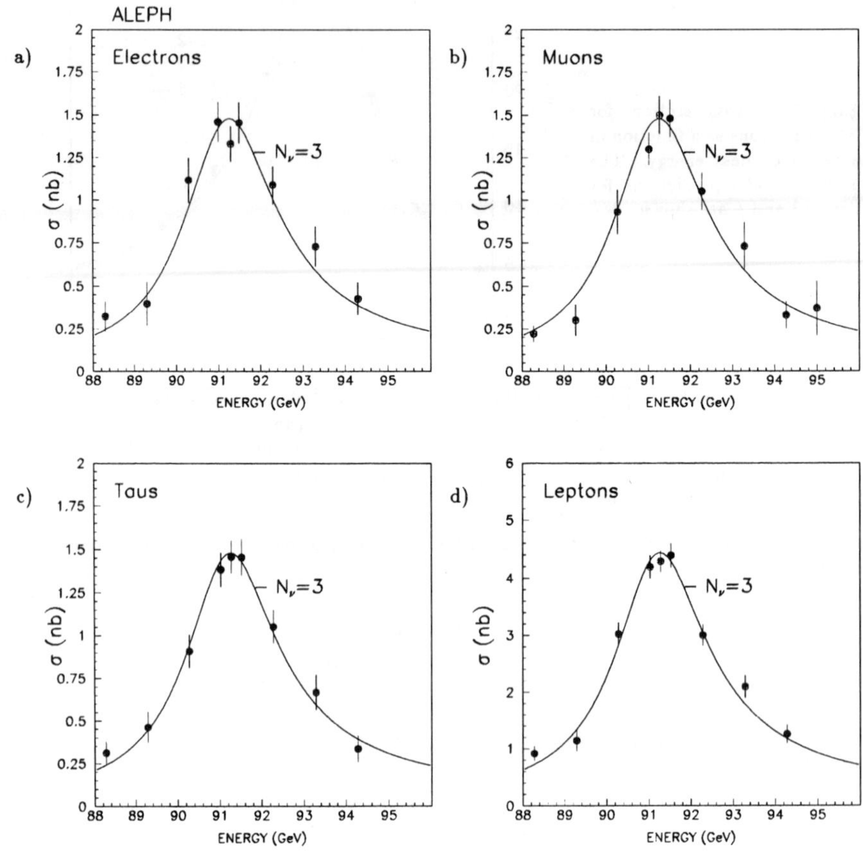

Figure 8: Cross sections for $e^+e^- \rightarrow$ lepton pairs as a function of the centre-of-mass energy. The Standard-Model prediction for $N_\nu = 3$ is shown. a) $e^+e^- \rightarrow e^+e^-$, b) $e^+e^- \rightarrow \mu^+\mu^-$, c) $e^+e^- \rightarrow \tau^+\tau^-$ and d) $e^+e^- \rightarrow$ lepton pairs.

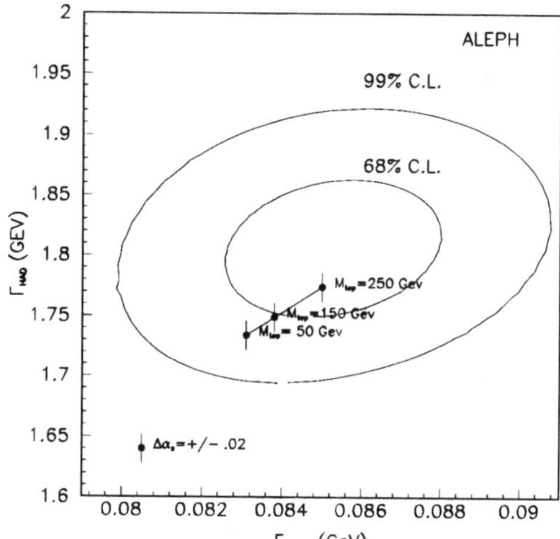

Figure 9: Contours of constant χ^2 for Γ_{had} versus Γ_{ll}, together with the Standard-Model predictions for various top-quark masses. The uncertainty due to the QCD correction is indicated by the error bars.

and

$$N_\nu = \frac{\Gamma_{ll}}{\Gamma_\nu} \left(\sqrt{\frac{12\pi}{M_Z^2 \sigma_{\mathrm{had}}^0} R} - R - 3 \right) , \tag{7}$$

where R stands for $\Gamma_{\mathrm{had}}/\Gamma_{ll}$.

Using the Standard-Model prediction for $\Gamma_{ll}/\Gamma_\nu = 0.5010 \pm 0.0005$, $R = 20.93 \pm 0.21$ [13] and our measured values of M_Z and σ_{had}^0, one finds

$$N_\nu = 2.99 \pm 0.14_{\mathrm{exp}} \pm 0.03_{\mathrm{theory}} . \tag{8}$$

The second error comes from the uncertainty in the prediction of R, which is dominated by the uncertainty in the QCD correction. If we use instead our measured value of $R = 21.10 \pm 0.58$, we obtain

$$N_\nu = 2.97 \pm 0.16 . \tag{9}$$

The importance of this last result compared to the previous one, to which it is completely correlated, is that here the only assumption from the Standard Model is the ratio Γ_{ll}/Γ_ν. This result is still valid if unexpected states yielding hadrons are present in the Z decay. The second determination of N_ν can also be obtained in an equivalent way using our measured value of $\Gamma_{\mathrm{inv}}/\Gamma_{ll}$ and multiplying it by the predicted value of Γ_{ll}/Γ_ν. These results are in good agreement with our earlier measurements [14].

In the Standard Model both Γ_{had} and Γ_{ll} depend on the top-quark mass while the QCD correction only affects Γ_{had}. This prediction is shown in Fig. 9 for top-quark masses from $50\,\mathrm{GeV}$ to $250\,\mathrm{GeV}$. Our measurements are consistent with the Standard Model, favoring a larger value for the top-quark mass.

The ratios $\Gamma_{\mathrm{had}}/\Gamma_{ll}$ and $\Gamma_{\mathrm{inv}}/\Gamma_{ll}$ are almost independent of the top-quark and Higgs masses and therefore provides a powerful test of the Standard Model. The only uncertainty in the prediction comes from the QCD correction to the hadronic cross section. In Fig. 10 the correlation between the two ratios $\Gamma_{\mathrm{had}}/\Gamma_{ll}$ and $\Gamma_{\mathrm{inv}}/\Gamma_{ll}$ is shown together with the prediction. Again the data are consistent with the Standard-Model prediction for 3 neutrinos, including QCD corrections.

374

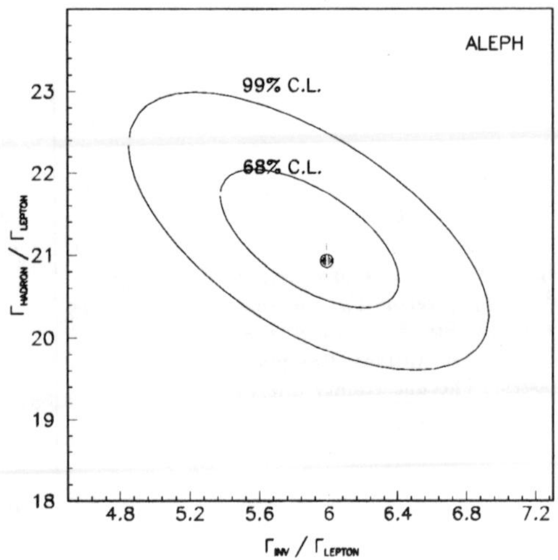

Figure 10: Contours of constant χ^2 for $\Gamma_{\rm had}/\Gamma_{ll}$ versus $\Gamma_{\rm inv}/\Gamma_{ll}$, together with the Standard-Model prediction. The uncertainty in the QCD correction is indicated by the error bar.

The leptonic partial width can be related to the effective weak mixing angle $\sin^2\theta_W$ in the following manner:

$$\Gamma_{ll} = (1+\kappa)\ \frac{\alpha(M_Z)\cdot M_Z\cdot\left[1+(1-4\sin^2\theta_W(M_Z))^2\right]}{48\sin^2\theta_W(M_Z)\cos^2\theta_W(M_Z)}\,, \tag{10}$$

where $\kappa = (0.2\pm0.3)\%$ represents additional electroweak effects [15] and $\alpha(M_Z)$ is the QED coupling constant. From our measurement of Γ_{ll} we find

$$\sin^2\theta_W(M_Z^2) = 0.227 \pm 0.005\,. \tag{11}$$

This determination is insensitive to assumptions on the top-quark mass, the Higgs boson mass or the Higgs structure of the theory.

7. Conclusions

From the data collected by ALEPH in 1989 we have measured the mass of the Z boson with a precision of 0.04% and its partial widths into hadrons and leptons with a precision of 2%, obtaining

$$\begin{aligned}
M_Z &= (91.197 \pm 0.021_{\rm exp} \pm 0.030_{\rm LEP})\ {\rm GeV}\\
\Gamma_{\rm had} &= (1803 \pm 37)\ {\rm MeV}\\
\Gamma_{ll} &= (85.1 \pm 1.7)\ {\rm MeV}\\
\Gamma_{\rm inv} &= (508 \pm 33)\ {\rm MeV}\,.
\end{aligned} \tag{12}$$

The number of light neutrinos, $N_\nu = 2.97 \pm 0.16$, has been determined from the measured ratio $\Gamma_{\rm inv}/\Gamma_{ll}$. Using the measured value of Γ_{ll} we have determined $\sin^2\theta_W(M_Z^2) = 0.227 \pm 0.005$. This result is, within the experimental error, free from assumptions on the top-quark mass and on the Higgs structure of the theory.

All of these results are in agreement with the Standard Model and favor a large top-quark mass. They are also in good agreement with our previous measurements [14] and the measurements of the other LEP and SLC experiments [16].

Acknowledgements

The results presented here are some of the fruits of a cooperative effort of about 360 physicists from 30 institutions, forming the ALEPH collaboration. We wish to express our appreciation to the technical staffs of CERN and the outside institutions for their support in constructing ALEPH and to the LEP division for the construction and rapid commissioning of the accelerator.

References

[1] D. Decamp et al. (ALEPH Collab.), CERN-EP/90-25, submitted to Nucl. Inst. Meth.

[2] D. Decamp et al. (ALEPH Collab.), Phys. Lett. B **234** (1990) 209.

[3] M. Bengtsson and T. Sjöstrand, Phys. Lett. B **185** (1987) 435; G. Marchesini and B.R. Webber, Nucl. Phys. **B310** (1988) 461.

[4] H. Burkhardt, F. Jegerlehner, G. Penso and C. Verzegnassi, Z. Phys. **C43** (1989) 497.

[5] D.Y. Bardin et al., Monte Carlo group in *Proceedings of the Workshop of Z Physics at LEP*, CERN Report 89-08.

[6] F.A. Berends and R. Kleiss, Nucl. Phys. **B228** (1983) 737; M. Böhm, A. Denner and W. Hollik, Nucl. Phys. **B304** (1988) 687; F.A. Berends, R. Kleiss and W. Hollik, Nucl. Phys. **B304** (1988) 712; The computer program BABAMC, courtesy of R. Kleiss.

[7] S. Jadach and Z. Was, Comp. Phys. Commun. **36** (1985) 191; Monte Carlo Group in *Proceedings of the Workshop of Z Physics at LEP*, CERN Report 89-08 (1989) Vol. III.

[8] A. Borelli, M. Consoli, L. Maiani and R. Sisto, CERN-TH-5441 (1989).

[9] F.A. Berends et al., Z Line Shape group in *Proceedings of the Workshop of Z Physics at LEP*, CERN Report 89-08 (1989) Vol. I.

[10] F.A. Berends, G. Burgers and W.L. van Neerven, Nucl. Phys. **B297** (1988) 429; Nucl. Phys. **B304** (1988) 921; The computer program ZAPP, courtesy of G. Burgers.

[11] E.A. Kuraev, V.S. Fadin, Sov. J. Nucl. Phys. **41** (1985) 466; G. Altarelli, G. Martinelli, in *Physics at LEP*, CERN 86-02 (1986) Vol. I, 47; O. Nicrosini, L. Trentadue, Phys. Lett. B **196** (1987) 551.

[12] A. Hofmann, private communication, LEP note to appear.

[13] This value of R and its error are obtained as the prediction of the Standard Model [9] for $\alpha_s = 0$, multiplied by the QCD correction computed in G. D'Agostini, W. deBoer and G. Grindhammer, Phys. Lett. B **229** (1989) 160.

[14] D. Decamp et al. (ALEPH Collab.), Phy. Lett. B **231** (1989) 519; Phys. Lett. B **234** (1990) 399; Phys. Lett. B **235** (1990) 339.

[15] D.C. Kennedy, B.W. Lynn, J.C. IM and R.G. Stuart, Nucl. Phys. **B321** (1989) 83.

[16] G.S. Abrams et al. (MARK II Collab.), Phys. Rev. Lett. **63** (1989) 724, 2173; B. Adeva et al. (L3 Collab.), Phys. Lett. B **231** (1989) 509; Phys. lett. B **237** (1990) 136; L3 preprint #005, submitted to Phys. Lett. B; M.Z. Akrawy et al. (OPAL Collab.), Phys. Lett. B **231** (1989) 530; Phys. Lett. B **240** (1990); P. Aarnio et al. (DELPHI Collab.), Phys. Lett. B **231** (1989) 539; P. Abreu et al. (DELPHI Collab.), CERN-EP/90-31; CERN-EP/90-32.

Z^0 Mass and Width and the Number of Neutrinos

The OPAL Collaboration [1]

presented by: **Gordon J. VanDalen**

University of California at Riverside

Abstract

We report a measurement of the mass and total width of the Z^0 boson. Based on 25,801 hadronic decays at 11 energy points, we obtain a mass of $M_Z = 91.145\pm 0.022$ (exp)$\pm$ 0.030 (LEP) GeV, and a total width of $\Gamma_Z = 2.526\pm0.047$ GeV.

INTRODUCTION

We present here a measurement of the hadronic decays of the Z^0, based on the entire data sample collected in 1989 with the OPAL detector. This corresponds to a sixfold increase in luminosity compared with our previous publication [1]. An improved understanding of the luminosity monitor has led to a significant reduction in the systematic uncertainties of the measurements.

The cross sections were measured at 11 centre-of-mass energies between 88.28 and 95.04 GeV by repeatedly scanning across the Z^0 resonance. Measurements at nearby energies, within 10 MeV, were combined into one data point at the luminosity-weighted average energy. The fractional error in the energy of each beam was $3 \cdot 10^{-4}$, corresponding to 30 MeV in the centre-of-mass energy. The fractional point-to-point error in the energy was $1 \cdot 10^{-4}$.

THE OPAL DETECTOR

The data were recorded with the OPAL detector [3] at the CERN e^+e^- collider LEP. The detector consists of a vertex detector, a jet chamber, and a z-chamber positioned inside a solenoidal coil, which is surrounded by a time-of-flight counter array, a lead glass electromagnetic calorimeter with a presampler, an instrumented magnet return yoke serving as a hadron calorimeter and four layers of outer muon chambers. Forward detectors serve as a luminosity monitor.

The electromagnetic calorimeter, which played a central role in this analysis, consists of a cylindrical array and endcaps of lead glass blocks, covering the region $|\cos\theta| < 0.98$. The blocks each subtend a solid angle of approximately 40×40 mrad^2. The time-of-flight system (TOF) covers the region $|\cos\theta| < 0.82$. More details about the detector and the trigger have been given in [1].

For Monte Carlo studies the OPAL detector was simulated using a computer program [4], which includes the detector geometry and material as well as effects of resolution and efficiencies.

THE LUMINOSITY MEASUREMENT

The integrated luminosity of the colliding beams was determined by the measurement of small angle Bhabha scattering. The measurement used the forward detector, consisting of two identical elements placed around the beam pipe at either end of the central tracking chambers. Two components of this detector were used in a complementary manner: (i) a calorimeter provided a high statistics measurement of the relative luminosity at each energy; (ii) proportional tube chambers with a well defined acceptance provided the absolute luminosity calibration.

The energy resolution of the calorimeter was measured to be $19\%\sqrt{E}$. The acceptance of the calorimeter extends from 39 to 155 mrad, and is essentially complete in azimuth. The proportional tube chambers are positioned between the presampler and main sections of the calorimeter. These

[1] Birmingham, Bologna, Bonn, U.C.Riverside, Cambridge, Carleton, CERN, Chicago, Freiburg, Heidelberg, Queen Mary College, Birkbeck College, University College London, Manchester, Maryland, Montréal, NRCC-Ottawa, Rutherford, Saclay, Technion-Israel, Tel Aviv, Tokyo, Kobe, Brunel, Weizmann.

chambers each consist of a vertical, horizontal, and diagonal plane of proportional tubes of 1 cm^2 cross section. The tube chamber acceptance extends from 50 mrad to 135 mrad in polar angle, and covers 95% in azimuth. The positions of the tube chambers were surveyed to 1.0 mm and were checked with electron tracks measured in drift chambers in front of the calorimeter.

Absolute luminosity calibration is determined within a fiducial region extending from 58 mrad to 124 mrad in polar angle, and excluding azimuthal angles within 10 degrees of the horizontal and vertical planes. Particles in this angular region traversed less than 0.2 X_0 before reaching the forward detectors. The difference in the azimuthal angles between the two ends, was required to be $160° < \Delta\phi < 200°$, to reject background due to off-momentum beam particles. Finally, the average of the energies of the largest cluster in each end was required to be larger than 2/3 of the beam energy. The overall trigger efficiency was $99.0 \pm 0.3\%$ for the events selected by these cuts.

The acceptance of the luminosity monitor was determined using the BABAMC Monte Carlo program [5] to generate $e^+e^- \rightarrow e^+e^-(\gamma)$ events. A theoretical uncertainty of 1% was assigned to this calculation [6]. Throughout the fiducial region the measured average polar angles agreed well with the simulated distribution, which corresponds approximately to $1/\theta^3$, as shown in figure 1.

The sources of error in the absolute luminosity normalisation are detailed in table 1. The final error of 2.2% includes a statistical error of 0.7% corresponding to 17,379 events in this sample.

A check was made on the accuracy of the tube chamber luminosity calibration by using the shadow of the beam pipe support ring to establish a well defined inner edge for the calorimeter. A measurement using this method, which is independent of the tube chambers, agreed with the tube chamber measurement to within $0.8 \pm 2.6\%$.

The relative luminosity between points of different beam energy was measured using the main calorimeters only. Events were selected in which the average energy of the largest clusters seen in the main sections of each calorimeter exceeded 70% of the beam energy and $160° < \Delta\phi < 200°$. A total of 58,124 events were selected by these cuts, of which less than 0.1% were background.

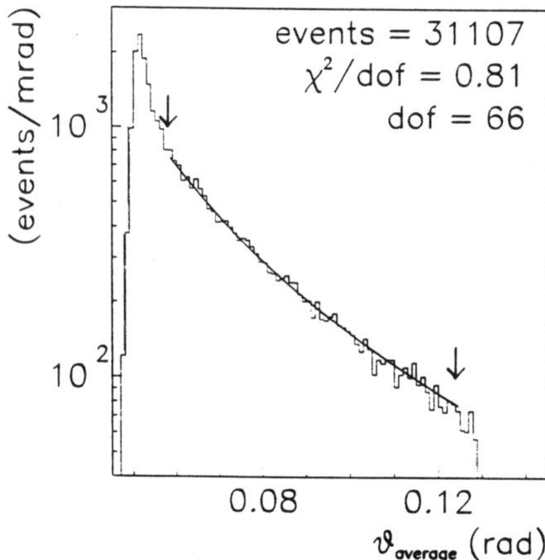

Source of Error	Magnitude
Theory	1%
Radial survey	1.5%
Efficiency of tube chambers	0.5%
Trigger efficiency	0.3%
Energy cut	0.5%
Radial cut	0.5%
Background subtraction	0.5%
Total systematic error	2.1%
Statistical error	0.7%
Total uncertainty	2.2%

Figure 1: Angular distribution of luminosity event candidates, as measured by the tube chambers. The arrows show the bounderies of the fiducial region, inside of which the data follow the expected $1/\theta^3$ distribution.

Table 1: Contributions to the absolute luminosity measurement uncertainty. The systematic errors contributing to the overall uncertainty are listed and then combined with the statistical error.

The energy calibration of the main calorimeter was maintained to within 0.5% over the entire period of data taking, so that the luminosity defined by these events was stable with time to within

0.8%. The calorimeter luminosity measurement was compared to the tube chamber acceptance for 70% of the data taking period. No statistically significant systematic differences were observed. The beam energy dependent systematic error of the integrated luminosity was estimated to be 1% and was taken into account as a point-to-point systematic error in the cross section measurements.

THE HADRONIC DECAYS

The criteria used to select hadronic Z^0 decays were nearly identical to those described in our previous publication [1]. The selection was mainly based on energy clusters in the electromagnetic calorimeter. Clusters in the barrel region were required to have an energy of at least 100 MeV, and clusters in the end cap were required to contain at least two adjacent lead glass blocks and have an energy of at least 200 MeV. The following three requirements defined a multihadron candidate: (i) at least 8 clusters, (ii) a total energy deposited in the lead glass of at least 10% of the centre-of-mass energy

$$R_{vis} = \Sigma E_{clus}/\sqrt{s} > 0.1,$$

where E_{clus} is the energy of each cluster, and (iii) an energy imbalance along the beam direction

$$R_{bal} = \mid \Sigma(E_{clus} \cdot \cos\theta) \mid /\Sigma E_{clus} < 0.65.$$

The measured distributions of these variables are shown in figure 2. The cut on the number of clusters efficiently eliminated Z^0 decays into e^+e^- and $\tau^+\tau^-$. The R_{vis} cut discarded two-photon and beam-gas events. The cut in R_{bal} rejected beam-wall, beam-gas and beam-halo events, and cosmic rays in the end caps.

In order to reject cosmic ray background information from the TOF counters was used. All events with at least 4 TOF counters fired within 8 ns of the expected time were accepted. Events with less than 4 TOF counters for which at least 50% of the observed energy was seen in the barrel lead glass were rejected. All remaining events with $N_{TOF} < 4$ were visually inspected. The background from cosmic-rays, beam-wall and beam-halo was estimated to be less than 0.1%. A total of 25,801 hadronic Z^0 decays remained after all these cuts, corresponding to an integrated luminosity of 1.25 pb^{-1}.

The main contamination in the hadronic data sample came from $\tau^+\tau^-$ events; with a background fraction of $0.33 \pm 0.04\%$ estimated using Monte Carlo, and checked with the data for low charged track multiplicity. The contribution of $e^+e^- \rightarrow e^+e^-$ to the hadronic event sample was found to be less than 0.1%. The background from two-photon processes was estimated by a Monte Carlo calculation using a quark-parton model [7] and was checked by measuring the ratio of the numbers of events with high and low R_{vis} as function of beam energy. Both estimates gave consistent results; the background was $0.03 \pm 0.03\%$ under the Z^0 peak and $0.2 \pm 0.2\%$ in the tail region.

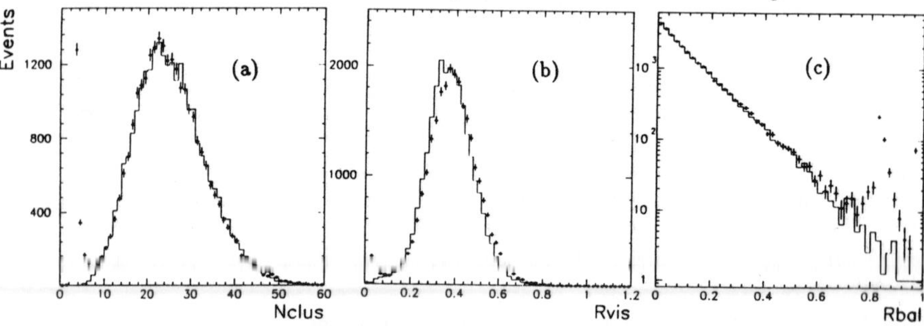

Figure 2: Distributions of variables used in the selection of hadronic events, each shown after applying all other cuts: (a) N_{clus}, (b) R_{vis}, (c) R_{bal}. QCD parton shower model distributions are also shown.

To calculate the acceptance of the event selection procedure, the process $e^+e^- \rightarrow q\bar{q}$ with subsequent hadronisation was simulated using the JETSET parton shower model with five flavours and string fragmentation [8]. There is good agreement between the model predictions and the properties of the hadronic event sample (see for example figure 2 and reference [9]). An acceptance of 97.7% with a negligible statistical error was calculated for multihadronic events. The effect of uncertainties in the fragmentation model was estimated to be 0.5% by varying the parameters of the JETSET model and by using the HERWIG hadronisation model [10]. Imperfections in the detector simulation lead to a variation of the acceptance of 0.5%. The energy dependence of the acceptance in the region of the scan was 0.2%. The error due to the specific choice of the hadronic selection cuts was estimated to be 0.2%. The resulting total systematic error on the selection and acceptance of hadronic events was 0.8%.

The overall trigger efficiency for hadronic decays of the Z^0 that would pass the acceptance criteria was determined from the redundancy between the three main trigger modes: the TOF trigger, the track trigger and the electromagnetic energy trigger. The overall trigger inefficiency was less than 0.1% and therefore also negligible.

In table 2 the numbers of events, the integrated luminosities and the corresponding cross sections are listed as a function of the centre-of-mass energy. The error on the cross section includes the statistical errors of the hadronic event sample and of the luminosity as well as a 1% point-to-point systematic error from the luminosity. In the analysis of the cross section an overall normalisation error of 2.3% is taken into account which includes both the systematic error of 2.2% on the luminosity and the systematic error on the acceptance of 0.8%.

$\sqrt{s}$ (GeV)	$\mathcal{L}_{\text{int}}(\text{nb}^{-1})$	N_{had}	σ_{had} (nb)
88.278	115.1±1.6	569	5.04±0.23
89.283	80.7±1.4	766	9.68±0.40
90.284	103.7±1.6	1990	19.56±0.56
91.034	210.9±2.3	6192	29.94±0.58
91.289	186.2±2.1	5633	30.86±0.62
91.529	230.8±2.4	6612	29.21±0.55
92.282	85.5±1.5	1781	21.24±0.66
92.562	9.2±0.5	150	16.66±1.62
93.286	111.4±1.7	1286	11.77±0.39
94.277	95.4±1.6	710	7.59±0.32
95.036	17.7±0.7	112	6.44±0.66
Total	1246.6	25801	

Table 2: The hadronic cross section, σ_{had}, the integrated luminosity $\mathcal{L}_{\text{int}}$ and the number of observed hadronic events N_{had}, as a function of the luminosity-weighted centre-of-mass energy, $\sqrt{s}$.

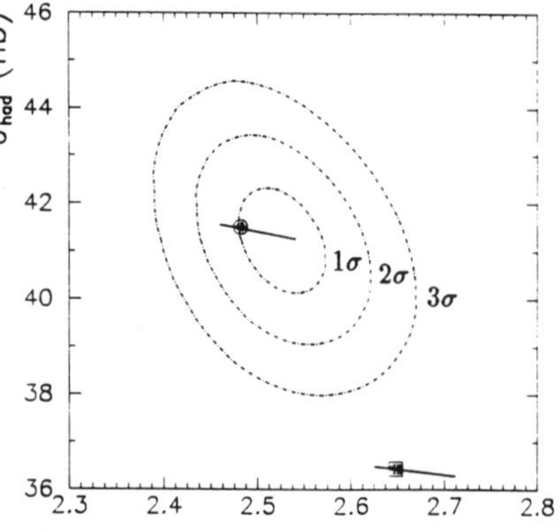

Figure 3: Confidence contours in the Γ_Z - $\sigma_{\text{had}}^{\text{pole}}$ plane. Standard Model predictions for 3 and 4 light ν generations are also shown.

ANALYSIS OF THE HADRONIC CROSS SECTION

To extract the Z^0 resonance parameters from the cross sections given in table 2 we perform a fit to the line shape parametrisation given in reference [11].

We perform a model independent fit to the data based on reference [11], treating M_Z, Γ_Z and $\sigma_{\text{had}}^{\text{pole}}$ as free parameters where $\sigma_{\text{had}}^{\text{pole}}$ represents the resonance hadronic cross section at $s = M_Z^2$ without photonic corrections. The parameter values obtained are

$$M_Z = 91.145 \pm 0.022 \text{ (exp)} \pm 0.030 \text{ (LEP) GeV}$$

$$\Gamma_Z = 2.526 \pm 0.047 \text{ GeV} \qquad \sigma_{\text{had}}^{\text{pole}} = 41.2 \pm 1.1 \text{ nb}.$$

380

The χ^2 value for this fit is 4.5 for 8 degrees of freedom. Figure 3 shows the central values and the confidence contours for our measurement of Γ_Z and $\sigma_{\text{had}}^{\text{pole}}$.

The Standard Model predictions for the pole cross section $\sigma_{\text{had}}^{\text{pole}}$, the width Γ_Z and the partial decay widths are a function of M_Z and depend on the values assumed for the top quark mass m_t, the Higgs mass m_H and the strong coupling constant α_s. In order to compare our measurement with the Standard Model predictions we calculate those assuming $m_t = m_H = 100$ GeV, and setting $\alpha_s = 0.12$. The errors on the predictions are derived by allowing a variation of m_t from 50 to 250 GeV, m_H from 20 to 1000 GeV and α_s from 0.09 to 0.15.

Figure 3 shows that the effect of varying the number of light neutrinos from 3 to 4 is large compared to the uncertainty in the fit values of $\sigma_{\text{had}}^{\text{pole}}$ and Γ_Z. The probability for obtaining our measurement assuming three neutrinos is 54% whereas the assumption of four neutrinos would result in a probability of only 0.001%.

The number of light neutrinos, N_ν, can be found by using the Standard Model in a fit where the only other free parameter is M_Z. The total width and hadronic pole cross section are given by

$$\Gamma_Z = \Gamma_{\text{had}}^{\text{SM}} + 3\Gamma_e^{\text{SM}} + N_\nu \Gamma_\nu^{\text{SM}} \quad \text{and} \quad \sigma_{\text{had}}^{\text{pole}} = \frac{12\pi}{M_Z^2} \frac{\Gamma_e^{\text{SM}} \Gamma_{\text{had}}^{\text{SM}}}{\Gamma_Z^2}.$$

Here, $\Gamma_{\text{had}}^{\text{SM}} = 1734^{+47}_{-21}$ MeV, $\Gamma_e^{\text{SM}} = 83.4^{+1.3}_{-0.6}$ MeV and $\Gamma_\nu^{\text{SM}} = 166.2^{+2.7}_{-0.7}$ MeV are the Standard Model predictions for the partial width for hadrons, electrons and for each light neutrino. The two parameter fit has a χ^2 of 5.5 for 9 degrees of freedom, and yields

$$M_Z = 91.141 \pm 0.022 \,(\text{exp}) \pm 0.030 \,(\text{LEP}) \text{ GeV}$$

$$N_\nu = 3.09 \pm 0.19 \,(\text{exp})^{+0.06}_{-0.12} \,(\text{theor}).$$

ACKNOWLEDGEMENTS

It is a pleasure to thank the LEP Division for the efficient operation of the machine, the precise information on the absolute energy, and their continuing cooperation with our experimental group.

References

[1] OPAL Collaboration, M.Z. Akrawy et al., Phys. Lett. B231 (1989) 530

[2] OPAL Collaboration, M.Z. Akrawy et al., Phys. Lett. B235 (1990) 379

[3] OPAL Technical proposal (1983) and CERN/LEPC/83-4, and
OPAL Collaboration, K. Ahmet et al., The OPAL Detector at LEP, to be submitted to Nucl. Instr. and Meth.

[4] J. Allison et al., Comp. Phys. Comm. 47 (1987) 55;
R. Brun et al., GEANT 3, Report DD/EE/84-1, CERN (1989).

[5] M. Böhm, A. Denner and W. Hollik, Nucl. Phys. B304 (1988) 687;
F.A. Berends, R. Kleiss, W. Hollik, Nucl. Phys. B304 (1988) 712

[6] R. Kleiss et al., Z Physics at LEP1, CERN 89-08, ed. G. Altarelli et al., Vol 3 (1989) 126

[7] F.A.Berends, P.H.Daverveldt and R.Kleiss, Comp. Phys. Comm. 40 (1986) 271, 285, 309.

[8] T. Sjöstrand, Comp. Phys. Comm. 39 (1986) 347; JETSET, Version 7.1.

[9] OPAL Collaboration, M.Z. Akrawy et al., Phys. Lett. B236 (1990) 364

[10] G. Marchesini and B.R. Webber, Nucl. Phys. B310 (1988) 461; HERWIG, Version 3.2.

[11] D. Bardin et al., Z Physics at LEP1, CERN 89-08, ed. G.Altarelli et al., Vol.1 (1989) 89.

ELECTROWEAK PHYSICS

Conveners: J. Dorfan, E. Eichten, S. Smith, & D. Treille

Recent Semileptonic Charm Decay Results from E691

Jean E. Duboscq
Dept. of Physics, UCSB
Santa Barbara, CA 93106
for The Fermilab Tagged Photon Spectrometer Collaboration (E691)

We present two results. The first is a direct measurement of the form factors in $D^+ \to \overline{K}^{*0} e^+ \nu_e$. The second is an upper limit on $B(D_s^+ \to \phi e^+ \nu_e)/B(D_s^+ \to \phi \pi^+)$. Using this along with a simple prediction for $B(D_s^+ \to \phi e^+ \nu_e)$ we obtain a lower limit on $B(D_s^+ \to \phi \pi+)$.

I. Form Factors in $D^+ \to \overline{K}^{*0} e^+ \nu_e$ decays:

The Feynman diagram for a spectator semileptonic decay is shown at right. In it, a meson (Qq) decays into (eν) via a virtual W and into a (q'q) meson. The differential decay rate is $d\Gamma \propto (G_F V_{Qq'} L_\mu H^\mu)^2 d(PS)$. The first three elements are relatively simple to write down. The hadronic current H^μ is were the difficulty lies, and is paramatrized by the use of form factors. These functions describe the evolution of the initial and final hadronic state overlaps as a function of momentum transfer in the decay.

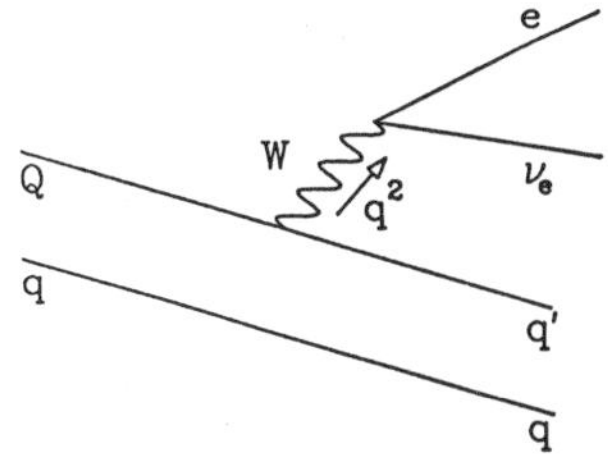

The number of different form factors in a decay depends on the relative spins of the hadrons involved. In the decay $D^0 \to K^- e^+ \nu_e$, there is only one relevant form factor, called f_+. In the decay $D^+ \to \overline{K}^{*0} e^+ \nu_e$, there are three form factors A_1, A_2, and V.

In Ref 1, our group reported a measurement of the branching ratio $D^+ \to \overline{K}^{*0} e^+ \nu_e$. Since then, we have been reanalyzing the same sample with greater Monte Carlo statistics in order to isolate the form factors A_1, A_2, and V. We report here on our initial results for these form factors at zero momentum transfer. In order to extract 3 functions from one decay, we use the fact that each form factor in the squared matrix element has a different coefficient multiplying it dependent on the $\overline{K}^{*0}$ spin, the virtual W spin and the mismatch of these two spins. There is also the particular q^2 dependence of each form factor. In the $\overline{K}^{*0}$ rest frame, we name the angle between the Kaon direction and the D direction θ_V. In the W rest frame, the angle between the e and the D direction is called θ_e. The angle between the (K^-, π^+) and the (e^+, ν_e) decay planes is termed χ. Lorentz invariance dictates that:

$$d\Gamma \propto cos^2\theta_V \, sin^2\theta_e \, |H_0|^2 + sin^2\theta_V \left((1+cos\theta_e)^2 |H_-|^2 + (1-cos\theta_e)^2 |H_+|^2 \right) + I.T.$$

H_+, H_-, and H_0 represent the helicity of the W and are functions of the form factors:

$$H_0 \propto 1/q \left((...)A_1(q^2) - (...)A_2(q^2) \right) \qquad H_\pm \propto e^{\pm i\chi} \left((...)A_1(q^2) \mp (...)V(q^2) \right)$$

where the (...) hide coefficients depending on the various masses and energies in the problem. $I.T.$ represents interference terms between these helicity states and, since we integrate out the χ dependence, this term can be ignored. As found in Ref 2, the form factor in $D^0 \to K^- e^+ \nu_e$ does have a particular q^2 dependence termed "nearest pole dominance." This is expected for all semileptonic decays, and so we take the same kind of q^2 evolution for A_1, A_2 and V: $A_{1,2} = A_{1,2}(0)/(1-q^2/M_A)$, $V = V(0)/(1-q^2/M_V)$ where $M_A = 2.11$ GeV and $M_V = 2.53$ GeV are $(c\bar{s})$ resonances. While the exact details of q^2 dependence are still being debated, this approximation should suffice for our study since we are in a regime where $q^2 < 1 \ GeV^2 << M_{A,V}^2$.

In Ref 1, by examining only the θ_V dependence, we found that the ratio of longitudinal to transverse polarization of the K^* (Γ_L/Γ_T) in the decay was $2.4 + 1.7 - .9 \pm .2$, as opposed to various models' expectations of a value near 1. The rate for $D^+ \to \overline{K}^{*0} e^+ \nu_e$ was half that of $D^0 \to K^- e^+ \nu_e$. These rates were also expected to be equal. A detailed look at the form factors should help theorists understand this and other semileptonic decays.

The individual events from the original $\overline{K}^{*0} e^+ \nu_e$ analysis were binned into a 3 dimensional θ_V, θ_e, q^2 space (as mentionned above, we integrate out the χ dependence.) We then threw many monte carlo phase space weighted $D^+ \to \overline{K}^{*0} e^+ \nu_e$ events to account for our detector's acceptance. Each such event is characterized by its value of θ_V, θ_e and q^2. The maximum likelihood fit is a 3 step algorithm: a) pick values of $A_1(0)$, $A_2(0)$, $V(0)$. b) for each MC event with (θ_V, θ_e, q^2), form the weight of the event according to the matrix element and the above form factors. c) Form the likelihood of the resulting MC distribution reproducing the data about each data point. Repeat a), b), c), until this likelihood is maximized (this method reproduces the results of the 1d fit in ref 1.)

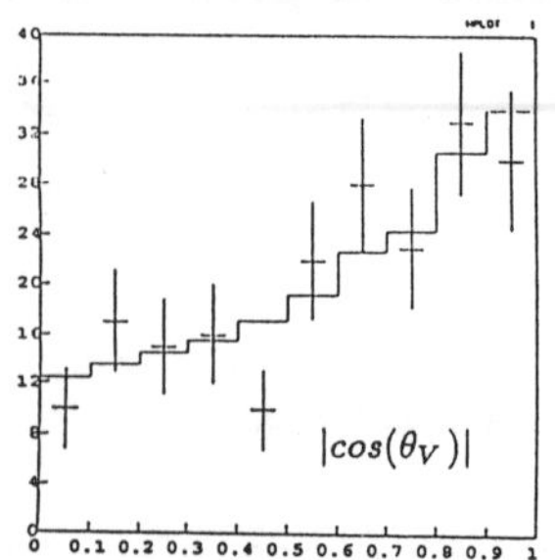

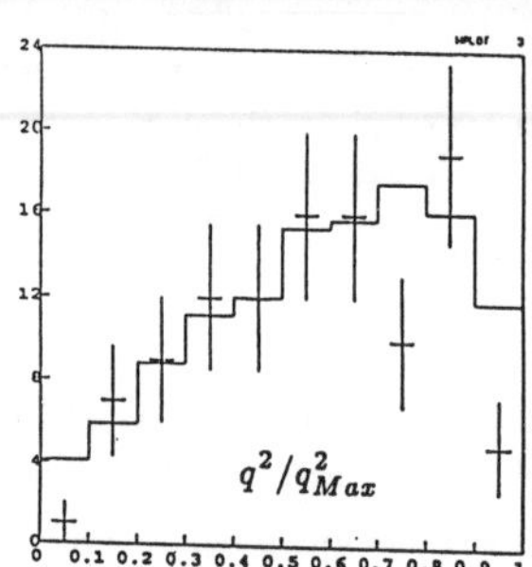

Above right is shown the projection of our fit onto $|cos(\theta_V)|$. The smooth curve is the Monte Carlo. At first glance, this fit seems less good than the PRL fit. Remember however that in Ref 1 we fit to only one variable while here we do a 3-d fit and project out. Above left is a projection with cuts on θ_V and θ_e that emphasize regions where $|H_\pm(q^2)|$ dominate the $|H_0(q^2)|$ contribution.

Below are our preliminary results, along with a sample of theoretical estimates:

Form Factor	E691	ISGW[3,a]	BSW[4]	GS[5,a]	KS[6]
$A_1(0)$	$.50 \pm .05 \pm .05$	.8	.9	.8	1.
$A_2(0)$	$.00 \pm .25 \pm .10$	.9	1.15	.6	1.
$V(0)$	$.93 \pm .40 \pm .09$	1.1	.9	1.	1.

An added bonus of this fit is that we obtain a new and improved value of $\Gamma_L/\Gamma_T = 1.8 \pm .5$ (the theoretical estimates are respectively 1.09, .91, 1.2 and 1.2) There are two notable facts here. First, A_2 is about zero. Recall that $H_0 \propto (...)A_1 - (...)A_2$, and that Γ_L depends on H_0. Hence, with a vanishing A_2, we can have a large H_0 and large Γ_L/Γ_T. Second, very roughly speaking, the three form factors are almost $1/\sqrt{2}$ times their predicted values. Hence the mismatch between the predicted and observed decay rates.

II. Study of $D_s \to \phi e \nu$:

Due to the lack of any knowledge of the absolute branching fractions of the D_s, all D_s decays are normalized to the mode $D_s^+ \to \phi\pi^+$. Thus, if one were able to measure a mode whose branching fraction is theoretically predictable, one could obtain absolute numbers for all seen D_s decays. Since leptons do not interact strongly, the best candidate for a reliable calculation of a decay rate are semileptonic decays. In this spirit, we present an analysis in which we set an upper limit on the ratio of branching fractions $B(D_s^+ \to \phi e^+ \nu_e)/B(D_s^+ \to \phi\pi^+)$, estimate what $B(D_s^+ \to \phi e^+ \nu_e)$ should be, and thus derive a lower limit on the branching fraction of $D_s^+ \to \phi\pi^+$.

This analysis is similar in many respects to that reported in Ref 1 for $D^+ \to \overline{K}^{*0} e^+ \nu_e$. In contrast to the $\overline{K}^{*0}$ which decays into a (K^-,π^+) pair, the ϕ decays into a (K^+,K^-) pair, and also has a very sharp mass distribution. Thus by simply identifying one Kaon and looking for a particle that is not a confirmed pion we have a pool of events that should be enriched with $D_s^+ \to \phi e^+ \nu_e$ decays.

Possible backgrounds include $D^+ \to \overline{K}^{*0} e^+ \nu_e$ $K^* \to K^- \pi^+$ events. By using our previous measurement of this mode and our well tested monte carlo, we can estimate the size of this feedthrough and find that it should be the dominant background. Its contribution to the ϕ mass region (1.012 GeV to 1.027 GeV) is however very small. That this is our background can be verified by requiring that the K and the e have the same sign eletrical charge. In this case there should be very little $D^+ \to \overline{K}^{*0} e^+ \nu_e$ feedthrough which is what we find. This sample is not used however because the decrease in background under the phi does not increase the significance of the observed number of phis sufficiently.

The resultant (KK) mass plot (not shown here) contains a very obvious and large ϕ peak. We must establish whether these ϕs all come from D_s decays. We found that the χ^2/dof of the decay vertex for these events does not track our $D_s \to \phi e \nu$ monte carlo, nor does it follow the geometrically similar $D^+ \to \overline{K}^{*0} e^+ \nu_e$ signal or monte carlo. Also, the proper decay time distribution of these events shows that many have lifetimes much greater than the D_s (.48 ps). Hence, we conclude that not all of our ϕs come from $D_s \to \phi e \nu$ decays. At right is the mass plot for events with lifetimes below 0.8 ps. Evidently there are very few events in the phi peak. Below it, we plot the proper decay time of the 11 events between 1.012 GeV and 1.027 GeV. The dashed curve is the monte carlo distribution of a pure signal of 11 events, while the solid line is the best fit to a D_s signal and an exponentially decreasing background distribution (where the lifetime is allowed to float.) From this fit, we find 0 $D_s \to \phi e \nu$ events with less than 7.8 events at a 90% confidence level. Factoring in our efficiency of $0.2 \pm .02\%$, and a total of 8900 ± 1000 $D_s \to \phi\pi$ events, we find that:

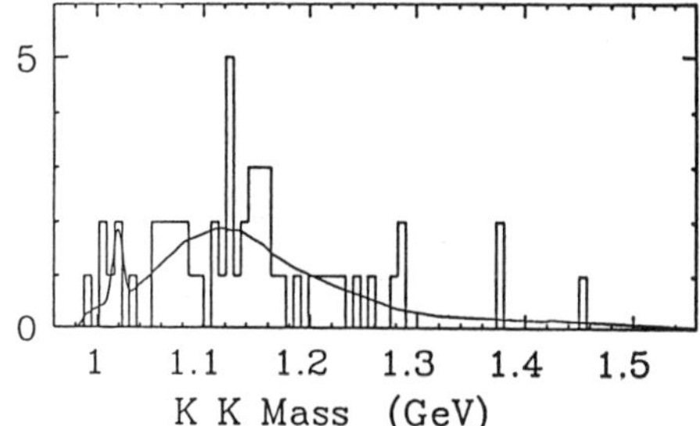

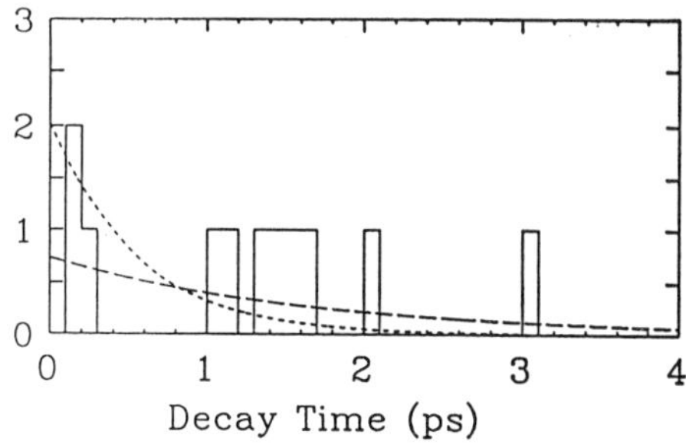

$$B(D_s^+ \to \phi e^+ \nu_e)/B(D_s^+ \to \phi\pi^+) < .45 \text{ at } 90\% \text{ C.L.}$$

The next ingredient in our analysis is an estimate of what $B(D_s \to \phi e \nu)$ should be. By SU(3) symmetry, the rates of $D^+ \to \overline{K}^{*0} e^+ \nu_e$ and $D_s^+ \to \phi e^+ \nu_e$ should be approximately equal. Hence by scaling $D^+ \to \overline{K}^{*0} e^+ \nu_e = 4.5 \pm .7 \pm .3\%$, we expect $B(D_s^+ \to \phi e^+ \nu_e) = 1.7 + / - .3\%$ (we have included an extra 11 % error on the decay rate for scaling the form factors between the two decays.) Using the above measured number, we obtain:

$$B(D_s^+ \to \phi \pi^+) > 3.4\% \text{ at } 90\% \text{ C.L.}$$

Note that MARKIII has found, by looking for $D_s^{*-} D_s^+$ events, that $B(D_s \to \phi \pi) < 4.1\%$, while CLEO has found $B(D_s \to \phi \pi) = 2 \pm 1\%$ by subtracting out D^0, D^+, Λ_c production from continuum charm production in $e^+ e^-$ collisions. Our number rules out only the lower end of this last value and is consistant with the MARKIII upper limit.

In summary, we have directly measured the 3 form factors in $D^+ \to K^* e \nu$ decays, have set a limit of $B(D_s \to \phi e \nu)/B(D_s \to \phi \pi) < .45$, and obtained $B(D_s \to \phi \pi) > 3.4\%$.

REFERENCES

(1) J.C. Anjos *et al.*, Phys. Rev. Lett. **62**, 722 (1989).

(2) J.C. Anjos *et al.*, Phys. Rev. Lett. **62**, 1587 (1989).

(3) N. Isgur and D. Scora, Phys. Rev. D **40**, 1491 (1989).

(4) M. Bauer and M. Wirbel, Z. Phys. C **42**, 671 (1989).

(5) F. J. Gilman and R. L. Singleton, Jr., Phys. Rev. D **41**, 142 (1990).

(6) J. G. Korner and G. A. Schuler, Z. Phys. C **38**, 511 (1988).

(7) J. Adler *et al.*, "An Upper Limit on the Absolute Branching Fraction for $D_s \to \phi \pi^+$", SLAC PUB 5044 (1989), submitted to Phys Rev Lett.

(8) W.Y. Chen *et al.*, Phys. Lett. B **226** , 192 (1989).

(a) These form factors are calculated at $q^2 = q^2_{max}$ and extrapolated down to $q^2 = 0$. using nearest pole dominance.

HADRONIC CHARMED MESON DECAY RESULTS FROM E691

A. BEAN for <u>Fermilab Tagged Photon Spectrometer Collaboration (E691)</u>

We report a series of results on the hadronic decays of charmed mesons from the charm photoproduction experiment Fermilab E691. These include a limit on the decay $D_s^+ \to \eta' \pi^+$ which is much lower than previously reported measurements.[1] We also have made a series of measurements on $D \to K4\pi$ modes, including $D^0 \to \overline{K^{*0}}\eta$ and $D^0 \to \overline{K^{*0}}\omega$. These are among the largest modes previously unmeasured. We also report a number of Cabibbo-suppressed D decays.

The D samples used in this study were obtained from the analysis of the full data sample of 10^8 events from Fermilab phoproduction experiment E-691.[2] This experiment collected data using the Tagged Photon Spectrometer, which is a large acceptance two-magnet spectrometer with silicon microstrip detectors (SMDs), drift chambers, Cerenkov counters and both electromagnetic and hadronic calorimetry. Each of the candidate charged decay products was required to traverse both spectrometer magnets and to have a momentum of at least 2.5 GeV/c. We required the charged tracks to give a good fit to a single secondary vertex with the total momentum of the secondary vertex products pointing back to the primary vertex. A cut on the significance of the separation between the primary and secondary vertices was applied as well as an isolation cut requiring no other tracks near the secondary vertex. A Cerenkov cut was applied for charged particle identification while the K_s^0 was identified by finding a two prong vertex in the downstream drift chambers. In addition, for the D^0 modes discussed the decay was required to be consistent with $D^{*+} \to D^0\pi^+$.

Figure 1 shows the invariant $K_s^0\pi^+\pi^+\pi^-\pi^-$ mass spectrum. The curve shown represents the result of a fit which gives a signal of 6.2 ± 2.1 events. One finds the branching ratio shown in Table 1. For the decay $D^+ \to K^-\pi^+\pi^+\pi^+\pi^-$, the invariant mass is shown in Figure 2. The fit found 113 ± 12 events over a background of 46 events. The relative and absolute branching fractions are given in Table 1. The absolute branching fractions in this paper are calculated using information in reference 3. The resonant subcomponent fractions found are shown in Table 2.

For the $D^0 \to K^-\pi^+\pi^+\pi^-\pi^0$ channel, the analysis used the satellite peak, a peak about 200 MeV/c^2 in width in the $K^-\pi^+\pi^+\pi^-$ mass spectrum when the π^0 is ignored. The $K^-\pi^+\pi^+\pi^-$ mass spectrum is shown in Figure 3, showing a clear peak at the D^0 mass with 427 ± 15 events. The fit also included terms for: the $K^-\pi^+\pi^+\pi^-\pi^0$ signal which has a sharp turn on below 1.72 GeV/c^2, radiative $D^0 \to K^-\pi^+\pi^+\pi^-\gamma$ decays, and background. There is a large unambiguous signal from the $D^0 \to K^-\pi^+\pi^-\pi^+\pi^0$ decay containing 180 ± 16 events. The branching ratios found are given in Table 1. Since the $K^-\pi^+\pi^-\pi^+\pi^0$ mode is much larger than the two $K4\pi$ channels discussed above, it is natural to think that this might be due to two-body modes such as $\overline{K^{*0}}\eta$ and $\overline{K^{*0}}\omega$. In the $D^0 \to K^-\pi^+\pi^+\pi^-\pi^0$ sample we searched for resonant contributions from $\overline{K^{*0}} \to K^-\pi^+$ and $\eta, \omega \to \pi^+\pi^-\pi^0$. The results of these searches are shown in Table 2. The branching fractions at the 1% level found for the $D^0 \to \overline{K^0}\pi^+\pi^+\pi^-\pi^-$ and $D^+ \to K^-\pi^+\pi^+\pi^+\pi^-$ are consistent with the theoretical expectations.[4] The larger branching fraction for the $D^0 \to K^-\pi^+\pi^+\pi^-\pi^0$ could be due to the possible inclusion of the vector $\omega \to \pi^+\pi^-\pi^0$ decay although our results limit this contribution in the two-body $\overline{K^{*0}}\omega$ channel. The $D^0 \to \overline{K^{*0}}\eta$ limits strongly constrain the theoretical predictions.

The $D^0 \to K_s^0 K^-\pi^+$ sample (as distinct from the $D^0 \to K_s^0 K^+\pi^-$ sample) was divided into three exclusive sets according to consistency with the hypotheses $K^0\overline{K^{*0}}$, $K^{*+}K^-$, and non-resonant $K^0 K^-\pi^+$. The $D^0 \to K_s^0 K^+\pi^-$ sample was divided and fit similarly. The results of these fits are shown in Table 3. We can draw a fair conclusion that $D^0 \to K^+ K^{*-}$ is observed while $D^0 \to K^0\overline{K^{*0}}$ is not. We also observe no $D^0 \to K^+ K^{*-}$. This assymmetry between the branching ratios to $K^{*+}K^-$ and $K^+ K^{*-}$ distinguishes $W^+ \to$vector decays from spectator$\to$vector decays.[4,5] The $K^+ K^-\pi^+\pi^-$ data was divided into four mutually exclusive sets. The first contains those events with $\phi \to K^+ K^-$ candidates, and the remaining three contain those with two, one, or zero K^* candidates. The results of a maximum likelihood fit are collected in Table 4. There is a clear signal in the $\overline{K^{*0}}K^{*0}$ sample which may include some feedthrough from $K^*K\pi$ and non-resonant $KK\pi$ in addition to the $\overline{K^*}K^*$ contribution. The 2.1σ signal provides evidence for hadronic final state interactions.[4,5]

The measurement of $D_s^+ \to \eta'\pi^+$ was conducted by searching for the decay $D_s^+ \to \eta'\pi^+$, $\eta' \to \eta\pi^+\pi^-$, $eta \to \pi^+\pi^-\pi^0$. The kinematics of this decay chain allow us to search for a signature of the $D_s^+ \to \eta'\pi^+$ decay without reconstructing the π^0. The $\pi^+\pi^-\pi^+\pi^-$ mass from the η', $(M(4\pi))$, must be between 0.56 and 0.82 GeV, and the $\pi^+\pi^-\pi^+\pi^-\pi^+$ mass spectrum $(M(5\pi))$ from $D_s^+ \to \eta'\pi^+$ events is peaked near the upper limit of 1.83 GeV. To extract a possible signal, a two-dimensional fit was performed using $M(4\pi)$ and $M(5\pi)$. The fit yielded 5.8 ± 3.9 D_s^+ events and 7.0 ± 4.5 D^+ events. The projection of data and fit onto the $M(5\pi)$ axis is shown in Figure 5. We find $B(D_s^+ \to \eta'\pi^+)/B(D_s^+ \to \phi\pi^+) = 0.7 \pm 0.5 \pm 0.4 \leq 1.6$. This measurement is consistent with theoretical predictions[4] but contrasts with previous experimental measurements.[1]

REFERENCES

(1) G. Wormser *et al.*, Phys. Rev. Lett. **61**, 1057 (1988) and G. Wormser, International Symposium on Heavy Quark Physics, Ithaca, NY (1989).

(2) J.C. Anjos *et al.*, Phys. Rev. Lett. **58**, 311 (1987). J.R. Raab, Ph.D. thesis, Univ. of California at Santa Barbara (1987, unpublished). T.E. Browder, Ph.D. thesis, Univ. of California at Santa Barbara (1988, unpublished).

(3) J. Adler *et al.*, Phys. Rev. Lett. **60**, 89 (1988).

(4) L.L. Chau and H.Y. Cheng, Phys. Rev. **D36**, 137 (1987); B.Y. Blok and M.A. Shifman, Sov. J. Phys. **45**, 522 (1987); M. Bauer, B. Stech, and M. Wirbel, Z. Phys. **C34**, 103 (1987); P. Rosen, Los Alamos preprint LA-UR-89-2175, July 1989; J.F. Donoghue, Phys. Rev. **D33**, 1516 (1986); A.N. Kamal, Phys. Rev. **D33**, 1344 (1986).

(5) X.-Y. Pham, Phys. Lett. **B193**, 331 (1987).

Table 1

Mode A	Mode B	B(Mode A)/B(Mode B)	B(Mode A)
$D^0 \to \overline{K^0}\pi^+\pi^+\pi^-\pi^-$	$D^0 \to \overline{K^0}\pi^+\pi^-$	$0.18 \pm 0.05 \pm 0.04$	$(1.2 \pm 0.4 \pm 0.3)\%$
$D^0 \to K^-\pi^+\pi^+\pi^-\pi^0$	$D^0 \to K^-\pi^+\pi^+\pi^-$	$0.57 \pm 0.06 \pm 0.05$	$(5.2 \pm 0.7 \pm 0.6)\%$
$D^+ \to K^-\pi^+\pi^+\pi^+\pi^-$	$D^+ \to K^-\pi^+\pi^+$	$0.09 \pm 0.01 \pm 0.01$	$(0.83 \pm 0.09 \pm 0.16)\%$

Table 2

Mode A	Mode B	Fraction of Mode B	B(Mode A)
$D^0 \to \overline{K^{*0}}\pi^+\pi^-\pi^0$	$D^0 \to K^-\pi^+\pi^+\pi^-\pi^0$	$0.30 \pm 0.10 \pm 0.10$	$(2.3 \pm 0.8 \pm 0.8)\%$
$D^0 \to \overline{K^{*0}}\eta$	$D^0 \to K^-\pi^+\pi^+\pi^-\pi^0$	$0.00 \pm 0.02 \pm 0.02$	$< 1.4\%$
$D^0 \to \overline{K^{*0}}\omega$	$D^0 \to K^-\pi^+\pi^+\pi^-\pi^0$	$0.00 \pm 0.10 \pm 0.10$	$< 2.3\%$
$D^+ \to \overline{K^{*0}}\pi^+\pi^-\pi^+$	$D^+ \to K^-\pi^+\pi^+\pi^+\pi^-$	$0.83 \pm 0.08 \pm 0.15$	$(1.0 \pm 0.2 \pm 0.3)\%$
$D^+ \to \overline{K^{*0}}\rho^0\pi^+$	$D^+ \to \overline{K^{*0}}\pi^+\pi^+\pi^-$	$0.75 \pm 0.17 \pm 0.19$	$(0.8 \pm 0.2 \pm 0.3)\%$

Table 3

Decay Mode f	$\dfrac{B(D^0 \to f)}{B(D^0 \to K^-\pi^+)} \times 100$	Absolute Branching Ratio (%)
$\overline{K^{*0}}K^0$	$.000^{+.020}_{-.000}$	$< .11$
$K^{*+}K^-$	$.160^{+.077}_{-.063}$	$.70^{+.33}_{-.27} \pm .15$
$(K^0 K^-\pi^+)_{nr}$	$.056^{+.070}_{-.056}$	$.23^{+.29}_{-.23} \pm .05$
$K^{*0}\overline{K^0}$	$.000^{+.040}_{-.000}$	$< .24$
K^+K^{*-}	$.000^{+.026}_{-.000}$	$< .14$
$(\overline{K^0}K^+\pi^-)_{nr}$	$.101^{+.056}_{-.050}$	$.43^{+.23}_{-.21} \pm .09$

Table 4

Decay Mode f	$\dfrac{B(D^0 \to f)}{B(D^0 \to K^-\pi^+\pi^-\pi^+)} \times 100$	Absolute Branching Ratio (%)
inclusive $K^-K^+\pi^-\pi^+$	3.6 ± 0.8	$.33 \pm .07 \pm .07$
$\phi\pi^+\pi^-$	$.83^{+.65}_{-.49}$	$.08^{+.06}_{-.04} \pm .02$
$K^{*0}\overline{K^{*0}}$	$3.7^{+2.2}_{-2.8}$	$.34^{+.20}_{-.17} \pm .07$
$K^{*0}K^-\pi^+ + \overline{K^{*0}}K^+\pi^-$	$1.63^{+2.26}_{-1.63}$	$< .41$
$(K^-K^+\pi^-\pi^+)_{nr}$	$0.85^{+1.90}_{-0.85}$	$< .31$

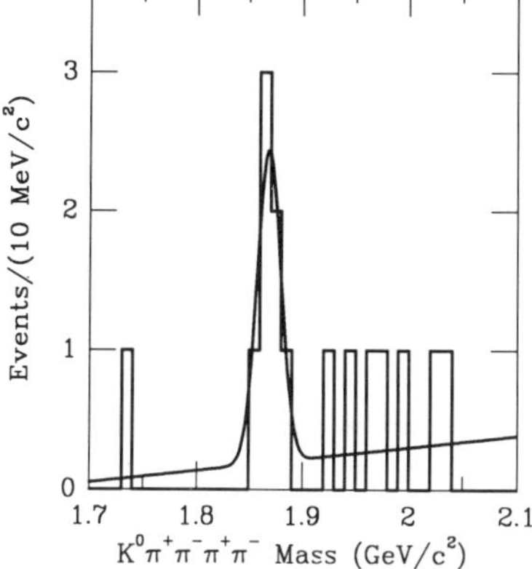

Figure 1. Invariant $K^0\pi^+\pi^+\pi^-\pi^-$ mass.

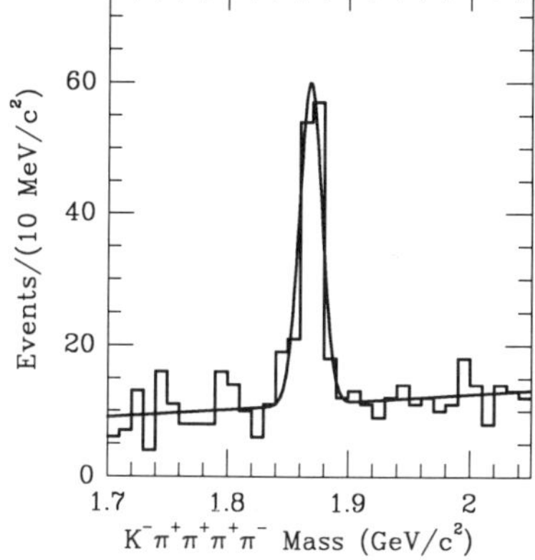

Figure 2. Invariant $K^-\pi^+\pi^+\pi^-\pi^0$ mass.

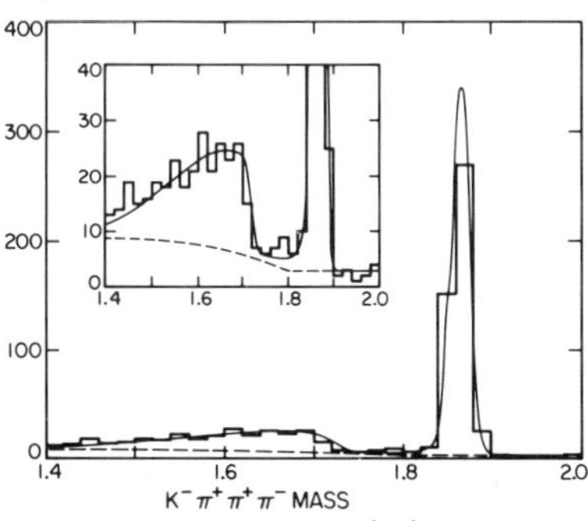

Figure 3. Invariant $K^-\pi^+\pi^+\pi^-$ mass. The solid line is the fit and the dashed line is the background contribution. The inset shows the same plot on a finer scale.

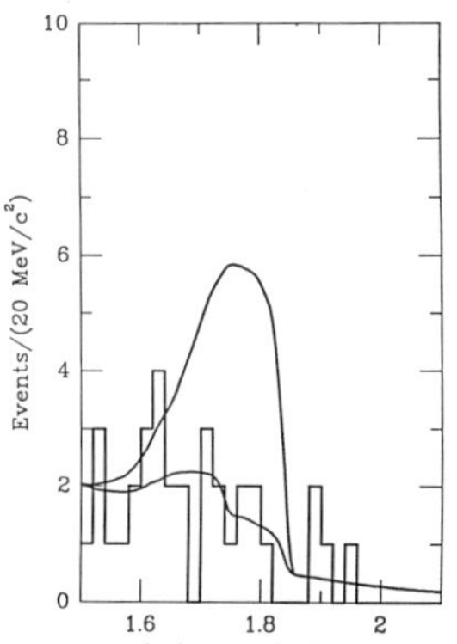

Figure 4. The $\pi^+\pi^-\pi^+\pi^-\pi^+$ mass spectrum for events with $\pi^+\pi^-\pi^+\pi^-$ mass less than 0.82 GeV.

Measurement of the Decays $D^0 \to \pi^- \pi^+$ and $D^0 \to K^- K^+$

William R. Ross

Department of Physics, Yale University
New Haven, Connecticut
06511
Representing the E691 Collaboration at Fermilab[1]

Abstract

If SU(3) flavor symmetry is preserved in weak decays, then the ratio of branching fractions of Cabibbo suppressed decays, $B(D^0 \to K^- K^+) \,/\, B(D^0 \to \pi^- \pi^+)$, should be on the order of unity. The central value of the existing world average of 3.5 (± 1.1) is hard to accommodate in present models, even those which include strong corrections to weak decays. A preliminary measurement of this ratio, giving 2.14 ± 0.49, as well as the relative branching fractions between Cabibbo suppressed and allowed modes, $B(D^0 \to K^- K^+) \,/\, B(D^0 \to K^- \pi^+)$ and $B(D^0 \to \pi^- \pi^+) \,/\, B(D^0 \to K^- \pi^+)$, is reported from Fermilab Experiment E691, based on signals of approximately 2400 $D^0 \to K^- \pi^+$ 70 $D^0 \to \pi^- \pi^+$ and 130 $D^0 \to K^- K^+$ events.

Introduction

The Cabibbo suppressed decays $D^0 \to K^- K^+$ and $D^0 \to \pi^- \pi^+$ proceed by external spectator diagrams. If there were perfect SU(3) flavor symmetry in the final state, then the amplitudes of these diagrams would be the same. The relative rate between the two decays would be determined by the difference in phase space. In that case, the ratio,

$$R(KK/\pi\pi) = \frac{B(D^0 \to K^- K^+)}{B(D^0 \to \pi^- \pi^+)}$$

would be .86. (The decays of the charge conjugate states are included implicitly throughout). The existing measurements of this ratio from the Mark II [2] (1979) and Mark III [5] (1985), experiments are given in Table 1. The weighted average is 3.5 ± 1.1. Although the error is large, the central value suggests a rather big SU(3) flavor symmetry breakdown. There are models for hadronic charm decays [7,8,10,11] which take into account the effect of strong interactions causing SU(3) breaking. Predictions [9,15] for $R(KK/\pi\pi)$ are in the range of 1.4 to 2.7. The discrepancy between the theoretical predictions and the central value of the measurement has remained a puzzle.

Apparatus and Analysis

The analysis presented here is based on the 10^8 photon-beryllium interactions recorded in Fermilab experiment E691. The major elements of the apparatus [14] were a silicon microstrip vertex detector, a magnetic spectrometer with drift chamber tracking, two multi-cell threshold Čerenkov counters, and a calorimeter measuring both electro-magnetic and hadronic energy deposition.

In the off-line analysis, events were selected which had candidates for secondary vertices measured to be more than 5.0 standard deviations downstream of the primary vertex. From these events, a further selection was made for events with secondary vertices with two oppositely charged tracks. The line of flight of the resultant D^0 candidate was determined from the summed 3-momentum of the two tracks and the position of the two-track vertex. Only those candidates were selected whose line of flight passed within 80 microns of the primary vertex.

The range of momenta of the two tracks was restricted to regions where particle identification was possible using the Čerenkov counters and where there was significant production and acceptance into our spectrometer.

Probabilities for particle identification were assigned to each charged track based on the light seen in the Čerenkov counter photo-tubes [6]. To extract the signals for $D^0 \to K^- K^+$ and $D^0 \to \pi^- \pi^+$, only those decay candidates were selected where both tracks had pion probabilities $> .84$ or where both tracks had kaon probabilities > 0.4.

Results

The resulting mass plots are shown in Figures 1 and 2. In the KK mass plot there is a clean peak at the D^0 mass. The number of signal events was determined by a least squares fit for a Gaussian signal and a linear background. The D^0 mass was fixed at the established value of 1.865 GeV/c^2 and the standard deviation, σ, of the Gaussian (due to the detector resolution) was fixed at 9.6 Mev/c^2 as determined from a Monte Carlo simulation. The fit gives 129 ± 14 signal events.

In the $\pi\pi$ mass plot shown in Figure 2, there are two prominent structures. The broad peak at about 1.75 GeV/c^2 is due to the decays $D^0 \to K^- \pi^+$ where the kaon is mistaken as a pion. There is also a Gaussian peak at the D^0 mass due to $D^0 \to \pi^-\pi^+$. The function used to fit the data includes terms with a Gaussian for the false peak, a term of the form $e^{A+Bm+Cm^2}$ for the combinatoric background and Gaussian signal with fixed mass and $\sigma = 13.0$ Mev/c^2. The fit gives a signal of 71 ± 14 $\pi\pi$ events.

To test the sensitivity of the signal estimate on the parameterization of the false peak and combinatoric background, different functional forms were tried without any significant ($< 1 \, \sigma$) change in the signal estimate. The form of the false peak is well understood from the large number of $D^0 \to K^-\pi^+$ events and from the Monte Carlo.

The calculation of the relative branching fractions between the three decay modes of interest ($K^-\pi^+$, $\pi^-\pi^+$, and K^-K^+) requires knowledge of the relative detection efficiency, ϵ, for each mode, eg.,

$$R(KK/\pi\pi) = \frac{N(D^0 \to K^-K^+)}{\epsilon(K^-K^+)} \frac{\epsilon(\pi^-\pi^+)}{N(D^0 \to \pi^-\pi^+)}.$$

The efficiencies are determined as the product of two contributions, eg.,

$$\epsilon(K^-K^+) = \epsilon_G(KK)\epsilon_C(KK)$$

where ϵ_G is due to the geometric acceptance and track reconstruction efficiency of the apparatus and ϵ_C is due to the Čerenkov cuts. The efficiencies $\epsilon_G(K\pi)$ and $\epsilon_C(K\pi)$ are unity by definition. The terms ϵ_G are estimated from the Monte Carlo simulation to be $.99 \pm .03$ for KK and $.86 \pm .03$ for $\pi\pi$. The difference between $\epsilon_G(KK)$ and $\epsilon_G(\pi\pi)$ is due to the different Q- values in the decays.

The efficiency of Čerenkov detection for pions and kaons, as a function of particle momentum, was determined from the signals for $D^{*+} \to \pi^+ D^0$, $D^0 \to K^-\pi^+$. These efficiencies agree with those determined from a Monte Carlo for momentum ranges where ϕ data could be used to tune the simulation parameters.

The overall Čerenkov efficiencies are, $\epsilon_C(\pi\pi) = .59\pm.03$ and $\epsilon_C(KK) = .44\pm.02$. The efficiency $\epsilon_C(K\pi) = 1$ since no Čerenkov cuts are applied. Combining the results for efficiency and the number of signal events, the relative branching fractions are given in Table 1. Using the corrected Mark III [16] (1988) result for the absolute branching fraction for $D^0 \to K^-\pi^+$, the absolute branching fractions for $D^0 \to K^-K^+$ and $D^0 \to \pi^-\pi^+$ are measured to be 0.48 ± 0.09 % and 0.23 ± 0.05 % respectively. For the E691 measurement, the statistical error is given first followed by the systematic error due to the uncertainty in the relative efficiencies. The results presented here are consistent with preliminary results [13] from ARGUS [3] and CLEO [12] as shown in Table 1. All three new measurements of $R(KK/\pi\pi)$ are smaller than the previous world average and are in a range compatible with existing models of charm decay. This gives more confidence in the ability of these models to incorporate the effects of SU(3) flavor symmetry breaking in charmed particle decays.

393

References

[1] The members of the E691 Collaboration are:
J.C. Anjos, J.A. Appel, A. Bean, S.B. Bracker, T.E. Browder,
L.M. Cremaldi, J.R. Elliott, C.O. Escobar, P. Estabrooks, M.C. Gibney,
G.F. Hartner, P.E. Karchin, B.R. Kumar, M.J. Losty, G.J. Luste,
P.M. Mantsch, J.F. Martin, S. McHugh, S.R. Menary, R.J. Morrison,
T. Nash, P. Ong, J. Pinfold, G. Punkar, M.V. Purohit,
W.R. Ross, A.F.S. Santoro, J.S. Sidhu, K. Sliwa,
M.D. Sokoloff, M.H.G. Souza, W.J. Spalding, M.E. Streetman, A.B. Stundžia,
M.S. Witherell

[2] G. S. Abrams et al., *Phys. Rev. Lett.* **43** (1979) 481.

[3] H. Albrecht et al., "A Study of Cabibbo Suppressed D^0 Decays", contributed paper to the XIV International Symposium on Lepton and Photon Interactions, Stanford Univ., 1989.

[4] J.C. Anjos et al, *Phys. Rev. Lett.* **62** (1989) 125.

[5] R.M. Baltrusaitis et al., *Phys. Rev. Lett.* **55** (1985).

[6] D. Bartlett et al., *Nucl. Inst. Meth.* **A260** (1987) p. 71.

[7] M. Bauer, B. Stech and M. Wirbel, *Z. Phys.* **C34** (1987) 103.

[8] I. Bigi, to appear in proceedings of *1989 International Symposium on Heavy Quark Physics*, Cornell University.

[9] A. Buras et al., *Phys. Lett.* **B268** (1986) 16.

[10] L.L. Chau and H-Y. Cheng, *Phys. Rev. Lett.* **56** (1986) 1655.

[11] L.L. Chau and H-Y. Cheng, *Phys. Rev.* **D 36** (1987) 137.

[12] CLEO Collaboration, "Study of Rare D^0 Decays", contributed paper to the XIV International Symposium on Lepton and Photon Interactions, Stanford Univ., 1989.

[13] P.E. Karchin, to appear in proceedings of the XIV International Symposium on Lepton and Photon Interactions, Stanford Univ., August, 1989, World Scientific, Singapore.

[14] J.R. Raab et al., *Phys. Rev.* **D 37**, 2391 (1988).

[15] K. Terasaki and S. Oneda, *Phys. Rev.* **D38** (1988) 132.

[16] J. Adler et al., *Phys. Rev. Lett.* **60** (1988) 89.

expt.	$K^-K^+/K^-\pi^+$	$\pi^-\pi^+/K^-\pi^+$	$K^-K^+/\pi^-\pi^+$
E691	$.116 \pm .012 \pm .008$	$.055 \pm .010 \pm .002$	$2.14 \pm 0.48 \pm 0.13$
Mark II	$.113 \pm .030$	$.033 \pm .015$	3.4 ± 1.8
Mark III	$.122 \pm .018$	$.033 \pm .010$	3.7 ± 1.4
ARGUS	$.10 \pm .02 \pm .01$	$.04 \pm .007 \pm .006$	2.5 ± 0.7
CLEO	$.11 \pm .008 \pm .007$	$.046 \pm .006 \pm .006$	2.2 ± 0.5

Table 1: Comparison of relative branching fractions.

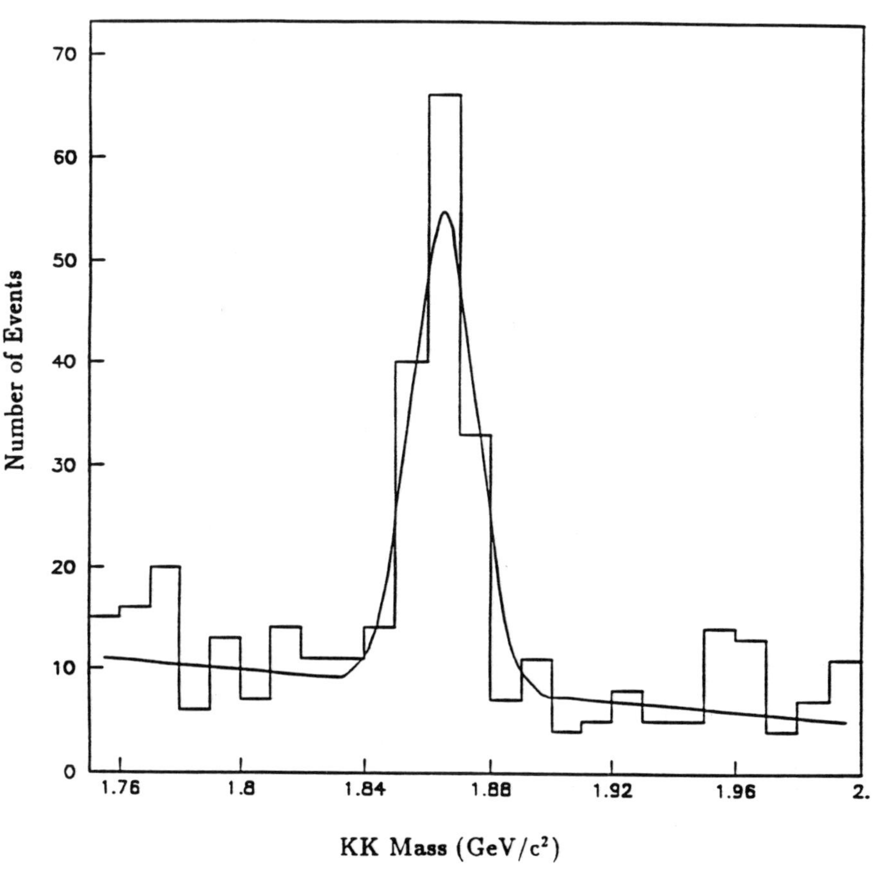

Figure 1: Signal for $D^0 \rightarrow K^-K^+$. The solid line is the fit to signal and background.

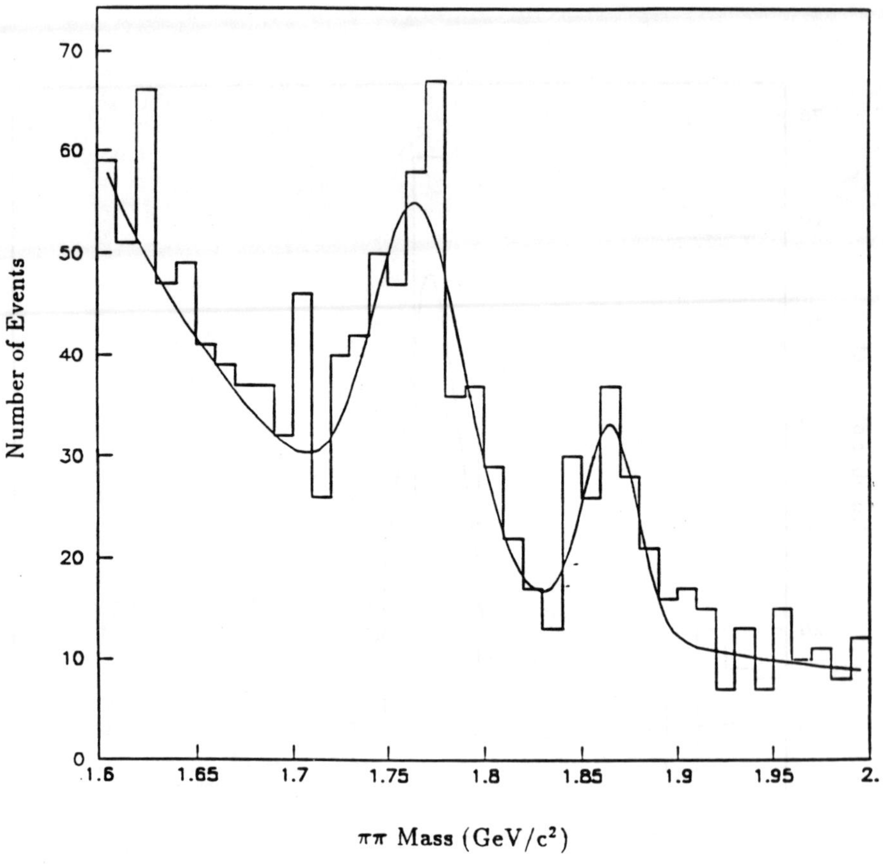

Figure 2: Signal for $D^0 \rightarrow \pi^-\pi^+$. The solid
line is the fit to signal and background.

Early Charm Results from E687[*]

J. R. Wilson

Department of Physics

University of Illinois, Urbana, IL, 61801

for the E687 COLLABORATION[†]

Abstract

Fermilab experiment E687 has observed sizable charm signals in several decay modes using a partial sample of a dataset which consists of 60 million high energy photon-Be triggers.

1 Introduction

E687 took data with the highest energy photon beam in the world during a run which began in July 1987. The general purpose spectrometer emphasized vertex detection and particle identification. We expect to reconstruct $\sim 10^4$ charm particles in the full data sample.

2 The Wide Band Photon Beamline

The Fermilab wide band beam line is a bremsstrahlung beam line with a large momentum bite (13%). Primary protons of 800 GeV were used to generate an intense beam of photons from π^0 decay. These photons passed through a radiator and the electrons produced were steered around a neutral beam dump and directed onto a 20% lead radiator. The scattered electron passed through a sweeping magnet and struck a recoil shower counter (RESH), the bremsstrahlung photons produced continued on to the experimental target. Photons which did not interact in the target deposited their energy in a beam dump calorimeter at the back of the spectrometer. Figure 1 shows the inferred energy distribution of photons impinging on our target, also shown is the acceptance for detecting the scattered electron in the RESH. The mean energy of the photon spectrum with the RESH requirement was about 220 GeV.

3 The E687 Spectrometer

The E687 spectrometer is a general purpose magnetic spectrometer with a high resolution silicon vertex detector, Cerenkov particle identification, muon tagging and electromagnetic calorimetry (the muon sytem and calorimetry will not be discussed here). Figure 2 is a plan view of the spectrometer, with an insert showing a blowup of the target region.

---

[*]Supported by US DOE and NSF and by INFN

[†]E687 COLLABORATION: INFN Bologna; University of Colorado; Fermilab; INFN Frascati; University of Illinois; University of Notre Dame; INFN Milano; University of Milan; Northwestern Universiy; INFN Pavia; University of Pavia; University of Puerto Rico — Mayaguez

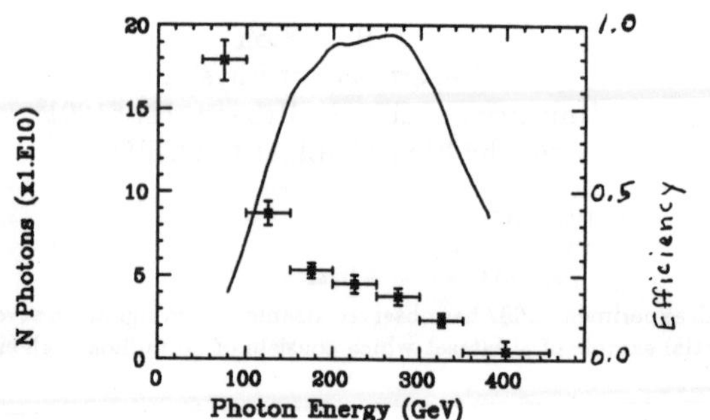

Figure 1: Photon energy spectrum (points, left scale). RESH acceptance (solid line, right scale).

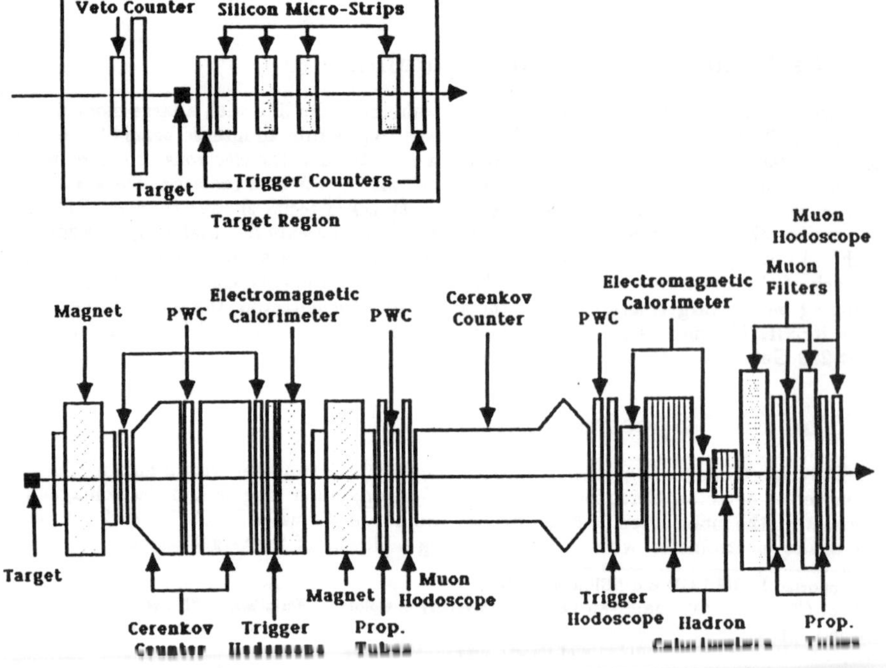

Figure 2: Plan view of the E687 spectrometer. The insert is a blowup of the target region. The spectrometer is approximately 35 meters long and the target region occupies about 40 cm.

The silicon microvertex detector (SSD) consisted of 12 planes, organized into 4 stations. Each station contained a plane with strips parallel to the y-axis of the spectrometer and two planes which were rotated by $\pm 45^0$ with respect to the first. The pitch of the planes in the first station was $25\mu m$, all other planes were $50\mu m$. The outer 2/3 of each plane had twice the pitch of the inner 1/3 (by area) which decreased the channel count by a factor of two. Since a large fraction of the tracks passed through the high resolution region, the pitch doubling had little effect on the average track resolution.

The resolution of the detector is best expressed as the transverse resolution of tracks extrapolated to the center of the target. This resolution function has two terms: an asymptotic term calculated from the geometry and a term which scales with momentum and is due to multiple coulomb scattering (MCS). The silicon in the SSD caused most of the scattering with small contributions from the 5% Be target and a trigger counter. In the y projection the resolution function looked like:

$$\sigma_y = 7.7 \ \mu m \quad \sqrt{1 + \left(\frac{25 \ GeV}{p} \right)^2} \tag{1}$$

In the x-view the asymptotic resolution is 11 μm and the MCS term is 17.5 GeV. These resolutions are verified in the data by looking at normalized impact parameters of single tracks with vertices.

We do our momentum analysis with two vertical bending dipole magnets (M1 and M2) and a system of 5 multiwire proportional chambers (P0-P4), in addition to the SSD system. These chambers each had 4 planes, oriented to measure x, y and $y\pm 11^0$, with wire spacings of $2mm$ in P0 and P3 and $3mm$ in P1, P2 and P4. The primary use of the chamber system was in determing the momentum of PWC tracks which linked to tracks found in the SSD. For tracks which traversed the whole spectrometer we obtained a momentum resolution of:

$$\frac{\sigma_p}{p} = 1.4\% \quad \left(\frac{p}{100 \ GeV} \right) \quad \sqrt{1 + \left(\frac{23 \ GeV}{p} \right)^2} \tag{2}$$

Again, the $1/p^2$ term is due to the contribution of multiple scattering to the resolution function. Tracks which only went through M1 had as asymptotic resolution of 3.4% and a MCS turnover point of 17 GeV. We check these resolutions by examining the difference in the momentum measured in each magnet normalized by the error prediction.

The final aspect that we are going to consider is the Cerenkov particle identification system. This system consisted of three gas filled threshold Cerenkov counters denoted as C1-C3 in Figure 2. The gas types and π thresholds of the three counters were: HeN_2, N_2O, He and 6.7, 4.4 and 17.0 GeV respectively.

We used a digital algorithm that treated each counter as either on, off, or confused for each charged track passing through its fiducial region. The patterns were then checked for consistency with each of the possible particle types: e, π, K, and P. Typically K's were required to be K or K/P ambiguous (consistent with both K and P hypothesis). This gave good identification up to a momentum of 60 GeV, the range could be extended to 115 GeV by allowing the track to be K/π/e ambiguous above 60 GeV. This is the K cut used in this analysis.

Cerenkov performance was checked through Monte Carlo and by looking at specific states in data with and without identification cuts (K_s, Λ, ϕ, and D^{*+}). K's were found to be correctly identified 80% of the time and the probability of misidentifying a π as a K was about 4%.

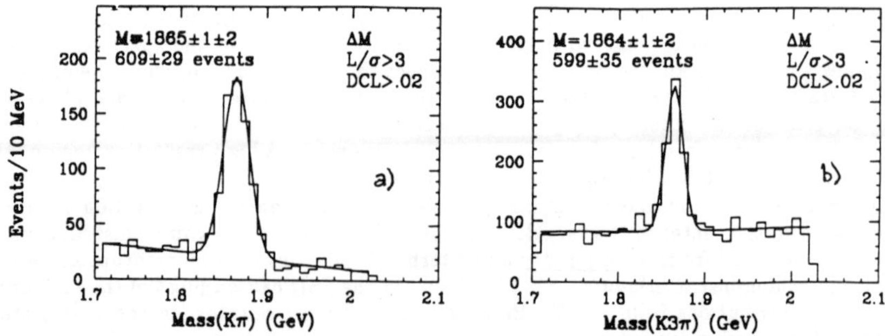

Figure 3: D^0 from $D*$ decay with cuts as described in the text. a) $D^0 \rightarrow K^-\pi^+$, b) $D^0 \rightarrow K^-\pi^+\pi^+\pi^-$

4   A Candidate Driven Vertex Algorithm

The signals presented in this paper were found using a candidate driven vertex algorithm (CDVA). This algorithm took the charm candidate tracks, fit them to form a secondary vertex (with a fit confidence level denoted DCL) and then formed a track vector for the candidate. This track vector was then used as a seed for finding the primary vertex. Basically, tracks were added to the primary vertex until a confidence level cut of 2% was failed. Full covariance matrices with multiple scattering contributions as decribed earlier were used in doing the vertex fits. We also performed a global fit to the candidate tracks which required the candidate to point back to the primary vertex taking correct cognizance of all errors. This fit produced a confidence level of the decay (denoted LCL) and a decay length with its associated error (L, σ_L). The candidate tracks were usually chosen kinematically with mild particle id. requirements. The CDVA itself was extremely efficient and unbiased until specific cuts were put on the confidence level (DCL, LCL) and L/σ_L.

5   Decays of the D^{*+}

We have obtained a sizable sample of $D^{*+} \rightarrow D^0\pi^+$ using two D^0 decay modes: $D^0 \rightarrow K^-\pi^+$ and $D^0 \rightarrow K^-\pi^+\pi^+\pi^-$. In all discussion of signals it should be assumed that we are including charge conjugate states. In Figure 3a we show the invariant mass plot for the $K^-\pi^+$ hypothesis, for 90% of our data, with the following requirements: 1) the track assigned as a K must be identified by the K cut defined above, 2) M(D^{*+}) - M(D^0) must be in the range 142.5—148.5 MeV, 3) The secondary vertex must have a good confidence level (4) The secondary and primary must be separated by $L/\sigma_L > 3$. Figure 3b is the same plot for the $K^-\pi^+\pi^+\pi^-$ decay mode. Gaussian fits have been superimposed on both histograms, with the mass coming in consistent with current world averages. We are reconstructing inclusive D^0's (D^* tag is not checked) in a ratio of about 4:1 to the D^* tagged events, this puts our projected sample of D^0's in the range of 4000-5000.

Since $K\pi$ mode is reasonably clean without any vertex requirements at all (just particle id.), we can use this signal to measure the performance of the vertex algorithm. The amount of signal lost by requiring a primary vertex to be found with no cuts on separation is about 10%. In Figure 4a we plot the number of D^*'s remaining after a separation

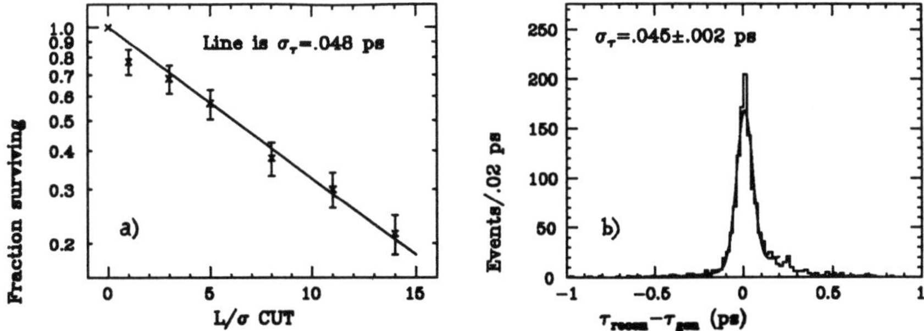

Figure 4: a) Number of $D^0 \to K^-\pi^+$ surviving an L/σ_L cut. b) Difference between generated proper time and reconstructed proper time in Monte Carlo events

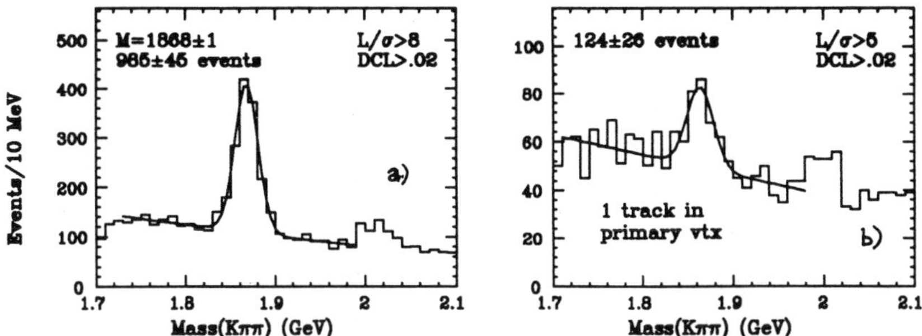

Figure 5: $D^+ \to K^-\pi^+\pi^+$ with cuts as described in the text. a) All D^+'s , b) D^+'s with only 1 track found in the primary vertex.

cut of N sigma. A line is superimposed which shows the effect of a resolution of $\sigma_L =$ .048ps assuming that the D^0 lifetime is .43ps. The second plot (4b) shows the difference between the generated lifetime and reconstructed lifetime for monte carlo events which should be a direct measurement of our lifetime resolution. Both of these techniques give results consistent with a lifetime resolution of $\approx$ .048ps.

6 Decay of the D^+

We have no mass difference tricks for the D^+ and must rely on lifetime cuts only, fortunately these have greater effect due to the longer lifetime of the D^+. In Figure 5a we show the decay $D^+ \to K^-\pi^+\pi^+$ for 30% of our data. This has the same cuts as the D^* plots except that $L/\sigma_L > 8$. To illustrate the power of the CDVA Figure 5b shows the D^+ signal for the case where only one other track in addition to the D^+ was found in the primary vertex. Projecting the D^+ yield to the full data sample and after accounting for known skim efficiency improvements we expect to see 3000-4000 D^+'s in the final sample.

A Preliminary Measurement of the Lifetime of the D_s^+ in Fermilab Experiment E687 *

John D. Cunningham
Department of Physics, University of Notre Dame, Notre Dame, IN 46556
The E687 Collaboration**

I. SIGNAL EXTRACTION - Nearly 60 million photoproduced events have been analyzed to extract the D_s^+ signal via the $\phi\pi^+$ mode in Fermilab Experiment E687. (Charge conjugate states are included.) The beamline and spectrometer are discussed elsewhere in these proceedings.[1] The data set was reduced to 400K events by retaining all events with K^+K^- pairs Cerenkov consistent with the kaon hypothesis. The D_s^+ signal was then extracted requiring that:

1. the invariant mass of the K^+K^- combination be between 1.01 Gev/c^2 and 1.03 Gev/c^2;

2. the primary vertex be in the target region;

3. the secondary vertices formed from the $K^+K^-\pi^+$ system have χ^2/DOF < 3;

4. the secondary vertex be downstream of the primary vertex and be separated from the primary vertex by at least a distance L such that $L/\sigma > n$. Here σ is the error in L and n is chosen to maximize the signal-to-noise ratio;

and 5. the angle θ between the K^+ and the π in the rest frame of the ϕ must be such that $|\cos(\theta)| > 0.3$.

In fig. 1 are shown four plots of the $\phi\pi$ invariant mass distribution with different values of n. Two prominent peaks are present: one corresponds to the D_s^+ and the other to the Cabibbo-suppressed decay of the D^+. Table I gives the fitted mass, width, and number of events for n > 3.

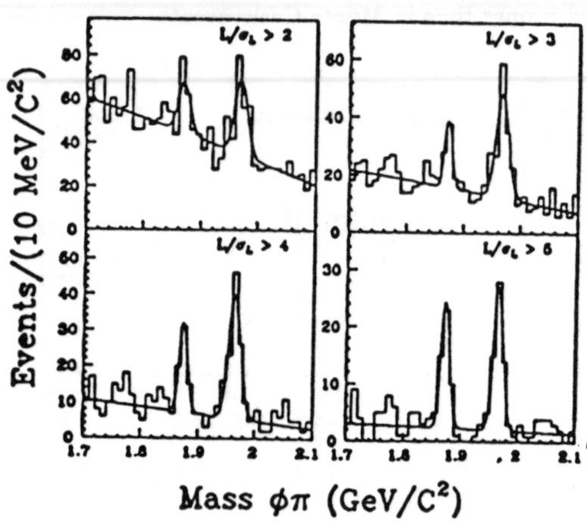

Figure 1

II. LIFETIME MEASUREMENT - For the D_s^+ lifetime measurement, the cut $L/\sigma > 3$ was chosen. The D_s^+ lifetime was determined using a mass and proper time maximum likelihood fit to the parameters m and t' where ct' = ct - 3σ.

Fitted Quantity	D_s^+	D^+
Mass (MeV)	1967±2	1871±3
σ (MeV)	11.1± 2.6	7.4±3.7
Number of Events	104±15	50±10

TABLE I

The likelihood function L was then maximized. Here $L = \sum_{i=1}^{events} \log(PS_i + PB_i)$,

$$PS_i = \frac{N_{D_s}}{\sigma_{D_s}\sqrt{2\pi}}\exp\left(\frac{(m-\overline{m_D})^2}{2\sigma_{D_s}^2}\right)\frac{1}{\tau_{D_s}}\exp\left(\frac{-t'}{\tau_{D_s}}\right) + \frac{N_D}{\sigma_D\sqrt{2\pi}}\exp\left(\frac{(m-\overline{m_D})^2}{2\sigma_D^2}\right)\frac{1}{\tau_D}\exp\left(\frac{-t'}{\tau_{D_s}}\right),$$

and PB_i (the background term) is given by:

$$PB_i = A(1+Bm)(\exp(\frac{-t'}{\tau_a}) + C\exp(\frac{-t'}{\tau_b})).$$

Background parameters τ_a, τ_b and C were obtained from the data and were tested on Monte Carlo generated events. Their values were 0.069 ps, 0.501 ps, and 5.41 respectively. The function L was normalized over mass and proper time. The fitted mass region was from 1.7 Gev/c^2 to 2.1Gev/c^2. The proper time t' was normalized on an event-by-event basis over the range from zero to τ_{max}.

The value of τ_{max} was determined from hardware constraints. (It was not possible to accurately determine the vertex positions downstream of the first microstrip plane.) This fit resulted in a D_s^+ lifetime of **0.53 ± 0.07** ps and a D^+ lifetime of **1.09 ± 0.19** ps. Figure 2 is a plot of the D_s^+ proper time (1.94 Gev/c^2 to 1.99 Gev/c^2 mass cut) showing contributions from both signal and background.

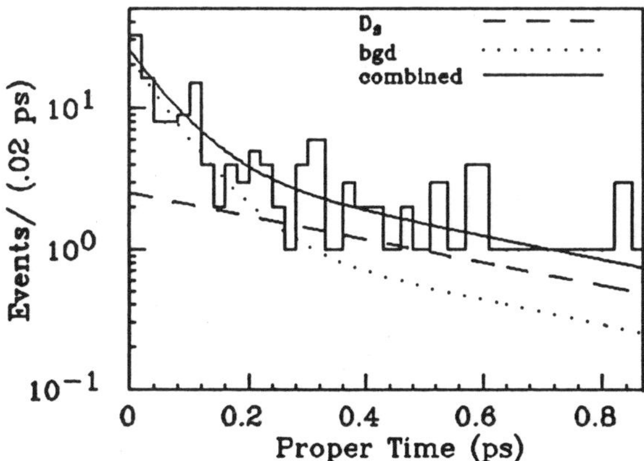

Figure 2

III. FURTHER WORK IN PROGRESS - A correction function, f(t), which incorporates absorption in the target and acceptance as a function of proper time are being studied and will modify the function L. The systematic and statistical errors are also being studied in detail.

[1]J. Wilson *et al*, these proceedings.
*Supported by US DOE, NSF and INFN
**E687 Collaboration: Univ. of Colorado; Fermilab; Univ. of Illinois; INFN Bologna; INFN Frascati; INFN Milano; Northwestern Univ.; Univ. of Notre Dame; INFN Pavia

Fermilab Experiment 687
Preliminary Charm Baryon Lifetime Measurements[*]

C. W. Bogart
Department of Physics
University of Colorado, Boulder, CO 80309
for the E687 Collaboration[†]

January 4, 1990

1 INTRODUCTION

We present preliminary results from Fermilab Experiment 687 on the lifetime of the charmed baryon Λ_c^+ in the decay mode $\Lambda_c^+ \to pK^-\pi^+$. These results are based on a sample of approximately 47 million photoproduced hadronic interaction events. We first discuss the techniques used in data reduction and analysis, and then present our preliminary lifetime result.

2 DATA REDUCTION

The E687 spectrometer has been described elsewhere.[1] We will discuss the charged track selection techniques used to obtain the $\Lambda_c^+ \to pK^-\pi^+$ signal.

The first stage of data reduction involves the use of several charged track selection criteria based on characteristics of the $\Lambda_c^+ \to pK^-\pi^+$ daughter tracks. First, the putative kaon daughter track is required to be of opposite signed charge to the putative pion and proton tracks; this reflects the charge-sign correlation in the charm to strange quark decay. The three daughter tracks are required to be linked from the silicon microstrip vertex detector to the main spectrometer (proportional wire chambers) to take full advantage of the microstrip vertex resolution and momentum information provided by the analysis magnets. The Cerenkov system criteria are then applied to the daughter tracks. The proton, pion, and kaon tracks are required to match the respective Cerenkov hypotheses for these particles. Finally no daughter track must identify as consistent with being a muon.

The second set of criteria involves the Λ_c^+ production and decay vertices. The former is termed the "primary" vertex; the latter is termed the "secondary" vertex. The secondary

[*]Supported by USDOE, NSF, and INFN, Italy

[†]University of Colorado; Fermilab; University of Illinois; INFN Bologna; INFN Frascati; INFN Milano; Northwestern University, Notre Dame University, INFN Pavia

[1]See paper by J.Wilson, these Proceedings.

vertex position is determined by a fit to a vertex hypothesis of the three daughter tracks selected as described above. If the χ^2 of this fit is acceptable, the Λ_c^+ momentum vector is formed and used as a "seed" for the primary vertex. Remaining tracks in the event are added to this momentum vector to form the primary and the vertex is fit; addition of tracks ceases when a good fit can no longer be obtained. A selected $\Lambda_c^+ \to pK^-\pi^+$ event is required to have both a good primary and secondary vertex. To further reduce the background, a cut is made on the significance of separation L between the vertices. The quantity $\frac{L}{\sigma}$, where σ is the total error on the vertex positions, is required to be greater than 3.0.

3 LIFETIME MEASUREMENT

The lifetime measurement technique involved a maximum likelihood fit to the combined mass and time distribution of the $\Lambda_c^+ \to pK^-\pi^+$ signal and background events (of mass 2.10 to 2.45 GeV/c^2) obtained with the above cuts. The proper decay length $\frac{L}{\beta\gamma}$ or ct was determined for each event. The distibution functions for the events used in the likelihood fit are PS for the signal and PB for the background, where

$$PS = \frac{A}{\sigma\sqrt{2\pi}} \exp\frac{(m-\bar{m})^2}{2\sigma^2}\frac{1}{\tau}(\exp\frac{t}{\tau})\frac{1}{(1-\exp\frac{t}{t_{max}})}$$

$$PB = \frac{1}{N}(B + C(m-\bar{m}) + D(m-\bar{m})^2)\frac{1}{\tau_b}(\exp\frac{t}{\tau_b})\frac{1}{(1-\exp\frac{t}{t_{max}})}$$

where the constant N is given by

$$B(0.35) + \frac{C}{2}[(2.45-\bar{m})^2 - (2.10-\bar{m})^2)] + \frac{D}{3}[(2.45-\bar{m})^3 - (2.10-\bar{m})^3]$$

and m is the Λ_c^+ mass, t the Λ_c^+ proper decay time, for each $\Lambda_c^+ \to pK^-\pi^+$ event. The constant N assures the proper normalization for the background; t_{max}, the maximum decay time, is derived from the distance from the event primary vertex to the first vertex detector plane.

The lifetime obtained by means of this fit for the Λ_c^+ is 0.22$\pm$0.03 picoseconds for a signal of 90 $\pm$ 19 events. The quoted error is the statistical error of the fit. The systematic errors on the lifetime due to such factors as spectrometer acceptance vs. lifetime and absorption in the target are still under study. Hence this result should be considered preliminary. The result is in agreement, within error, with the results obtained by experiments E691[2] and NA-32[3] for the Λ_c^+ lifetime.

4 CONCLUSIONS

Preliminary results for the lifetime of the Λ_c^+ have been obtained by Fermilab Experiment 687. These results agree within error with results obtained by other recent experiments. Study of the systematic errors in lifetime measurement are in progress.

---

[2] J.Anjos *et. al.*, Phys. Rev. Lett. **60**, 1379(1988)
[3] S. Barlag *et.al.*, Phys. Lett. **218B**, 374 (1989)

CHARM PHOTOPRODUCTION FROM THE NA14/2 EXPERIMENT

J.-P. Engel

CENTRE DE RECHERCHES NUCLEAIRES
IN2P3-CNRS/UNIVERSITE LOUIS PASTEUR
BP 20, F-67037 STRASBOURG CEDEX, France

NA14/2 Collaboration : National Technical University (Athens) – Universidad Autonoma (Barcelona) – CERN – Imperial College (London) – LAL (Orsay) – Collège de France (Paris) – C.E.N. Saclay – University of Southampton – C.R.N. (Strasbourg) – University of Warsaw

Abstract

We present results on charm photoproduction in the energy interval 40 to 160 GeV, obtained from charm samples of the NA14/2 experiment at CERN. These measurements include cross-section, x_F and p_T^2 distributions testing QCD in its perturbative behaviour, and various production ratios and asymmetries testing its non-perturbative behaviour through the hadronization scheme described by a phenomenological model.

1. INTRODUCTION

The photoproduction of charmed particles is of interest because to first order in QCD only a single diagram contributes, the photon-gluon fusion. Higher order corrections are expected to be small.

The hadronization scheme is also simple and is expected to be dominated by a single topological graph in the framework adopted, that of the Dual Parton Model.

2. THE NA14/2 EXPERIMENT

The NA14 spectrometer[1] is characterized by a large acceptance for photons and charged particles with 73 MWPC planes and three electromagnetic calorimeters. Two Cherenkov counters allow the separation of pion/kaon/proton. For the study of charmed particles the most important component is a a high resolution silicon vertex detector (Fig 1) It consists of a segmented silicon target[2], 31 planes of 24 slabs, with analogue readout. It gives a measurement of the ionization deposited around the vertex position. It is followed by a telescope of ten planes of silicon microstrips[3] of 1000 channels each with a 50 μm pitch read in a digital mode and gives an accuracy of about 17 μm in the transverse plane at the vertex.

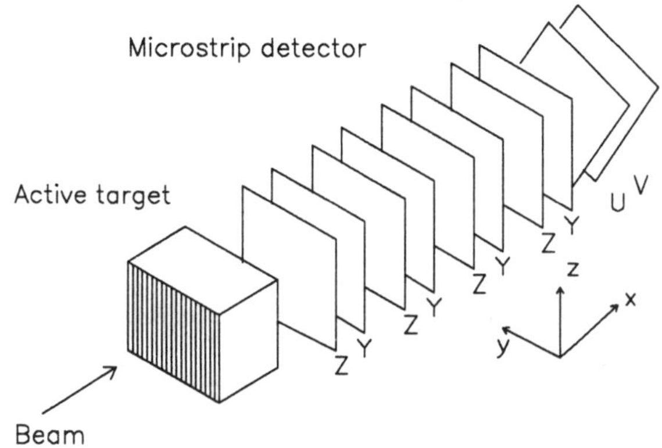

Fig. 1 : The vertex detector.

The trigger is designed to select hadronic interactions of the incident photon beam. It requires at least one charged particle above and below the horizontal plane and also at least 2.5 mip of charge deposited in each of the last three planes of the active target. The trigger efficiency is 30% for non charmed hadronic interactions. For charmed events it is approximately two times higher.

The incident electron energy is measured before and after radiation. The accuracy of the reconstruction of the radiated energy is about 2 GeV. The efficiency of the tagging spectrometer is 40%, limited by the high intensity of the beam ($\sim 10^7$ per burst, $E_\gamma > 50\,\mathrm{GeV}$).

About 60 events were registered on tape per 2 second burst and a total of 17 million events were recorded over the period 1985-1986. Events of interest for particular lines of analysis were selected using fast filter algorithms. Part of the initial sample, i.e. 4.3 million events, was processed directly through the full reconstruction program using the emulator farm at CERN. The charm signals are extracted by imposing a cut on the separation of the production and decay vertices reconstructed with tracks fitted in the microstrip detector.

3. INCLUSIVE CHARM PHOTOPRODUCTION

The mass of the charmed quark is high in comparison to the hadronization scale given by Λ_{QCD} and this allow to compute the charm photoproduction cross-section in a perturbative QCD approach. At first order in QCD a single diagram remains, $\gamma - g$ fusion process and higher order diagrams are expected to give a relatively small contribution. Information will be gained about the mass of the charmed quark m_c and the gluon structure function $G(x)$ by measuring the cross-section and the p_T^2 and x_F dependence of charmed hadron production. However the x_F distribution depends also on the choice of the charm fragmentation function.

The cross-section is measured using the decay modes $D^0 \to K^-\pi^+$ and $D^+ \to K^-\pi^+\pi^+$ extracted from the directly processed data. As we measure only the total radiated energy, the radiation of several photons is taken into account by transforming the tagging answer distribution to the real photon by means of a Monte Carlo derived matrix. The data are corrected for trigger acceptance and efficiencies. The cross-section is estimated by comparing our charm data spectrum with the hadronic spectrum using a hadronic photoproduction cross-section of $115\,\mu$b at $120\,$GeV. Correcting for the branching fractions of the decays used and for the average D multiplicity and assuming that the total contribution of the D mesons to the total incoherent charm protoproduction is 83%, we obtain

$$\sigma_{c_{tot}} = 414 \pm 52 \pm 45 \,\text{nb}$$

at a mean energy of $100\,$GeV and the energy dependence is shown in figure 2.

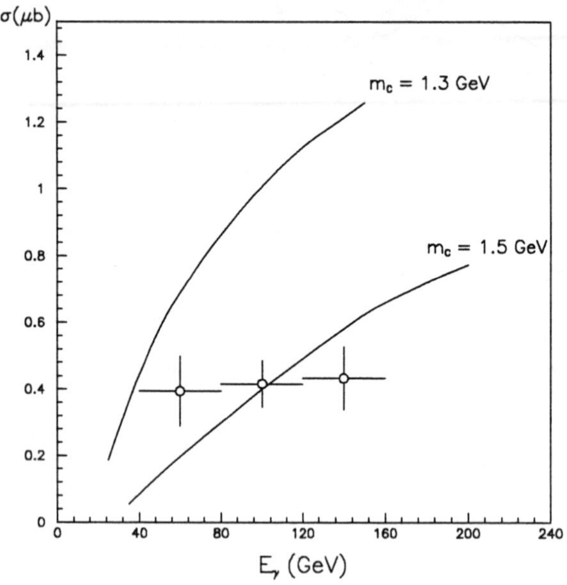

Fig. 2 : Energy dependence of the charm photoproduction cross-section.

For the p_T^2 and x_F distributions we used the filtered samples for D mesons. Because of its mass, the D meson is expected to have a higher mean transverse momentum than a π, a K or a proton produced in an inclusive way. It is observed experimentally that inclusive particle production follows the parametrization e^{-6m_T} (m_T is the transverse mass). The p_T^2 distribution for the signal (Fig.3) shows a clear excess at high p_T (over $2\,\text{GeV}^2$) above the background estimated from the wings on each side of the signal and also from events in the signal region after relaxing the cuts imposed by the vertex detector. The background is well described by e^{-6m_T}.

The x_F distribution obtained from the $D^0 \to K^-\pi^+$ and $D^0 \to K^-\pi^+\pi^+\pi^-$ samples is presented in figure 4. It is corrected for efficiencies and acceptances for these two modes and for multiple photon radiation.

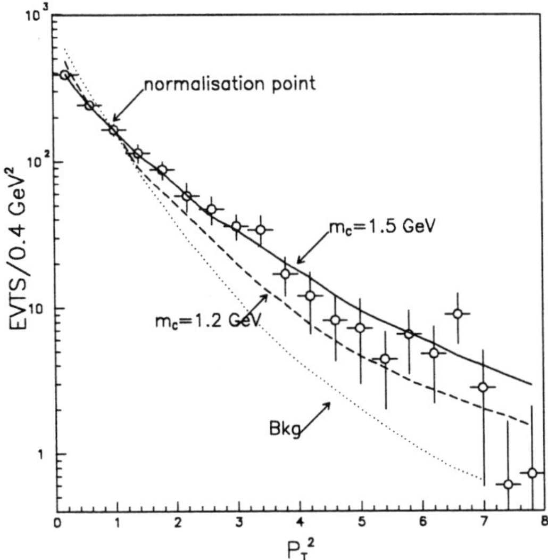

Fig. 3 : p_T^2 distribution of D-mesons.

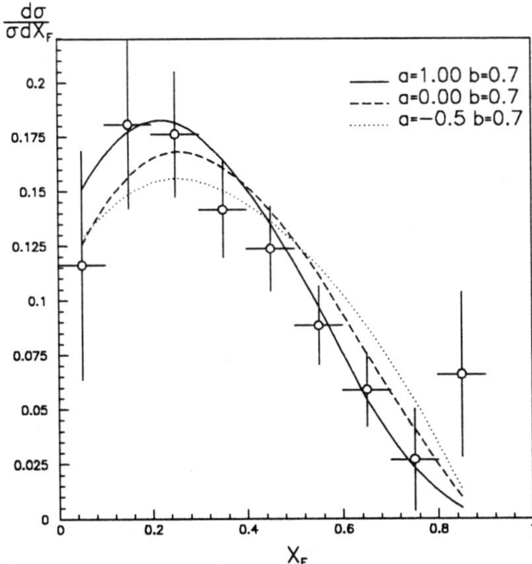

Fig. 4 : x_F distribution of D^0 mesons normalized for $x_F > 0$. The curves are the predictions for various fragmentation functions $zf(z) = (1 - z)^a e^{-bm_T^2/z}$.

The D meson production has been simulated using the $\gamma - g$ fusion QCD diagram, with a "naive" gluon structure function $xG(x) \sim (1-x)^5$, no Q^2 evolution, $Q^2 = m_c^2$, and a fixed mass for the charmed quark. The predictions for the p_T^2 distribution are presented in figure 3. Sensitive to the charmed quark mass, the data favours $m_c = 1.5$ GeV. The predictions for the x_F distribution are presented for $m_c = 1.5$ GeV and with different fragmentation functions, $zf(z) = (1 - z)^a e^{-bm_T^2/z}$, in figure 4. The data are well described by the default Lund values $a = 1.$ and $b = -0.7$. Using a softer gluon structure function $xG(x) \sim (1-x)^{10}$ gives a narrower distribution but its parametrization cannot be extracted from the data owing to the fragmentation of the charmed quark.

4. THE HADRONIZATION SCHEME

At the parton level, the charmed quark and antiquark are produced through the simple $\gamma-g$ fusion mechanism. The hadronization produces then diffractive and non-diffractive events. The diffractive component is expected to be small. We are thus in a situation where essentially all events are coming from a single QCD diagram. In the Dual Parton Model[4] used, hadrons are produced along a mesonic string stretched between the anti-charmed quark and a target spectator quark and a baryonic string stretched between the charmed quark and the remaining target spectator diquark (Fig.5). The structure functions for the spectator partons are obtained from low p_T physics. For quarks, at small x, a $x^{-1/2}$ behaviour is expected, whereas for diquarks we get $x^{1.5}$. This implies that most of the target energy is taken by the diquark. The masses of the two strings are very different and we expect particle-antiparticle asymmetry for charmed particles produced. Hadrons are created along the two independant strings using the LUND scheme.

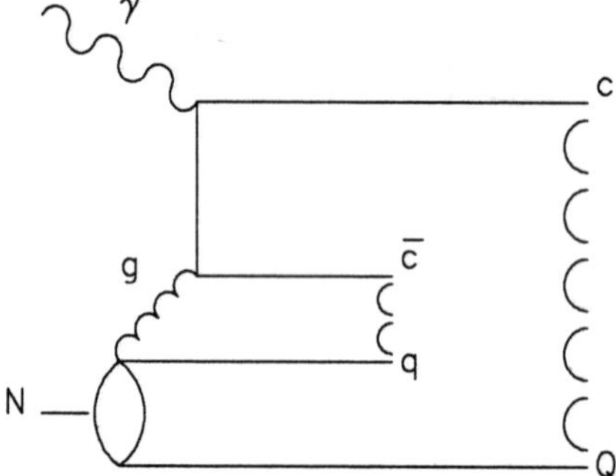

Fig. 5 : The hadronization scheme of the mesonic ($\bar{c}q$) and baryonic string (cQ).

A key issue is the Λ_c production. Photoproduction of the Λ_c ($\overline{\Lambda}_c$) is generally considered to result from the association of a c ($\bar{c}$) quark with a Q ($\overline{Q}$) diquark. At high incident photon energy $Q\overline{Q}$ pairs are easily produced in mesonic or baryonic string and $\Lambda_c/\overline{\Lambda}_c$ production is symmetric. At low energy, below about 50 GeV, insufficient phase space is available for $Q\overline{Q}$ production but the c quark can still form a Λ_c through the association with the spectator diquark Q from the nucleon. Thus Λ_c production is favoured compared with $\overline{\Lambda}_c$.

The $D/\overline{D}$ asymmetry reflects the $\Lambda_c/\overline{\Lambda}_c$ asymmetry. At high incident photon energy, the asymmetries are expected to be small.

For the D_S^+/D_S^- asymmetry, even at low energy, we do not expect more D_S^- than D_S^+ because of phase space limitation in the mesonic string which prevents $D_S^- K X$ production. There may be even a small excess of D_S^+ mesons.

The D^+/D^0 production ratio depends mainly on the amount of D^* produced, expressed as the ratio of vector (V) to pseudo-scalar (PS) charmed mesons $P_V = V/(PS + V)$.

The various charmed particle rates versus photon energy are presented in figure 6. In our Monte Carlo simulation we use the default LUND parameters and the branching fraction $B_* = \mathrm{Br}\left(D^{*+} \to D^0 \pi^+\right) = 0.49 \pm 0.08^{(5)}$.

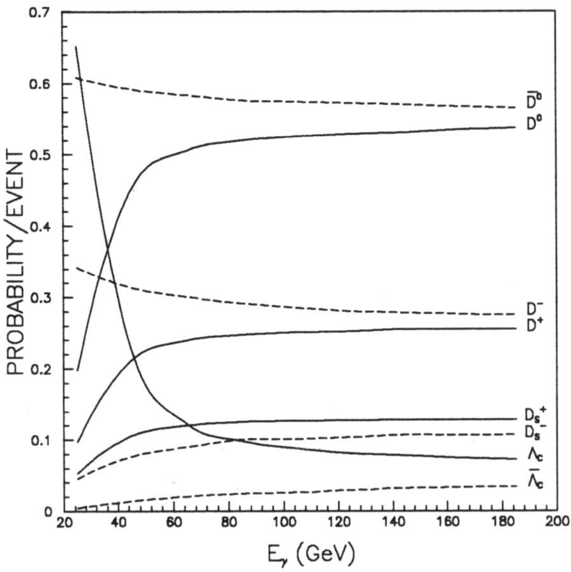

Fig. 6 : Production rate predictions for charmed particles.

5. PRODUCTION RATIOS AND CHARGE ASYMMETRIES

Production ratios or asymmetries of various charmed particles expected from the hadronization scheme are compared with measurements obtained from reconstructed events from D^0, D^+, D_S and Λ_c filtered samples.

The measurement of D^*/D and D^+/D^0 ratios gives information on the vector to pseudo-scalar charmed mesons ratio P_V. To derive this parameter we make the important assumption of an equal production of respectively D^0 and D^+, D^{*0} and D^{*+} at the parton level. The fraction R_1 of D^0 produced through the D^{*+} decay gives a determination of $P_V B_* = R_1/(1 - R_1)$. From our $D^0 \to K^- \pi^+$ and $D^0 \to K^- \pi^+ \pi^+ \pi^-$ samples we measure $R_1 = 0.26 \pm 0.04$ and obtain $P_V B_* = 0.35 \pm 0.07$. The measurement of the D^+/D^0 production ratio R_2 gives another determination $P_V B_* = (1 - R_2)/(1 + R_2)$. It is obtained using our $D^0 \to K^- \pi^+$ and $D^+ \to K^- \pi^+ \pi^+$ samples and the branching ratio $\text{Br}(D^+ \to K^- \pi^+ \pi^+)/\text{Br}(D^0 \to K^- \pi^+) = 2.17 \pm 0.37^{[5]}$. We measure $R_2 = 0.45 \pm 0.09$, giving $P_V B_* = 0.38 \pm 0.09$. The two values are combined and we obtain $P_V B = 0.36 \pm 0.06$. Using $B_* = 0.49 \pm 0.08$ we get $P_V = 0.72 \pm 0.10 \pm 0.12$, in agreement with the value expected from simple spin counting rule, 0.75, which is in the LUND model.

The prediction of our simulation for the ratio $\frac{\sigma(D^+)}{\sigma(D^0)} = 0.49$ is in agreement with our measurement 0.45 ± 0.09.

For charmed mesons, defining the asymmetry as $A_D = (N_D - N_{\overline{D}})/(N_D + N_{\overline{D}})$ results are presented in the following table :

	Measured	Predicted with our beam
$\overline{D}/D$	1.07 ± 0.09	1.17
A_D	-0.03 ± 0.05	-0.07
D_S^+/D_S^-	1.6 ± 0.6	1.26
A_{D_S}	0.24 ± 0.17	0.12

We observe small asymmetries that are expected by the simulation.

The production ratio of D_s^+ over D^+ provides also information on the hadronization since a charmed quark must combine with a stange quark of the sea, thus suppressing D_S^+ production relative to D^+. Measurement of the decay channels $D_S^+ \to \varphi \pi^+$, $D^+ \to K^- \pi^+ \pi^+$ and $D^+ \to \varphi \pi^+$ allows us to extract the ratio $\frac{\sigma(D_s^+)}{\sigma(D^+)}\text{Br}(D_S^+ \to \phi \pi^+) = (1.81 \pm 0.63)\%$, using known branching fractions$^{[5]}$. If there is a measurement of the absolute D_S^+ branching fraction, we can infer the relative cross-section. Alternatively, assuming the value $\frac{\sigma(D_s^+)}{\sigma(D^+)} = 0.38 \pm 0.15$ obtained from our simulation, we get $\text{Br}(D_S^+ \to \varphi \pi^+) = (4.8 \pm 1.7 \pm 1.9)\%$.

The Λ_c production has been studied with the sample of reconstructed $pK\pi$ decays. The $\Lambda_c/\overline{\Lambda}_c$ ratio measured is 0.6 ± 0.3. Taking into account the experimental acceptance the Monte Carlo simulation predits 0.7. The photoproduction cross-section can be estimated and with the identification of the Λ_c limited to the 30–80 GeV range we quote $\sigma_{\Lambda_c}(E > 30\,\text{GeV})\,\text{Br}(\Lambda_c \to pK\pi) = 2.5 \pm 0.7 \pm 0.7\,\text{nb}$. The charmed baryon production relative to charmed mesons is found higher than the predictions of the Monte Carlo simulation, as expressed by the experimental and theoretical Λ_c/D production ratios : $(\Lambda_c/D^0)_{\text{exp}}/(\Lambda_c/D^0)_{\text{Lund}} = 3.4 \pm 0.9$ and $(\Lambda_c/D^+)_{\text{exp}}/(\Lambda_c/D^+)_{\text{Lund}} = 3.1 \pm 0.8$ using a branching fraction $\text{Br}(\Lambda_c \to pK\pi)$ as high as 5%.

6. SUMMARY

We have measured the cross-section for photoproduction of charm at $< E_\gamma >=$ 100 GeV to be $\sigma_{c_{tot}} = 414\pm52\pm45$ nb. The p_T^2 and x_F distributions for D mesons are compared to those expected by the simple γ-g process. These measurements favour $m_c = 1.5$ GeV. Various production ratios and charge asymmetries were obtained and found in agreement with Monte Carlo predictions using the Dual Parton Model in which hadrons are generated along the mesonic and baryonic strings using the Lund scheme. However the Λ_c/D ratio is measured higher than the prediction of the Lund model.

REFERENCES

1. M.P. Alvarez et al. — CERN–EP/90–26
2. R. Barate et al. — Nucl. Inst. Meth. A235 (1985)235
3. G. Barber et al. — Nucl. Inst. Meth. A253 (1985)530
4. A. Capella and J. Tran Thanh Van — Phys. Lett. 93B (1980)146
   P. Roudeau — Nucl. Phys. B (Proc. Suppl.) 1B (1988)33
5. Particle Data Group — Phys. Lett. B204 (1988)1

A new determination of the electroweak mixing angle from muon-neutrino electron scattering

The CHARM II Collaboration

D.Geiregat, P.Vilain and G.Wilquet
IIHE, Brussels, Belgium

F.Bergsma, U.Binder, H.Burkard, Y.Eisenberg, W.Flegel, H.Grote, T.Mouthuy
H.Overas, J.Panman, R.Santacesaria, K.Winter, G.Zacek and V.Zacek
CERN, Geneva, Switzerland

R.Beyer, V.Blobel, F.W.Büsser, C.Foos, L.Gerland, T.Layda, F.Niebergall
G.Rädel, P.Stähelin, A.Tadsen and T.Voss
University of Hamburg, Germany

Th.Delbar, D.Favart, G.Gregoire and E.Knoops
Catholic University of Louvain, Belgium

P.Gorbunov, E Grigoriev, V.Khovansky, A.Maslennikov and A.Rozanov
ITEP, Moscow, USSR

W.Lippich, A.Nathaniel and A.Staude
University of Munich, Germany

M.Caria, B.Eckart, A.Ereditato, E.Gorini, F.Grancagnolo, R.Iasevoli
V.Palladino and P.Strolin
University/INFN, Naples, Italy

A.Capone, D.de Pedis, E.Di Capua, U.Dore, A.Frenkel-Rambaldi
P.F.Loverre, G.Piredda and D.Zanello
University/INFN, Rome, Italy

Presented by T. Mouthuy (CERN)

Abstract

We are reporting on a new determination of $sin^2\theta_w$ from the ratio of $\nu_\mu e$ to $\bar{\nu}_\mu e$ scattering cross sections. 762 $\nu_\mu e$ and 1017 $\bar{\nu}_\mu e$ have been observed in the 1987 and 1988 exposure of the CHARM II detector to the Wide Band Beam of the 450 GeV CERN SPS. From the ratio of cross sections we determined $sin^2\theta^0 = 0.233 \pm 0.012(stat) \pm 0.008(syst)$ without radiative corrections. With radiative corrections for $m_t = m_H = 100\ GeV$ we find $sin^2\theta_w = 0.232 \pm 0.012(stat) \pm 0.008(syst)$.

Introduction

A precise measurement of the electroweak mixing angle $sin^2\theta_w$ provides a test of the standard model by comparing experimental values at different Q^2 and the theoretically calculated radiative corrections.

We are reporting here on the new determination of $sin^2\theta_w$ from a measurement of the ratio of the cross sections of elastic $\nu_\mu e$ and $\bar{\nu}_\mu e$ scattering. The ratio depends on the value of $sin^2\theta_w$ in the Born approximation by:

$$R = \frac{\sigma(\nu_\mu e)}{\sigma(\bar{\nu}_\mu e)} = 3\,\frac{1 - 4sin^2\theta_w + (16/3)sin^4\theta_w}{1 - 4sin^2\theta_w + 16sin^4\theta_w}$$

The method of the ratio has many advantages. Only the flux ratio is needed, the method is independent of the value of ρ and most systematic uncertainties in the efficiencies cancel. QED corrections cancel to a large extent in the ratio and can be neglected. Electroweak corrections depend weakly on the values of m_t and m_H [1,2]. Experimentally one forms an empirical ratio:

$$R_{vis} = \frac{N(\nu e)}{N(\bar{\nu} e)}\,F$$

with F being the energy weighted flux ratio of the main component of the beam.

Experimental setup

The leptonic scattering of neutrinos on electrons is a rare process compared with the semileptonic scattering (about 10^{-4}). A large detector and an intense neutrino beam are needed to achieve a statistically significant result. The CHARM II Collaboration has built a 692 tons detector [3] made of 420 planes of 4.8 cm thick glass interspaced with 4 by 4 m^2 streamer tubes with digital and analogue readout. This type of detector showed a good angular resolution ($\sigma(\theta) = \frac{22mrad}{\sqrt{E}}$) and good energy resolution ($\sigma(E) = \frac{0.3}{\sqrt{E}}$). This is essential to reconstruct the scattered electron emitted at small angle ($E_e\theta_e^2 = 2m_e(1 - y) < 2m_e$). The calorimeter is further equipped with scintillation planes used in the e/π^0 discrimination by a dE/dx method.

The detector was exposed to the SPS Wide Band Beam providing two 6 ms long pulses of 450 GeV protons for a total intensity of $1.8\ 10^{13}$ protons per cycle. The total amount of protons used in 1987 and 1988 is $0.9\ 10^{19}$ (statistics presented here) and $0.6\ 10^{19}$ in a 1989 run. In order to ensure the symmetry between ν_μ and $\bar{\nu}_\mu$ running, the beam polarity is inverted approximately every 2 days.

Measurement of neutrino electron scattering

After the rejection of cosmic events, a series of separators e.g. based on width and density of the shower have been applied to reject the ten thousand times more abundant nucleon scattering events. Showers with additional charged particle tracks or

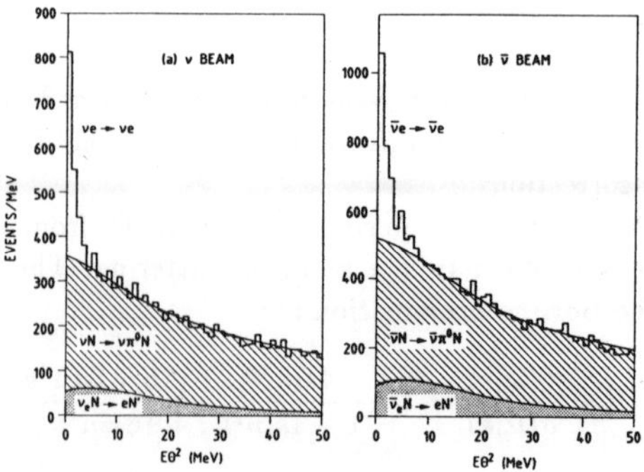

Figure 1: Distribution of selected events as function of $E\theta^2$

with backscattering have also been rejected. Finally only events with a single hit in the vertex plane have been kept for further analysis. The selection efficiency was determined from a test beam and found to be 0.76 without apparent energy dependance. The $E\theta^2$ distributions observed in the neutrino and antineutrino beam are shown in figure 1. The pronounced peak ($E\theta^2 < 3MeV$) is attributed to neutrino electron scattering. The remaining background is composed of semileptonic interactions with an electromagnetic shower in the final state, mainly due to coherent and diffractive π^0 production, and to quasielastic scattering of the ν_e and $\bar{\nu}_e$ components of the beam [4]. This background was studied into a reference region ($5 \leq E\theta^2 \leq 72MeV$).

Events due to coherent and diffractive π^0 production have been simulated by Monte-Carlo techniques following the theoretical description of Rein and Seghal [5] and Bel'kov and Kopeliovitch [6]. The events were then fed through the same analysis chain as the candidates.

The distributions for the reactions $\nu_e N \rightarrow e^- N'$ and $\bar{\nu}_e N \rightarrow e^+ N'$ have been contructed starting from observed $\nu_\mu N \rightarrow \mu^- N'$ and $\bar{\nu}_\mu N \rightarrow \mu^+ N'$ events. The muon was replaced using Monte-Carlo techniques by an electron of the same direction and energy.

The relative abundance of these two types of background was determined for a subsample of events by measuring the energy released in the scintillation counter following the vertex. In the case of an electron induced shower a energy loss typical of one charged particle is observed whereas π^0 induced showers starting with $e^+ e^-$ pairs show a larger energy deposition. Table 1 gives the observed background composition in comparison with the relative abundances obtained by fitting the background description to the data.

Beam	E_{first}	$E - E\theta^2$ fit
ν beam	0.20 ± 0.03	0.185 ± 0.020
$\bar{\nu}$ beam	0.25 ± 0.03	0.250 ± 0.020

Table 1: Relative abundance of ν_e and π^0

After subtraction of the background (see figure 2) we found $N(\nu e) = 762 \pm 43(stat)$ events in the neutrino sample and $N(\bar{\nu}e) = 1017 \pm 51(stat)$ events in the antineutrino

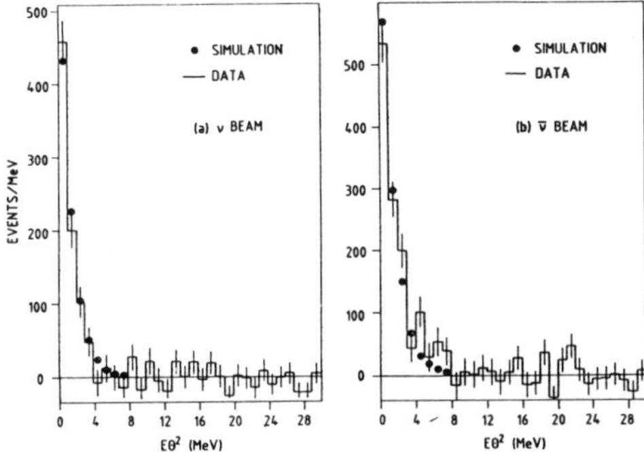

Figure 2: Distribution of neutrino electron scattering events as a function of $E\theta^2$ after background subtraction. The points represent the expected distributions.

sample, with $E\theta^2 < 3MeV$ and $3 < E < 24\ GeV$. The observed shape is in good agreement with that expected from experimental resolution and kinematics.

Flux normalisation

We determined the flux ratio in four independent ways.

Method	Flux ratio
Muon flux	1.207 ± 0.049
Quasi-elastic	1.270 ± 0.063
π^0 production	1.220 ± 0.061
Inclusive	1.166 ± 0.058
Mean value	1.215 ± 0.028

Table 2: Energy-weighted flux ratios

The *flux of muons* produced by the decay of pions and kaons in the beam has been measured by solid state detectors inserted into slots of the iron shield following the decay tunnel. The relation between the neutrino flux and the measured muon flux is obtained by a Monte-Carlo simulation. The dominating error is due to instabilities of the beam profile and uncertainties in the beam composition.

The cross sections of neutrino and antineutrino *quasi-elastic scattering* are almost equal at $Q^2 = 0$. A correction taking into account this small difference as well as the difference in the reconstruction efficiencies due to the final state (Δ and N^* resonances included) has been applied. Events have been further selected for a small Q^2 ($0.05 < Q^2 < 0.15\ GeV^2$).

The cross sections for *coherent π^0 production* on nuclei are equal for neutrinos and antineutrinos. The number of $\overset{(-)}{\nu_e}\ N \to eN'$ events has been subtracted from the observed number of events in a reference region ($5 \leq E\theta^2 \leq 72MeV$).

The *total number of interactions* has been estimated by using our minimum-bias sample of events. The comparison with the known ratio of inclusive cross sections [7] allows us to extract the relative fluxes of the beams. Corrections have been made to take into account the wrong-helicity neutrinos and the $\overset{(-)}{\nu_e}$ component of the beam. The main uncertainty comes from the poor knowledge of the cross section below $E_\nu < 30\ GeV$.

Determination of $sin^2\theta_w$

Multiplying the ratio of the number of $\nu_\mu e$ and $\bar{\nu}_\mu e$ scattering events observed with the flux ratio F, we obtain [8]:

$$\mathbf{R_{vis} = 0.910 \pm 0.068(stat) \pm 0.045(syst)}$$

which leads to:

$$\mathbf{sin^2\theta^0 = 0.233 \pm 0.012(stat) \pm 0.008(syst)}$$

Applying electroweak radiative corrections [2] with $m_t = m_H = 100\ GeV$ gives:

$$\mathbf{sin^2\theta_w = 0.232 \pm 0.012(stat) \pm 0.008(syst)}$$

Flux normalisation	± 0.003
$\bar{\nu}_\mu/\nu_\mu$ composition	± 0.002
$\nu_e, \bar{\nu}_e$ admixture	± 0.002
Beam spectra	± 0.002
Background subtraction	± 0.007
Energy calibration	± 0.002

Table 3: Sources of systematic errors

A comparison between the result presented here and the average value of $sin^2\theta_w$ [9] gives an upper limit on m_t of 257 GeV (90 % CL).
The comparison of $sin^2\theta_w$ derived from neutrino-electron scattering and from the new precise measurement of the Z^0 mass shows good agreement. To evaluate the radiative corrections we used the relation [10]:

$$\delta_z = 1 - \left(\frac{A_0}{sin\theta^0 cos\theta^0 m_Z}\right)^2$$

where $sin^2\theta^0$ is the value derived from neutrino-electron scattering and $A_0 = 37.281\ GeV$. With $M_Z = 91.09 \pm 0.16\ GeV$,

$$\delta_z(exp) = 0.003 \pm 0.039$$

Theoretically [11] we obtain for $m_t = m_H = 100\ GeV$,

$$\delta_z(theor) = 0.0592 \pm 0.0013$$

Conclusions and prospects

In conclusion the new determination of $sin^2\theta_w$ based on high statistics of $\nu_\mu e$ and $\bar{\nu}_\mu e$ events shows a good agreement with $sin^2\theta_w$ obtained from deep inelastic ν_μ scattering and with m_Z. The radiative corrections observed are in good agreement with the predictions of the Standard Model.

More data are expected from the 1989 and 1990 runs. 2000 events in each channels are expected, leading to a statistical error of 0.007. More strict event selection will also improve the signal/backgound ratio. The systematic error can be reduced using the results of a calibration run in electron and pion beams.

Acknowledgements

The experiment has been made possible by grants from the IIHE (Belgium), CERN (Switzerland), the Bundesministerium für Forschung und Technologie (Germany), ITEP (USSR) and INFN (Italy). We would like to thank the skilful crew operating team of the CERN SPS.

References

[1] S.Sarantakos, A.Sirlin and W.J.Marciano, *Nucl.Phys. B217 (1983) 84*
M.J.Marciano and S.Sirlin *Phys.Rev. D22 (1980) 2695, D29 (1984) 945, D31 (1985) 213 E*

[2] D.Y.Bardin and O.M.Dokuchaeva *Nucl.Phys. B246 (1984) 221*

[3] K.De Winter et al. CHARM II Collaboration *Nucl. Instr. Methods A278 (1986) 670*

[4] F.Bergsma et al. CHARM Collaboration *Phys.Lett. 157B (1985) 469*

[5] D.Rein and L.M.Sehgal *Nucl.Phys. B223 (1983) 29*

[6] A.A.Belk'kov and B.Z.Kopeliovich *Sov..Nucl.Phys. 46 (1987) 499*

[7] J.V.Allaby et al. CHARM Collaboration *Z.Phys. C38 (1988) 403*
Particle Data Group *Phys.Lett. 204B (1988) 1*

[8] D.Geiregat et al. CHARM II Collaboration *Phys.Lett. B232 (1989) 539*

[9] J.V.Allaby et al CHARM Collaboration *Z.Phys C36 (1987) 611*
H.Abramovitch et al. CDHSW Collaboration *Phys.Rev.Lett. 57 (1986) 298*

[10] U.Amaldi et al. *Phys.Rev. D36 (1987) 1385*

[11] F.Jegerlehner *Z.Phys. C32 (1986) 195*

A Report on the measurement of the Weak Radiative Decay $\Lambda \to n + \gamma$.[*]

A.J. NOBLE, M.D. HASINOFF, D.F. MEASDAY, S. STANISLAUS, C.E. WALTHAM,
Univ. of British Columbia, Vancouver, B.C., Canada V6T 2A6;

P.G. JONES, J. LOWE,
Univ. of Birmingham, U.K. B15 2TT, and Univ. of New Mexico, Albuquerque, NM 87131;

E.K. MCINTYRE, J.P. MILLER, B.L. ROBERTS, T.M. WARNER, D.A. WHITEHOUSE,
Boston Univ., Boston, MA, USA 02215;

M. SAKITT,
Brookhaven National Laboratory, Upton, NY, USA 11973;

W.J. FICKINGER, D.K. ROBINSON,
Case Western Reserve Univ., Cleveland, OH, USA 44106;

D. HORVÁTH,
KFKI, Budapest, Hungary and TRIUMF, Vancouver, B.C.,Canada V6T 2A3;

B. BASSALLECK, J.R. HALL, K.D. LARSON, D.M. WOLFE,
Univ. of New Mexico, Albuquerque, NM 87131;

A.L. HALLIN,
Princeton Univ., Princeton, NJ 08544;

M. SALOMON,
TRIUMF, Vancouver, B.C., Canada V6T 2A3.

Abstract

We report on the measurement of the Weak Radiative Decay $\Lambda \to n + \gamma$. An enhancement in the region of interest has been seen and all indications are consistent with a branching ratio higher than previously measured.

Introduction.

The weak radiative decays of hyperons, $B \to B' + \gamma$, are of particular interest as they provide a relatively simple system combining facets of the weak and electromagnetic interactions, and of QCD (through chiral symmetry breaking effects). The general framework of weak interaction theory has been well established but there are still a number of mechanisms, such as the empirically observed $\Delta I = \frac{1}{2}$ rule and the f/d ratio, which do not have satisfactory explanations.

Much of the motivation for studying radiative decays is the hope that since the electromagnetic vertex can be calculated with some confidence compared with pion emission, the weak interaction vertex can be isolated and examined in some detail. However, present theoretical predictions are inconsistent amongst themselves and with the limited data. Theoretical models attempt to calculate the experimentally accessible parameters; the branching ratio and the asymmetry. These are given by

$$\alpha = \frac{2Re(a^*b)}{|a|^2 + |b|^2} \qquad \Gamma = \frac{1}{\pi}(\frac{m_i^2 - m_f^2}{2m_i})^3 \, (|a|^2 + |b|^2)$$

where a and b are the parity conserving and parity violating amplitudes and m_i and m_f are the masses of the initial and final baryons. The asymmetry is predicted to be zero in

[*] This work was supported by the Canadian NSERC, the U.S. DOE and NSF, and the U.K. SERC.

the chiral limit of SU(3)$_f$ and most models have a hard time to make the asymmetry large, (as has been demonstrated experimentally in the case of the decay $\Sigma^+ \to p + \gamma$), while maintaining the branching ratio at a reasonable limit.

The decay $\Sigma^+ \to p + \gamma$ is the most well known experimentally and can be used to constrain free parameters of the models. Then measurements of other decays can provide sensitive tests of the theory. Of these, the decay $\Lambda \to n + \gamma$ has been singled out as being particularly sensitive. (Clos 80). For a review of the theoretical status see Żenczykowski and references therein. (Zenc 89).

Apparatus.

The experimental setup is shown in figure 1. We used the Low Energy Separated Beam line, LESB-II, at Brookhaven National Lab as a source of K^-'s. Under typical running conditions a π/K ratio of less than 6:1 was possible. These K^-, at an initial momentum of 680 MeV/C, were stopped in a LH_2 target after being degraded by beam counters, particle identification Čerenkov counters, dE/dx counters, a 6×6 hodoscope array and typically 5" of copper.

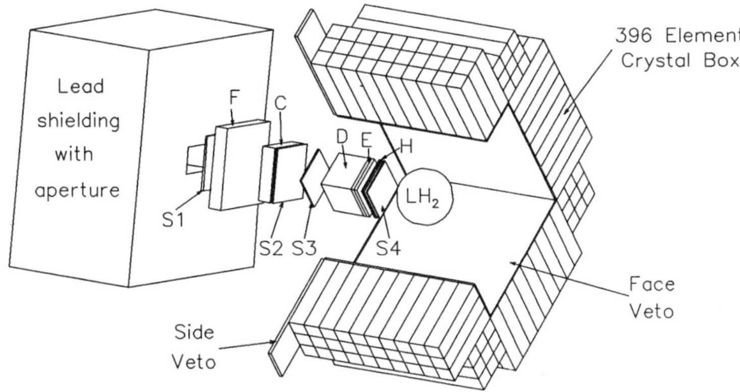

Figure 1: Schematic cutaway view of Crystal Box and beam counters S_i, Čerenkov counters F and C, dE/dx counters E, hodoscope H and degrader D.

The Λ's are generated in association with a monoenergetic π^0 by the stopped K^- reaction, $K^- p \to \Lambda + \pi^0$, and we use the decay of this π^0 as the signature for Λ production. The hydrogen target is centered within the Crystal Box, a 396 element NaI array with about 2π solid angle coverage. Each face crystal is 2.5" × 2.5" × 12" deep and is optically decoupled from its neighbors. The hydrogen target is encircled with scintillators, and there is a second layer covering the faces of the Crystal Box, enabling us to reject charged particles. The Crystal Box is an excellent multi-particle spectrometer and it is used to detect the photons from the π^0 decay in addition to the decay products of the Λ.

The hard-wired trigger required a K^- incident on the LH_2 target, sufficient energy deposited in the Crystal Box and no charged particles amongst the decay products. Suitable events were passed to tape via a system of two CAMAC CAB preprocessors, a microprogramable branch driver and a PDP 11/44. We used 400 channels of FERA ADC's to digitize the analog energy signals for each crystal, and another 400 to search for pileup by sampling a slice of the analog signal.

Analysis.

A clump finding routine was used to determine the position and energy of each particle entering the Crystal Box. This information combined with the hodoscope position information allowed us to reconstruct the event. The program requires that two of three photons reconstruct to be a π^0 and the third photon is then Doppler corrected in energy to the rest frame of the Λ. In the rest frame of the Λ we expect a peak around 162 MeV, broadened by the position and energy resolutions of the detector.

In figure 2 we show such a spectrum. The region from 50 to 140 MeV is dominated by γ's from the much more prolific decay $\Lambda \to n + \pi^0$ with only one γ being detected from the subsequent π^0 decay. Also contributing in this region are γ's from the reaction $K^-p \to \Sigma + \pi^0$ which can have up to 5 γ's in the final state with some accidental combination of γ's conspiring to look like a 288 MeV π^0.

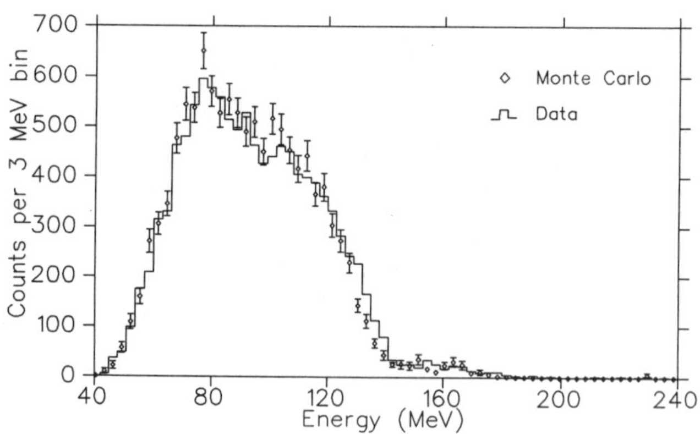

Figure 2: Weak radiative decay Doppler corrected energy spectrum.

The region from 150 to 170 MeV is where the signal is expected. With the present state of analysis a clear enhancement is seen, but the peak has not been clearly isolated. Investigations continue to determine the nature of the backgrounds beneath the peak. At present all fits to the data are indicative of a branching ratio higher than the value 1.02±0.33 previously reported. (Biag 86).

Bibliography.

Biag 86   S.F. Biagi et al., Z. Phys. **C30**, 201 (1986).
Clos 80   F.E. Close and H.R. Rubinstein, Nucl. Phys. **B173**,477 (1980)
Zenc 89   P. Żenczykowski, Phys Rev. **D40** 2290 (1989).

Rare Z Decays and New Physics

E. W. N. Glover

Fermi National Accelerator Laboratory

P. O. Box 500, Batavia, Illinois 60510

Abstract

Although the signatures for rare Z decays are often spectacular, the predicted standard model rates are usually extremely small. In many cases, however, rare decays are very sensitive to new phenomena and may lead to an observable rate. In this talk, I select some interesting rare decays and discuss how new physics might be identified.

1. Introduction

In the summer of 1989, experiments at LEP commenced and, in a few short months, the four experiments observed around 10^5 Z events between them. It is anticipated that each experiment will collect 10^6 events in 1990. With this data sample, it is possible to test the standard model both at the level of electroweak radiative corrections and by searching for rare decays of the Z. Although the signatures for rare decays are in many cases spectacular, the predicted branching rates are usually extremely small. On the other hand, rare decays are very sensitive to phenomena beyond the standard model, and attempts to isolate them may yield valuable information on new physics. Rather than catalogue the expectations for rare decays within the standard model [1], I will focus on a few decays, which, if observed, may lead to new physics.

2. Higgs Bosons

The most copious source of Higgs bosons in Z decay is the Bjorken process [2],

$$Z \to HZ^* \to Hf\bar{f}, \tag{1}$$

which, because of the large HZZ coupling, may have a branching ratio as large as 1% (see Fig. 1). For relatively light Higgs bosons, the rate for (1) summed over the fermions f is large enough that the few tens of thousands of events collected in 1989 at LEP are sufficient to set limits of $m_H > 25$ GeV [3] and $m_H > 24$ GeV [4] from OPAL and ALEPH respectively. Both of these limits make use of the $Z \to H\nu\bar{\nu}$ channel in addition to the smaller $Z \to H\ell^+\ell^-$ channel. Due to the Yukawa coupling with the fermions, the Higgs boson preferentially decays into the heaviest available fermion pair, which for $m_H \geq 10$ GeV means $H \to b\bar{b}$.

For light Higgs bosons, the $Z \to H\nu\bar{\nu}$ decay leads to a clean signal containing large amounts of missing energy accompanied by two jets. As m_H increases however, the missing energy decreases and the event appears more like a two jet event with some energy imbalance. Ordinary two jet events with missing energy generated by either the semileptonic

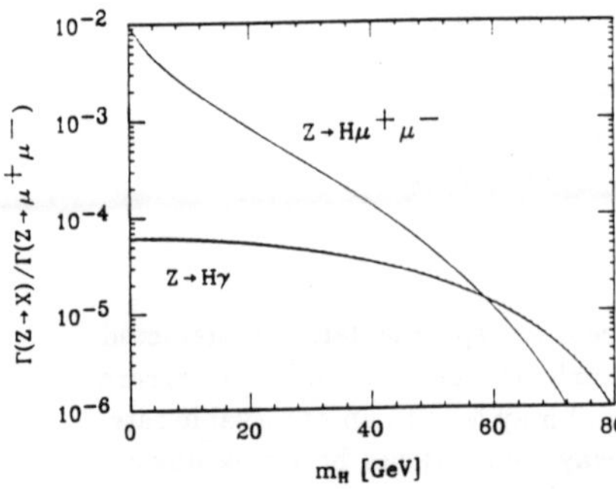

Figure 1: The decay rate, normalised to $\Gamma(Z \to \mu^+\mu^-)$ for (a) $Z \to H\mu^+\mu^-$, (b) $Z \to H\gamma$ for $m_t = 90$ (solid line) and 200 GeV (dotted line) as a function of the Higgs boson mass, m_H.

decay of a heavy quark, long lived neutrals or energy leakage then provide a significant background, leading to a reduced detection efficiency.

The $Z \to H\mu^+\mu^-$ decay channel is much cleaner and, in principle, the Higgs boson mass may be reconstructed from the muon four momenta,

$$m_H^2 = \hat{s} - 2\sqrt{\hat{s}}(E_{\mu^+} + E_{\mu^-}) + m_{\mu^+\mu^-}^2. \tag{2}$$

As m_H increases, the mass determination improves since the cancellations on the right hand side of (2) become smaller and less sensitive to the experimental resolution. Although the $Z \to \mu^+\mu^-$ branching rate is quite small, the event rate is significant (see Table 1).

m_H (GeV)	25	35	45	55	65
$Z \to H\mu\mu$	168	67	24	7.1	1.4
$Z \to H\gamma$	16.5 (16.7)	13.0 (13.3)	9.3 (9.6)	5.7 (5.9)	2.7 (2.9)

Table 1: Number of Higgs boson events in 10^7 Z events. We use $m_t = 90$ (200) GeV.

The signal consists of a pair of b quarks from the Higgs boson decay with $m_{b\bar{b}} \sim m_H$ and a muon pair from the decay of the virtual Z boson. Due to the effect of the Z propagator, $m_{\mu^+\mu^-}$ is as large as possible (see Fig. 2). The dominant background is the four fermion decay, $Z \to b\bar{b}\mu^+\mu^-$, or, since it is difficult to efficiently distinguish quark jets, $Z \to q\bar{q}\mu^+\mu^-$ [5]. However, as shown in Figs. 2 and 3, the background has a different structure in $m_{\mu^+\mu^-}$ and $m_{q\bar{q}}$ allowing a clean separation of signal and background for $m_H \lesssim 45$ GeV. Imposing an invariant mass cut on the muon pair, $m_{\mu^+\mu^-} > 20$ GeV, considerably reduces the background for heavy Higgs bosons while leaving the signal essentially untouched [5]. Provided that the experimental resolution on $m_{q\bar{q}}$ is O(few GeV), the discovery potential is only limited by the number of Z boson events obtained. With 10^6 (10^7) Z events, LEP can probe Higgs boson masses up to ≈ 40 (60) GeV.

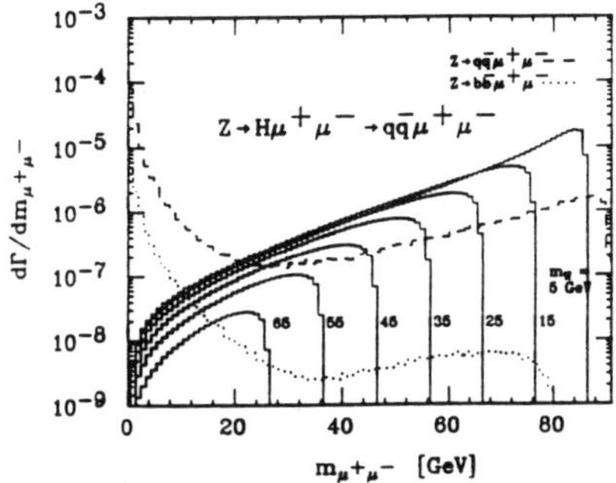

Figure 2: The invariant mass distribution of the muon pair, $d\Gamma/dm_{\mu^+\mu^-}$, produced in both $Z \to q\bar{q}\mu^+\mu^-$ decay and $Z \to H\mu^+\mu^- \to q\bar{q}\mu^+\mu^-$ decay for $m_H = $ 5, 15, 25, 35, 45, 55 and 65 GeV. The branching ratio for Higgs decay into heavy quarks, $H \to b\bar{b}$ and $H \to c\bar{c}$, has been folded into the signal, while the background is summed over all quark flavours. The contribution from $Z \to b\bar{b}\mu^+\mu^-$ is shown separately.

The Higgs particle may also be produced in association with a photon, $Z \to H\gamma$, which, because both the Z and H are neutral, occurs via top quark and W boson loops [6]. The branching fraction is therefore small (see Fig. 1), however, the phase space is larger than in the three body $Z \to H\mu^+\mu^-$ decay, and $Z \to H\gamma$ becomes important for $m_H \gtrsim 60$ GeV. Due to charge conjugation, the contribution from the top quark loop is proportional to the product of the electric charge and the small vector coupling with the Z. Therefore, the W loops dominate, leading to the very small top quark mass dependence shown in Table 1 and Fig. 1.

In principle, this decay is sensitive to the untested $WW\gamma$, WWZ and $WWZ\gamma$ vertices. Furthermore, because the Higgs boson couples to the mass of the particle in the loop, heavy particles do not decouple and the decay rate is sensitive to the existence of heavy particles that couple to both Z and γ. For example, in supersymmetric models, the scalar top quark and chargino loops contribute, and may change the decay rate appreciably [7]. Although, the $Z \to H\gamma$ rate is too small to effectively search for the Higgs, it would be an interesting

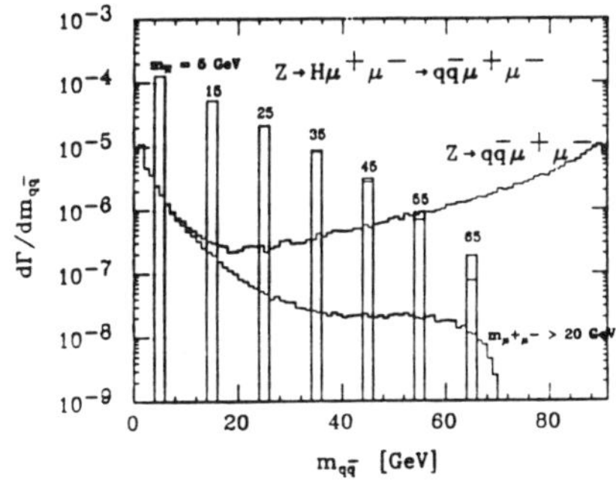

Figure 3: The invariant mass distribution of the quark pair, $d\Gamma/dm_{q\bar{q}}$ produced in both $Z \to q\bar{q}\mu^+\mu^-$ decay and $Z \to H\mu^+\mu^- \to q\bar{q}\mu^+\mu^-$ decay for $m_H = $ 5, 15, 25, 35, 45, 55 and 65 GeV. Both $H \to b\bar{b}$ and $H \to c\bar{c}$ decays are included in the signal, while the background is summed over all quark flavours. We also show the effect of making a cut on the invariant mass of the muon pair, $m_{\mu^+\mu^-} > 20$ GeV.

reaction to study once the Higgs boson has been found.

Finally, the Higgs boson may also be pair produced, $Z \to HHZ^* \to HHf\bar{f}$, [8] or produced with gluons, $Z \to Hgg$, [9]. Both of these decays are too rare to be seen at LEP.

3. W-Boson Production

The decay $Z \to Wf\bar{f}'$ was first proposed as a possible source of W-bosons [10], however, the branching rate for on-shell W production is rather small,

$$Br(Z \to W^{\pm}X) = 1.6 \ 10^{-7}. \tag{3}$$

There are two reasons for this. Firstly, because of the relatively small $W - Z$ mass difference the available phase space is very small and secondly, there is a significant cancellation between diagrams containing the ZWW vertex and those which don't. Recently, Barger and Han [11] have investigated the contributions from off-shell W production in the $Z \to \ell^{\pm}\nu q\bar{q}'$ decay and find,

$$Br(Z \to \ell^{\pm}\nu q\bar{q}') = 1.5 \ 10^{-7}, \tag{4}$$

three times larger than in the on-shell case.

The major backgrounds from $Z \to \tau^+\tau^-$, $b\bar{b}$ or $c\bar{c}$ followed by one leptonic and one hadronic decay are easily eliminated by isolation cuts on the charged lepton and missing energy vector and by an invariant mass cut on the invariant mass of the hadrons, $m_{hadron} > m_{\tau}$. Nevertheless, because of the gauge cancellation, the rate is too small to observe. On the other hand, if the ZWW vertex were to deviate from the standard model, the cancellations might be spoiled leading to an enhanced rate.

The most general Lorentz invariant effective Lagrangian for ZWW interactions may be described in terms of seven form-factors [12]. In models where mixing of the $SU(2)$ and $U(1)$ neutral gauge bosons occurs, electromagnetic gauge invariance eliminates three of these form-factors [13] while a fourth is heavily constrained by measurements of the neutron electric dipole moment [14]. The effective Lagrangian may then be written in terms of the

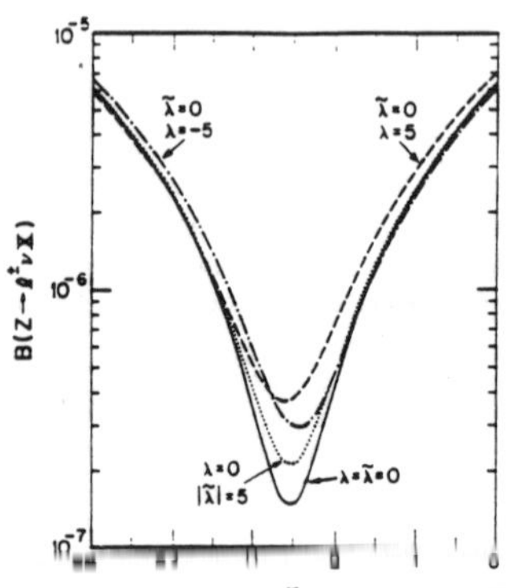

Figure 4: The $Z \to \ell^{\pm}\nu X$ branching fraction as a function of κ. The solid curve is for $\lambda = \bar{\lambda} = 0$; the dotted curve is for $\lambda = 0$, $|\bar{\lambda}| = 5$; the dashed curve is for $\lambda = 5$, $\bar{\lambda} = 0$; the dot-dashed curve is for $\lambda = -5$, $\bar{\lambda} = 0$. This figure is taken from ref. [11].

three remaining form-factors, κ, λ and the CP violating $\tilde{\lambda}$,

$$\frac{\mathcal{L}_{ZWW}}{g\cos\theta_W} = W^\dagger_{\mu\nu}W^\mu Z^\nu - W^\dagger_\mu Z_\nu W^{\mu\nu} + \kappa W^\dagger_\mu W_\nu Z^{\mu\nu}$$
$$+ \frac{\lambda}{M_W^2}W^\dagger_{\lambda\mu}W^\mu_\nu Z^{\nu\lambda} + \frac{\tilde{\lambda}}{2M_W^2}W^\dagger_{\lambda\mu}W^\mu_\nu \epsilon^{\nu\lambda\alpha\beta}Z_{\alpha\beta}. \tag{5}$$

In the standard model, $\kappa = 1$, while $\lambda = \tilde{\lambda} = 0$. As shown in Fig. 4 [11], deviations from the standard model values can lead to a significant increase in the $Z \to \ell^\pm q\bar{q}'$ rate. After cuts to eliminate backgrounds, Barger and Han [11] find that with $5\ 10^7$ Z events, one could test deviations of $\Delta\kappa = \pm 1$ at the 95% confidence level.

4. Lepton Flavour Violation

In the standard model lepton flavour is absolutely conserved. At LEP, it is possible to search for lepton flavour violating decays of the Z, the observation of which would then be a clear indication of physics beyond the standard model. Alternatively, the absence of such decays places potentially much more stringent bounds on lepton flavour violation than those from low energy data, and may restrict models which contain lepton flavour violation.

The most general lepton flavour violating (LFV) effective Lagrangian for interaction between leptons l_i and $\bar{l}_j$ is,

$$\mathcal{L}_{ij}^{LFV} = -i\,g_Z\,\bar{l}_i\,\gamma^\mu\left(a_L^{ij}\left(\frac{1-\gamma_5}{2}\right) + a_R^{ij}\left(\frac{1+\gamma_5}{2}\right)\right)Z_\mu\,l_j$$
$$+ g_Z\,\bar{l}_i\,\sigma^{\mu\nu}\frac{k_\nu}{M_Z}\left(b_L^{ij}\left(\frac{1-\gamma_5}{2}\right) + b_R^{ij}\left(\frac{1+\gamma_5}{2}\right)\right)Z_\mu\,l_j + \text{h.c.}, \tag{6}$$

where k_ν is the Z boson four momentum. Unlike the $b_L(b_R)$ terms, which must be generated via loop interactions, the $a_L(a_R)$ terms can arise either at tree level through mixing in models with extra exotic particles or at the one loop level, if the tree level couplings are flavour diagonal.

The branching ratio for lepton flavour violating Z decay into massless leptons is,

$$\frac{Br(Z \to l_i^+ l_j^- + l_i^- l_j^+)}{Br(Z \to \mu^+\mu^-)} = \frac{2\,(a_L^{ij2} + a_R^{ij2}) + b_L^{ij2} + b_R^{ij2}}{c_L^2 + c_R^2}, \tag{7}$$

where $c_L = -\frac{1}{2} + \sin^2\theta_W$ and $c_R = \sin^2\theta_W$. Although the lepton violating couplings a_L^{ij}, a_R^{ij}, b_L^{ij} and b_R^{ij} are a priori unknown, some limits may be extracted from low energy data (see Table 2). We note that at low energy $k^2 \ll M_Z^2$, and therefore, low energy data is essentially insensitive to both b_L^{ij} and b_R^{ij}, and only a_L^{ij} and a_R^{ij} are constrained. Non zero values for b_L^{ij} and b_R^{ij} will increase the lepton flavour violating decay of the Z, and the observation of such decays at rates larger than the limits quoted below would indicate the presence of lepton flavour violating $\sigma^{\mu\nu}$ terms in the Lagrangian.

The event signature for the processes $Z \to e\tau$ or $\mu\tau$ is very distinctive. An energetic electron (muon) of beam energy recoils against a τ which then provides a well-defined

$l_i l_j$	Upper Limit	Process giving limit
$e\mu$	2.2×10^{-11}	$\mu \to eee$
$e\tau$	5.0×10^{-3}	$\tau \to e\mu$
$\mu\tau$	3.6×10^{-3}	$\tau \to \mu\mu\mu$

Table 2: Limits on $Br(Z \to l_i l_j)$ from low energy data.

signature of one or three charged prongs plus missing energy and momentum carried off by undetected neutrinos. Event selection and background suppression as well as the influence of the detector (resolution, inefficiencies, etc) have been discussed in ref. [15].

The limiting background is $Z \to \tau^+ \tau^-$, where one $\tau^\pm \to e^\pm \nu_e \nu_\tau$ or $\tau^\pm \to \mu^\pm \nu_\mu \nu_\tau$, with electron (muon) energy close to the end-point of its spectrum. To suppress this background, one has to take advantage of the fact that the energy distribution of the electron (muon) produced in τ decays is a smooth linear distribution near the end-point, while the expected signal for the electron (muon) produced in the process $Z \to e\tau$ ($\mu\tau$) would be (after convoluting with the detector resolution) roughly a Gaussian distribution with a radiative tail. For a detector of resolution $\sigma_E/E = 1\%$, the prediction [15] is,

$$Br(Z \to l^\pm \tau^\mp) < 7 \times 10^{-5} \ (l = e, \mu) \text{ for } 10^6 \ Z \text{ events,}$$
$$Br(Z \to l^\pm \tau^\mp) < 7 \times 10^{-6} \ (l = e, \mu) \text{ for } 10^7 \ Z \text{ events,} \tag{8}$$

which improves the limits given in Table 2 by at least two orders of magnitude and begins to constrain exotic models.

5. Triple Photon Decays

Although the decay of an on-shell Z-boson into two photons is forbidden by Yang's theorem, the Z-boson may decay into three photons. In the standard model this is achieved through charged fermion or W boson loops.

Both triple gauge boson vertices ZWW and γWW contribute as do the quadruple gauge boson vertices $Z\gamma WW$ and $\gamma\gamma WW$. No estimate of the W loop contribution exists at present, however, due to the fact that $M_W > M_Z/2$, the W loops have already decoupled and their contribution is expected to be small.

As in the case of the $Z \to H\gamma$ decay, only the vector coupling contributes in the fermion loop and, ignoring the W-boson loop contributions, the three photon decay width is,

$$\Gamma(Z \to \gamma\gamma\gamma) = 0.7 \text{ eV}. \tag{9}$$

This is clearly unobservable at LEP.

On the other hand, composite models often generate a large $Z \to \gamma\gamma\gamma$ rate [16]. For example, if the Z boson is a bound state of constituents with electric charge Q, then [17],

$$Br(Z \to \gamma\gamma\gamma) = 2 \ 10^{-4} <Q^3>^2 . \tag{10}$$

Alternatively, residual four boson contact terms at a scale Λ can also lead to a sizeable branching ratio [16],

$$Br(Z \to \gamma\gamma\gamma) = 7 \; 10^{-9} \left(\frac{m_Z}{\Lambda}\right)^8 , \tag{11}$$

provided $\Lambda < m_Z$. Other possibilities also exist [18].

The limit of observability is determined by the pure QED process,

$$e^+e^- \to \gamma\gamma\gamma, \tag{12}$$

where, in general, two photons are energetic and approximately back-to-back while the third photon is relatively soft. This contrasts with the signal which is more 'Mercedes'-like with three well separated energetic photons. Nevertheless, current estimates [19] indicate that (12) provides an irreducible background corresponding to a $Z \to \gamma\gamma\gamma$ branching ratio of 10^{-5}. OPAL have placed a limit on the $Z \to \gamma\gamma\gamma$ branching ratio [20],

$$Br(Z \to \gamma\gamma\gamma) < 2.8 \; 10^{-4}, \tag{13}$$

which still leaves a significant window for new physics.

Finally, there has been recent interest in the $Z \to \pi^0\gamma$ decay. Early estimates found the extremely small branching fraction of $\sim 10^{-11}$ [21] due mainly to the neutral current form-factor $\sim m_\pi^2/m_Z^2$ [22]. Jacob and Wu [23] have claimed that the axial anomaly prevents this form-factor dependence and find $Br(Z \to \pi^0\gamma) = 1.7 \; 10^{-3}$. However, their arguments, based on soft pion results, are almost certainly wrong [24] and the $Z \to \pi^0\gamma$ decay is expected to be unobservable at LEP. Based on the absence of events of this nature in the 1989 data, OPAL and ALEPH place the limits,

$$\begin{aligned} Br(Z \to \pi^0\gamma) &< 3.9 \; 10^{-4}, \quad [20] \\ Br(Z \to \pi^0\gamma) &< 4.9 \; 10^{-4}. \quad [25] \end{aligned} \tag{14}$$

6. Summary

With a data sample of 10^6 Z boson events, experiments at LEP should be able to probe the Higgs sector up to $m_H \sim 40-45$ GeV, improve limits on lepton flavour violating decays by nearly two orders of magnitude and saturate the possibility of observing an anomalously large $Z \to \gamma\gamma\gamma$ rate. The observation of such events is a clear indication of new physics.

Acknowledgements

I thank W. Bernreuther, M. Duncan, J. Gomez-Cadenas, C. Heusch and in particular J. van der Bij and R. Kleiss for an enjoyable collaboration. I am also happy to thank C. Verzegnassi and G. Altarelli for several stimulating discussions.

References

1. E. W. N. Glover and J. J. van der Bij, in *Z Physics at LEP I*, CERN 89-08, Vol. 2, p1 (1989).

2. J. D. Bjorken, in *Weak Interactions at High Energy and the Production of New Particles*, Proceedings of the SLAC Summer Institute on Particle Physics, 1976, p1 (1976); J. Finjord, Physica Scripta **21**, 143 (1980).

3. OPAL Collab., M. Z. Akrawy *et al.*, presented by R. Barlow at *Symposium on Z Physics*, Madison (1990).

4. ALEPH Collab., D. Decamp *et al.*, CERN preprint CERN-EP/90-16 (1990).

5. E. W. N. Glover, R. Kleiss and J. J. van der Bij, FNAL preprint FERMILAB-PUB-89/244-T.

6. R. N. Cahn, M. S. Chanowitz and N. Fleishon, Phys. Lett. **B82**, 113 (1979); L. Bergström and G. Hulth, Nucl. Phys. **B259**, 137 (1985).

7. G. Gamberini, G. F. Guidice and G. Ridolfi, Nucl. Phys. **B292**, 237 (1987); T. J. Weiler and T.-C. Yuan, Nucl. Phys. **B318**, 337 (1989).

8. E. W. N. Glover and A. D. Martin, Phys. Lett. **226B**, 393 (1989).

9. B. A. Kniehl, Madison preprint MAD/PH/552 (1990).

10. F. M. Renard and M. Talon, Phys. Lett. **B82**, 113 (1979); W. J. Marciano and D. Wyler, Z. Phys. **C3**, 181 (1979).

11. V. Barger and T. Han, Madison preprint MAD/PH/548 (1989).

12. K.-I. Hikasa, K. Hagiwara, R. D. Peccei, D. Zeppenfeld, Nucl. Phys. **B282**, 253 (1987).

13. M. Kuroda, F. M. Renard and D. Schildknecht, Phys. Lett. **183B**, 366 (1987).

14. W. J. Marciano and A. Queijeiro, Phys. Rev. **D33**, 3449 (1986).

15. J. J. Gomez-Cadenas and C. A. Heusch, Snowmass: DPF Summer Study 1988, p247, (1989).

16. F. Boudjema and F. M. Renard, in *Z Physics at LEP I*, CERN 89-08, Vol. 2, p182 (1989).

17. F. M. Renard, Phys. Lett. **116B**, 264 (1982), *ibid.* **116B**, 269 (1982).

18. F. M. Renard, Phys. Lett. **126B**, 59 (1983), *ibid.* **132B**, 449 (1983).

19. D. Treille *et al.*, in *ECFA Workshop on LEP 200*, CERN 87-08, p414 (1987).

20. OPAL Collab., M. Z. Akrawy *et al.*, CERN preprint CERN-EP/90-29 (1990).

21. L. Arnellos, W. J. Marciano and Z. Parsa, Nucl. Phys. **B196**, 378 (1982).

22. G. P. Lepage and S. J. Brodsky, Phys. Lett. **87B**, 359 (1979); A. Duncan and A. Mueller, Phys. Rev. **D21**, 1636 (1980).

23. M. Jacob and T. T. Wu, Phys. Lett. **B232**, 529 (1990).

24. N. Deshpande, P. Pal and F. I. Olness, Oregon preprint, OITS-433-R (1990); K.-I. Hikasa, KEK preprint, KEK-TH-246 (1990); T. Schröder, Heidelberg preprint, HD-HEP-90-16 (1990); T. N. Pham and X. Y. Pham, Paris preprint, PAR-LPTHE 90-16 (1990).

25. ALEPH Collab., D. Decamp *et al.*, CERN preprint CERN-EP/90-23 (1990).

Forward-Backward Charge Asymmetry in $e^+e^- \rightarrow$ Hadron Jets

Winston Ko, David Stuart, and the AMY Collaboration

ABSTRACT

The forward-backward asymmetry of quarks produced in e^+e^- annihilations, summed over all flavors, is measured at $\sqrt{s}$ between 50 and 60.8 GeV, near the maximum expected in the 5 flavor standard model, The asymmetry is found to be $(8.3 \pm 2.9 \pm 1.9)\%$ in agreement with the standard model.

Because of the interference of γ and Z exchange in e^+e^- annihilation into a quark pair, the angular distribution of the quarks is asymmetric. Since this is sensitive to the *interference* of the γ and Z, it is an excellent test of the Standard Electroweak theory. We do not know the flavor or direction of the original quarks, but we can sum over all flavors and use the thrust axis to define the direction of the original quarks. With the production angle defined between the incoming electron and the outgoing negatively charged jet, the u and c quarks will contribute with a positive asymmetry, due to their charge, while the d, s, and b quarks will contribute with a negative asymmetry. Since the u and c contributions dominate, the net asymmetry is positive. The figure below shows the expected asymmetry for u-type quarks (dot-dash curve), d-type quarks (dashed curve), and a combination of all 5 flavors (solid curve) as a function of the center of mass energy. The asymmetries peak between 55 and 75 GeV and then fall off gradually across the Z.

We measured the hadron jet charge asymmetry at the KEK e^+e^- collider, TRISTAN. We used a $27.4pb^{-1}$ sample of 3211 multihadronic events collected by the AMY detector at center-of-mass energies of 50 to 60.8 GeV ($< \sqrt{s} >= 56.6$ GeV). In order to assure a reasonably pure sample of two jet events, we select high thrust events ($T > 0.8$). Each event is split into two jets with the plane perpendicular to the thrust axis. Jet charge is calculated by summing the charge of all the particles in the jet, weighted by their rapidity with respect to the thrust axis. This is then divided by the total weighting to obtain an average charge. We require that the difference of the two jet charges be greater than 0.15 to improve the identification of the charge direction of the jet pair which is needed to define the production angle. Monte Carlo studies show that the charge direction is assigned correctly in about 75% of the events.

We measure an asymmetry of $(9.3 \pm 3.1 \pm 2.0)\%$ which agrees well with the standard model prediction of 9.7 %. This measurement depends on the method used to determine the jet charge. In order to compare to lower energy experiments, we unfold a method independent asymmetry by taking into account the probability of misidentifying the charge direction of the jet pair. In this manner, we unfold a method independent asymmetry of $(8.3 \pm 2.9 \pm 1.9)\%$. The standard model prediction is 8.7%. This method independent asymmetry is plotted along with values calculated from data published by MAC[1] and JADE[2] on the figure below.

Note added in proof: The details of this work were published after the meeting in Phys. Rev. Lett. **64**, 983 (1990).

[1] MAC Collaboration, W. Ash et al., Phys. Rev. Lett. **58**, 1080 (1987).

[2] JADE Collaboration, T. Greenshaw et al., Z. Phys. C **42**, 1 (1989).

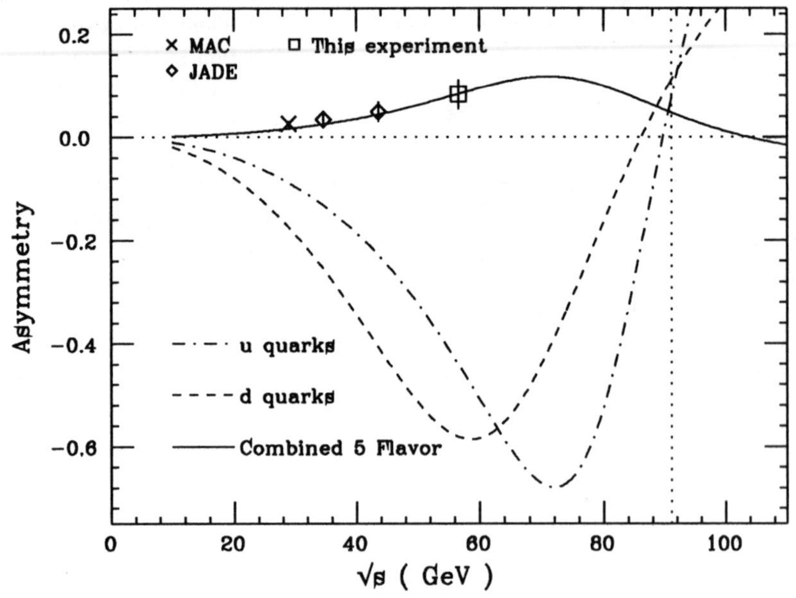

SEARCH FOR ISOLATED PHOTONS IN ELECTRON-POSITRON COLLISIONS

KANGPING HU
Physics Department, VPI & SU
Blacksburg, Virginia 24061

ABSTRACT

Studies have been made of the direct production of energetic
photons produced in multihadronic electron-positron annihilation events using the AMY detector at the KEK TRISTAN
collider.The energy spectrum of the photons is compared with
that expected from initial and final state radiation and from
quark fragmentation with five quark flavors.

1. INTRODUCTION

From conservation of 4-momentum, production of a Higgs
boson through electron-positron annihilation could be
accompanied by an isolated photon which would have a monochromatic peak in the recoil mass spectrum.
We have recorded data with integrated luminosities of 0.67,
4.0, 3.3, 6.0, 1.0, and 4.4 inverse picobarns at center-of-
mass energies of 50, 52, 55, 56, 56.5, and 57 GeV, respec-
tively.Studies of isolated photon production have been made
for these data. The event selection and background calculation will be discussed.

2. EVENT SELECTION CRITERIA

We select multi-hadronic annihilation events by requiring
five or more good charged tracks, a total visible energy
(Evis) greater than half of the C.M. energy, a momentum
imbalance along the beam direction with a magnitude less
than 0.4Evis and more than 3 (5) GeV energy deposited in
the barrel calorimeter at 50 and 52 (55 to 57) GeV.
From the multi-hadronic event sample, we select events
where an energetic photon (Ephoton/Ebeam greater or equal
to 0.32) is isolated by the requirement that less than 2GeV
of additional energy be contained in a cone of twenty
degrees about the photon direction.
Also in order to reduce events from radiative background,
we reject events with thrust greater than 0.95 and with the
angle between the isolated photon and thrust axis less than
five degrees.
The efficiency for event selection after the above cuts is
shown in Fig.1 as a function of the Higgs mass.

434

3. BACKGROUND CALCULATION

The total background is the sum of the radiative background
and hadronic background. The radiative background includes
the contribution from initial state radiation, final state
radiation and the interference between these two terms[1].
The hadronic background is from the contribution of the
quark fragmentation with five quark flavors, calculated with
the parton shower model in the Lund 6.3 program[2].

4. RESULT

We define recoil mass for the isolated photon events as
$$RM = sqrt[Ecm (1 - Ephoton/Ebeam)]$$
The recoil mass spectrum for the experimental data is shown
in Fig.2, also shown in the same graph is the recoil mass
spectrum expected from the background Monte Carlo. 62 events
survived all the cuts for the experimental data, but only
43.2(+/-)1.4 events survived for the background.

5. CONCLUSION

Experimental data is about 2.5-sigma higher than the background. No monochromatic peak in the recoil mass spectrum
is observed. More statistics in the data is needed to make
further investigations.

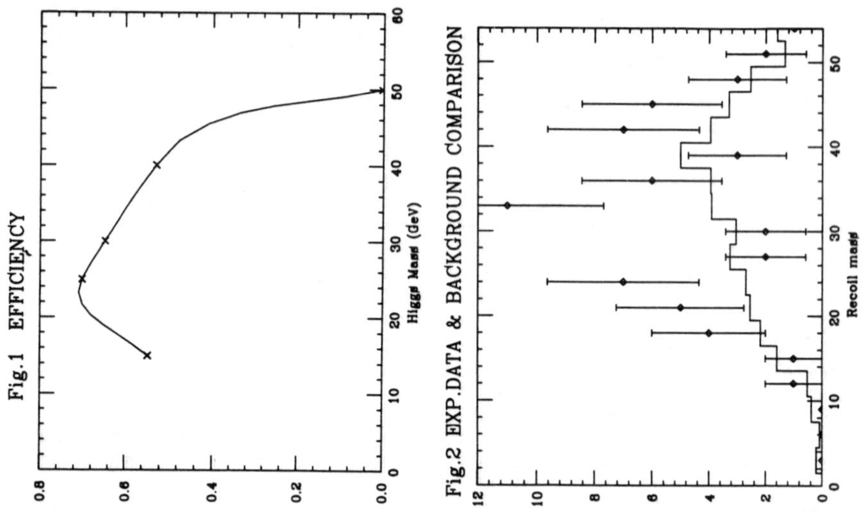

REFERENCES:

[1] J.Fujimoto, Y.Shimizu, Mod.Phys.Letters,A3(1988)581
[2] T.Sjostrand, M.Bengtsson, Comp.Phys.Commun.43(1987)367
    T.Sjostrand,Int.Journal of Mod.Phys.A,V.3,No.4,(1988)751

PRELIMINARY RESULTS ON $\sin^2\vartheta_W$ FROM DEEP INELASTIC NEUTRINO SCATTERING AT THE TEVATRON

R. BROCK

Department of Physics and Astronomy
Michigan State University
East Lansing, MI 48824

ABSTRACT

Preliminary determination of the ratio of deep inelastic neutral current to charged current events is presented with an emphasis on the experimental uncertainties associated with event selection. Preliminary extraction of $\sin^2\vartheta_W$ is provided with a discussion of the sensitivity to parton model uncertainties.

1. INTRODUCTION

There has been renewed interest in precision measurements of the weak mixing angle, $\sin^2\vartheta_W$[1]. While the top quark may continue to elude direct detection, it should be possible to bound the top quark mass from the comparison of a variety of processes which are precise and have different sensitivities in high order to a heavy fermion[2]. Pairs of measurements which do indeed have these characteristics are the precision measurements of the masses of the W and Z and the determination of $\sin^2\vartheta_W$ from the measurement of $R=\sigma_{NC}(\nu N \rightarrow \nu X)/\sigma_{CC}(\nu N \rightarrow \mu X)$ in deep-inelastic neutrino scattering[3] (DINS). Of course, inherent in the assumption of making a precision measurement is an adequate understanding of the experimental and theoretical uncertainties. The situation with regard to DINS is presently not clear due to the fact that the measurements which make up the world's data come from neutrino beams of low average energies of around 50GeV. As has been emphasized[1], these are energies in which the mass scale of the charm quark is not negligible and hence the parton model language requires substantial modification. The so-called slow-rescaling prescription is presently insufficient in its application to be totally reliable. While there are possible attempts to understand it better through future experimentation[4] and/or phenomenological effort[5], the only uncomplicated approach is to perform the measurement at neutrino energies which are not compromised by these non-scaling difficulties.

The FMMF collaboration[6] has performed such a measurement using data taken during the recent running of the Tevatron Quadrupole Triplet neutrino beam (QTB) and we report preliminary results here. The detector is the fine-grained Flash Chamber - Proportional Tube neutrino detector in Laboratory C at Fermilab. Details of this detector can be found elsewhere[7]. Briefly, it is a 300t (100t fiducial) digital flash-chamber calorimeter with fine sampling in both the transverse and longitudinal directions. The longitudinal sampling is $3\%\lambda$ and $22\%X_o$ and the density is 1.35g/cc. Typically, a minimum ionizing particle loses 3MeV/cm of energy as it traverses the calorimeter which is 1900cm long.

436

Downstream of the calorimeter is an iron toroid spectrometer with a drift plane readout in four stations. Detailed features of the Tevatron QTB have been described elsewhere[8]. The running period from which these data come was in the 1987/88 fixed target Tevatron run of 5×10^{17} proton on target at 800GeV/c. The restricted fiducial volume for this preliminary analysis results in 35000 fully reconstructed CC events (out of a total sample of 74000 analyzable CC events in the whole fiducial volume). The features of performing an experiment of this nature in the QTB relative to the more conventional Dichromatic beam include: an average interacting neutrino energy of 150 GeV; reduced misidentification of charged current (CC) events; minimal sensitivity to kinematical cuts; reduced background from cosmic rays; the presence of $\bar{\nu}$ in the ν beam; an increased background from decays in the hadronic showers leading to increased misclassification of neutral current (NC) events; and an increased background from ν_e-induced events.

2. EVENT CLASSIFICATION

There have been historically two methods of separating NC from CC events for the extraction of the observable, R: a determination of event "lengths" statistically and an event-by-event classification. The second method is employed by the fine-grained detectors (CHARM and FMMF). In our analysis, we have concentrated on developing pattern-recognition codes designed to find tracks and distinguish long hadrons from muons. Further, we have spent considerable effort in understanding the efficiency of these algorithms, often using the data itself to cross-check the codes.

We assign all events as falling into one of three categories: **CC** [an event which has an unambiguously identified μ due to a successful fit in the muon spectrometer downstream of the calorimeter]; **NC** [an event in which there is no track which satisfies the criteria for a μ (defined below)]; or **CC0** [an event which satisfies the μ criteria but does not have a successful fit]. Determination of R involves counting all of the events which fall into these categories, evaluating the number which are misidentified, and subtracting/adding events to account for these inefficiencies. A "μ" is <u>defined</u> to be a track which has a projected distance from the vertex of >1000cm or a track which leaves the side of the detector after traveling a projected distance of >500cm. Frequently, the data themselves are used to simulate desired final states. The most common "tricks" of this sort are the removal in software of tracks or the addition of "tracks" from an external data base.

NC events can be misinterpreted as CC events three different ways:

1. An event which has a π or K decay in the shower resulting in a muon which successfully penetrates the muon spectrometer iron and gives a successful fit. Such an event would be classified as **CC**.

2. An event which has a π or K decay in the shower sometimes results in a short, stopping track which goes a sufficient distance to be identified as a "μ" by the muon algorithm. Such an event is defined as ambiguous and given the designation **CC0**.

3. An event which has a long hadron track with insignificant showering at the stopping point, insignificant straggling, and still penetrates the calorimeter the required distance would also be classified as **CC0**.

The manner in which this source (NC→CC) of misidentification is handled is to measure the ways in which undesired "μ" are found by the software and correct for those losses. There are two, consistent, techniques used to measure this loss. The first technique assumes that the production of muons and long hadrons from decays in a hadron shower of E_h which would be found by the software would be essentially the same, whether the event was one instigated by an incoming hadron or by a neutrino. The second method assumes that, apart from charm-decay, extra muon candidates in neutrino events are all due to the same kind of background that would cause a true NC event to be misclassified as **CC**.

Throughout the running of the experiment there was continuous hadron beam exposure in the 15 seconds or so of flattop in which there was no fast neutrino extraction. This was for continuous energy calibration as well as for muon background determination. Therefore, there is a substantial body of data in which there are hadronic showers in the upstream third of the detector. Figure 1 shows the rate of muon production for various muon momenta at various E_h settings for this experiment as compared with a previous hadronic beam dump at Fermilab, E379[9].

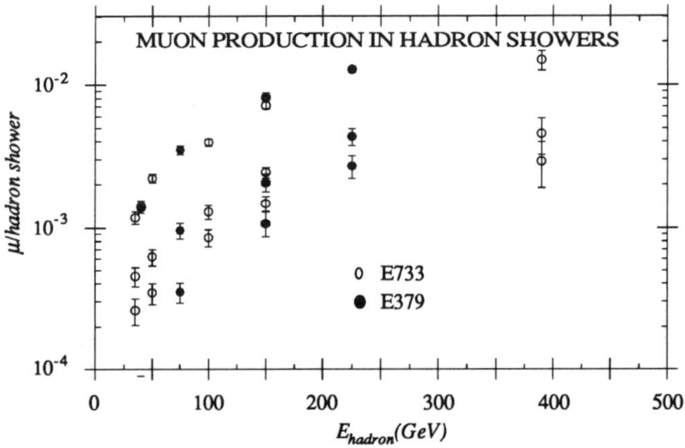

Figure 1. *Comparison of the production of muons from the Lab E Fermilab neutrino calorimeter from E379 (closed circles) and the Lab C calorimeter (open circles). The families of data are for (from top to bottom) $p_\mu>$8.45, 15.45, and 20.45 GeV/c. The E379 data have been scaled to the E733 density and atomic number.*

By removing in software the incoming hadron track, the hadron beam events are turned into "**NC**" candidate events. These events are then processed through the standard neutrino codes and classified as **NC**, **CC**, or **CC0**. However, since these data are all concentrated at the upstream end of the detector, this method, by itself, is insufficient to provide us with an acceptable background measurement. This is because of the differences in visibility of and acceptance for these muons, which travel the entire detector length as compared with those from neutrino data further downstream which have a shorter distance to travel before exiting the detector.

As we mentioned above, it is straightforward to find and ignore the presence of muons in neutrino events. This technique was applied to all **CC** events in order to determine how often there was a second "μ" found. The primary muon was ignored by the software (by isolating it within an angle/impact parameter window) and the remaining event was treated, in effect, as if it were a new event. In order to take into account the presence of semi-leptonic charm production and decay, a subtraction was calculated using the known $\nu N \rightarrow \mu\mu X$ cross section, our earlier analysis of charm production[10], and the assumption that all same-sign dimuon events originate from π/K decays[11]. In order to

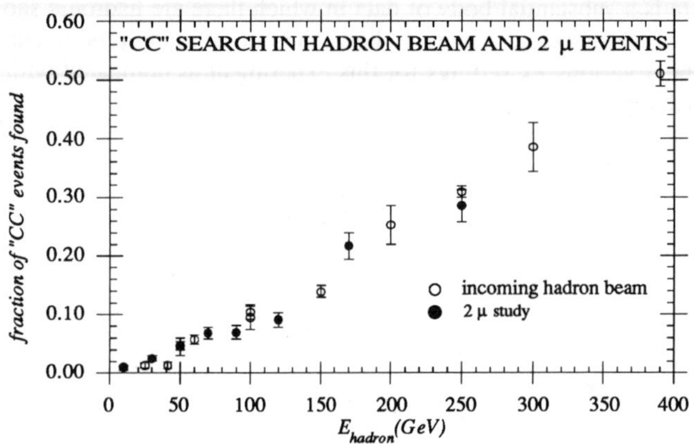

Figure 2. *The fraction of events with muons found by the pattern recognition software for incoming hadron beams (open circles) and dimuon events with the primary muon removed in software (closed circles). The dimuon data were limited to the region of the detector populated by the hadron beam for comparison.*

provide confidence in this method, the 2μ technique was compared to the test-beam technique *in that region of the detector in which the test beam is active.* Figure 2 shows the fraction of CC events found in the pseudo-NC samples for both techniques (after charm subtraction for the 2μ method) showing good agreement. When averaged over the observed E_h shape, the average fraction of "μ" found is 11.7±1.2%. The 2μ technique was used in the actual extraction of R.

The criteria for defining a muon track were not designed to find every muon, but rather to be easily and reliably simulated. The assumption in this analysis is that the kinematics of muons in DINS events is well known and, when combined with our understanding of the energy loss mechanisms in our calorimeter, will result in a reliable means of calculating how often real CC events will be misclassified as **NC** or **CC0** events. A Monte Carlo simulation of the physics of CC events was used which incorporated the μ criteria, the understanding of decay background, and the efficiency of the pattern recognition algorithm. The latter was determined by a scan of 5000 randomly chosen events in which three physicists analyzed each event to determine whether the algorithm would have classified the event as **NC**, **CC**, or **CC0**. Note that this is not the same as classifying the events as NC or CC according to the the scanner's judgement. From the scan it was determined that the algorithm is not consistent with its rules 0.92% of the time. This inefficiency was coded into the Monte Carlo. The net fractional CC→NC correction to the data can be described

in terms of the quantity $\Delta_v \equiv \dfrac{(CC \to NC) - (NC \to CC_{fit\ muon})}{CC}$. The result for Δ_v is 0.0382±0.0023 and constitutes the major uncertainty in R at the present time. A major contribution to the uncertainty in Δ_v is the statistical error on the scan, and so a sample which is three times the size of the original is presently being scanned.

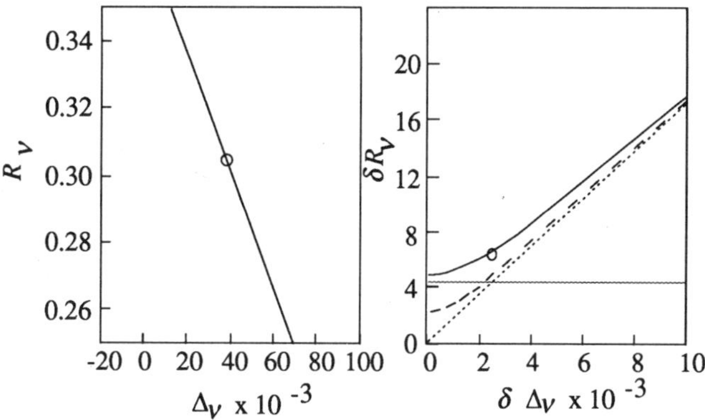

Figure 3. *(a) The contribution of Δ_V to R_V on the left. (b) The contribution to δR from the uncertainty in Δ_V on the right. For the latter, the contributions are: solid top curve, all experimental uncertainties; horizontal dotted line, statistical uncertainty; wide dashed curve, all experimental systematic uncertainties (exclusive of statistical contributions); and short dashed curve, the contribution due only to $\delta\Delta_V$. The open circles are the final values which contribute to the present result.*

This quantity is one of 12 separate systematic uncertainties which are considered in the analysis of δR. Figure 3a shows the variation of R *with* Δ_V showing why this is the most sensitive of all of our corrections. Figure 3b shows the contributions to δR as a function of $\delta\Delta_V$ from various sources.

3. PRELIMINARY RESULTS ON R AND EXTRACTION OF $\text{Sin}^2\vartheta_W$

From the above analysis, we preliminarily report that we find $R_V = 0.305\pm0.006$ where the error is a combination of statistical and experimental systematic uncertainties. We expect to be able to reduce the uncertainty in R by adding the complete data set and by more scanning. We hope to be able to get to the level of about ±0.003. We can extract $\sin^2\vartheta_W$

from this measurement and we find the preliminary result of $\sin^2\vartheta_W = 0.235\pm0.009$. The error is only experimental. This result is very preliminary and uses very simple assumptions about the parton densities, radiative corrections[12], isoscalar correction of 0.002, $s/\bar{u} = 0.5$, and $m_c = 1.5$ with the usual slow-rescaling prescription. We have checked that the sensitivity of $\sin^2\vartheta_W$ to the uncertainty on m_c is <u>half</u> of that of our earlier experiment at lower energies[13]. The subtraction for ν_e events is for the $K^\pm$ contribution only. This is not sufficient as the open geometry of the QTB leads to a contribution to this flux from K_L decay. The sensitivity to this is non-trivial: a 10% increase in the ν_e flux results in nearly a 1% change in $\sin^2\vartheta_W$. The actual effects for this uncertainty are being calculated at the present time.

[1] U. Amaldi *et al*, Phys. Rev. **D36** (1987) 1385 , R.Brock, "Deep Inelastic Neutrino Measurements of $\sin^2\vartheta_W$", published in Proceedings of the New Directions in Neutrino Physics at Fermilab, September, 1988.

[2] W. Marciano, Phys. Rev. **D20** (1979) 274, A. Sirlin, PRD22 (1980) 971, W. Marciano and A. Sirlin Phys. Rev. **D22** (1980) 2695, W. Marciano and A. Sirlin PR **D29** (1984) 945, M. Veltman, PL91B (1980) 95, F. Antonelli *et al.*, PL **91B** (1980) 90.

[3] G.Altarelli, "Theory of Precision Electroweak Experiments", Rapporteur's Talk at the XIVth International Symposium on Lepton and Photon Interactions at High Energy, Stanford, U.S.A., August 1989.

[4] N.W. Reay, P803, Fermilab Letter of Intent (April, 1989) and N.W. Reay, "Oscillation Search with an Emulsion Neutrino Detector", published in Proceedings of the New Directions in Neutrino Physics at Fermilab, September, 1988; N.Stanton, "What Can Be Learned from Charm Production by ν and ν Near Threshold?", to be published in Proceedings of Physics at Fermilab in the 1990s, August 1989.

[5] W.K. Tung, "Charm Production in ν and ν Deep Inelastic Scattering", to be published in Proceedings of Physics at Fermilab in the 1990s, August 1989.

[6] The FMMF group is a collaboration of Fermilab, Michigan State University, MIT, and the University of Florida. The subject of this talk is from the Ph.D. thesis work of George Perkins of Michigan State University.

[7] F.E. Taylor *et al.*, IEEE Trans. Nucl. Sci. **NS-27** (1980) 30, D. Bogert *et al.*, IEEE Trans. Nucl. Sci. **NS-29** (1982) 363, M. Tartaglia, Ph.D. Thesis, Massachusetts Institute of Technology (1984), unpublished.

[8] L. Stutte, " N-Center Wide Band Neutrino Beam", Fermilab TM-1305.

[9] Data from E379 are reproduced in: K. Lang, "An Experimental Study of Dimuons Produced in High Energy Neutrino Interactions", Ph.D. Thesis, University of Rochester, 1985.

[10] B. Strongin, "Opposite Sign Dimuon Production in High Energy Neutrino-Nucleon Interactions", Ph.D. Thesis, Massachusetts Institute of Technology, May 1989, unpublished.

[11] When the results of this method were extrapolated to the measurement of R, a 100% uncertainty on the charm subtraction was included in δR .

[12] A. Sirlin and W. Marciano, Nucl. Phys. **B189**, 442 (1981) 442..

[13] D. Bogert *et al.*, Phys. Rev. Lett. **55** (1985) 1969.

A Study of Semi-Muonic Charmed Particle Decays

Soren G. Frederiksen

(E653 Collaboration)

We present a study of the semi-leptonic decays of the D^0 using the Fermilab experiment E-653. 138 semi-muonic D^0 decays found in an emulsion target were used to determine the branching fraction ratio $B(D^0 \rightarrow K^- \mu^+ \nu_\mu)/B(D^0 \rightarrow \mu^+ X) = 0.32 \pm 0.07 \pm 0.02\%$. This ratio was then used to determine $\Gamma(D^0 \rightarrow K^- \mu^+ \nu_\mu) = (5.5 \pm 1.2(\text{stat.}) \pm 0.9(\text{syst.})) \times 10^{10} \text{ sec}^{-1}$.

Semi-leptonic charm decays have recently become of great interest because they are fairly easy to interpret theoretically [1]. One of the simplest decays is $D \rightarrow K l \nu$ which is usually treated as a spectator decay involving the emission of a virtual W, a process well understood and described in the standard model. Information about the hadronic portion of the decay, which is less well understood, can be obtained from the decay rate.

This analysis is based on data from the first run of Fermilab experiment E653, a hybrid emulsion-spectrometer experiment extensively described elsewhere[2, 3]. An 800 GeV/c proton beam hit an emulsion target upstream of a a large aperture spectrometer. Particles leaving the emulsion target were detected in 18 planes of silicon strip detectors, and were then bent by a dipole magnet and tracked by 10 planes of drift-chambers. Muons were tagged using a system consisting of drift chambers on either side of a toroid magnet, and two planes of muon hodoscopes, all of which were imbeded in 5 GeV of range steel. The detector was triggered by an interaction in the emulsion target with at least one muon, and offline cuts (which included in 8 GeV momentum cut on the muons) were then applied to select semi-muonic charm decays.

Out of 1206 charmed particle candidate events found in the emulsion 153 muonic two-prong decays (D^0) had both emulsion tracks matched to spectrometer tracks. In order to eliminate K^0 and Λ^0 decays, at least one of the two decay tracks had to have a $P_\perp$ with respect to the charm direction greater than 0.250 GeV/c. The application of this cut resulted in 138 semi-muonic two-prong decays, along with three four-prong decays consistent with charm.

The two main sources of background to the D^0 decays were secondary interactions, estimated to be 2 events, on the basis of observed 2 prong interactions with visible breakup stubs, and hadronic D^0 decays in which one of the hadrons decayed into a muon and was thus tagged as a semi-muonic decay. Using a complete Monte Carlo simulation of the experiment it was estimated that 16 ± 3 % of the events are due to this hadronic feedthrough.

Since semimuonic D^0 decays always involve an unseen neutrino, it is impossible to separate completely the decays which are $D^0 \rightarrow K^- \mu^+ \nu_\mu$, $D^0 \rightarrow \pi^- \mu^+ \nu_\mu X^0$ or $D^0 \rightarrow K^- \mu^+ \nu_\mu X^0$ (where X^0 is a non-zero number of π^0 and/or K^0). However, it is possible to

442

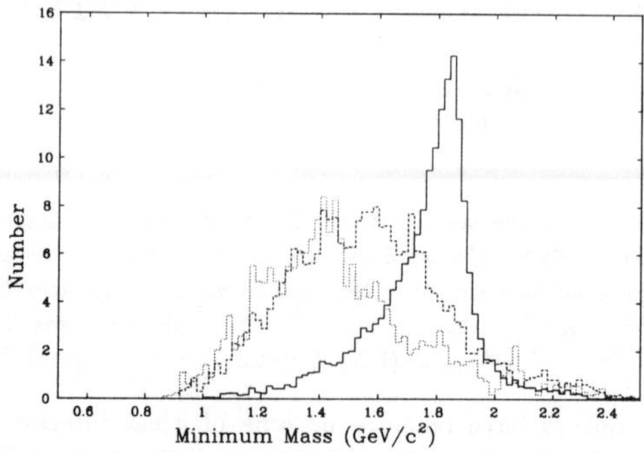

Figure 1: Monte Carlo $M_{\min}$ distribution. Assuming the decay $D^0 \to K^- \mu^+ \nu_\mu$.

determine the relative number of events for these decays on a statistical basis by using the "Minimum Effective Mass" ($M_{\min}$) technique. $M_{\min}$ is defined as

$$M_{\min} = \sqrt{m_{vis}^2 + p_\perp^2} + \sqrt{m_\nu^2 + p_\perp^2} \tag{1}$$

where m_{vis} is the effective mass of the observed particles from the decay, $p_\perp$ is the transverse momentum of the observed tracks with respect to the charm direction, and m_ν^2 is the assumed mass of the missing neutral in the decay, taken to be zero.

Figure 1 shows the $M_{\min}$ distribution for muonic two-prong decays generated by a Monte Carlo program which included the effects of experimental resolution and the detection efficiencies. The solid line is for the decay $D^0 \to K^- \mu^+ \nu_\mu$, while the dashed line is for the decay $D^0 \to K^{*-} \mu^+ \nu_\mu$, and the dotted line is for $D^0 \to K^{*-} \mu^+ \nu_\mu \pi^0$. The $M_{\min}$ distributions for the decay $D^0 \to K^- \mu^+ \nu_\mu$ has a sharp peak at the charm mass, and the peak of the distribution gets lower and broader as the number of neutrals in the decay increases.

A maximum likelihood fit was done to the experimental $M_{\min}$ data, to determine the relative fraction of the various decay modes. The fitted value must then be corrected for the relative finding efficiencies of the two modes, and for the hadronic D^0 background. Assuming the only multi-neutral mode is $D^0 \to K^{*-} \mu^+ \nu_\mu$, we find $\Gamma(D^0 \to K^- \mu^+ \nu_\mu)/\Gamma(D^0 \to \mu^+ X) = 0.31 \pm 0.09 \pm 0.03$, and if we assume $D^0 \to K^{*-} \mu^+ \nu_\mu \pi^0$, we find $0.38 \pm 0.05 \pm 0.02$. The most representative fit is one involving an equal mixture of the two mutli-neutral modes, and for that we find $0.32 \pm 0.07 \pm 0.02$.

Fig. 2 shows the $M_{\min}$ distribution for the 138 data events, the dashed histogram shows the fit for $\Gamma(D^0 \to K^- \mu^+ \nu_\mu)/\Gamma(D^0 \to \mu^+ X) = 0.32$. Using this ratio, assuming that [4] $B(D^0 \to \mu^+ X) = 0.96 \times B(D^0 \to e^+ X) = 7.4 \pm 1.1$, we determined $\Gamma(D^0 \to K^- \mu^+ \nu_\mu) = (5.5 \pm 1.2(\text{stat.}) \pm 0.9(\text{syst.})) \times 10^{10} \text{ sec}^{-1}$. This value can be compared to the Mark III decay width of [5] $(7.9 \pm 1.2 \pm 0.9) \times 10^{10} \text{ sec}^{-1}$, or to the result from Fermilab experiment E691 [6], $(8.9 \pm 1.2 \pm 1.4) \times 10^{10} \text{ sec}^{-1}$. Our value is lower than that found by Mark III and E691, but still consistent within errors.

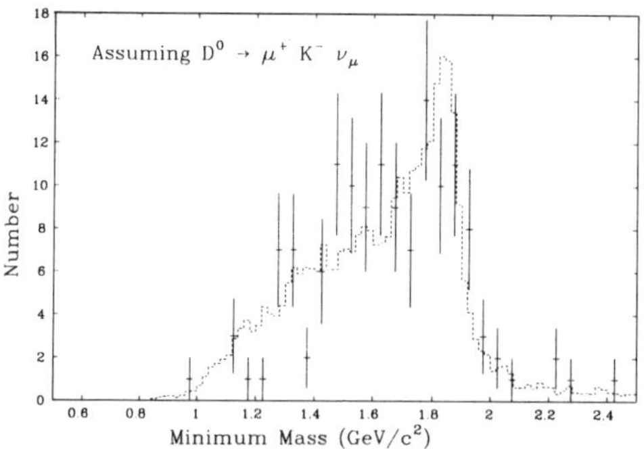

Figure 2: M_{min} distribution for Muonic V's assuming the decay $D^0 \to K^-\mu^+\nu_\mu$. The dashed line shows the Maximum Likelihood fit to the data.

We wish to acknowledge the efforts of the Fermi National Accelerator Laboratory staff. This work was supported in part by the US Department of Energy; the US National Science Foundation; The Japan Society for the Promotion of Science; the Japan-US Cooperative Research Program for High Energy Physics; the Ministry of Education, Science and Culture of Japan; the Korea Science and Engineering Foundation; and the Basic Science Research Institute Program, Ministry of Education, Republic of Korea.

References

[1] M. Wirbel, B. Stech, and M. Bauer, Z. Phys. C **29**, 637 (1985). T. Altomari and L. Wolfenstein, Phys. Rev. bf D37 681 (1988). M. Bauer and W. Wirbel, U. Heidelberg preprint HD-THEP 88-22 (1988). J.M. Cline, W.F. Palmer and G. Kramer, Phys. Rev. **D40** 793 (1989). D. Scora and N. Isgur, Phys. Rev. **D40** 1491 (1989). F.J. Gilman and R.L. Singleton, Stnaford Linear Accelerator preprint SLAC-PUB-5065 (1989).

[2] K. Kodama *et al.*, Nucl. Instr. & Meth. A **289**, 146 (1990).

[3] J.M. Dunlea, Ph.D. Thesis, The Ohio State University (1987). T.S. Jaffery, M.S. Thesis, University of Oklahoma (1988). A. Mokhtarani, Ph.D. Thesis, University of California, Davis (1988). G.A. Oleynik, Ph.D. Thesis, The Ohio State University (1987). K. Taruma, Ph.D. Thesis, Kobe University (1987) (in Japanese).

[4] G.P.Yost *et al.* (Particle Data Group), Phys. Lett. B **204**, 1 (1988).

[5] J. Adler *et al.*, Phys. Rev. Lett. **62**, 1821 (1989).

[6] J.C. Anjos *et al.*, Phys. Rev. Lett. **62**, 1587 (1989).

A MEASUREMENT OF Re(ε'/ε) WITH DOUBLE-BEAM TECHNIQUE

Hitoshi Yamamoto
(representing the E731 collaboration)

Enrico Fermi Institute, University of Chicago
Chicago, Illinois 60637

ABSTRACT

A recent measurement of the direct CP violation parameter Re(e'/e) by the E731 collaboration at Fermilab is reported. The emphasis is placed on control of systematic errors using the double-beam method.

1. INTRODUCTION

If the CP violation in 2π decays of neutral kaon is restricted to the CP contaminations in K_L and K_S states, then the rate of CP violation is expected to be independent of the isospin of final state; namely $\eta_{+-}=\eta_{00}$, where,

$$\eta_{+-} \equiv \frac{a(K_L \to \pi^+\pi^-)}{a(K_S \to \pi^+\pi^-)} = \varepsilon + \varepsilon', \quad \eta_{00} \equiv \frac{a(K_L \to \pi^0\pi^0)}{a(K_S \to \pi^0\pi^0)} = \varepsilon - 2\varepsilon'.$$

On the other hand, if there is non-vanishing CP violating decay amplitude $K_2(CP-) \to 2\pi$ (direct CP violation), then in general it would depend on the isospin structure of final state; thus, $\eta_{+-} \neq \eta_{00}$, or equivalently, $\varepsilon' \neq 0$. Such is the case for the standard model[1], while in the superweak model of CP violation[2] the direct CP violation is virtually zero.

In practice, the difference between η_{+-} and η_{00} is measured through the following double ratio:

$$R_d \equiv \frac{\Gamma(K_L \to \pi^+\pi^-)/\Gamma(K_S \to \pi^+\pi^-)}{\Gamma(K_L \to \pi^0\pi^0)/\Gamma(K_S \to \pi^0\pi^0)} = 1 + 6\mathrm{Re}\frac{\varepsilon'}{\varepsilon} \ .$$

Thus, when the double ratio is measured with an accuracy of 0.006, then Re(ε'/ε) is measured to 0.001.

2. BASIC DESIGN

The double ratio above involves four decay modes to be measured, and in order to achieve the desired accuracy, several problems have to be overcome. They are: 1) deadtimes, 2) gain and efficiency shifts in drift chambers and calorimeters, 3) understanding of acceptances as functions of P_K and decay position, and 4) background subtractions. Here, we will focus our attention on items 1), 2) and 3).[3]

2.1 Deadtimes

These are what can be called "passive inefficiencies", and primarily consist of deadtimes caused by data acquisition and accidental hits in veto counters (either at trigger stage or at offline analysis stage). These inefficiencies are independent of rate of signals themselves (thus "passive"). If each of the four modes are measured separately in time, these inefficiencies should be understood to much better than 1%, which is not an easy task since instabilities of beam intensity can readily cause tens of percents of variation in the absolute inefficiencies. The problem, however, can be effectively solved by taking the four modes in pairs. There are two ways:

I)   take two K_L modes simultaneously and two K_S modes simultaneously, or

II) take two $\pi^+\pi^-$ modes simultaneously and two $\pi^0\pi^0$ modes simultaneously.

In either case, the probability to lose signals due to the deadtime effects are identical for the modes taken at the same time, thus cancel in the double ratio.

2.2 Gain Shifts, Efficiency Shifts in Drift Chambers and Calorimeter

These are the instabilities in the parts of the detector which actively detect the signals, thus these inefficiencies can be called "active inefficiencies". Some of the drift chamber wires may die during the course of the experiment and gains of calorimeter blocks may change due to exposure to radiation, and these can directly affect the detection efficiencies. Since the variation in the drift chamber efficiency and that of the calorimeter are usually independent of each other, drifts in the relative efficiencies between the charged and neutral modes do not cancel when two K_L modes and two K_S modes are taken in pairs (choice I above).

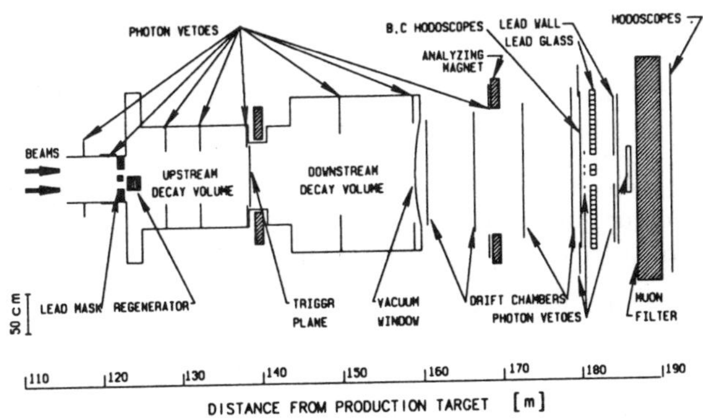

Figure 1. Side view of E731 detector.

The way around it is to take the four modes in pairs in such a way that two modes that use the same part of the detector are tken at the same time (choice II above). This way, for a given kaon momentum and decay position, the shift in the detection efficiencies cancel in the double ratio.

In order to take K_L and K_S decays simultaneously, we utilize a double K_L beam where in one of the beams a regenerator is placed to provide K_S. The regenerator alternates between the beams every spill (every minute) to equalize the various asymmetries. The cancellations described above are first order effects; higher order effects together with more precise accounts of the first order effects will be discussed later.

2.3 Detector and Data

Figure 1 shows the side view of the E731 detector. Neutral kaons are generated by 800 GeV protons hitting a Be target which is formed into a double beam through a two-hole collimator. The regenerator is a 2-interaction-length B_4C block implemented with veto counters to reject inelastic scatterings and is placed at z=123m (z is the distance from the target). Neutral decays are reconstructed by a 804-block lead-glass array with photon energy resolution of $2.5+5/\sqrt{E(GeV)}\%$, and charged decays are reconstructed by a 200 MeV-kick magnet and 16 layers of drift chambers with single-hit position resolution of 100 μm. In addition, 11 layers of photon veto counters are employed to detect photons escaping the calorimeter.

The data collection lasted from August of 1987 to February of 1988, and data analysis have been completed for 20% of the data where all four modes are taken simultaneously, which gives additional benefit of, among others, being able to use high statistics charged mode events (such as Ke3 decays) for calibration of the calorimeter and aperture surveys.

The statistics for the 20% of data is: $K_L\rightarrow\pi^+\pi^-$ 43357 (0.42%), $K_S\rightarrow\pi^+\pi^-$ 178803 (0.12%), $K_L\rightarrow\pi^0\pi^0$ 52226 (5.15%), and $K_S\rightarrow\pi^0\pi^0$ 201332 (2.57%), where numbers are before background subtractions while the percentages in the parentheses are background fractions. With small corrections due to acceptances and the K_L component in the K_S beam, the result for $Re(\varepsilon'/\varepsilon)$ is -0.0005 ± 0.0014 where the error is statistical only. In the following, we will analyze the double beam method in detail and attempt to obtain insights into systematic errors involved.

3. ANALYSIS OF DOUBLE-BEAM METHOD

3.1 Mathematical Formulation

The double beam is taken at a horizontal targeting angle of 5 mrad, and the two beams are separated vertically so that their momentum distributions are similar. Yet small differences in momentum distribution and intensity remain due mainly to a non-vanishing vertical targeting angle. In fact, the mean momentum was found about 1% higher for the bottom beam.

We denote the K_L flux (number of kaons per second) in the two beams by $f_i(P)dP$, where i = 1 (top beam) or 2 (bottom beam). Right after the regenerator, the K_L beam is attenuated by a factor s(P), and the regenerated K_S amplitude is given by $\rho(P) \equiv amp(K_S)/amp(K_L)$. The regeneration amplitude ρ varies as $P^{-\alpha}$ with $\alpha \approx 0.6$, while the attenuation s is consistent with being independent of P. Both s and ρ are specific to the regenerator block and do not depend on weather it is in the top beam or in the bottom beam.

For a given momentum P and decay position z, the probability for a decay to be detected (detection efficiency ε) depends not only in which beam the decay occurred but also in which beam the regenerator is in at that time. The detection efficiency could depend on the beam of the decay when, for example, there are more dead drift chamber wires in the top half or if the energy resolution is systematically worse in the top half of the calorimeter. It could also depend on the position of the regenerator if, for example, some veto counters near the regenerator are more efficient for the top half and the probability of accidental veto is greater when the regenerator is in the top beam. Thus, the number of events detected per second Y(P,z)dPdz for the top or bottom beam (subscripts 1 and 2) and for the case regenerator is up or down (subscripts u and d) is (Figure 2)

$$Y_{1u}(P,z) = f_1(P)\, s(P)\, |\rho(P)|^2\, g_S(P,z)\, \varepsilon_{1u}$$

$$Y_{2u}(P,z) = f_2(P)\, |\eta|^2\, g_L(P,z)\, \varepsilon_{2u}$$

$$Y_{1d}(P,z) = f_1(P)\, |\eta|^2\, g_L(P,z)\, \varepsilon_{1d}$$

$$Y_{2d}(P,z) = f_2(P)\, s(P)\, |\rho(P)|^2\, g_S(P,z)\, \varepsilon_{2d},$$

where $\eta = \eta_{+-}$ or η_{00} depending on the final state, and the geometric factors g_L and g_S depend only on whether it is K_L or K_S and normalized as $g_L=g_S=1$ at the regenerator.[4] Here, the 2π decays from the transmitted K_L beam in

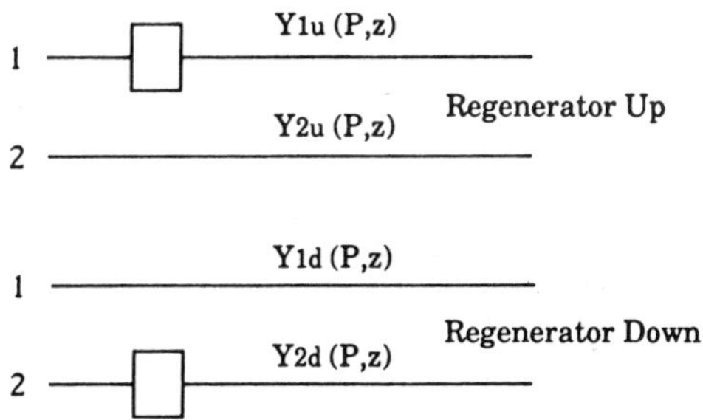

Figure 2. Two beams with two configurations of regenerator position.

the K_S beam is neglected (it is included in the final analysis). This is adequate for the discussion of systematics.

In order to calculate the double ratio R_d, one has to combine the regenerator-up data with the regenerator-down data to obtain the K_L-K_S yield ratio r for each final state; there are two ways:

a) simple mean

$$r = \frac{Y_{2u}+Y_{1d}}{Y_{1u}+Y_{2d}} = \frac{f_2\, \varepsilon_{2u} + f_1\, \varepsilon_{1d}}{f_1\, \varepsilon_{1u} + f_2\, \varepsilon_{2d}} \frac{|\eta|^2}{|\rho|^2} \frac{g_L}{s\, g_S} \,, \text{ or}$$

b) geometric mean

$$r = \left(\frac{Y_{2u}Y_{1d}}{Y_{1u}Y_{2d}}\right)^{1/2} = \left(\frac{\varepsilon_{2u}\,\varepsilon_{1d}}{\varepsilon_{1u}\,\varepsilon_{2d}}\right)^{1/2} \frac{|\eta|^2}{|\rho|^2} \frac{g_L}{s\, g_S} \,.$$

Note that the flux difference between the top and bottom beam canceled in the geometric mean. Also, it is straightforward to see that the statistical power is the same for both methods as long as Y_{2u} and Y_{1d} are comparable and Y_{1u} and Y_{2d} are comparable. The parameters ρ, s, and $g_{L,S}$ are common to both charged and neutral final states; thus, if one uses the geometric mean for both final states and form the double ratio, then one obtains $\mathrm{Re}(\varepsilon'/\varepsilon)$ provided that $\varepsilon_{2u}\,\varepsilon_{1d} = \varepsilon_{1u}\,\varepsilon_{2d}$.

$$R_d = r_{+-}/r_{00} = |\eta_{+-}|^2/|\eta_{00}|^2 = 1 + 6\mathrm{Re}(\varepsilon'/\varepsilon), \quad \text{if } \varepsilon_{2u}\,\varepsilon_{1d} = \varepsilon_{1u}\,\varepsilon_{2d}\,. \quad (1)$$

This condition is equivalent to saying that the detection efficiencies can be decomposed as below:

$$\varepsilon_{1u} = \varepsilon_1 \cdot \varepsilon_u, \quad \varepsilon_{2u} = \varepsilon_2 \cdot \varepsilon_u,$$

$$\varepsilon_{1d} = \varepsilon_1 \cdot \varepsilon_d, \quad \varepsilon_{2d} = \varepsilon_2 \cdot \varepsilon_d\,.$$

Before examining actual meaning of the above mathematical arguments, it is worth noting that even though the cancellation is not complete for the simple mean method it still is highly effective. For example, $f_{1,2}$ drop out if $\varepsilon_1=\varepsilon_2$ and $\varepsilon_u=\varepsilon_d$, $\varepsilon_{1,2}$ drop out if $\varepsilon_u=\varepsilon_d$ and $f_1=f_2$, and $\varepsilon_{1,2}$ drop out if $\varepsilon_u=\varepsilon_d$ and $f_1=f_2$ (assuming the decomposability). In fact, both methods were tried in the final analysis and they gave consistent results within 0.0001 in $\mathrm{Re}(\varepsilon'/\varepsilon)$.

3.2 Check of the Cancellations and Non-cancelling biases

The effects that can cause $\varepsilon_1 \neq \varepsilon_2$ are up-down asymmetries in the detector such as aforementioned dead wires in the top half of detector. The effect that can cause $\varepsilon_u \neq \varepsilon_d$ is the asymmetry in the intensity of the two beams that results in different accidental veto rate depending on which beam the regenerator is in. Or it can also be caused by an up-down asymmetry of veto counters coupled to the position of the regenerator as described earlier. As shown above, however, these asymmetries do not cause biases if the detection efficiencies are decomposable as in (1). Figure 3 shows the accidental veto rate of a photon veto counter for $K_{L,S} \to \pi^0\pi^0$ plotted as a function of time (for the 20% of data). The cut is tightend to kill about 10 times more than the one used in the actual analysis to enhance possible biasing effects. The veto counter is located right downstream of the regenerator, and thus could cause a bias through $\varepsilon_u \neq \varepsilon_d$. Even though the veto rate changes substantially with time, the ratio of K_L to K_S is constant within the statistical error (the plot is equivalent to the simple mean). The plot demonstrates the first order cancellations described in 2.1 rather than the higher order effects discussed here. If the second order effects are large, however, it could certainly show up in the plot.

What could invalidate the decomposability of the detection efficiencies? The condition (1) is rewritten as

$$\frac{\varepsilon_{1u}}{\varepsilon_{2u}} = \frac{\varepsilon_{1d}}{\varepsilon_{2d}}, \quad \text{or equivalently,} \quad \frac{\varepsilon_{1u}}{\varepsilon_{1d}} = \frac{\varepsilon_{2u}}{\varepsilon_{2d}}.$$

Namely, the necessary and sufficient condition for the method of geometric mean to cancel the efficiency asymmetries is that the ratio of the detection efficiency of the top beam to that of the bottom beam does not depend on the position of the regenerator (or equivalently, the ratio of efficiencies for the regenerator up to down does not depend on whether it is for the top or bottom beam). The difference between the sizes of regenerator-up data and regenerator-down data, which can be caused by fluctuation of beam intensity, has the same effect as $\varepsilon_u \neq \varepsilon_d$, and thus causes no bias.

In order for the ratio of top-beam to bottom-beam efficiencies to depend on the position of the regenerator, it is necessary for the decay products to interact with some effects caused by the regenerator. Thus, they should

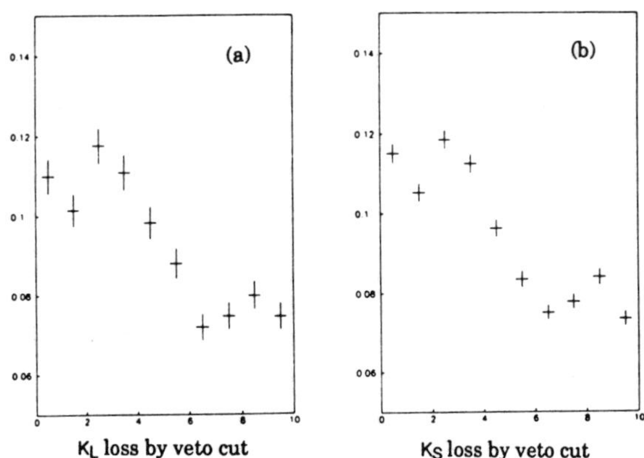

Figure 3. Fraction of good events killed by a photon veto for $K_L \to 2\pi^0$ (a) and $K_S \to 2\pi^0$ (b) plotted as a function of time.

have to do with "active inefficiencies". Possibilities are: 1) charged mode trigger plane counters, and 2) accidental overlaps both in neutral and charged modes.

The charged mode trigger hodoscope is located at the end of decay volume. The scintillators are arranged vertically so that each counter sees the entire vertical fiducial distance covering both beams. Individual counter is looked on each end by a phototube and pulse heights from two ends are analog-summed before discriminated to generate a trigger. This design combined with a high efficiency (99.7%) ensured that inefficiency is not caused preferentially to the decays in the K_S beam.

The accidental hits in the tracking chambers and the calorimeters are found more around the K_L beam (only in the regions close to the regenerator the accidental hits are more in the neighborhood of the K_S beam). Since the chamber hits or the photon clusters from kaon are centered around the beam of origin, this could cause the K_L decays to be killed more than K_S decays due to accidental overlap of hits. In order to study these effects, "accidental events" are taken throughout the run triggering randomly on RF buckets.[5] After proper veto cuts are applied, the accidental events had an average of 0.027 clusters and 8.5 drift chamber hits. These accidental events were overlaid on Monte Carlo 2π events to find out possible biases. The losses were about 3% for each mode and there was no bias observed within a statistical error of 0.07% in R_d which we take as the systematic error due to accidental overlap effects.

3.3 Momentum and Vertex Distributions

In the calculation of the double ratio, the factor $g_L / (|\rho|^2 s\, g_S)$ dropped out because it is common to both neutral and charged mode, which is not the case if P and z are not measured correctly. The result, however, is not affected if $g_L / (|\rho|^2 s\, g_S)$ is a constant, or the true P,z distribution of K_S is the same as that of K_L. In general, it is desirable to have similar P,z distribution for K_L and K_S.

When integrated over z, the P distribution for K_L is quite similar to that of K_S (Figure 4). This is because the boost factor 1/P for K_L is closely matched by the momentum dependence of $|\rho|^2 \propto P^{-1.2}$. However, the vertex distribution is quite different for K_L and K_S (Figure 5). This makes it important to understand the acceptance as a function of z. The acceptance is obtained by a detailed Monte Carlo simulation of beam and detector; thus, a good check of acceptance is to see the agreement of z vertex distribution between data and the Monte Carlo. Figure 6 shows the comparison of z distributions of data vs Monte Carlo for $K_L \rightarrow \pi^+\pi^-$ decays. The dotted line corresponds to an error that would cause the $Re(\epsilon'/\epsilon)$ value to increase by 0.0033, which appears to be unlikely. The agreement is equally good in the neutral mode. Also, the acceptance was studied using 1×10^7 Ke3 decays and 6×10^6 $3\pi^0$ decays, both of which showed excellent agreements between data and Monte Carlo. Together with the fit to τ_S and Δm which gave values consistent with accepted values, the systematic error due to acceptance is estimated to be 0.25% in R_d.

3.4 Moving the Regenerator or Target

A possible solution to the different z distribution for K_L and K_S is to move the regenerator (or a K_S target as in the case of the NA31 detector[6]) along

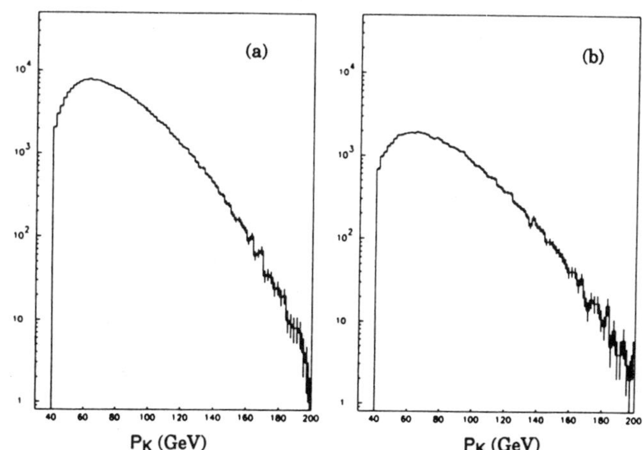

Figure 4. Kaon momentum distributions for $K_S \rightarrow 2\pi^0$ (a) and $K_L \rightarrow 2\pi^0$ (b).

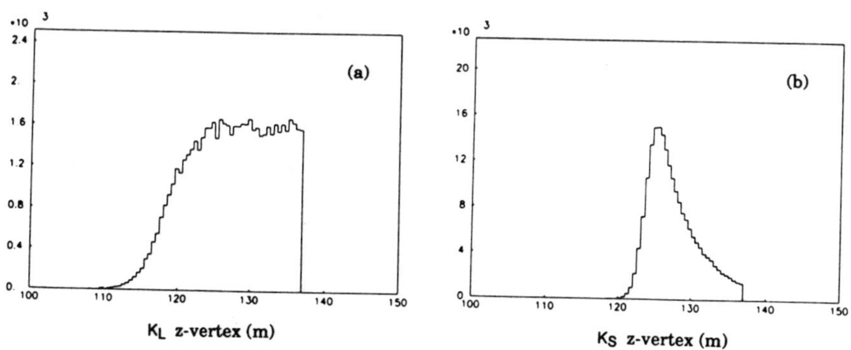

Figure 5. Decay vertex distributions for for $K_L \rightarrow 2\pi^0$ (a) and $K_S \rightarrow 2\pi^0$ (b).

the beam to make K_S decays mimic the flat K_L distribution. There are, however, rather fundamental incompatibilities between the double beam method and the moving target/regenerator method.

First, in the double beam method, the distance between the two beams has to be as small as possible in order to reduce the biases discussed above; thus, when the target or regenerator moves over a substantial distance longitudinally, it comes in the way of decays upstream of it. Second, the target/regenerator motion method partially defeats the purpose of the double beam method in the following way: Suppose that the detection efficiency becomes small toward the down stream end of the fiducial region. Then, when the target/regenerator is near the downstream end it will spray the nearby veto counter with debris causing an inefficiency. This does not affect K_S yield much since the detection efficiency for K_S in this configuration is small to begin with. The K_L yield, however, is directly affected since K_L is taken in the whole fiducial volume in this configuration. Thus, it becomes necessary to understand the relative inefficiency caused by the veto as a function of the target/regenerator position.

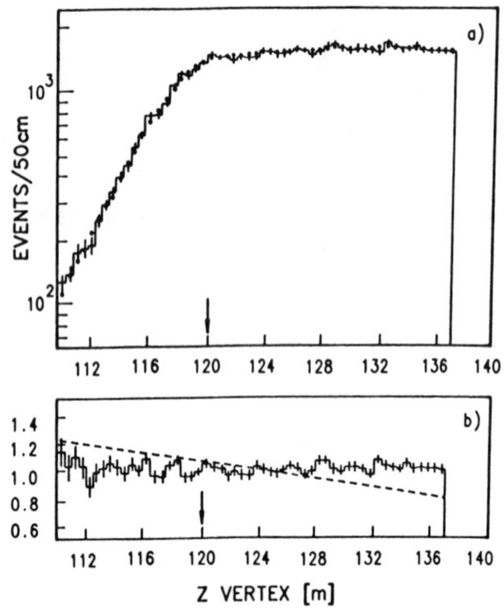

Figure 6. Vertex distributions for data and Monte Carlo for $K_L \rightarrow \pi^+\pi^-$ decays (a). The histogram is data and dots are Monte Carlo. Ratio of data to Monte Carlo is shown in (b). The dashed line corresponds to a 2% error in the double ratio.

4. CONCLUSION

Together with an systematic error due to background subtraction of 0.18% in R_d, the final result is[7] $Re(\varepsilon'/\varepsilon) = -0.0004\pm0.0014(\text{stat})\pm0.0006(\text{syst})$. This result is consistent with the superweak model and does not confirm the NA31 experiment which reported[8] an evidence for direct CP violation $(Re(\varepsilon'/\varepsilon) = 0.0033\pm0.0007(\text{stat})\pm0.0008(\text{syst}))$. The standard model of CP violation, however, can still account for such a small value of $Re(\varepsilon'/\varepsilon)$[1].

REFERENCES

1. See, for example, J. M. Flynn and L. Randall, Phys. Lett. **B224**, 221 (1989).

2. L. Wolfenstein, Phys. Rev. Lett. **13**, 569 (1964).

3. For background subtractions see, for example, B. Winstein, Proceedings of the 1989 International Symposium on Lepton and Photon Interactions at High Energies. For a detailed account of analysis in general, see J. R. Patterson, PhD thesis, University of Chicago, unpublished.

4. The exact expression is: $g_{L,S} = \exp(-\Gamma_{L,S}\, t)M_K/Pc$ where $\Gamma_{L,S}$ is the $K_{L,S}$ decay rate, M_K the kaon mass, and t the proper time corresponding to z with respect to the regenerator position.

5. The accidental trigger was actually given by a muon hodoscope looking at the target. Thus, the rate of the trigger was proportional to the number of protons in each bucket. This was important in reproducing the true amount of accidental overlap on the kaon decays.

6. H. Burkhardt et al., Nucl. Instrum. Methods. **A268**, 116 (1988).

7. J. R. Patterson et al., Phys. Rev. Lett. **64**, 1491 (1990).

8. H. Burkhardt et al., Phys. Lett. **B206**, 169 (1988).

454

MEASUREMENT OF THE PHASES Φ_{+-} AND Φ_{00}
a precise test of CPT symmetry in the $K^\circ - \bar{K}^\circ$ system

NA31 collaboration
presented by Etienne Augé - Laboratoire de l'Accélérateur Linéaire - Orsay

The NA31 collaboration at CERN [1] has performed a detailed study of CP violation in the $K^\circ \to 2\pi$ decays. One usually defines the two ratios :

$$\eta_{+-} = \frac{Amplitude(K_L \to \pi^+\pi^-)}{Amplitude(K_S \to \pi^+\pi^-)} \quad , and \quad \eta_{00} = \frac{Amplitude(K_L \to \pi^\circ\pi^\circ)}{Amplitude(K_S \to \pi^\circ\pi^\circ)}$$

A first set of data-taking periods has been devoted to the measurement of

$$R = \frac{\Gamma(K_L \to \pi^\circ\pi^\circ)}{\Gamma(K_S \to \pi^\circ\pi^\circ)} \cdot \frac{\Gamma(K_S \to \pi^+\pi^-)}{\Gamma(K_L \to \pi^+\pi^-)} = \left| \frac{\eta_{00}}{\eta_{+-}} \right|^2$$

Any deviation of R from 1 provides an evidence that a part of the CP violation observed in the decay $K_L \to 2\pi$ depends on the final state, and therefore occurs in the decay process itself. This part is called Direct CP Violation, as opposed to the violation coming from the fact that K_L and K_S are not exact CP eigenstates.

In 1987, another data taking period was devoted to the measurement of the phases Φ_{00} and Φ_{+-} of the two ratios η_{00} and η_{+-}. This measurement provides a stringent test that the CPT symmetry is not violated in the neutral kaon decays.

Given the CP eigenstates :

$$|CP = +1 >= \frac{K^\circ + \bar{K}^\circ}{\sqrt{2}} \quad , and \quad |CP = -1 >= \frac{K^\circ - \bar{K}^\circ}{\sqrt{2}}$$

one can in general write the expansion of the physical states :

$K_S \approx |CP = +1 > +(\epsilon + \delta)|CP = -1 >$

$K_L \approx |CP = -1 > +(\epsilon - \delta)|CP = +1 >$

where the complex numbers ϵ and δ have a modulus much smaller than 1.

From unitarity, one can write [2]:

$$\delta = \frac{1}{2} \frac{< \bar{K}^\circ|H|\bar{K}^\circ > - < K^\circ|H|K^\circ >}{(M_L - M_S) + i/2(\Gamma_S - \Gamma_L)} = \frac{1}{2} \frac{(M_{K^\circ} - M_{\bar{K}^\circ}) - i/2(\Gamma_{K^\circ} - \Gamma_{\bar{K}^\circ})}{(M_L - M_S)(1 + i \cdot cotg\Phi_{,w})}$$

where $\Phi_{,w} = Atan(2\frac{M_L - M_S}{\Gamma_S})$, neglecting Γ_L in front of Γ_S.

If CPT is a good symmetry, the mass and the total width are the same for K° and $\bar{K}^\circ$, and δ is therefore zero.

From unitarity, one also gets :

$$-i[(M_L - M_S) + i(\frac{\Gamma_L + \Gamma_S}{2})] < K_L|K_S >= \sum_f < f|T|K_L >^* < f|T|K_S >$$

The contribution from the final state f = 2π with isospin 0 is dominant. Neglecting the contribution from other final states in the above formula, one gets :

$Im(\delta)/Re(\epsilon) = \Phi_{,w} - 2/3 \cdot \Phi_{+-} - 1/3 \cdot \Phi_{00}$

and therefore

$$\delta_T = \frac{M_{K^\circ} - M_{\bar{K}^\circ}}{2(M_L - M_S)} = |\eta_{+-}| \cdot (\Phi_{sw} - \Phi_{+-} - \frac{\Phi_{00} - \Phi_{+-}}{3})$$

We now introduce the two amplitudes :

$Amplitude(K^\circ \rightarrow 2\pi, isospin\ 2) = A_2 \cdot e^{i\delta_2}$, and $Amplitude(K^\circ \rightarrow 2\pi, isospin\ 0) = A_0 \cdot e^{i\delta_0}$ and the ratio $\epsilon' = \frac{1}{\sqrt{2}} Im\left(\frac{A_2}{A_0}\right) \cdot e^{i(\pi/2 + \delta_2 - \delta_0)}$

If K_L and K_S are CPT eigenstates $(\delta = 0)$, we can easily write :

$\eta_{+-} = \epsilon + \epsilon'$, $\eta_{00} = \epsilon - 2 \cdot \epsilon'$, and $Re(\epsilon'/\epsilon) = (1 - R)/6$

Moreover, the unitarity relation above now reads : $\Phi_\epsilon = 2/3 \cdot \Phi_{+-} + 1/3 \cdot \Phi_{00} = \Phi_{sw}$

Given the measured values of R [3] and $(\delta_2 - \delta_0)$ [4] , the phases Φ_{+-}, Φ_{00} and Φ_ϵ should be equal to Φ_{sw} within 0.1 degree. Conversely, a difference between Φ_{00} and Φ_{+-} would indicate that CPT is not a good symmetry of the decay process itself.

The measurements available up to 1987 were not sufficiently accurate to exclude a large violation of CPT. This was especially true for Φ_{00}, which precision was much poorer than the one on Φ_{+-}.

We extract the phases from the distributions of the proper time of the decays for an incoherent mixture of K° and $\bar{K}^\circ$. If K° and $\bar{K}^\circ$ are not produced in equal amounts at the target, then the interference between the K_L and the K_S components of the beam can be measured (figure 1). In NA31, the asymmetry between the numbers of K° and $\bar{K}^\circ$ at the production target, $D = (N_{K^\circ} - N_{\bar{K}^\circ})/(N_{K^\circ} + N_{\bar{K}^\circ})$, increases from 0.2 to 0.4 with increasing Kaon energy.

The proper time distributions for $\pi^+\pi^-$ and $\pi^\circ\pi^\circ$ are shown on figure 2. The cosine term of the interference has been extracted and is clearly visible over a full oscillation. However, a measurement of the phases based on these distributions suffers from large systematic uncertainties because of acceptance calculation. We have chosen instead a method where the effect of acceptance cancels in first approximation.

We use two production targets ("Near" and "Far"), producing two collinear beams in alternance (period 16 hours). The distance between the targets corresponds to a phase shift of $\pi/2$ of the cosine term. We fit the ratio of the number of observed decays coming from the two targets : N(near)/N(far), in bins according to the Kaon energy and to the proper time of the decay. The proper time τ is computed from the mid-point between the two targets. Therefore, the acceptance is nearly identical in each energy and τ bin for the events coming from one or the other target. The corresponding systematic uncertainty is estimated to be 0.1° on Φ_{+-}, Φ_{00} and $\Delta\Phi$.

We have measured $1.8 \cdot 10^6$ $K \rightarrow \pi^\circ\pi^\circ$ decays from the "near" target, and $0.3 \cdot 10^6$ from the "far" target. For $\pi^+\pi^-$, the numbers are $2.2 \cdot 10^6$ from the "near" target, and $0.6 \cdot 10^6$ from the "far" target. These data have been divided in ten 5 GeV energy bins and thirty-two $\tau_S/2$ life-time bins. A global fit was done to compute D, a charged and a neutral "Near/Far" normalisation factor in each 10 GeV interval, and Φ_{00} and Φ_{+-}.

The K_S and K_L life times, the $K_L - K_S$ mass difference and $|\eta_{+-}|$ have been taken from other experiments (see table 1). The corresponding value of Φ_{sw} is $43.7^\circ \pm 0.2^\circ$. The value of $|\eta_{00}|$ has been extracted from the measurement of ϵ'/ϵ by NA31 [3]. One gets $|\eta_{00}/\eta_{+-}| = 0.990 \pm 0.003$.

Typical fits are shown in figure 3. A plateau is seen at high τ where the K_L component dominates in both beams. For the lowest energies, the plateau at small τ corresponding to the K_S component is also seen. The total χ^2 is 796 for 768 degrees of freedom.

The fraction of residual background (coming from $K\pi_3$ and Kl_3 decays) is of the order of 10^{-3}. Ignoring this background completely causes a shift of $-0.6°$ for Φ_{00} and of $-0.2°$ for Φ_{+-}. We estimate the systematic uncertainty due to the substraction of the residual background to be $0.1°$ on Φ_{00} and Φ_{+-}.

The dominant systematic uncertainty comes from the uncertainty on the absolute energy calibration ("energy scale"), since the proper time measurement is based on the measured Kaon energy and position of the decay vertex. The problem is especially serious for $\pi^0\pi^0$ final states, since an error on the energy scale also causes an error on the vertex position which adds to the first one when computing the proper time. The absolute calibration is extracted from special runs every period (16 hours), where we use a special target in the decay volume with a counter in the beam to veto early decays. The position of the counter is fitted from the reconstructed position of the observed decays. If this fitted counter position coincides with its nominal position, the longitudinal distance scale and, therefore, the energy scale are correct. We achieve a precision of 0.08% on the energy scale, which translates into a systematic uncertainty of $0.8°$ on Φ_{00}. The energy scale estimates for various Kaon energies and various positions of the target are in good agreement. We have therefore no evidence of any non-linearity in the energy measurement. Nevertheless, we cannot exclude an error of $0.5°$ on Φ_{00}, coming from non-linearity.

In the case of $\pi^+\pi^-$ final states, we estimate the Kaon energy from the angle between the two tracks and the ratio of their energy (measured by calorimetry). The energy scale is essentially given by the "angle scale", itself coming from the distance between the two wire chambers, and from the wire pitch. We estimate the uncertainty on Φ_{+-} to be $0.6°$. This measurement is much less affected than Φ_{00} by non-linearity in the energy measurement: the corresponding uncertainty is estimated to be $0.1°$.

To estimate the total uncertainty on our result, we also have to consider the uncertainty coming from the external parameters. Especially relevant are the K_S life time τ_S and the $K_S - K_L$ mass difference. They however do not influence the phase difference. We have varied these quantities by one standard deviation and redone the fit. As a consequence, our result reads [5]:

$$\Phi_{+-} = 46.9° \pm 1.4°(stat) \pm 0.7°(syst) \pm 0.6° \cdot \sigma_{\tau_s} \pm 1.4° \cdot \sigma_{\Delta m}$$

$$\Phi_{00} = 47.1° \pm 2.1°(stat) \pm 1.0°(syst) \pm 0.5° \cdot \sigma_{\tau_s} \pm 1.4° \cdot \sigma_{\Delta m}$$

$$\Delta\Phi = 0.2° \pm 2.6°(stat) \pm 1.2°(syst)$$

We have improved the precision on Φ_{00} by a factor 2 with respect to previous measurements. $\Delta\Phi$ is compatible with 0 (no Direct CPT violation). This is confirmed by the recent result from the Fermilab E731 collaboration [6]. However, both Φ_{+-} and Φ_{00} seem slightly higher than $\Phi_{s,w}$. This small discrepancy might entirely be blamed on the uncertainty on $M_L - M_S$ and τ_S. We therefore observe no CPT violation from state mixing :

$$\delta_T = (1.3 \pm 0.8) \cdot 10^{-4} < 2.0 \cdot 10^{-4} \quad (90\%C.L.)$$

Given the small mass difference $M_L - M_S$, this upper limit translates into a stringent limit on the relative mass difference

$$\left|\frac{M_{K^0} - M_{\bar{K}^0}}{M_{K^0}}\right| < 4 \cdot 10^{-18} \quad (90\%C.L.)$$

Parameter	Value	Reference
τ_S	$(0.8922 \pm 0.020) \cdot 10^{-10}$ s	[7]
τ_L	$(5.18 \pm 0.04) \cdot 10^{-8}$ s	[7]
$M_L - M_S$	$(3.521 \pm 0.014) \cdot 10^{-15}$ GeV	[8]
$|\eta_{+-}|$	$(2.272 \pm 0.021) \cdot 10^{-3}$	[8]

Table 1 : External parameters of the fits

I thank A. Schaffer for a critical reading of the text.

References

[1] : NA31 collaboration : CERN, Edimburgh, Mainz, Orsay, Pisa, Siegen

[2] : Bell, Steinberger : Int. Conf. on High Energy Physics (Oxford,1965)
Barmin et al : Nucl. Phys. B247 (1984) 296

[3] : Burkhardt et al (NA31 collaboration) : Phys. Lett. 206B (1988) 169
: Patterson et al (E731 collaboration) : submitted to Phys. Rev. Lett.

[4] : Devlin and Dickey : Rev. Mod. Phys. 51 (1979) 237
Biswas et al : Phys. Rev. Lett. 47 (1981) 1378
Kamal et al : Journ. Phys. G12 (1986) L43
Cheng : Phys. Lett. B201 (1988) 155

[5] : Carosi et al (NA31 collaboration) : Phys. Lett. 273B (1990) 303

[6] : Karlsson et al (E731 collaboration) : submitted to Phys. Rev. Lett.

[7] : Particle Data Group Phys. Lett. B204 (1988) 1

[8] : Wahl CERN Pre-print EP 89/86 (july 1989)

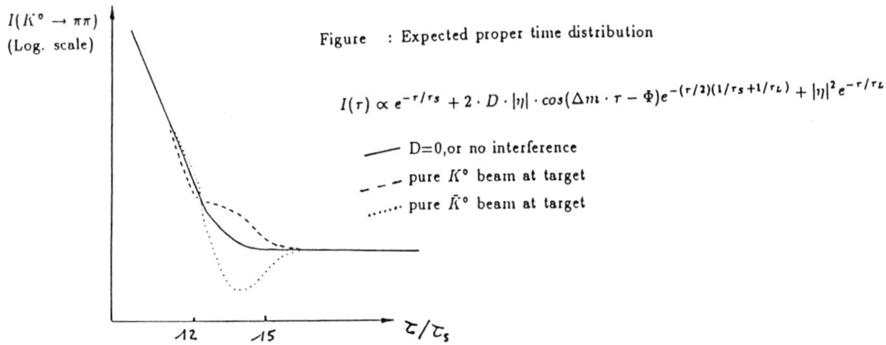

458

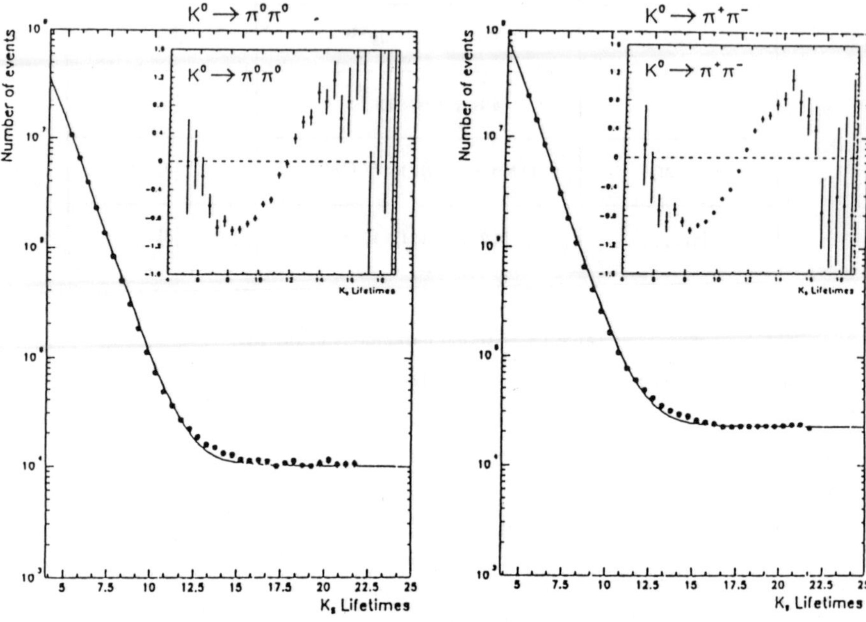

Fig. 2   Lifetime distribution, acceptance corrected. The line is a fit without the interference term. Shown in insets is the difference between this fit and the data, averaged over energy.

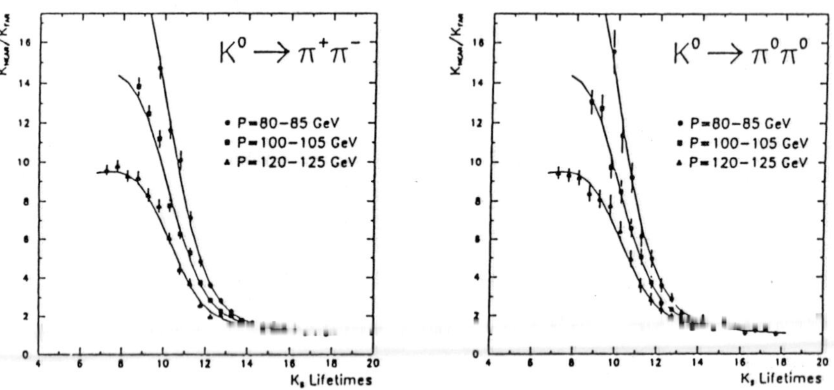

Fig. 3   Ratio of 'lifetime' distributions from the K_N and K_F targets.

Weak Transition Form Factors between Heavy Mesons

Nathan Isgur
Department of Physics, University of Toronto
Toronto, Canada M5S 1A7

Mark B. Wise
California Institute of Technology
Pasadena, California, USA 91125

ABSTRACT

We present an extension of our earlier observations on symmetries in operation in the weak decays of heavy mesons containing a single heavy quark. The new symmetries allow us to obtain absolutely normalized model-independent predictions in the heavy quark limit of all of the form factors for the $Q_1 \to Q_2$ induced weak pseudoscalar to pseudoscalar and pseudoscalar to vector transitions in terms of a single universal function $\xi(t)$ with $\xi(0) = 1$.

The properties of hadrons containing a single heavy quark Q ($m_Q \gg \Lambda_{QCD}$) along with light degrees of freedom are constrained by symmetries which are not manifest in QCD [1]. The first of these is a flavor symmetry which arises from the fact that the long wavelength properties of the light degrees of freedom in such a hadron become independent of m_Q for $m_Q \gg \Lambda_{QCD}$. Thus, for example, the *light* degrees of freedom of a $\bar{B}$ and D meson can be related by an (approximate) $b \leftrightarrow c$ SU(2) symmetry even though m_b and m_c are very different. The second symmetry pointed out in Ref. [1] is a related spacetime symmetry which arises in QCD because the spin of a heavy quark decouples from the gluon field [2]. This makes $\vec{S}_Q$, the heavy quark spin operator, the generator of another SU(2) group of symmetries of the light degrees of freedom in a meson containing a single heavy quark. Thus, for example, the *light* degrees of freedom in the $\bar{B}$ and $\bar{B}^*$ mesons are in (approximately) the same state since the spin orientation of the b quark does not affect their dynamics. These symmetries are manifest in an effective theory where the heavy quark acts in its hadron's rest frame like a spatially static triplet source of (short distance) color electric field [2,3]. In the effective theory the heavy quark's couplings to the gluon degrees of freedom are independent of its mass and described by a Wilson line [3].

To apply these symmetries to the matrix elements of weak currents, we must know these currents in the effective theory. For the currents of interest here (color indices are suppressed)

$$V_\nu^{ji} \equiv \bar{Q}_j \gamma_\nu Q_i \tag{1}$$

$$A_\nu^{ji} \equiv \bar{Q}_j \gamma_\nu \gamma_5 Q_i \tag{2}$$

one has [1,4] for $J_\nu = V_\nu$ or A_ν

$$J_\nu^{ji} = C_{ji} \mathbf{J}_\nu^{ji} + ... \tag{3}$$

where the ellipses denote higher dimension operators suppressed by powers of m_Q, and where on the right hand side of Eq. (3) $\mathbf{J}_\nu^{ji}$ is the naive weak current in the effective theory. In the leading logarithmic approximation with $m_j < m_i$

$$C_{ji} = \left[\frac{\alpha_s(m_i)}{\alpha_s(m_j)} \right]^{-\frac{6}{33-2N_f}}. \tag{4}$$

Note that it is the charges associated with the currents $\bar{Q}\gamma_\nu \frac{\vec{\tau}}{2}Q$ with

$$Q \equiv \begin{pmatrix} Q_1 \\ Q_2 \end{pmatrix}, \tag{5}$$

the heavy quark field operators in the effective theory, which lead to the flavor symmetry.

In our earlier paper [1] we applied this symmetry for static quarks, where a $Q_1 \to Q_2$ transition simply substituted one static heavy quark for another. Here we exploit a powerful extension of this method which makes use of the fact that when Q_1 at *velocity* $\vec{v}$ makes a weak transition into Q_2 at *velocity* $\vec{v}\,'$, the amplitude for the light degrees of freedom to make any associated transition is independent of m_1 and m_2 if they are sufficiently large. *I.e.*, the light degrees of freedom interact only with the (moving) color fields of Q_1 and Q_2 which depend only on the Lorentz boosts required of the (mass independent) rest frame color fields. In the effective theory the mass of a heavy quark Q is taken to infinity in such a way that p_Q^μ/m_Q is held fixed, but the four momenta of the light degrees of freedom are neglected compared to m_Q. In this limit the interactions of the gluons with the heavy quark do not alter its straight world line and are independent of its mass and spin. In the matrix elements of the currents in Eqs. (1) and (2), changes in the heavy quark velocity and spin occur only due to the actions of the currents. In a typical transition $H_1(\vec{v} = \vec{0}) \to H_2(\vec{v}\,')$ (where H_i is a hadron containing the single heavy quark Q_i), the form factors will therefore be determined by the product of the amplitude for the heavy quarks to make the transition $Q_1(\vec{v} = \vec{0}) \to Q_2(\vec{v}\,')$ and for the light quarks to be "excited" by the transition from the hadron H_1 at rest into the hadron H_2 moving with velocity $\vec{v}\,'$. The resulting symmetry is somewhat unusual in that it relates states of equal *velocity* but different mass, and therefore of different momentum.

Given this peculiarity, it is convenient to remove insofar as possible the dependence of the heavy quark and hadron formalism on the heavy quark mass m_Q.[†] We therefore work with heavy boson fields $\tilde{\phi} \equiv m_Q^{-1}\phi$ and heavy fermion fields $\tilde{\psi} \equiv m_Q^{-\frac{3}{2}}\psi$ with creation operators normalized according to

$$[\tilde{a}(\vec{v}\,', s'), \tilde{a}^\dagger(\vec{v}, s)] = 2\gamma\delta_{s', s}\delta^3(\gamma'\vec{v}\,' - \gamma\vec{v}) \tag{6}$$

[†] Henceforth we will ignore differences between the heavy quark mass and the mass of the hadron of which it is a part.

and

$$[\tilde{b}(\vec{v}\,',s'),\tilde{b}^\dagger(\vec{v},s)] = \gamma\delta_{s's}\delta^3(\gamma'\vec{v}\,' - \gamma\vec{v}) \tag{7}$$

respectively, where $\gamma = E/m_Q = (1 - \vec{v}^2)^{-\frac{1}{2}}$. In contrast to the usual flavor symmetry we then define symmetry operators $\mathbf{U}$ by

$$\mathbf{U}^\dagger\tilde{a}_i^\dagger(\vec{v},s)\mathbf{U} = R_{ij}\tilde{a}_j^\dagger(\vec{v},s) \tag{8}$$

etc., where $R = exp(i\frac{\vec{\tau}}{2}\cdot\theta)$.

Let's now work out the consequences of this symmetry for pseudoscalar to pseudoscalar transitions via the heavy quark vector current (1). Since in the effective theory

$$\langle\tilde{P}_j(\vec{v}\,')|\bar{\tilde{Q}}_j(0)\gamma_\nu\tilde{Q}_i(0)|\tilde{P}_i(\vec{v})\rangle$$

is independent of i and j, in terms of the usual meson states with normalizations

$$\langle M(\vec{p}\,',s')|M(\vec{p},s)\rangle = 2E\delta_{s's}\delta^3(\vec{p}\,' - \vec{p})$$

and the full vector current we have that

$$\rho_{ji} \equiv \frac{\langle P_j(\vec{p}_j\,')|\bar{Q}_j(0)\gamma_\nu Q_i(0)|P_i(\vec{p}_i)\rangle}{C_{ji}\sqrt{m_j m_i}} \tag{9}$$

is independent of i,j. By definition

$$\langle P_1(\vec{p}_1\,')|\bar{Q}_1(0)\gamma_\nu Q_1(0)|P_1(\vec{p}_1)\rangle = f_{11}(t_{11})(p_1 + p_1')_\nu \tag{10}$$

$$\langle P_2(\vec{p}_2\,')|\bar{Q}_2(0)\gamma_\nu Q_2(0)|P_2(\vec{p}_2)\rangle = f_{22}(t_{22})(p_2 + p_2')_\nu \tag{11}$$

$$\langle P_2(\vec{p}_2\,')|\bar{Q}_2(0)\gamma_\nu Q_1(0)|P_1(\vec{p}_1)\rangle = f_+(t_{21})(p_1 + p_2')_\nu + f_-(t_{21})(p_1 - p_2')_\nu \tag{12}$$

where $t_{rs} = (p_s - p_r')^2$. Thus using $\rho_{11} = \rho_{22} = \rho_{21}$ and equating coefficients of the velocities $\vec{v}\,'$ and $\vec{v}$ separately gives

$$f_{11}(t_{11}) \equiv \xi(t_{11}) \tag{13}$$

$$f_{22}(t_{22}) = \xi(\frac{m_1^2}{m_2^2}t_{22}) \tag{14}$$

$$f_\pm(t_{21}) = \pm\left[\frac{\alpha_s(m_1)}{\alpha_s(m_2)}\right]^{-\frac{6}{33-2N_f}}\left[\frac{m_1 \pm m_2}{\sqrt{4m_1 m_2}}\right]\xi(\frac{m_1}{m_2}[t_{21} - t_m]) \tag{15}$$

where $t_m \equiv (m_1 - m_2)^2$ is the maximum momentum transfer in the $P_1 \to P_2 e\bar{\nu}_e$ transition. All four form factors are thus determined by a single universal function $\xi(t)$. Moreover, since $\bar{Q}_1\gamma_\nu Q_1$ is a conserved current, $\xi(0) = 1$. In our earlier work [1], we found only the value of $f_+(m_1+m_2)+f_-(m_1-m_2)$ at $t = t_m$. Eq. (15) is, to the order we are working, consistent with the Ademollo-Gatto theorem which states that as $m_2 \to m_1$, $f_+(t_m)$ deviates from unity by terms quadratic in $\Delta m \equiv m_1 - m_2$. When m_2 is close to m_1, the linear term in Δm which arises from expanding the ratio of strong interaction fine structure constants in C_{21} is, with $\bar{m} \equiv \frac{1}{2}(m_1 + m_2)$, $\alpha_s(\bar{m})\Delta m/\pi\bar{m}$, and so is a higher order effect. It can be compensated by one loop contributions to the matching condition between the full field theory and the effective theory. Although there are other corrections of order $\alpha_s(\bar{m})/\pi$, in our numerical results to be quoted below we will use $C'_{21} = C_{21}(1 - \frac{\alpha_s(\bar{m})}{\pi}\frac{\Delta m}{\bar{m}})$ to ensure compliance with the Ademollo-Gatto theorem in the approach to the equal mass limit. For $b \to c$ transitions, the resulting compensation of C_{21} is almost complete.

By using the symmetry of the light degrees of freedom under a rotation of the heavy quark spin, analogous relations can be derived for $P_1 \to V_2$ transitions, where V_2 is the vector meson degenerate with P_2 as $m_2 \to \infty$. To use this symmetry we note that

$$h_Q|P_Q(\vec{p})\rangle = \frac{1}{2}|V_Q(\vec{p},0)\rangle \tag{16}$$

where h_Q is the helicity operator of the heavy quark Q, and $|V_Q(\vec{p},0)\rangle$ is the helicity zero state of V_Q. Thus for Γ any product of γ-matrices,

$$\langle V_2(\vec{p}\,',0)|\bar{Q}_2(0)\Gamma Q_1(0)|P_1(\vec{p})\rangle = 2\langle P_2(\vec{p}\,')|[h_{Q_2},\bar{Q}_2(0)\Gamma Q_1(0)]|P_1(\vec{p})\rangle. \tag{17}$$

Since

$$[\hat{n}\cdot\vec{J}_Q, Q^\dagger(0)] = [\hat{n}\cdot\vec{S}_Q, Q^\dagger(0)] = Q^\dagger(0)\hat{n}\cdot\vec{\Sigma} \tag{18}$$

where $\Sigma^i = \frac{i}{8}\epsilon^{ijk}[\gamma^j,\gamma^k]$, one has many useful commutation relations, e.g.,

$$[S^z_{Q_2}, A^{21}_0] = -\frac{1}{2}V^{21}_3, \quad [S^z_{Q_2}, A^{21}_3] = -\frac{1}{2}V^{21}_0, \quad [S^z_{Q_2}, A^{21}_\pm] = \mp\frac{1}{2}A^{21}_\mp, \tag{19,20,21}$$

and

$$[S^z_{Q_2}, V^{21}_0] = -\frac{1}{2}A^{21}_3, \quad [S^z_{Q_2}, V^{21}_3] = -\frac{1}{2}A^{21}_0, \quad [S^z_{Q_2}, V^{21}_\pm] = \mp\frac{1}{2}V^{21}_\mp, \tag{22,23,24}$$

which are more than sufficient to determine all of the form factors in $P_1 \to V_2$ transitions. With the definitions

$$\langle V_2(\vec{p}\,',\vec{\epsilon})|A^{21}_\nu(0)|P_1(\vec{p})\rangle = f\epsilon^*_\nu + a_+(\epsilon^*\cdot p)(p+p')_\nu + a_-(\epsilon^*\cdot p)(p-p')_\nu \tag{25}$$

$$\langle V_2(\vec{p}\,',\vec{\epsilon})|V^{21}_\nu(0)|P_1(\vec{p})\rangle = ig\epsilon_{\nu\rho\tau\sigma}\epsilon^{*\rho}(p+p')^\sigma(p-p')^\tau \tag{26}$$

the $P_1 \to V_2$ form factors can be related to the $P_1 \to P_2$ form factors and thereby to the universal function $\xi(t)$. The results, which can be most easily derived in the frame where $\vec{p}\,' = \vec{0}$, are

$$a_+(t_{21}) = -a_-(t_{21}) = -g(t_{21}) = -\left[\frac{\alpha_s(m_1)}{\alpha_s(m_2)}\right]^{-\frac{6}{33-2N_f}}\frac{\xi(\frac{m_1}{m_2}[t_{21}-t_m])}{\sqrt{4m_1m_2}} \tag{27}$$

$$f(t_{21}) = \left[\frac{\alpha_s(m_1)}{\alpha_s(m_2)}\right]^{-\frac{6}{33-2N_f}} [4m_1m_2 + (t_m - t_{21})]\frac{\xi(\frac{m_1}{m_2}[t_{21} - t_m])}{\sqrt{4m_1m_2}}. \tag{28}$$

We previously found [1] only the result

$$f(t_m) = \sqrt{4m_1m_2}\left[\frac{\alpha_s(m_1)}{\alpha_s(m_2)}\right]^{-\frac{6}{33-2N_f}}.$$

That $a_+ + a_- = 0$ in the heavy quark limit is transparent in the $\vec{p}\,' = \vec{0}$ frame: this combination of form factors is proportional to the amplitude for A_ν^{21} acting on $\langle V_2(\vec{0}, \vec{\epsilon})|$ to produce P_1 in a D-wave, but the axial current acting at the origin can only produce Q_1 in an S- or P-wave.

These results are all consistent with those of Refs. [5,6] in the appropriate limit, apart from the short distance factor C_{21} which these authors neglected, and given that they worked to leading order in $\vec{v}$ and $\vec{v}\,'$. However, unlike this previous work based on the nonrelativistic quark model, the results of this paper are systematic consequences of QCD, corrections to them being suppressed by $\alpha_s(m_Q)/\pi$ and Λ_{QCD}/m_Q. The results of Ref. [5] may now be used to provide estimates for the departures of real systems from the $m_Q \to \infty$ limit of this paper. In Table 1 we compare our model-independent, absolutely normalized predictions for $b \to c$ form factors at $t = t_m$ with the quark model results of Ref. [5]. The deviations are all less than about 5 percent, which strongly suggests that the large m_Q limit is applicable to these decays. It should also be noted that present experimental evidence on the rates for $\bar{B} \to De\bar{\nu}_e$ and $\bar{B} \to D^*e\bar{\nu}_e$, and on the D^* polarization in the latter decay are consistent with the predictions of this limit.

We have so far not directly addressed the question of the range of validity in $(t_m - t)$ of our results. This range is determined by two approximations. Since the Fock states of the heavy mesons are only independent of the heavy quark mass on length scales greater than m_Q^{-1}, the universality of the function ξ will break down when these states are probed at distances smaller than this. This restricts the range of validity of our results to

$$\frac{t_m - t}{2m_1m_2} \ll \left[\frac{m_2}{\Lambda_{QCD}}\right]^2 \tag{29}$$

and gives rise to $\frac{\Lambda_{QCD}}{m_Q}$ corrections to our results at $t = t_m$. Even if valid up to such $t_m - t$ as an evaluation of the effects of the lowest dimension operator, our results may be *inapplicable* if higher order terms in Eq. (3) become important at lower values of $t_m - t$. This could happen since hard transverse gluon exchange effects suppressed by $\frac{\Lambda_{QCD}}{m_Q} \cdot \frac{\alpha_s(m_Q)}{\pi}$ produce an asymptotically [†] dominant power law tail to the form factors [7] not present in $\xi(t)$. However, since $t_m - t = (\frac{m_1 m_2}{m_\ell^2})Q_\ell^2$, where

---

[†] See Ref. [8] for a discussion of the momentum transfers required to reach asymptotia.

m_ℓ and Q_ℓ are the effective mass and momentum transfer of the light degrees of freedom, and since asymptopia is surely above $Q_\ell^2 = 1\ GeV^2$, we can confidently apply our results over the full kinematic range explored by the $b \to c$ transition.

Table 1: zero recoil values of $\bar{B} \to D, D^*$ form factors of Ref. [5] compared to the results of this paper

	f_+	$f/\sqrt{4m_B m_{D^*}}$	$\sqrt{4m_B m_{D^*}}\, g$	$\sqrt{4m_B m_{D^*}}\, a_+$
Ref. [5] (corrected*)	1.15	1.06	1.08	-0.97
heavy quark limit	1.16	1.01	1.01	-1.01

* We have multiplied the Ref. [5] results by $C'_{cb} \simeq 1.02$ to make this comparison meaningful.

REFERENCES

[1] N. Isgur and M.B. Wise, Phys. Lett. **B232**(1989)113.

[2] G.P. Lepage and B.A. Thacker, in *Field Theory on the Lattice*, proceedings of the International Symposium, Seillac, France, 1987, ed. A. Billoire *et al.* , Nucl. Phys. **B**, Proc. Suppl. **4** (1988) 199; W.E. Caswell and G.P. Lepage, Phys. Lett. **B167** (1986) 437.

[3] E. Eichten and F.L. Feinberg, Phys. Rev. Lett. **43** (1979) 1205; Phys. Rev. **D23** (1981) 2724; E. Eichten in *Field Theory on the Lattice*, proceedings of the International Symposium, Seillac, France, 1987, ed. A. Billoire *et al.* , Nucl. Phys. **B**, Proc. Suppl. **4** (1988) 170.

[4] M.B. Voloshin and M.A. Shifman, Yad. Fiz. **45** (1987) 463 [Sov. J. Nucl. Phys. **45** (1987) 292; H.D. Politzer and M.B. Wise, Phys. Lett. **B206** (1988) 681; **B208** (1988) 504.

[5] N. Isgur, D. Scora, B. Grinstein, and M.B. Wise, Phys. Rev. **D39** (1989) 799.

[6] T. Altomari and L. Wolfenstein, Phys. Rev. Lett. **58** (1987) 1583.

[7] G.P. Lepage and S.J. Brodsky, Phys. Rev. **D22** (1980) 2157.

[8] N. Isgur and C.H. Llewellyn Smith, Nucl. Phys. **B317** (1989) 526.

DISCOVERING b' THROUGH ITS FCNC DECAYS

George Wei-Shu Hou
Paul Scherrer Institute
CH-5232 Villigen PSI, Switzerland

ABSTRACT

The fourth generation b' quark, if it exists, may have unusual decay properties. Within the Standard Model framework, for a wide range of the parameters, $m_{b'}$, m_t, $m_{t'}$, M_H, $V_{cb'}$ and $V_{tb'}$, FCNC $b' \to b$ transitions could be substantial, even dominating over CC decays. Interesting FCNC decays would be $b' \to b\gamma$, $b' \to bg$, $b' \to bZ^*$, $b' \to bZ$ and $b' \to bH^0$. Search strategies should be broadened to take into account these decay modes.

1. INTRODUCTION

Two questions have to be addressed right away: Why is a fourth generation still interesting? Why is this talk put in a session on B physics? The neutrino counting result reported[1] at this Conference indicates that Nature has only 3 light neutrino species. For the conservative, one would think that there is no fourth generation. But for those who are interested in extra heavy fermions, the situation actually becomes more interesting: if it exists, the fourth generation is "non-sequential", namely the fourth neutrino has to be rather heavy, unlike the first three families. As can be seen from the talks by MARK II, ALEPH and OPAL[2], the experimenters who found no evidence for a fourth light neutrino have not themselves given up on the quest for fourth generation quarks. Why this talk is scheduled in a B physics session is beyond my control. However, the phenomenology I will discuss has many similarities to the B system. In fact, I will try to argue that the b' system may be an even more exciting system than the B, even if $m_{b'}$ is very heavy.

One can learn from the K and B systems. Due to kinematics ($m_K \approx 3m_\pi$) and the smallness of the Cabibbo angle (i.e. V_{us}), the K lifetime is rather prolonged. We therefore had the fortune to discover the marvelous phenomena of $K^0 - \bar{K}^0$ mixing and CP violation, while loop suppressed FCNC decays such as $K \to \pi\nu\bar{\nu}$ become more interesting. The B lifetime turned out to be even more prolonged, implying $V_{cb} \ll V_{us}$. This lead to the subsequent surprise of large $B^0 - \bar{B}^0$ mixing, good prospects for CP violation studies, and the result that loop induced FCNC decays have much larger BR's than in the K system[3]. Thus, down type heavy quarks (s, b, ...) have been great, because they are the lighter components of weak doublets, and because of the existence of hierarchy of quark mixing angles.

A "parable" regarding the B system prepares us for the b' quark. Imagine a world in which everything is the same as ours, *except* $m_c > m_b$. This seemingly innocent deviation leads to dramatic consequences: $b \to c$ decays are forbidden. In our world, $b \to u$ transitions are KM suppressed ($V_{ub} \ll V_{cb}$), while loop induced (FCNC) $b \to s$ decays are loop suppressed; both decay modes are at the per cent level, while the $b \to u$ mode vanishes with $V_{ub} \to 0$. Thus, in the world where $m_c > m_b$ and $V_{ub} \to 0$, the B meson will mostly decay through FCNC $b \to s$ decays, with τ_B a hundred times longer than in our world. The "rare" decays become dominant!

The above of course was not realized in Nature, and $b \to c$ is the major b decay mode. However, if a fourth generation exists, one could well have an analogous scenario realized in the b' system. Indeed, through detailed study[4,5,6], it is found that the b' decay phenomenology could be very rich. In this talk, all details regarding computation, *etc.*, are suppressed. Only the qualitative features are discussed.

2. CHARGED CURRENT DECAY MODES

The charged current decay rate $b' \to (c,\, t) + W^{(*)} \cong 9 \cdot \Gamma(b' \to (c,\, t) + e\bar{\nu})$ is plotted w.r.t. $m_{b'}$ for three m_t values, with the assumption that $m_{b'} < m_{t'}$. The dotted,

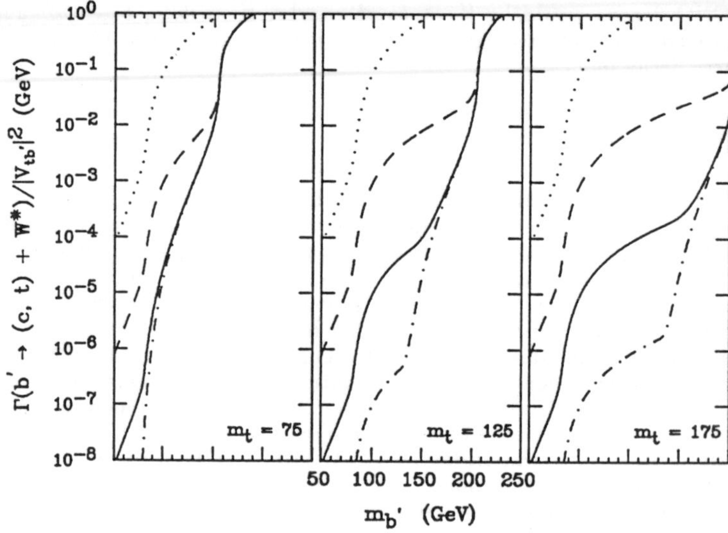

Fig. 1

dashed, solid and dotdashed curves corresponds to $V_{cb'}/V_{tb'} = 1,\ 0.1,\ 0.01,\ 0.001$, respectively. Both real and virtual W effects are included. One sees that the CC rate could vary over a spectacular range, depending on $m_{b'}$, m_t and $V_{cb'}/V_{tb'}$. Above the tW threshold, the b' width is at the 100 MeV level or higher. For $m_t < m_{b'} < m_t + M_W$, the dominant CC mode is $b' \to cW$, unless it is rather suppressed by the KM factor $V_{cb'}$, and $b' \to tW^*$ becomes important. For $m_{b'} < m_t$, the major threshold is whether the on-shell $b' \to cW$ is allowed. The CC rate can get very suppressed if $V_{cb'}$ is small. This deep valley of very suppressed CC decay partial widths for small $m_{b'}$ and $V_{cb'}/V_{tb'}$ forms the basis for our scenario of FCNC b' decay dominance. As m_t is raised, this valley extends towards higher $m_{b'}$ values. The plots could be extended from the lower left corner to cover in more detail the case when $m_{b'} < M_Z$.

3. FLAVOR-CHANGING NEUTRAL CURRENT DECAY MODES

The FCNC decay modes $b' \rightarrow bZ$ (dashed line) and $b' \rightarrow bH^0$ (solid line) for $m_{t'} = 250$ GeV are plotted in a similar fashion as the CC modes for ease of

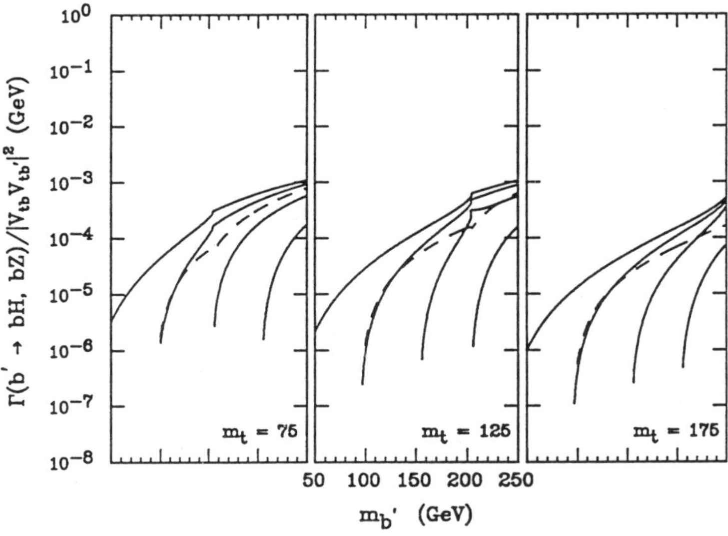

Fig. 2

comparison[6]. The Higgs mode is given for the M_H values of 20, M_Z, 150, 200 GeV, respectively, from left to right. Given the present LEP limit on M_H, the first curve serves as an upper bound on $b' \rightarrow bH^0$ decay. The very strong (m_Q^4 in rate) dependence on heavy internal quark masses are not exhibited. Note that $b' \rightarrow bH^0$ and $b' \rightarrow bZ$ are comparable in rate when $M_H \approx M_Z$. This is easy to understand, since the strong m_Q dependence for $b' \rightarrow bZ$ is due to the "longitudinal" component of the Z boson, which is actually the would-be Goldstone boson that was eaten, and belonged to the same complex neutral scalar field as the Higgs boson itself. If $M_H < M_Z$ the $b' \rightarrow bH^0$ mode dominates, while if $M_H > M_Z$, the $b' \rightarrow bZ$ mode tends to dominate. The two modes never differ by more than a factor of ten in rate, unless $b' \rightarrow bH^0$ is kinematically forbidden. Comparing with the CC b' decay rate in Fig. 1, one observes the following qualitative features:

- If $m_{b'} > m_t + M_W$, then CC $b' \rightarrow tW$ dominates over FCNC $b' \rightarrow b$ modes, the latter having branching ratios below 10^{-2}, independent of $V_{cb'}/V_{tb'}$.
- The region $M_Z < m_{b'} < m_t + M_W$ is rich. Depending somewhat on $V_{cb'}/V_{tb'}$ the three decay modes $b' \rightarrow cW$, $b' \rightarrow bZ$ and $b' \rightarrow bH^0$ could all be of similar order of magnitude (i.e. not differing by more than a factor of ten in rate).

The FCNC results in Fig. 2 have been plotted in the narrow Z and H width approximation, which is valid for $m_{b'} > m_b + M_Z$ or M_H. As one goes to the lower left corner, i.e. for lower $m_{b'}$, the $b' \rightarrow bZ$ mode is readily continued into $b' \rightarrow bZ^*$, where the Z is now virtual, by incorporating the finite Z width. For the complete electroweak contributions to $b' \rightarrow b\ell^+\ell^-$, $b\nu\bar{\nu}$, $bq\bar{q}$, one also needs to take into account the γ^* and the box diagram contributions[5]. In contrast, when H · becomes virtual, the $b' \to bH^*$ mode is rather suppressed by the small Higgs width (dominated by $H \to b\bar{b}$ in our scenario). In this lower mass region, which is of more immediate experimental interest, one should consider also the decay modes[4] $b' \to b\gamma$ and $b' \to bg$, which have decay rates (modulo $|V_{tb}V_{tb'}|^2$) in the $10^{-10} - 10^{-6}$ GeV range as m_t, $m_{t'}$ is varied over a plausible range. These FCNC $b' \to b$ decay modes populate the low $m_{b'}$ valley, giving a lower bound on the b' width which is suppressed, but not so suppressed so as to provide an extra detection handle in itself. Thus, the main decay modes are $b' \to b\gamma$, $b' \to bg$, $b' \to bZ^*$, and the Higgs induced decay mode $b' \to bH^0$. I do not give numerical details, but summarize the qualitative features (for $m_{b'} < m_b + M_Z$) as follows:

- $b' \to bH^0 \gg$ other $b' \to b$ modes, when allowed[6,7]. It may even dominate over CC $b' \to cW^{(*)}$ so long that $V_{cb'}/V_{tb'}$ is not much larger than $\mathcal{O}(10^{-1})$.
- $b' \to bg > b' \to bg^*$ $(g^* \to q\bar{q}, gg)$. In the B system the reverse is true, so this is non-trivial. One may say that the simple parton model picture works here.
- $\Gamma(b' \to b\gamma)/\Gamma(b' \to bg)$ is of order 20% or more[4]!
- $\Gamma(b' \to bZ^*)$ is comparable to or even greater than[5] $\Gamma(b' \to bg)$!
- $b' \to bV^0 > b' \to c$ if $V_{cb'}/V_{tb'} < 10^{-2}$. This is not small compared to the analogous ratio $V_{ub}/V_{us} < 0.02$. $V_{cb'}$ is expected to be small, whereas $V_{tb'}$ could well be sizeable.
- If $b' \to b$ transitions dominate but $b' \to bH^0$ is forbidden, the branching ratios are roughly 10%, 45%, 45% for $b' \to b\gamma$, $b' \to bg$, $b' \to bZ^*$, respectively.
- In general, the lifetime of b' is still too short to be used as handle.

4. EXPERIMENTAL SIGNATURES AND PROSPECTS

One would always have b' pair-produced, and the signatures depend on the b' mass region. For $m_{b'} < m_b + M_Z$, one has the possible signatures:

$$
\begin{array}{ll}
b'\bar{b}' \to bg\bar{b}g, \, bg\bar{b}Z^*, \, bZ^*\bar{b}Z^* & \text{4-5-6 jets, roughly equal in } BR. \\
\quad\quad b\gamma\bar{b}g, \, b\gamma\bar{b}Z^* & \text{(isolated) } \gamma + \text{(3-4)-jets } (BR \text{ up to } 20\%). \\
\quad\quad bH^0\bar{b}H^0 & \text{BAD SIGNATURE.} \\
\quad\quad cW^*\bar{b}H^0 & \text{isolated } \ell^{\pm} \text{ events may contain Higgs signal!}
\end{array}
$$

This region can be studied at LEP II and the TEVATRON, i.e. at existing machines. Since $b' \to bH^0$ dominates the $b' \to b$ rate in this region whenever the decay is allowed, a powerful corollary[6] is that from the mere observation of, say, the $b' \to b\gamma$ mode, one could immediately draw the conclusion that $M_H > m_{b'} - m_b$. This could allow hadronic machines to set more stringent limits on M_H ahead of LEP I or II. Of course, there is also hope to discover the Higgs boson directly in tagged CC events (via $b'\bar{b}' \to cW^*\bar{b}H$), but a detailed Monte Carlo study is needed. If the Higgs mode is absent, the isolated photon plus jets decay mode provides an interesting signature, with the added incentive cited above. At e^+e^- machines one could also look for multijet events with up to 6 clearly separated jets. If there are no CC signal and no photonic signal, but one knows that a heavy flavor threshold has been crossed, it is likely that $b' \to bH^0$ dominates the b' rate. Fortunately, in this mass range one should be able to find the Higgs at LEP II via conventional means.

The mass region $m_b + M_Z < m_{b'} < m_t + M_W$ is beyond the reach of existing e^+e^- machines, and one has to rely on hadronic colliders in the foreseeable future. Even the TEVATRON cannot fully explore this range, and the physics signature reported here is relevant for SSC/LHC. The signatures are:

$b'\bar{b}' \to b Z \bar{b} Z$ $2Z$ (+2-jets, perhaps soft).

     $b H^0 \bar{b} H^0$ BAD SIGNATURE.

     $c W \bar{b} H^0$ isolated $\ell^{\pm}$ events may contain Higgs signal!

     $b Z \bar{b} H^0$ isolated $\ell^{\pm}\ell^{\mp}$ ($M_{\ell\ell} = M_Z$) events may contain Higgs signal!

If $b' \to bZ$ dominates, there should be no problem in finding the signal in a hadronic environment. What is perhaps more intriguing is the case when $b' \to cW$ and/or $b' \to bZ$ has BR of the same order as $b' \to bH^0$. In this case one may be hopeful to discover both the b' and the Higgs boson at the same time. It is exciting that this may help cover[6] the notoriously difficult (for hadronic machines) "intermediate mass Higgs" region $M_Z < M_H < 2M_W$. One may therefore hope that the b' is heavy enough.

5. CONCLUSIONS

It is exciting that not only the b' quark decay scenario may be different from naive expectations, the new signals may also be quite spectacular. The present limit $m_t > 89$ GeV from CDF leaves much room for our scenario to work, not only for the light b' scenario ($m_{b'} < M_Z$), but also for the "intermediate" b' scenario, $M_Z < m_{b'} < m_t + M_W$. The latter may allow the Higgs boson to be discovered together with b' even if the Higgs falls in the "intermediate mass" region of $M_Z < M_H < 2M_W$. SLC and LEP have reported[2] null results from their search following the scenario presented here, with a rather solid limit of $m_{b'} > M_Z/2$. While dissappointing, clearly the book is not closed yet. The search has to continue, even if there are only 3 light neutrinos.

I conclude that much (experimental and Monte Carlo) work remains to be done, and some excitement may await us in the years ahead, not just at e^+e^- machines, but hadronic machines as well. In particular, I stress the possibility that SSC/LHC may cover the intermediate mass Higgs region if b' is rather heavy. Higgs search in b' decays is analogous to that in B decays, except that the b' may well be as heavy as 200 GeV. What is at stake may be beyond just spectacular signatures. Discovering a fourth generation heavy quark, from studying its decays, inferring the masses and mixing angles involved, *etc.*, we may learn enough to get a hint at how to crack the age old problem of why the fermions have generations, and why they exhibit the observed mass-mixing patterns. At least a window would be closed.

ACKNOWLEDGEMENT:
I thank Robin Stuart for collaboration.

REFERENCES

1. Talks by B. Harral, W. Wallraff, F. Bird and G. Vandalen, this proceedings.
2. Talks by R. Van Kooten, V. Sharma and W. Gary, this proceedings.
3. See my talk on rare K and B decays, this proceedings.
4. W.S Hou and R.G. Stuart, Phys. Rev. Lett. **62** (1989) 617; Nucl. Phys. **B320** (1989) 277.
5. W.S Hou and R.G. Stuart, CERN-TH.5601/89; CERN-TH.5650/90, to appear in Phys. Lett. **B**.
6. W.S Hou and R.G. Stuart, Phys. Lett. **B233**(1989)485; CERN-TH.5686/90.
7. B. Haeri, A. Soni and G. Eilam, Phys. Rev. Lett. **62** (1989) 719; Phys. Rev. **D41** (1990) 875.

MEASUREMENT OF THE B MESON SEMILEPTONIC BRANCHING RATIO

MARK A. WORRIS
Cornell University
Ithaca, New York, 14850

ABSTRACT

Using data from the CLEO detector at the Cornell Electron Storage Ring (CESR), we fit the inclusive charged lepton momentum spectrum from the $\Upsilon(4S)$ to theoretical models to determine the branching ratio for semileptonic B decay. The preliminary result[1] is $(10.3\pm0.1\pm0.2\pm0.3)\%$ where the errors are the statistical and systematic errors in the measurement, and the systematic error associated with the theoretical models used. Electroweak radiative corrections are applied to the models.

The procedure followed is the same as has been used in past CLEO measurements.[2] The data used in this analysis consists of 212 pb^{-1} at the $\Upsilon(4S)$ resonance and 101 pb^{-1} from the continuum just below $B\overline{B}$ threshold. The CLEO detector, our standard hadronic event selection and lepton identification criteria have been described elsewhere.[3] For this analysis, we select hadronic events with five or more charged tracks and an identified muon or electron. From the lepton momentum spectra of the $\Upsilon(4S)$ data we scale and subtract the continuum data. Corrections are made for hadrons misidentified as leptons, and for event selection and lepton identification efficiencies. The result is normalized to the number of hadronic events, which we assume to be composed almost entirely of $B\overline{B}$ pairs.[4] Finally we subtract the lepton spectra from $B\to\psi X$; $\psi\to\ell^+\ell^-$, and $B\to\tau X$; $\tau\to\nu_\tau\ell\nu$.

The leptons from the $\Upsilon(4S)$ yield the momentum spectrum shown in fig. 1. The predominant components are the direct semileptonic decays $B\to X\ell\nu$, ($b\to c$ and $b\to u$), and, for electrons, the indirect semileptonic decays $B\to DX$; $D\to Y\ell\nu$, ($b\to c\to s$). The shape of the direct decays are predicted by various theoretical models. We fit the models of Altarelli[5] and Isgur.[6] The small $b\to u$ contribution[7] above 2.3 GeV/c will have no effect within the errors of this measurement. The shape of the secondary spectrum is obtained from our spectrum of $B\to D$ folded with the experimentally measured semileptonic spectrum of D decays.[8]

Electroweak radiative corrections to these models are significant. We use a prescription[9] which allows corrections to be applied to these models externally. This prescription includes the effects of short distance loop corrections and soft virtual and real photons.

The results of the fits are as follows:

Fit	BR($b\rightarrow c\ell\nu$)	BR($b\rightarrow c\rightarrow s\ell\nu$)
Altarelli average	$(10.4\pm0.1\pm0.3)\%$	$(9.4\pm0.4\ \pm0.8)\%$
IGSW average	$(10.1\pm0.1\pm0.3)\%$	$(10.0\pm0.3\pm0.8)\%$
Overall average	$(10.3\pm0.1\pm0.2)\%$	$(9.7\pm0.2\pm0.6)\%$

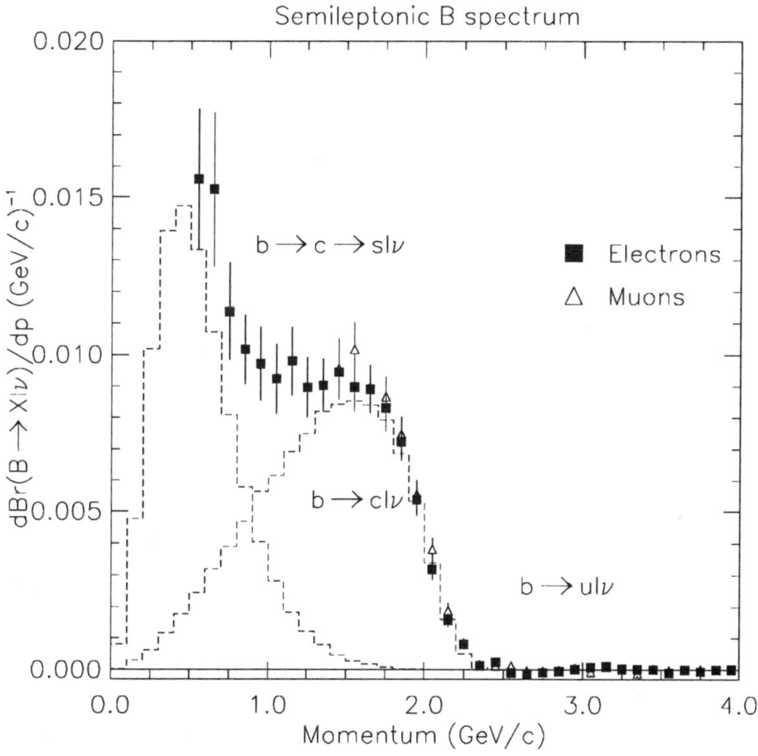

Fig. 1. Semileptonic B spectrum

REFERENCES

1. A final result described in detail will be submitted for publication soon.

2. CLEO Collaboration, S. Behrends *et al.*, *Phys. Rev. Lett.* **59**, 407 (1987). R. Kowalewski, Thesis, Cornell University, unpublished UMI 88-21246-m (microfiche) (1988).

3. CLEO Collaboration, D. Andrews *et al.*, *Nucl. Instr. Methods* **211**, 47 (1983). CLEO Collaboration, S. Behrends *et al.*, *Phys. Rev. D* **31**, 2161 (1985).

4. J. Alexander *et al.*, "Observation of $\Upsilon(4S)$ decays Into non-$B\overline{B}$ Final States Containing ψ Mesons", CLNS 90/975, CLEO 90-2, (to be published).

5. G. Altarelli *et. al*, *Nucl. Phys. B* **208**, 365 (1982).

6. N. Isgur *et. al*, *Phys. Rev. D* **39**, 799 (1989).

7. CLEO Collaboration, R. Fulton *et al.*, *Phys. Rev. Lett.* **64**, 16 (1990).

8. DELCO Collaboration, Bacino *et al.*, *Phys. Rev. Lett.* **43**, 1073 (1979).

9. D. Atwood and W. J. Marciano, Brookhaven Report No. BNL-43638, (1989) (to be published).

Exclusive Semileptonic Decays of B Mesons

GLEN CRAWFORD
Cornell University
Ithaca, New York, 14853

ABSTRACT

We report preliminary measurements of the branching fractions $B(B^- \to D^0 \ell^- \bar{\nu})$, $B(\overline{B}^0 \to D^+ \ell^- \bar{\nu})$, and $B(B^- \to D^{*0} \ell^- \bar{\nu})$. From these results, we extract the ratio of semileptonic widths for vector to pseudoscalar final states, the ratio of charged to neutral B meson lifetimes, and a model-dependent value for $|V_{cb}|$.

The data used in this analysis were collected with the improved CLEO detector at the Cornell Electron Storage Ring (CESR) in 1987. The data sample consists of an integrated luminosity of 212 pb^{-1} on the $\Upsilon(4S)$ and 102 pb^{-1} of e^+e^- data taken at an energy below the threshold of $B\overline{B}$ production (referred to as the continuum data sample). The data sample contains approximately 480,000 B meson decays. The CLEO detector and the recent modifications to its central tracking system are described in detail elsewhere.[1,2]

All events under consideration in this analysis pass our semileptonic event selection criteria.[3] To reduce the background from continuum e^+e^- annihilation we cut on an event–shape parameter[4] to reject "jet-like" events while retaining "spherical" events.[5] We also require that all charged particles be well-measured by our central tracking system. Lepton identification at CLEO is described elsewhere.[6] The candidate leptons are required to have a momentum p_ℓ between 1.4 and 2.4 GeV/c. The upper momentum limit is close to the maximum allowable for leptons from B decay to charmed particles; the lower momentum limit suppresses leptons that are not primary B decay products. We also reject leptons arising from the process $B \to \psi X$, $\psi \to \ell^+ \ell^-$.

We have already reported[7] a measurement of the exclusive rate $B(\overline{B}^0 \to D^{*+} \ell^- \bar{\nu})$, where we identified the exclusive final state by examining the spectrum of invariant missing mass (M_m^2) values in events containing a D^{*+} and an identified lepton. In this analysis we use the same technique but reconstruct only the D^0 or D^+ in leptonic events. We search for D^0 by observing its decay into $K^- \pi^+$, and for D^+ via its decay to $K^- \pi^+ \pi^+$. Since there is significant combinatorial background under the D^0 and D^+ peaks, we fit the $K\pi(\pi)$ mass spectra in several missing mass squared bins from -4.5 GeV^2/c^4 < M_m^2 < 3.0 GeV^2/c^4. Here we require the laboratory momentum of D^+ candidates to be between 1.5 and 2.5 GeV/c to improve the signal-to-background ratio. To reduce the background from events which contain a D and a ℓ^- which do not come from the same B, we require that the angle between the D and ℓ^-, $\theta_{D\ell}$, satisfy $\cos\theta_{D\ell} < 0$. Since the angle between an uncorrelated $D\ell^-$ pair is random, the $\cos\theta_{D\ell}$ cut eliminates half of this background.

There are several physical processes that could result in the observation of a $D\ell^-$ final state; we account for the ones not arising from semileptonic B decays as follows. The contribution from continuum e^+e^- annihilation events is subtracted directly using our continuum data. The contribution to the missing mass spectrum from events with fake leptons is calculated by examining the missing mass distribution for all events containing a D candidate and a track in the lepton fiducial volume; the contributions to the spectrum from tracks of different momenta are weighted by the momentum–dependent fake probability. The background to the missing mass squared spectra when the D comes from one B and the lepton from the other was estimated by studying the missing mass squared distribution of "wrong-sign" $D\ell^+$ events.

After subtracting these backgrounds, we fit the missing mass squared spectra to a function determined using our Monte-Carlo generator of semileptonic B decay. The function includes components for direct $\overline{B}\to D\ell^-\overline{\nu}$ decays, as well as feed-down from $\overline{B}\to D^{*(*)}\ell^-\overline{\nu}$, $D^{*(*)}\to DX$, or non-resonant $\overline{B}\to (D\pi)_{nr}\,\ell^-\overline{\nu}$. From the results of our analysis of $\overline{B}^0\to D^{*+}\ell^-\overline{\nu}$,[7] we fix the amount of $D^{*(*)}$ and non-resonant $(D\pi)_{nr}$ feed-down.

The results of the fits to the missing mass spectra yield $B(B^-\to D^0\ell^-\overline{\nu}) = (1.6 \pm 0.6\ ^{+0.9}_{-0.5})\%$, $B(\overline{B}^0\to D^+\ell^-\overline{\nu}) = (2.0 \pm 0.4 \pm 0.4)\%$, and $B(B^-\to D^{*0}\ell^-\overline{\nu}) = (4.6 \pm 0.5 \pm 0.7)\%$, where we have assumed that the fraction of $\Upsilon(4S)$ decays to B^+B^- and $B^0\overline{B}^0$ are equal . We use our previous result[7] of $B(\overline{B}^0\to D^{*+}\ell^-\overline{\nu}) = (4.6 \pm 0.5 \pm 0.7)\%$ in what follows. From these results we extract the ratio of vector to pseudoscalar decay widths in semileptonic B decay. We find

$$\frac{\Gamma(\overline{B}\to D^*\ell^-\overline{\nu})}{\Gamma(\overline{B}\to D\ell^-\overline{\nu})} = 2.4\ ^{+0.7\ +0.8}_{-0.4\ -0.5},$$

where we have taken a weighted average of our results for $\frac{\Gamma_{SL}(D^{*+})}{\Gamma_{SL}(D^+)}$ and $\frac{\Gamma_{SL}(D^{*0})}{\Gamma_{SL}(D^0)}$.

Isospin symmetry implies $\Gamma(\overline{B}^0\to D^{*+}\ell^-\overline{\nu}) = \Gamma(B^-\to D^{*0}\ell^-\overline{\nu})$.[8] Therefore the lifetime ratio of charged to neutral B mesons can be calculated from the ratio of semileptonic branching fractions:

$$\frac{\tau(B^-)}{\tau(B^0)} = \frac{B(B^-\to D^{*0}\ell^-\overline{\nu})}{B(\overline{B}^0\to D^{*+}\ell^-\overline{\nu})} = \frac{B(B^-\to D^0\ell^-\overline{\nu})}{B(\overline{B}^0\to D^+\ell^-\overline{\nu})} = (0.96 \pm 0.15 \pm 0.30),$$

where this result is a weighted average of our ratios of vector and pseudoscalar widths.

Using theoretical models,[9–11] which relate the exclusive width $\Gamma(\overline{B}\to D\ell^-\overline{\nu})$ to $|V_{cb}|^2$, and the average B lifetime, $\tau_B = 1.17 \pm 0.14$ ps,[12] our results imply $|V_{cb}| = 0.038 \pm 0.004$ in the ISGW model and 0.045 ± 0.005 in the WSB and KS models.

In conclusion, we have measured the exclusive semileptonic branching ratios of B mesons to the lowest-lying vector and pseudoscalar charmed mesons.

REFERENCES

1. D. Andrews *et al.*, *Nucl. Instr. and Methods* **211**(1983), 47.

2. D.G. Cassel *et al.*, *Nucl. Instr. and Methods* **A252**(1986), 325.

3. M. Artuso *et al.*, *Phys. Rev. Lett.* **62**(1989), 2233.

4. G.C. Fox and S. Wolfram, *Phys. Rev. Lett.* **41**(1978), 1581.

5. R. Fulton *et al.*, *Phys. Rev. Lett.* **64**(1990), 16.

6. R. Kowalewski, PhD. dissertation, Cornell University 1988 (unpublished).

7. D. Bortoletto *et al.*, *Phys. Rev. Lett.* **63**(1989), 1667.

8. J.L. Rosner, in *Proceedings of the Banff Summer Institute*, August 1988, Banff, Alberta, Canada, (World Scientific, Singapore, 1989).

9. (WSB): M. Wirbel, B. Stech and M. Bauer, *Z. Phys.* **C29**(1985), 269.

10. (ISGW): N. Isgur, D. Scora, B. Grinstein and M. Wise, *Phys. Rev.* **D39**(1989), 799.

11. (KS): J.C. Körner and G.A. Schuler, *Z. Phys.* **C38**(1988), 511.

12. H. Schröder, plenary talk at the *XXIV International Conference on High Energy Physics*, Munich (1988).

Anomalous ψ production in the $\Upsilon(4S)$ Energy Region

WEI-MING YAO
Purdue University
West Lafayette, Indiana, 47907

ABSTRACT

We report on the observation of ψ mesons from the $\Upsilon(4S)$ decays which are too energetic to come from B mesons. These events provide evidence for non-$B\bar{B}$ decays of the $\Upsilon(4S)$. The measured rate is $B(\Upsilon(4S) \to \psi X) = (0.22 \pm 0.06 \pm 0.04)\%$ for ψ momentum above 2 GeV/c.

The $\Upsilon(4S)$ resonance is the third radial excitation of $b\bar{b}$ system. It is massive enough to be above threshold for decay into $B\bar{B}$ and is thought to decay dominantly into these modes. In this analysis we investigate the production of ψ mesons from the $\Upsilon(4S)$ in the momentum range above the kinematic limit allowed for $\psi's$ from B decay. Low momentum $\psi's$ have previously been seen in $\Upsilon(4S)$ decay and were assumed to arise solely from B decay.[1] The inclusive branching ratio for $B \to \psi X$ was found to be 1.1%. Some fully reconstructed $B \to \psi K$ and $B \to \psi K^*$ events have been seen.

We use data taken with the CLEO detector using the Cornell Electron Storage Ring (CESR). The luminosities used consist of 212 pb^{-1} accumulated at the $\Upsilon(4S)$ resonance, 102 pb^{-1} taken at a center of mass energy 60 MeV below the $\Upsilon(4S)$. The CLEO detector is described in detail elsewhere.[2] This analysis uses the charged particle tracking system and the electron and muon identification system. The r.m.s. momentum resolution is $(\delta p/p)^2 = (0.007)^2 + (0.0023p)^2$ (where p is in GeV/c) and the dE/dx resolution is 6.5%.

The Cabbibo suppressed reaction $B \to \psi\pi$ gives the most energetic $\psi's$ possible from B decay. The ψ momentum is less than 2 GeV/c including Doppler broadening and momentum resolution. This translates into a kinematic limit $x_B = 0.378$ at the $\Upsilon(4S)$ where x is the ψ momentum divided by the beam energy.

The l^+l^- mass spectrum for the $\Upsilon(4S)$ sample is shown in Fig. 1a for $x > x_B$ and in Fig. 1b for $x < x_B$. We find 150 ± 14 events in the $x < x_B$ and $15.2^{+4.9}_{-4.5}$ for $x > x_B$. In latter sample we have 17 total events in the two bins centered on the ψ mass including a background of 5.0 events. The poisson probability of the signal at the ψ mass being caused by a background fluctuation is 2×10^{-5}. There are approximately equal number of di-electron and di-muon candidates. As these events can't be from B decay, they are either evidence of non-$B\bar{B}$ decay of the $\Upsilon(4S)$ or arise from the continuum under the $\Upsilon(4S)$ resonance.

The l^+l^- mass plot for continuum sample is shown in Fig. 1c for $x > x_B$ and in Fig. 1d for $x < x_B$. There is no signal in either x range.

We find $B(\Upsilon(4S) \to \psi X) = (0.22 \pm 0.06 \pm 0.04)\%$ for $x > 0.378$. The high momentum ψ events in the $\Upsilon(4S)$ sample have a flat momentum distribution and a spherical shape.

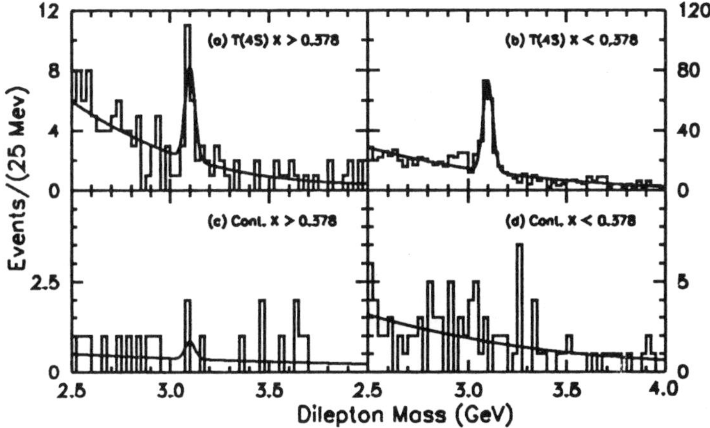

Figure 1. $m(l^+l^-)$ for various data samples. The curves are fits to the data using a Gaussian with fixed mass and width and a third order polynominal to describe the background. For $\Upsilon(4S)$ data (a) $x > x_B$ and (b) $x < x_B$. For continuum data (c) $x > x_B$ and (d) $x < x_B$.

The high momentum ψ events provide direct evidence for non-$B\bar{B}$ decays of the $\Upsilon(4S)$. Other such decays must be present. We have tried to find $D^{*\pm}$, ϕ and $\Upsilon(1S)$ signals without success. Lipkin[3] has argued that non-$B\bar{B}$ decays of the $\Upsilon(4S)$ are to be expected. However, Lipkin's predictions are not quantitative and therefore it is difficult to ascertain the validity of his assertions. There are other intriguing explanations. For example, the $\Upsilon(4S)$ has a mixture of pure $b\bar{b}$ and $b\bar{b}g$ hybrid state in its wave function.[4] The $b\bar{b}g$ part can't decay into $B\bar{B}$.[5]

In conclusion, we have observed $\Upsilon(4S)$ decays into non-$B\bar{B}$ final states containing ψ mesons. The Branching ratio is measured with $B(\Upsilon(4S) \rightarrow \psi X) = (0.22 \pm 0.06 \pm 0.04)\%$ for $x > 0.378$. We still need to determine the decay mechanism.

REFERENCES

1) P. Haas *et al. Phys. Rev. Lett.* **55**, 1248 (191985)
2) D. Andrews *et al. Nucl. Inst. Methods* **211**, 47 (1983)
3) H. J. Lipkin, *Phys. Lett.* **B179**, 278 (191986)
4) S. Ono, A. I. Sanda, and N. A. Tornqvist, *Phys. Rev.* **D34**, 186 (191986)
5) A. Le Yaouanc *et al. Z. Phys.* **C28**, 309 (191985)

Recent Results in B and D Physics from ARGUS

Sally Seidel
Physics Department
University of Toronto
Toronto, Ontario
Canada M5S 1A7

Abstract

A measurement of the branching ratio for the decay $B^0 \to D^- l^+ \nu$ is used to determine the Kobayashi-Maskawa matrix element $|V_{cb}|$. Values calculated in the contexts of six models are consistent and lie in the range $(0.038 - 0.047) \pm 0.009$. The ratio of lifetimes of charged and neutral B mesons is found to be $1.00 \pm 0.23 \pm 0.14$. The matrix element $|V_{ub}|$ is determined to be nonzero through studies of inclusive semileptonic B meson decays. The discoveries of the $D^*(2469)^+$, $D^*(2414)^+$, and $D_{sJ}(2536)^+$ are reported. The $D^*(2420)^0$ is resolved into its constituent states, and the $D^*(2459)^0$ is confirmed.

1. Introduction

Knowledge of the couplings between different generations of quarks is crucial to understanding the nature of the electroweak interaction. Relationships between coupling constants are expressed by the Cabibbo-Kobayashi-Maskawa mixing matrix,[1] which has four free parameters. Presented here are new measurements[2,3] of two of the matrix elements, V_{cb} and V_{ub}. Also presented is the first measurement[4] of the ratio of lifetimes of charged and neutral B mesons. In the latter half of this article, several results in the field of D meson spectroscopy are announced. Spectroscopy provides a tool for understanding the spin dependence of the interquark potential at large distances. The existence of the $D^*(2459)^0$ is confirmed,[5] and the first observation of its isospin partner, the $D^*(2469)^+$,[6] is reported.[7] The multi-state enhancement hitherto known as the $D^*(2420)^0$ is resolved into its components.[8] The discovery of a new charmed-strange meson, the $D_{sJ}(2536)^+$, is announced.[9]

The ARGUS detector is a cylindrical 4π spectrometer located at the $e^+ e^-$ storage ring DORIS II at DESY. The detector is optimized for particle identification, for which are used dE/dx measurements, time of flight measurements, muon coincidence signals, and shower counter information. A complete description of the ARGUS detector is given in Reference 10.

2. Measurement of V_{cb}

The matrix element for $b \to c$ transitions, V_{cb}, is related to the width for semileptonic decays according to the expression $\Gamma(B^0 \to D^- l^+ \nu) = k \times 10^{12} |V_{cb}|^2$, where k includes form factor effects. Table 1 summarizes values of k calculated for six theoretical models. The narrow range of values, 7.2–11.1, means that a measurement of the decay width provides an almost model-independent measurement of the matrix element. In practice the decay width is determined by measuring the semileptonic branching ratio and dividing it by a reliable estimate of the mean B^0 lifetime.

Model	k	$|V_{cb}|$
WBS[11]	8.1	0.044 ± 0.009
SP[12]	7.2	0.047 ± 0.009
KS[13]	8.3	0.044 ± 0.009
GISW[14]	11.1	0.038 ± 0.009
CPK[15]	9.0	0.042 ± 0.009
OS[16]	9.4	0.041 ± 0.009

Table 1. Values of k and $|V_{cb}|$ for different models.

The ARGUS data sample consists of 172 pb^{-1} recorded at the $\Upsilon(4S)$. The desired decay channel is isolated by calculating, for each event, the invariant mass of the system which recoils against the D–lepton pair during the B meson decay. By neglecting the small momentum of the B^0 and substituting the beam energy for the B^0 energy, the recoil mass squared is defined as $M^2_{recoil} = (E_{beam} - E_{D^-} - E_{\ell^+})^2 - (\vec{p}_{D^-} + \vec{p}_{\ell^+})^2$. As M_{recoil} can be identified with the mass of the neutrino, we demand that candidate events have a value in the range $-0.5 < M^2_{recoil} < 0.5$ GeV^2/c^4. We find a 135 ± 18 event D^- peak in that range due to the decay $D^- \to K^+ \pi^- \pi^-$; no corresponding peaks are present in the sidebands. In order to maintain consistency with the kinematics of D and B meson decays, the total momentum of the D–lepton system is required to be less than 2.5 GeV/c^2, the lepton momentum must be greater than 1 GeV/c, the momentum of the $K\pi\pi$ system is required to be greater than 1.5 GeV/c, and the number of charged tracks must exceed 4.

Four sources of background to this analysis are continuum D meson production (estimated at 7% by Monte Carlo techniques), accidental combination of a D and a lepton from different B meson decays (estimated at 4% from Monte Carlo), inclusion of hadrons misidentified as leptons (estimated at 4% from data), and combination of the D^- with the lepton and neutrino from the cascade decay

$$B^0 \to D^*(2010)^- l^+ \nu$$
$$\hookrightarrow D^- (\pi^0 \text{ or } \gamma),$$
$$\hookrightarrow K^+ \pi^- \pi^-,$$

where the π^0 or photon is unobserved.

The cascade contribution to the background is estimated by considering the decay chain

$$B^0 \to D^*(2010)^- l^+ \nu$$
$$\hookrightarrow \overline{D}^0 \pi^-$$
$$\hookrightarrow K^+ \pi^-.$$

This sequence is kinematically equivalent to the background one but is easier to reconstruct due to the absence of neutral particles in the D^* decay. The quantity M^2_{recoil} is computed for the $\overline{D}^0 l^+$ final state and is shown in Figure 1. The peak in the resulting histogram is offset from 0 due to the presence of a π^-, as well as a neutrino, in the final state. The histogram is scaled by the ratio in reconstruction efficiencies of the charged and neutral D mesons and is fit to a smooth curve, which is then the cascade contribution estimate. Figure 2 shows the cascade contribution curve overlayed on the data spectrum with the first three backgrounds' contributions subtracted. An excess of 82 ± 27 $B^0 \to D^- l^+ \nu$ events remains. Similar procedures for estimating the contribution of other possible cascades indicate that their contribution is negligible. Assuming e-μ universality, the branching ratio for the decay $B^0 \to D^- l^+ \nu$ is determined to be $1.8 \pm 0.6 \pm 0.5\%$.

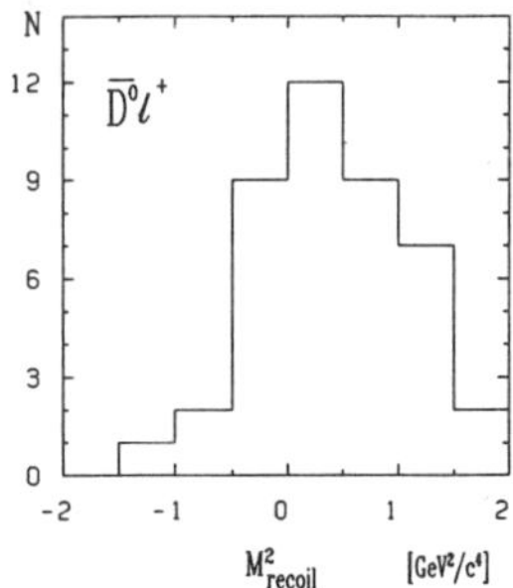

Figure 1. Distribution of M^2_{recoil} against the $\overline{D}^0 l^+$ system with $\overline{D}^0$ originating from cascade decays $B^0 \to D^*(2010)^- l^+ \nu$.

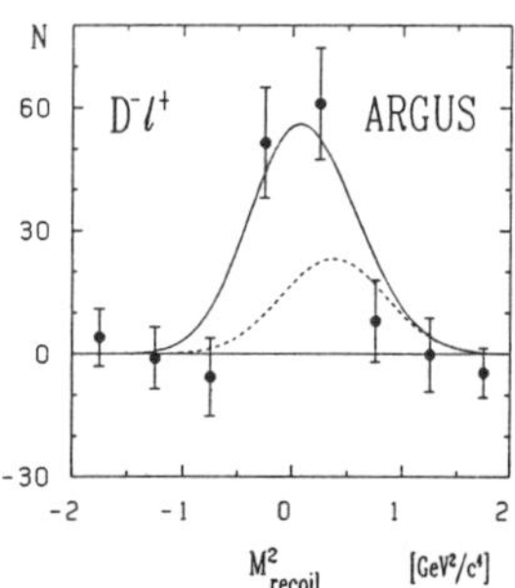

Figure 2. Distribution of M^2_{recoil} against the $D^- l^+$ system. The dashed curve shows the contribution from the cascade decay $B^0 \to D^*(2010)^- l^+ \nu$; $D^*(2010)^- \to D^- (\pi^0 \text{ or } \gamma)$.

We use the 1988 world average[17] value for the lifetime of the B meson, $\tau_B = (1.15 \pm 0.14) \times 10^{-12}$ s, to compute $|V_{cb}|$. Table 1 summarizes the results, which are mutually consistent and which lie in the range $(0.038 - 0.047) \pm 0.009$ for six models. This result supports the previous ARGUS measurement,[18] $|V_{cb}| = 0.046^{+0.008}_{-0.010}$, obtained by studying the polarization of the D^* in the decay $\overline{B}^0 \to D^{*+}l^-\nu$.

3. Measurement of τ_{B^+}/τ_{B^0}

In the context of the above measurement, it is important to consider the appropriateness of substituting the world average B lifetime, τ_B, for the lifetime of the neutral B meson. A new measurement by ARGUS of the ratio τ_{B^+}/τ_{B^0} supports this usage. Although the lifetime ratio of the charged and neutral D mesons is $2.58 \pm 0.09 \pm 0.09$,[19] several calculations suggest that replacement of the c quark by a b quark reduces the ratio.[20]

Assuming the absence of final state interactions, we expect the partial widths for semileptonic decay of charged and neutral B mesons to be equal:

$$\Gamma(B^+ \to \overline{D}^0 l^+\nu) = \Gamma(B^0 \to D^- l^+\nu) \quad \text{and} \quad \Gamma(B^+ \to \overline{D}^{*0}l^+\nu) = \Gamma(B^0 \to D^{*-}l^+\nu).$$

We use the fact that the width is given by the ratio of branching ratio to lifetime to rewrite the above expression as

$$\frac{\tau_{B^+}}{\tau_{B^0}} = \frac{BR(B^+ \to \overline{D}^0 l^+\nu) + BR(B^+ \to \overline{D}^{*0}l^+\nu)}{BR(B^0 \to D^- l^+\nu) + BR(B^0 \to D^{*-}l^+\nu)}.$$

We convert the righthand side of this expression to directly measurable quantities by using the fact that the charge of each observed D meson can be used to tag the charge of its parent B. The charged B mesons never decay to charged D mesons. Neutral B mesons decay both to charged D mesons and to neutral D mesons through the cascade $B^0 \to D^{*-}l^+\nu$; $D^{*-} \to \overline{D}^0\pi^-$. The cascade rate, however, is known. Using these relationships one can write

$$\frac{f_+}{f_0}\frac{\tau_{B^+}}{\tau_{B^0}} = \frac{N(\overline{D}^0 l^+) - N(D^{*-}l^+, D^{*-} \to \overline{D}^0\pi^-)/\epsilon_1}{\epsilon_2 N(D^- l^+) + N(D^{*-}l^+, D^{*-} \to \overline{D}^0\pi^-)/\epsilon_1}.$$

The expressions N represent the number of D mesons observed in a particular charge state. The factor ϵ_1 is the efficiency, 53%, for detection of the pion in the D^{*-} decay. The factor ϵ_2 represents the ratio in detection efficiencies for the neutral and charged D mesons, 58%. The ratio f_+/f_0 signifies the ratio in abundance of charged and neutral B mesons and is taken to be 1.

The event selection criteria[4] are similar to those used in the $B^0 \to D^-l^+\nu$ analysis above. Background due to event mixing, uncorrelated $\overline{D}l^+$ pairs, and continuum D production is considered. The resulting numbers of D species are given by $N(\overline{D}^0) = 325 \pm 28 \pm 9$, $N(D^-) = 183 \pm 37 \pm 12$, and $N(D^{*-}) = 58 \pm 9 \pm 3$. The ratio τ_{B^+}/τ_{B^0} is determined to be $1.00 \pm 0.23 \pm 0.14$. A study was performed to determine the effect upon this ratio of feeddown from as yet unobserved decays $B \to \overline{D}^*_J l^+\nu$. Denoting by R the righthand side of the above centered equation, we find that for R in the range 1.00 ± 0.23, feeddown from additional cascades changes τ_{B^+}/τ_{B^0} by less than 10% in the case where $\Gamma(B \to \overline{D}^*_J l^+\nu)/\Gamma(B \to (\overline{D} \text{ or } \overline{D}^*) l^+\nu) < 20\%$. The ratio in widths is guaranteed by existing inclusive semileptonic measurements by several groups.[3,21]

4. Measurement of $|V_{ub}|$

ARGUS has also measured a non-zero value for the matrix element V_{ub}. This result is of great interest because it is foundational to the Cabibbo-Kobayashi-Maskawa explanation[1] of CP violation. Inclusive semileptonic B meson decays are investigated for evidence of leptons having energies beyond the kinematic limit available from decays to charm. The ARGUS data sample used for this search consists of 201 pb^{-1} obtained at the $\Upsilon(4S)$ and 69 pb^{-1} obtained at the nearby continuum. Seven general selection requirements are applied to this sample: (1) The sum of the number of charged tracks and half the number of photon tracks must be greater than or equal to 6. (2) Photon candidates must have at least 80 MeV. (3) The likelihood coefficient[10] for a lepton track must exceed 70%. (4) Muon candidates must have a coincidence in the muon counters. (5) To reject converted photons, an electron candidate producing an invariant mass less than 100 MeV/c^2 when combined with any positron in its event is not considered. (6) Large angle two-photon events are rejected by requiring that $q_e \times \cos\theta < 0.9$, where q_e is the charge of the lepton being considered and θ is the angle between the missing momentum vector in the event and the detector z axis. (7) To guarantee good momentum resolution in the event, the magnitude of the cosine of the angle between the lepton and the detector z axis is required to be less than 0.68 for muons and less than 0.85 for electrons or positrons.

In addition to these general requirements, a set of special cuts is applied which favor spherical $\Upsilon(4S)$ events over the jet-producing continuum. For the special requirements, the data are partitioned into two independent sets: events containing exactly one lepton and events containing two leptons. We consider the single lepton case first.

The first special cut is placed on the scalar sum of the momenta perpendicular to the lepton track direction for particles having an angle A with respect to the lepton track such that $60° < A < 120°$. We demand that $\sum p_T > 2$ GeV/c, which saves 46% of $\Upsilon(4S)$ decays while removing 91% of the continuum background.

A second useful variable is the missing momentum in an event. From simple kinematics we expect the missing momentum of inclusive semileptonic B decays to be much larger on average than the missing momentum of hadronic B decays. Furthermore, differences in the product quark masses should lead to a missing momentum which is greater in $b \to u\ell\nu$ decays than in $b \to c\ell\nu$ cases. Figure 3 shows the WBS model's[11] predictions of the missing momentum spectra of $b \to u$ transitions and $b \to c$ transitions. Also plotted are ARGUS data for events with charged lepton momenta in the $b \to c$ regime. The excellent agreement between the data and the model in the $b \to c$ region provides support for the model's predictions in the $b \to u$ region. We use these histograms to select the momentum range where $b \to u$ transitions are most enhanced above $b \to c$ transitions and background: $p_{miss} > 1$ GeV/c. In order to reject two-photon events, p_{miss} must also be less than 3.5 GeV/c.

A third useful parameter is the cosine of the angle β between the lepton momentum vector and the vector associated with the event missing momentum. The value of $\cos\beta$ approaches -1 for signal events. Figure 4 shows the WBS model predictions for $\cos\beta$ in the $b \to u$ and $b \to c$ regions. It also shows the ARGUS data, which are consistent with the model and which clearly peak near -1. We demand that $\cos\beta < -0.5$, a cut which saves 81% of $b \to u$ transitions while rejecting 72% of the background.

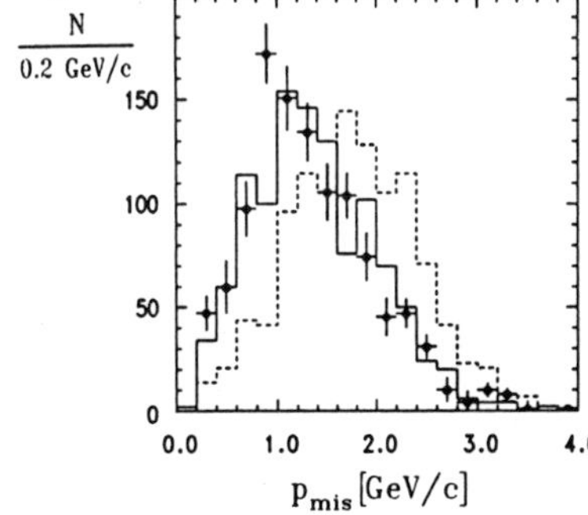

Figure 3. Distribution of missing momentum for direct $\Upsilon(4S)$ decays with leptons in the $b \to c$ range (crosses), Monte Carlo prediction for $b \to c$ (histogram), and Monte Carlo prediction for $b \to u$ (dotted histogram).

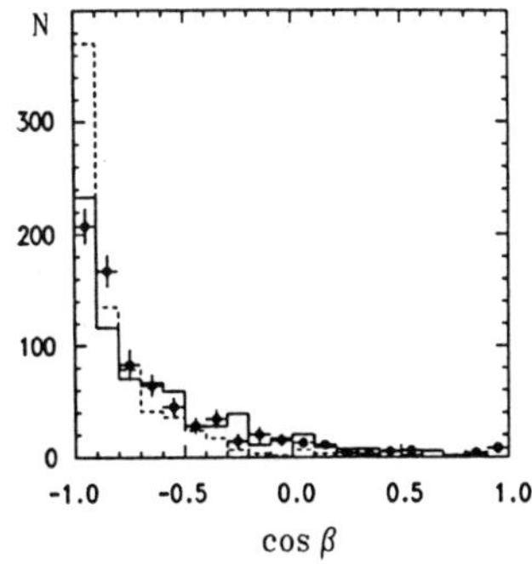

Figure 4. Distribution of $\cos\beta$ for direct $\Upsilon(4S)$ decays with leptons in the $b \to c$ range (crosses), Monte Carlo prediction for $b \to c$ (histogram), and Monte Carlo prediction for $b \to u$ (dotted histogram).

Two checks are performed to verify that the observed peak in the data is real. In the first check, we consider a sample of exclusive $B^0 \rightarrow D^{*-}l^+\nu$ events and for each event plot the cosine of the angle β' between its missing momentum vector and the direction of its approximately reconstructed neutrino track. This parameter peaks at 1 with approximately the same precision as is in the $\cos\beta$ plot, providing evidence that the $\cos\beta$ peak is indeed correlated with the neutrino direction. In a second check, the analysis is repeated for a sample of simulated events in which hadrons are substituted for the leptons. It is found that even if the fake rate in the detector were 100%–it is actually 0.5%–it could not account for the observed peak in $\cos\beta$.

Although continuum events account for the most background, other potential background sources are also considered. The first of these is resolution smearing in the high momentum tail of the $b \rightarrow c$ distribution. The momentum resolution in this regime is dominated by the resolution in the main drift chamber. That resolution has been closely studied in muon pair and Bhabha events, where it also dominates the event momentum resolution. In those cases no non-Gaussian tail has been observed. Furthermore, all of the events which contribute to the $b \rightarrow u$ signal to be discussed have been visually scanned and have tracks which are isolated and well measured. The model-dependent uncertainty in the exact shape of the $b \rightarrow c$ tail is estimated to be 20%. Taking all of the above into consideration, we estimate an 8.4 event background due to $b \rightarrow c$ events in the single lepton sample.

A second non-continuum background source is the production, via B meson decay, of J/ψ particles which decay to lepton pairs. In order to minimize this background, we require that the invariant mass of any observed lepton pair differ from the mass of the J/ψ by at least 100 MeV/c^2. The estimated residual background from this source is 1.0 event.

Finally, the inclusive $\Upsilon(4S) \rightarrow$ hadrons spectrum is folded with measured hadron misidentification probabilities in the ARGUS detector, permitting us to estimate 2.6 events of fake lepton background.

In the resultant momentum spectra for the single lepton $\Upsilon(4S)$ and continuum samples, the range $2.3 < p_l < 2.6$ GeV/c is sensitive to $b \rightarrow u$ transitions, while the range $2.0 < p_l < 2.3$ GeV/c, in which $b \rightarrow c$ transitions dominate, is used for normalization. Sixty events appear in the $b \rightarrow u$ region of the $\Upsilon(4S)$ sample; 3 events fall in the corresponding bins of the continuum sample. In order to estimate the continuum background conservatively, we calculate the average number of background events per bin for the range 2.0 to 2.9 GeV/c. This yields 7.3 estimated events in the signal region, which, when scaled by the difference in luminosities of the $\Upsilon(4S)$ and continuum samples, estimates 21.3 continuum events in the $b \rightarrow u$ region. Combining this number with the expectations for non-continuum background yields a total background estimate for the single lepton sample of 33.3 ± 5.2 events. Subtracting this value from the 60 events in the $\Upsilon(4S)$ sample leaves an excess of 26.7 ± 9.3 events in the $b \rightarrow u$ regime. There are 509 events in the normalization region.

A similar analysis is performed with the dilepton data sample. Four requirements are added to those described above. The momentum of the second lepton is required to lie between 1.2 and 2.3 GeV/c. The magnitude of the cosine of the angle between the second lepton track and the z axis is required to be less than 0.9. The cosine of the included angle between the two leptons is required to be greater than -0.8. In order to reject converted photons, the lepton pairs identified as a positron and an electron must have an included angle with a cosine less than 0.95. Non-continuum background estimates are 2.3 events from $b \rightarrow c$ decays, 0.3 event from leptonic J/ψ decays, and 2.1 events from fakes. In the dilepton sample analysis, 21 events appear in the $b \rightarrow u$ region of the $\Upsilon(4S)$ sample; no events appear in the corresponding window of the continuum sample. Continuum and non-continuum background totals 6.7 ± 2.2 events; hence there are 14.3 ± 4.9 excess events. The normalization region has 153.7 leptons.

Figure 5 shows the combined one and two lepton samples. An enhancement is clearly visible in the $b \rightarrow u$ transition region. We use the ratio of observed numbers of excess and normalization events, 0.062, as well as the ratio of reconstruction efficiencies for $b \rightarrow u$ and $b \rightarrow c$ events, 0.758, to calculate the ratio of semileptonic branching ratios in the $b \rightarrow u$ and $b \rightarrow c$ transition regions. The result is the value 0.047 ± 0.012. If the branching ratio for $\Upsilon(4S) \rightarrow B\overline{B}$ is 100%, this can be interpreted as a $b \to u$ signal. Table 2 summarizes the $|V_{ub}/V_{cb}|$ results for four models; these lie in the range 9–18%.

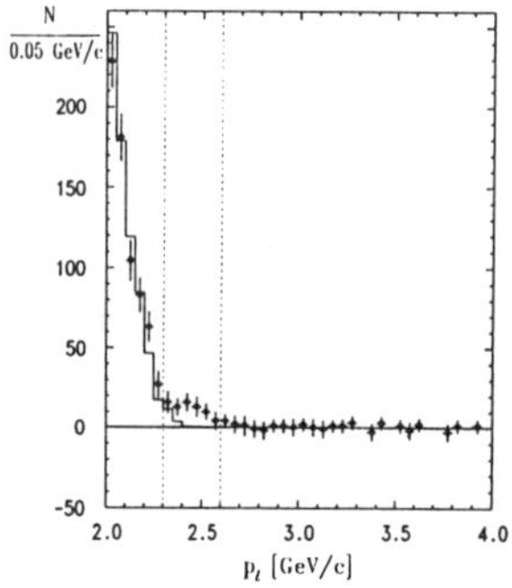

Figure 5. Combined lepton momentum spectrum for direct $\Upsilon(4S)$ decays; the histogram is a $b \to c$ contribution normalized in the region 2.0–2.3 GeV/c.

Model	$|V_{ub}|/|V_{cb}|$
ACM[22]	0.10 ± 0.01
WBS[11]	0.12 ± 0.02
KS[13]	0.09 ± 0.01
GISW[14,23]	0.18 ± 0.02

Table 2. $|V_{ub}|/|V_{cb}|$ for different models.

5. Observation of the $D^*(2469)^+$

The spectroscopy of the excited charmed mesons is the subject of considerable theoretical attention.[24–26] Information concerning new states is needed for better understanding both of the strong force at large distances and of the limits on the constituent quark masses.

The resonance $D^*(2459)^0$, first reported[27] by experiment E691, has now been seen by ARGUS in the channel $D^*(2459)^0 \to D^+\pi^-$. The state has mass $2455 \pm 3 \pm 5$ MeV/c^2 and width 15^{+13+5}_{-10-10} MeV/c^2. The Peterson parameter for its production in e^+e^- annihilation is $\epsilon = 0.06 \pm 0.03$. Its non-isotropic decay angular distribution makes this resonance a likely candidate for the $J^P = 2^+$ member of the $L = 1$ multiplet.

A search for evidence of the decay sequences $D^{*+} \to D^0\pi^+$; $D^0 \to K^+\pi^+(\pi^+\pi^-)$ has yielded the first observation of the isospin partner of the $D^*(2459)^0$, the $D^*(2469)^+$. The ARGUS data sample used in this search consists of 251 pb^{-1} recorded at the $\Upsilon(4S)$ and in the nearby continuum. In the single pion channel, the invariant mass of the reconstructed $K\pi$ pair is required to lie within 40 MeV/c^2 of the nominal D^0 mass. The cosine of the angle between the K^0 and the D^0 boost directions in the D^0 rest frame, $\cos\theta_K^*$, is required to be less than 0.7 in order to reduce the background which peaks strongly at 1. In the three-pion channel, the $K3\pi$ mass is permitted to differ from the D^0 mass by no more than 30 MeV/c^2, and the value of $\cos\theta_K^*$ must be less than 0.0. To favor charm quark fragmentation over the softer distributions of lighter quarks, combinations are also required to have a scaled momentum $p(D\pi)/p_{max}$ greater than 0.55, where $p_{max} = (E_{beam}^2 - m^2(D\pi))^{1/2}$. To reject the background due to uncorrelated slow pions, we demand that $\cos\theta_\pi^* > -0.9$, where θ_π^* is the angle, in the $D\pi$ rest frame, between their momentum vector and the direction of the $D\pi$ boost. A clear enhancement appears in the $D^0\pi - D^0$ mass difference spectrum with its peak around 600 MeV/c^2. Assuming that the observed state is the isospin partner of the $D^*(2459)^0$, we can predict that the two states have the same non-isotropic angular distribution. This suggests that the signal can be enhanced over the background by rejecting events with $|\cos\theta_\pi^*| \leq 0.4$. Investigation shows that the contribution of these rejected events to the signal is small. Applying this angular cut yields the distributions of Figure 6. We fit Breit-Wigners convoluted with Gaussians to the two peaks, the Gaussian widths being determined by simulation. The average mass difference for the two channels is 604.1 ± 3.7 MeV/c^2, corresponding to a mass of $2469 \pm 4 \pm 6$ MeV/c^2. The fit yields a natural width of 27.1 ± 11.8 MeV/c^2 and a statistical significance of 4.4σ. If the $D^*(2459)^0$ and the $D^*(2469)^+$ are true isospin partners, then neglecting small mass-related effects, the ratio of their production cross-sections multiplied by their branching ratios for decays to $D\pi$ should be equal. We find that this is the case:

$$\frac{\sigma(D^*(2469)^+) \times BR(D^*(2469)^+ \to D^0\pi^+)}{\sigma(D^*(2459)^0) \times BR(D^*(2459)^0 \to D^+\pi^-)} = 0.8 \pm 0.4 \pm 0.3.$$

The mass splitting between the states is $14 \pm 5 \pm 8$ MeV/c^2, consistent with theoretical prediction.[25]

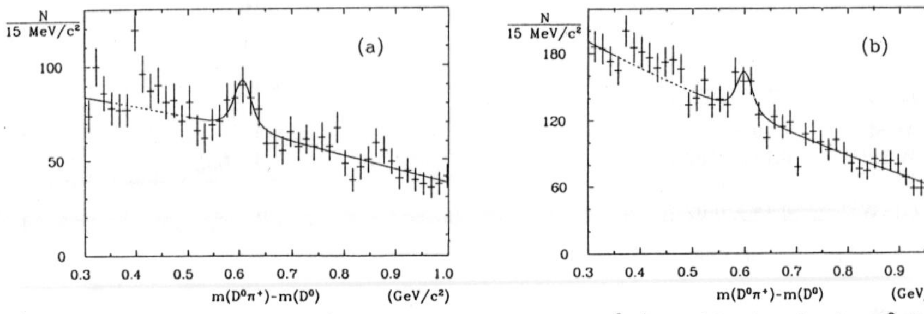

Figure 6. The $m(D^0\pi^+) - m(D^0)$ spectra for all accepted $D^0\pi^+$ combinations in the D^0 decay modes (a) $K\pi$ and (b) $K3\pi$ with the requirement that $|\cos\theta^*_\pi| > 0.4$.

6. Resonance Decomposition of the $D^*(2420)^0$

Studies of the state hitherto known as the $D^*(2420)^0$ have resulted in the observation of a new D^* state as well as a clearer picture of the $L = 1$ mesons. Table 3 lists some characteristics of the four expected states: allowed decay channels, decay angular distribution, and previous observational status. The decay angular distribution is measured with respect to the angle α, which is defined in the $D^*(2010)^+$ rest frame between the π^- from the $D_J^{(*)0}$ decay and the π^+ from the $D^*(2010)^+$ decay.

State	Decay Channels	$dN/d\cos\alpha$	Observed?
0^+	$D^+\pi^-$	—	No
1^+	$D^*(2010)^+\pi^-$	$A + B\cos^2\alpha$	No
1^+	$D^*(2010)^+\pi^-$	$A + B\cos^2\alpha$	No
2^+	$D^+\pi^-$	$\sin^2\alpha$	$D^*(2459)^0$
	$D^*(2010)^+\pi^-$		No

Table 3. Status of the $L = 1$ mesons prior to the analysis reported here.

A study was made to determine how the $D^*(2420)^0$, observed to decay to $D^*(2010)^+\pi^-$,[28] fits into this picture. The data sample used consists of 354 pb^{-1} recorded at the $\Upsilon(1S)$, $\Upsilon(2S)$, $\Upsilon(4S)$, and in the nearby continuum. A search was made for evidence of the decay chains

$$D_J^{(*)0} \to D^*(2010)^+\pi^-$$
$$\hookrightarrow D^0\pi^+$$
$$\hookrightarrow K^-\pi^+,\ K^0_s\pi^+\pi^-,\ K^-\pi^+\pi^+\pi^-.$$

Reconstruction methods[8] are similar to those used in the $D^*(2469)^+$ search. The observed enhancement is a broad structure centered at about 2420 MeV/c^2. The fact that it overlaps the known resonance at 2459 MeV/c^2 and the fact that its shape is difficult to fit with a single Breit-Wigner curve suggest that the peak may be a composite of several states. A possible way to separate the $J^P = 1^+$ and $J^P = 2^+$ states, if they are the components, is to place requirements on the decay angular spectra of the events observed. Figure 7(a) shows the peak produced by events having $|\cos\alpha| > 0.75$. This narrow state is composed principally of 1^+ events since the 2^+, proportional to $\sin^2\alpha$, is rejected with 91% efficiency. Figure 7(b) shows the broad enhancement peaking 13 MeV/c^2 higher produced by events having $|\cos\alpha| < 0.50$. As a variety of studies show that the detector's acceptance and resolution have no α-dependence, the variation of the peak is strong evidence for the presence of overlapping 1^+ and 2^+ states.

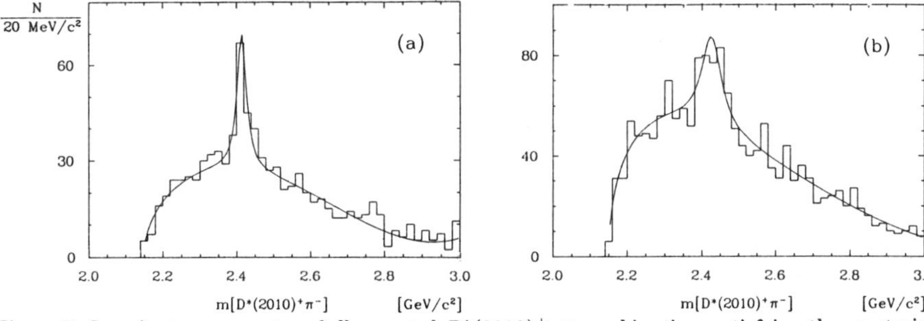

Figure 7. Invariant mass spectra of all accepted $D^*(2010)^+\pi^-$ combinations satisfying the constraint that (a) $|\cos\alpha| > 0.75$ and (b) $|\cos\alpha| < 0.50$.

To identify the lower-mass peak, the spectrum generated with the $|\cos\alpha| > 0.75$ cut is fit with two Breit-Wigners (each convoluted with a Gaussian), a cubic background, and a threshold function. The parameters of one of the Breit-Wigners are matched to those of the $D^*(2459)^0$: mass 2455 MeV/c^2 and width 15 MeV/c^2. A second peak of mass $M = 2414 \pm 2 \pm 5$ MeV/c^2 and width $\Gamma = 13 \pm 6^{+10}_{-5}$ MeV/c^2 is evident in the subsequent fit. The distribution is then refit using both peaks' masses and widths. The resultant distribution is shown in Figure 8. The $D^*(2414)^+$ decay angular distribution is well described by the function $1 + 2.8\cos^2\alpha$ and is definitely inconsistent with a $\sin^2\alpha$ distribution. This makes it a likely candidate for the $J^P = 1^+$. The $D^*(2459)^0$ decay angular distribution is consistent with $\sin^2\alpha$, suggesting that it is the 2^+. The fragmentation functions for the $D^*(2414)^+$ and $D^*(2459)^0$ have Peterson ϵ parameters 0.040 ± 0.010 and 0.054 ± 0.027, respectively. The ratio of branching ratios for decay of the $D^*(2459)^0$ to $D^+\pi^-$ and $D^*(2010)^+\pi^-$ is $3.0 \pm 1.1 \pm 1.5$. This falls within the range of theoretical predictions,[26] 1.5–3.5.

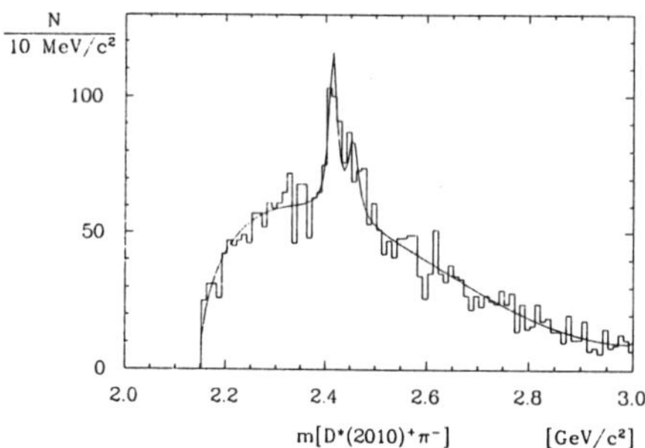

Figure 8. Invariant mass spectrum of all accepted $D^*(2010)^+\pi^-$ combinations.

7. Observation of the $D_{sJ}(2536)^+$

Recent studies show evidence for the existence of a new charmed-strange meson, the $D_{sJ}(2536)^+$. A 311 pb^{-1} data sample was searched for evidence of the decay

$$D^+_{sJ} \rightarrow D^{*+}K^0$$
$$\hookrightarrow D^0\pi^+$$
$$\hookrightarrow K^-\pi^+,\ K^-\pi^+\pi^+\pi^-,\ K^-\pi^+\pi^0.$$

Selection criteria[9] are similar to those used for the D^* searches. Figure 9 shows the result, a very narrow peak at mass $2535.9 \pm 0.6 \pm 2.0$ MeV/c^2 with statistical significance greater than 5σ.

486

The width is less than 4.6 MeV/c^2 at 90% confidence level. Studies of the wrong-charge spectra and of the sidebands in the range $160 \leq (m_{D^0\pi^+} - m_{D^0}) \leq 200$ MeV/c^2 show no peak. The Peterson distribution for the state has an ϵ parameter equal to $0.06^{+0.02}_{-0.01} \pm 0.02$. The product, $\sigma(e^+e^- \to D_{sJ}(2536)^+ X) \times BR(D_{sJ}(2536)^+ \to D^* K) = 32 \pm 9 \pm 6$ pb. It is noteworthy that no corresponding enhancement appears in the $D^+ K_S^0$ mass plot. In the absence of suppression by selection rules, phase space should favor this channel over the one reported. The lack of a signal for this decay mode supports the conjecture that the $D_{sJ}(2536)^+$ may be a state of unnatural parity and hence a candidate for the $J^P = 1^+$ state of the $L = 1$ multiplet.

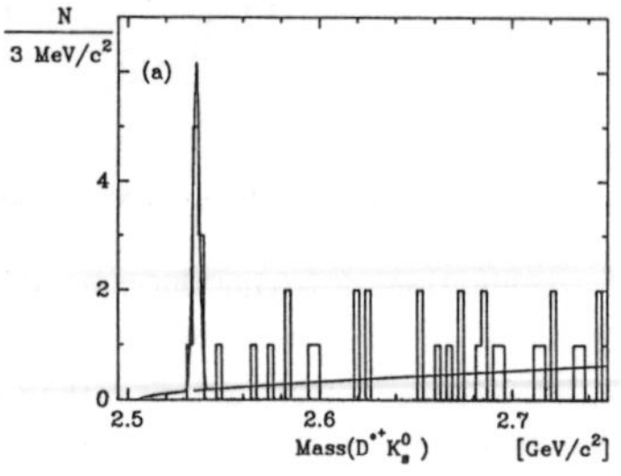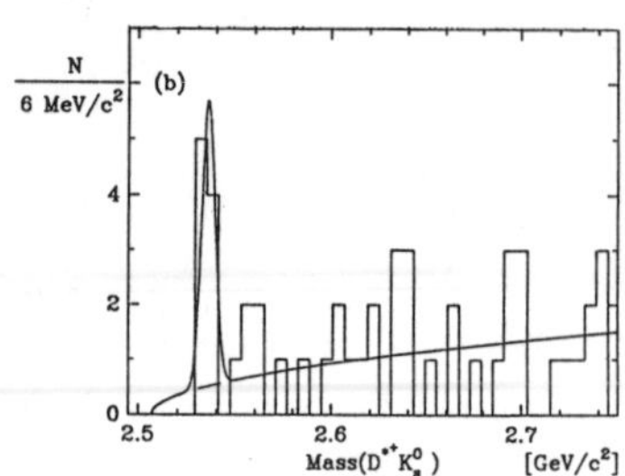

Figure 9. The $D^{*+} K_s^0$ invariant mass spectra for the D^0 decay modes (a) $K^-\pi^+$ and $K^-\pi^+\pi^+\pi^-$ and (b) $K^-\pi^+\pi^0$.

References

1. N. Cabibbo, Phys. Rev. Lett. **10** (1963) 531; M. Kobayashi and T. Maskawa, Prog. Theor. Phys. **49** (1973) 652.
2. ARGUS Collaboration, H. Albrecht et al., Phys. Lett. **229B** (1989) 175.
3. ARGUS Collaboration, H. Albrecht et al., Phys. Lett. **234B** (1990) 409.
4. ARGUS Collaboration, H. Albrecht et al., Phys. Lett. **232B** (1989) 554.
5. ARGUS Collaboration, H. Albrecht et al., Phys. Lett. **221B** (1989) 422.
6. References in this paper to a specific charge state imply the charge-conjugate state as well.
7. ARGUS Collaboration, H. Albrecht et al., Phys. Lett. **231B** (1989) 208.
8. ARGUS Collaboration, H. Albrecht et al., Phys. Lett. **232B** (1989) 398.
9. ARGUS Collaboration, H. Albrecht et al., Phys. Lett. **230B** (1989) 162.
10. ARGUS Collaboration, H. Albrecht et al., Nucl. Instr. Meth. **275A** (1989) 1.
11. M. Wirbel, B. Stech, and M. Bauer, Z. Phys. **29C** (1985) 637.
12. F. Schöberl and F. H. Pietschmann, Europhys. Lett. **2** (1986) 637.
13. G. Körner and G. A. Schuler, Z. Phys. **38C** (1988) 511.
14. B. Grinstein, N. Isgur, D. Scora, and M. Wise, Phys. Rev. **D39** (1989) 799.
15. J. M. Cline, W. F. Palmer, and G. Kramer, Phys. Rev. **D40** (1989) 793.
16. A. A. Ovchinnikov and V. A. Slobodenyuk, Z. Phys. **C44** (1989) 433.
17. D. Muller, in Proc. XXIV Int. Conf. on High Energy Phys., Munich, 1988, R. Kotthaus and J. H. Kühn, eds., Springer-Verlag, (1989) 886.
18. M. Danilov, in Proc. 1989 Int. Symp. on Lepton and Photon Int. at High Energy, SLAC, 1989, to be published.
19. Tagged Photon Spectrometer Collaboration, J. R. Raab et al., Phys. Rev. D **37** (1988) 2391.
20. A. Soni, Phys. Rev. Lett. **53** (1984) 1407; I. I. Bigi, Phys. Lett. **169B** (1986) 101.
21. ARGUS Collaboration, H. Albrecht et al., Phys. Lett. **210B** (1988) 121, D. Bortoletto et al., Phys. Rev. Lett. **63** (1989) 1667.
22. G. Altarelli et al., Nucl. Phys. **208B** (1982) 365.
23. B. Grinstein, M. B. Wise, and N. Isgur, Phys. Rev. Lett. **56** (1985) 298.
24. A. DeRújula, H. Georgi, and S. L. Glashow, Phys. Rev. Lett. **37** (1976) 785; R. Barbieri et al.,

Nucl. Phys. **B105** (1976) 125; D. Pignon and C. A. Piketty, Phys. Lett. **81B** (1979) 334; E. Eichten et al., Phys. Rev. **D21** (1980) 203; A. B. Kaidalov, Z. Phys. **C12** (1982) 63; A. B. Kaidalov and A. V. Nogteva, Sov. J. Nucl. Phys. **47** (1988) 32.

25. S. Godfrey and N. Isgur, Phys. Rev. **D34** (1986) 899.

26. J. L. Rosner, Comm. Nucl. Part. Phys. **16** (1986) 109; S. Godfrey and R. Kokoski, TRIUMF Preprint TRI-PP-86-51 (1986).

27. Tagged Photon Spectrometer Collaboration, J. C. Anjos et al., Phys. Rev. Lett. **62** (1989) 1717.

28. ARGUS Collaboration, H. Albrecht et al., Phys. Rev. Lett. **56** (1986) 549.

Neutrino Production of Opposite Sign Dimuons at Tevatron Energies

C.Foudas, K.T.Bachmann, [1] R.H.Bernstein, [2] R.E.Blair, [3] W.C.Lefmann, W.C.Leung,
S.R.Mishra, E.Oltman,[4] P.Z.Quintas, F.J.Sciulli, M.H.Shaevitz, W.H. Smith [5]
Columbia University, New York, NY 10027
F.S.Merritt, M.J.Oreglia, H.Schellman,[2] B.A.Schumm[4]
University of Chicago, Chicago, IL 60637
F. Borcherding, H.E.Fisk, M.J.Lamm, W.Marsh, K.W.B.Merritt, D.D.Yovanovitch
Fermilab, Batavia, IL 60510
A.Bodek, H.S.Budd, W.K.Sakumoto
University of Rochester, Rochester, NY 14627

Abstract

We have measured the strange quark content of the nucleon, $\eta_s = 0.057^{+0.012}_{-0.008}$, and the Kobayashi-Maskawa (KM) matrix element $|V_{cd}| = 0.220^{+0.015}_{-0.018}$ using a sample of 1797 ν_μ and $\bar{\nu}_\mu$-induced $\mu^-\mu^+$ events with $P_\mu \geq 9~GeV/c$ and $30 \leq E_\nu \leq 600$ GeV. The data are consistent with the slow rescaling hypothesis of charm production in ν-N scattering and within this formalism yield a value of the charm quark mass parameter $m_c = 1.31^{+0.64}_{-0.48}$ GeV/c^2.

Opposite-sign dimuon ($\mu^-\mu^+$) production in neutrino-nucleon interactions originates primarily from the production of charmed quarks. Since charmed quarks are mainly produced from valence down quarks and sea strange quarks, measurements of the strange content of the nucleon can be extracted from the production rate and kinematics of $\mu^-\mu^+$ events. In addition, these events can be used to extract the Kobayashi-Maskawa matrix element, V_{cd}, and a parameterization of the threshold behavior of charm production. (Knowledge of this threshold behavior is a dominant source of systematic error [1] in extracting $sin^2\theta_W$ from deep inelastic ν-N scattering.) The high statistics experiment presented here is the first measurement of opposite sign dimuon production above 300 GeV and yields the first measurement of the m_c parameter of slow rescaling [4], the first quantitative measurement of the momentum dependence of the strange sea, and a new measurment of V_{cd} with smaller errors than previous experiments.[2, 3]

The phenomenology of neutrino $\mu^-\mu^+$ production involves the slow-rescaling prescription[4] for the production of the heavy charm quark, and a fragmentation

---

[1] Present address: Widener Univ., Chester, PA 19013.

[2] Present address: Fermilab, Batavia, IL 60510.

[3] Present address: Argonne National Laboratory, Argonne, IL 60439.

[4] Present address: LBL, Berkeley, CA 94720

[5] Present address: Univ. of Wisconsin, Madison, WI 53706.

model for charmed meson production and decay. In the standard slow-rescaling parton model, the cross-section for producing a charm-quark of mass m_c from a light d or s quark in an isoscalar target is [3]:

$$\frac{d^2\sigma(\nu N \to cX)}{d\xi\, dy} = \frac{G^2 M E_\nu}{\pi}\left\{\xi[u(\xi) + d(\xi)]|V_{cd}|^2 + 2\xi s(\xi)|V_{cs}|^2\right\}[1 - \frac{m_c^2}{2M E_\nu \xi}] \quad (1)$$

where M is the nucleon mass, and V_{cd}, V_{cs} are the KM matrix elements[5]. In this expression, the usual momentum fraction $x = Q^2/2MEy$ is modified to $\xi = x(1 + m_c^2/Q^2)$. The function $q(\xi)$ ($q = u, d$ or s) is the probability distribution of a quark 'q' to have a fraction ξ of the nucleon momentum. For $\overline{\nu}$-N interactions, the cross section is also given by Eq. (1) where the functions q are replaced by $\overline{q}$.

The data were accumulated using the FNAL Tevatron Quadrupole Triplet Neutrino Beam (QTB) with the CCFR detector[6]. After fiducial and kinematic cuts, 670,000 ν_μ-N and 124,000 $\overline{\nu}_\mu$-N charged-current events remained with mean visible energies 160 GeV and 120 GeV respectively. From this charged-current sample, 1797 $\mu^-\mu^+$ events were extracted by searching for events with two tracks in the drift chambers. These events were required to have hadronic energy $E_{had} \geq 4$ GeV, both muons with momentum $P_\mu \geq 9$ GeV/c and an angle at the vertex $\theta_\mu \leq 0.250$ rad.

The $\mu^-\mu^+$ events were divided into those from incident ν or $\overline{\nu}$ by assuming that the leading muon has larger transverse momentum with respect to the direction of the hadron shower than the muon from the charm decay. This algorithm finds 1522 ν_μ-induced and 275 $\overline{\nu}_\mu$-induced $\mu^-\mu^+$ events. Monte Carlo studies, described below, indicate that the algorithm introduces a $2\pm.2\%$ ($26\pm3\%$) contamination in the ν_μ ($\overline{\nu}_\mu$) sample where the errors are due to uncertainties in the Monte Carlo modeling.

The background to the $\mu^-\mu^+$ sample from muonic decays of π and K mesons in the hadron shower[6] was calculated to be 101.7 ± 15.4 events for the ν_μ and 11.4 ± 1.7 for $\overline{\nu}_\mu$ events. All other background sources of $\mu^-\mu^+$ events (trimuons, b-quarks, J/ψ-production etc.) are calculated to be small ($< 1\%$). After the background subtraction and the contamination correction, the charm signal consists of 1460.4 ± 42.1 ν_μ and 223.5 ± 15.0 $\overline{\nu}_\mu$ induced $\mu^-\mu^+$ events.

A Monte Carlo was employed to simulate both single muon and $\mu^-\mu^+$ data in order to model the detector acceptance, resolution smearing and missing energy associated with the charm decay. In this simulation, the quark and antiquark momentum densities were taken from the CCFR structure functions[7]. The strange quark momentum density was parametrized as $x\overline{s}(x) = xs(x) = S_0(1 + \beta)(1 - x)^\beta$ with β and the normalization, S_0, left as free parameters. (Throughout the paper, $U_0, D_0, S_0, \overline{U}_0, \overline{D}_0,$ and $\overline{S}_0$ represent the integral of the quark momentum densities at the average $Q_0^2 = 16.85$ GeV^2/c^2 of the $\mu^-\mu^+$ data; i.e. $U_0 = \int_0^1 xu(x)dx$.)

The $\mu^-\mu^+$ Monte Carlo simulation follows Eq. 1, where the charm quark fragmentation into a charmed D-mesons is modeled using the Peterson function[8] with $\epsilon = .19 \pm .10$[9] and the transverse momentum distribution of the charmed mesons about the hadron shower direction by $dN/dp_t^2 \propto \exp(-ap_t^2)$ with $a = 1.1 \pm 0.3$ [10]. The resulting dimuon Monte Carlo and data E_{μ_2} and p_t distributions are in good agreement. The semileptonic decay kinematics of D-mesons follows Ref. [11] with the mean muonic branching ratio B_c left as a free parameter.

Information about the strange sea, the branching ratio, and the charm mass parameter is extracted by comparing the data and Monte Carlo x_{vis} and E_{vis} distributions where $E_{vis} = E_{\mu_1} + E_{\mu_2} + E_{had}$ and $x_{vis} = E_{vis}(E_{\mu_1} - P_{\mu_1}\cos\theta_{\mu_1})/M(E_{had} + E_{\mu_2})$ (μ_1 is the leading lepton vertex muon, and μ_2 the muon from the charm decay). The $\mu^-\mu^+$ data and Monte Carlo events are divided into ten x_{vis} bins and five E_{vis} bins and the $\mu^-\mu^+$ Monte Carlo normalization is fixed by the relative number of charged-current single muon events in the data and Monte Carlo simulation. A four parame-

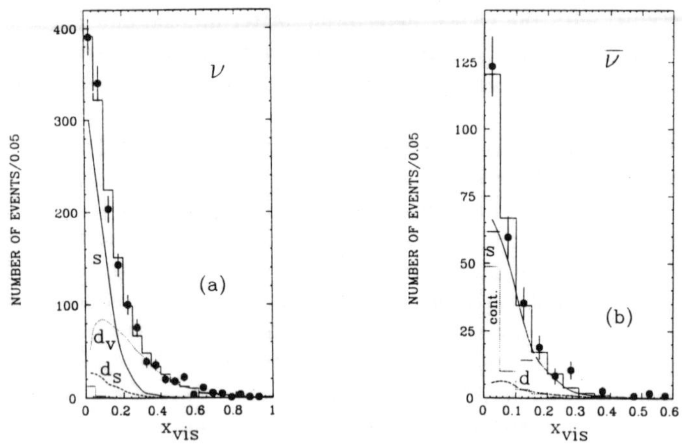

Fig. 1. Bjorken x_{vis} distributions of opposite sign background subtracted dimuon data with $P_\mu \geq 9.0\ GeV/c$ (circles) and the single charm Monte Carlo prediction (histogram) for ν_μ data (a) and $\overline{\nu}_\mu$ data (b). The Monte Carlo prediction is the sum of contributions from s ($\overline{s}$) quarks (solid curve), d-valence quarks (dotted curve), d-sea quarks (dashed curve), and contamination (dotted histogram).

ter χ^2 fit of the binned data compared to the predictions of the Monte Carlo yields: $\eta_s = 2S_0/(U_0 + D_0) = 0.057^{+0.010+0.007}_{-0.008-0.002}$, $B_c = 0.109^{+0.010+0.005}_{-0.010-0.001}$, $\beta = 10.8^{+1.1+0.7}_{-1.0-0.7}$, and $m_c = 1.31^{+0.31+0.56}_{-0.36-0.32}$ GeV/c^2, where the first error is statistical and the second systematic. (These parameters are extracted using $V_{cd} = 0.220 \pm 0.003$ and $V_{ss} = 0.9744 \pm 0.0011$ [5].) Combining η_s with the measured antiquark fraction, $R_{\overline{Q}} = (\overline{U}_0 + \overline{D}_0 + \overline{S}_0)/(U_0 + D_0 + S_0) = 0.153$ [7], we extract the more commonly quoted strange sea fraction, $\kappa = 2S_0/(\overline{U}_0 + \overline{D}_0) = 0.44^{+0.09+0.07}_{-0.07-0.02}$. The minimum χ^2 of the fit is 42 for 46 D.F., indicating good agreement between the data and Monte Carlo distributions(see Fig 1).

In order to investigate the shape differences implied by the difference between β and the exponent, α, of the total sea, $\alpha = 6.93[7]$, we compare the extracted $xs(x)$ and $x\bar{s}(x)$ distributions with $x\bar{q}(x) = x\bar{u}(x) + x\bar{d}(x) + x\bar{s}(x)$ distributions from Ref. [7] evaluated at the same Q^2 as the dimuon data. The $xs(x)$ and $x\bar{s}(x)$ are taken from the prompt dimuon x_{vis} distributions after corrections for acceptance, contamination, kinematic cuts, missing energy, threshold effects and the valence quark contribution (ν-data only). The comparison is shown in Figure 2 with $xs(x)$

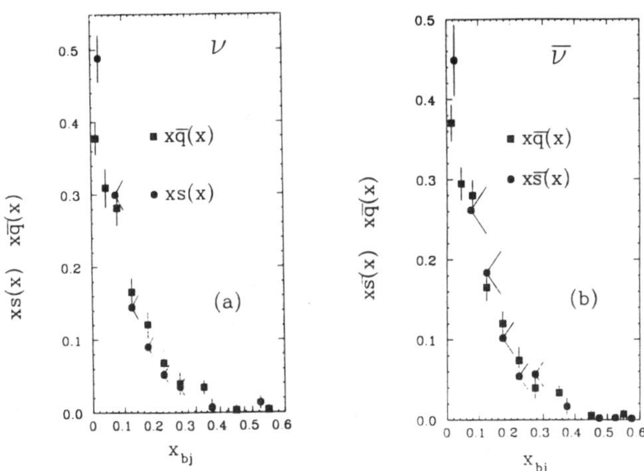

Fig. 2. Strange quark momentum distributions $xs(x)$ and $x\bar{s}(x)$ (circles) and $x\bar{q}(x) = x\bar{u}(x) + x\bar{d}(x) + x\bar{s}(x)$ (squares) evaluated at the mean Q^2 of the dimuon data for each x-bin, from **(a)** ν_μ data and **(b)** $\bar{\nu}_\mu$ data.

and $x\bar{s}(x)$ normalized to the same integral as the $x\bar{q}(x)$ distribution. The two distributions agree for $x > .05$ but differ in the first x bin indicating either a softer x distribution for s-quarks or deficiencies in the slow rescaling and QCD model at low x.

The KM coupling $|V_{cd}|$, assumed to be known in the above fits, can be determined from the dimuon data by fitting for different parameters. Three quantities, representing the contributions of s (or $\bar{s}$), d-sea, and d-valence quarks to the dimuon cross-section, are extracted assuming $m_c = 1.5^{+0.4}_{-1.5}\mathrm{GeV}/c^2$, $\beta = 10.1 \pm 2.0$, $\epsilon = 0.19 \pm 0.10$, and $R_{\bar{Q}} = 0.153$. This fit gives: $\dfrac{\kappa}{(\kappa + 2)}B_c|V_{sc}|^2 = (0.191^{+0.036+0.026}_{-0.039-0.074}) \times 10^{-1}$, $\dfrac{1}{(\kappa + 2)}B_c|V_{cd}|^2 = (0.318 \pm 0.430 \pm 0.332) \times 10^{-2}$, and $B_c|V_{cd}|^2 = (0.534 \pm 0.050^{+0.015}_{-0.060}) \times 10^{-2}$, where the first error is statistical and the second is systematic (the sensitivity to various model parameters is small because of the high energy and relatively flat energy spectrum of the quadrupole beam). Since from the $\mu^-\mu^+$ data we can only measure the product $B_c|V_{cd}|^2$, we extract V_{cd} using B_c=0.110±0.009 obtained from ν_μ-emulsion charmed particle fractions [12] combined with measured lifetimes[11]. This procedure yields a value of $|V_{cd}| = 0.220^{+0.015}_{-0.018}$, including statistical and systematic errors. This measurement has smaller errors than the lower value reported by CDHS[2] of $B_c|V_{cd}|^2 = (0.41 \pm 0.07) \times 10^{-2}$, where the errors given were statistical only.

We check the slow rescaling model, by comparing the rate of 2μ production versus single muons ($\sigma_{2\mu}/\sigma_{1\mu} |_{\nu,\bar{\nu}}$) as a function of E_ν with the predictions of the model. The dimuon rates versus E_ν, are shown in Figure 3 (squares) for ν and $\bar{\nu}$ data. The dependence on energy of the 2μ rates is attributed to the production of

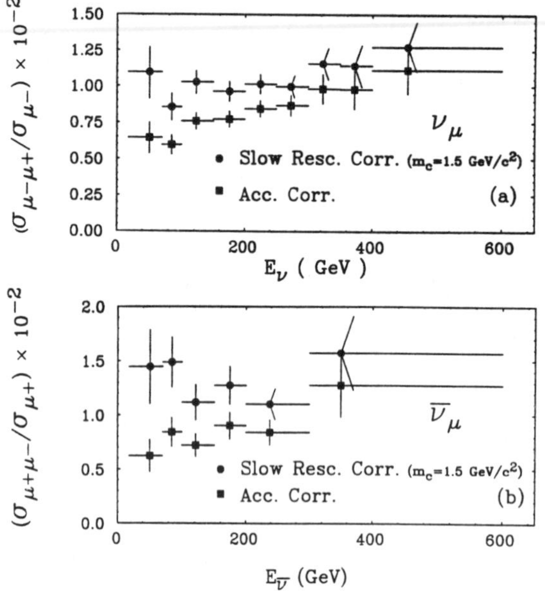

Fig. 3. Opposite sign dimuon rates as a function of E_ν for (a) ν_μ and (b) $\bar{\nu}_\mu$ data corrected for acceptance, kinematic cuts, missing energy and contamination (squares) and after corrections for slow rescaling (circles) assuming that m_c=1.5 GeV/c^2. The systematic uncertainties due to the fragmentation model are 15%, 15%, 10%, 10%, for the first four bins and 5% for the higher energy bins).

a heavy quark (charm) in the final state as described in Eq. 1. After the charm threshold effect was removed using the slow rescaling prescription with $m_c = 1.5$ GeV/c^2, the data (circles in Fig. 3) are consistent with the model predictions of no energy dependance.

From these studies of high energy dimuon production at the Tevatron, we conclude that the slow rescaling mechanism adequately describes our sample of single charm production in ν-N scattering with $m_c = 1.31^{+0.64}_{-0.48}$ GeV/c^2 for energies between 30 and 600 GeV. We have measured the Kobayashi-Maskawa coupling $|V_{cd}| = 0.220^{+0.015}_{-0.018}$. The momentum fraction carried by the strange quarks relative to non-strange quarks in the sea is found to be $\kappa = 0.44^{+0.09+0.07}_{-0.07-0.02}$, approximately half of that expected for an SU(3) flavor symmetric quark sea, and the strange quark momentum distribution is consistent with the non-strange sea for $x > .05$.

References

[1] U. Amaldi et al. Phys. Rev. **D36**, 1385(1987).

[2] CDHS: H.Abramowitz et al., Z. Phys. **C15**, 19 (1982).

[3] CCFR: K.Lang et al., Z. Phys. **C33**, 483 (1987).

[4] H.Georgi, H.D. Politzer, Phys. Rev. **D14**, 1829 (1976); R.M.Barnet, Phys. Rev. **D14**, 70 (1976); A.DeRujula et al., Rev. Mod. Phys. **46**, 391 (1974).

[5] Review of Particle Data Properties, Particle Data Group, Phys. Lett. **204B**, page 107,(1988). As described in this Ref. the values of V_{cd} and V_{cs} are determined by combining data from various experiments assuming three generations and unitarity of the KM matrix.

[6] B.A.Schumm et al., Phys. Rev. Lett. **60**, 1618 (1988).

[7] D.MacFarlane et al., Z. Phys., **C26**,1 (1984); E.Oltman, Proc. of the Storrs meeting of APS, World Scientific, June 1989. The functional forms of the parton densities, obtained from a Buras-Gaemers parametrization of CCFR data, were: $Sea(x) = 1.056(1-x)^{6.93}$, $xu_v(x) = 2.267x^{0.51}(1-x)^{2.61}$, $xd_v(x) = xu_v(x)(1-x)$ evaluated at the average Q^2=16.85 GeV^2/c^2 of our dimuon data. A. J. Buras, K. J. F. Gaemers, Nucl. Phys., **B132**, 249 (1978).

[8] C. Peterson et al., Phys. Rev. **D27**, 105(1983).

[9] H.Albrecht et al., Phys. Lett. **150B**, 235 (1985). The Argus e^+e^- fragmentation data is appropriate here since the mean E_{cm} energy is similar to the mean W^2 of the dimuon data. We use the Argus central value for ϵ but increase their reported error by a factor of three to account for possible differences in charm fragmentation between e^+e^- and νN production. (An analysis of the ν_μ-emulsion data of Ref. [12] with $W^2 > 30$ GeV^2 yielded $\epsilon = 0.18 \pm 0.06$.)

[10] M. Aguilar-Benitez et al., Phys. Lett. **123B**, 103 (1983). This value for the exponent, a, is consistent with 1.21 ± 0.34 coming from a fit to ν_μ-emulsion data [12] after imposing a $W^2 > 30$ GeV^2 cut appropriate for the dimuon data.

[11] J. Adler et al., Phys. Rev. Lett. **62**, 1821, (1989); J. Izen, Proc. of the Storrs meeting of APS, World Scientific, 1989. Also see SLAC-PUB-4753, Nov (1988); D. G. Hitlin Nucl. Phys. B (Procc. Suppl.) 3, (1988).

[12] N. Ushida et al., Phys. Lett. **121B**, 292 (1983); Phys. Lett. **206B**, 375(1988); Phys. Lett. **206B**, 380(1988); S.G.Fredriksen, Ph.D. Thesis (1987), University of Ottawa, Ontario, K1N 6N5, Canada)

SEARCH FOR THE FLAVOR-CHANGING NEUTRAL CURRENT DECAY
$K^+ \to \pi^+ \nu\bar{\nu}$

presented by M. M. Ito for the E787 Collaboration

M. S. Atiya, I-H. Chiang, J. S. Frank, J. S. Haggerty, T. F. Kycia,
K. K. Li, L. S. Littenberg, A. Sambamurti, A. J. Stevens, and R. C. Strand
Brookhaven National Laboratory, Upton, New York 11973

W. C. Louis
*Medium Energy Physics Division, Los Alamos National Laboratory,
Los Alamos, New Mexico 87545*

D. S. Akerib, M. M. Ito, D. R. Marlow, P. D. Meyers, M. A. Selen,
F. C. Shoemaker, and A. J. S. Smith
*Joseph Henry Laboratories, Princeton University,
Princeton, New Jersey 08544*

G. Azuelos, E. W. Blackmore, D. A. Bryman, L. Felawka, P. Kitching,
A. Konaka, Y. Kuno, J. A. Macdonald, T. Numao, P. Padley,
J.-M. Poutissou, R. Poutissou, J. Roy, R. A. Soluk, and A. S. Turcot
TRIUMF, Vancouver, British Columbia, Canada, V6T 2A3

1. MOTIVATION

The observation that flavor-changing neutral currents (FCNC) in weak decays are highly suppressed was first explained by Glashow, Iliopoulos and Maiani in 1970,[1] and their idea has since become a cornerstone of the Standard Model. They proposed a model of the weak interaction that included a then new fourth quark and which, in a natural way, allowed the existence of a neutral vector boson without inducing FCNC's at tree level. Thus "the couplings of the neutral intermediary...cause no embarrassment." In higher order though, decays like $K \to \pi l\bar{l}$ can proceed. Measurements of such processes provide a detailed test of the Standard Model since definite predictions for their rates can be calculated. Conversely, if no contradictions are found and the standard model is assumed to describe the physics, measurements limit the allowed values of the parameters of the model. As is often the case in studying processes that are highly suppressed and heretofore unseen, improvements in the sensitivity of experiments allow the possibility for the discovery of new physics that exhibits a similar experimental signature.

Fig. 1.1 shows diagrams that contribute to the decay $K^+ \to \pi^+ \nu\bar{\nu}$. Even at the one-loop level, the rate would vanish if not for the differences in the quark masses which upset the cancellation of the diagrams. Many authors have estimated the rate of $K^+ \to \pi^+ \nu\bar{\nu}$ in the context of the Standard Model.[2-8] From Ref. 4, each

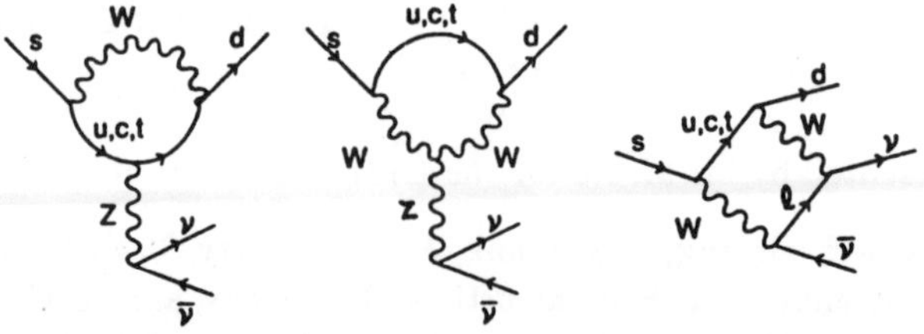

Figure 1.1: Diagrams for $K^+ \to \pi^+ \nu\bar{\nu}$.

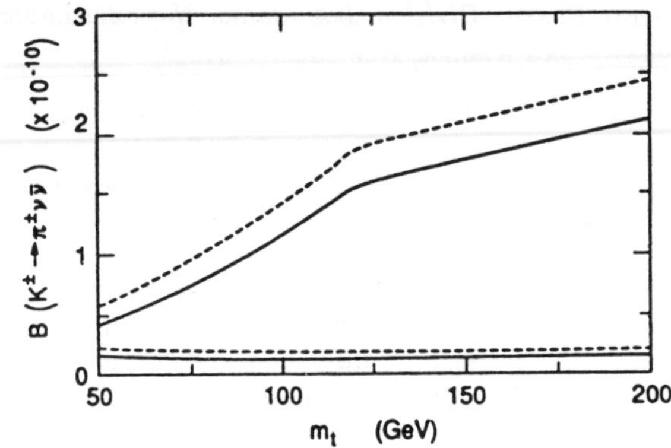

Figure 1.2: BR($K^+ \to \pi^+ \nu\bar{\nu}$) per neutrino family vs. top quark mass, from Ref. 7. Dashed line: no QCD corrections. Solid line: with QCD corrections.

flavor of neutrino contributes to the branching ratio an amount given by

$$\mathrm{BR}(K^+ \to \pi^+ \nu_i \bar{\nu}_i) = \frac{0.61 \times 10^{-6}}{|V_{us}|^2} \left| \sum_{j=c,t} V_{js}^* V_{jd} D(x_j) \right|^2 \tag{1.1}$$

$$= 0.61 \times 10^{-6} \left| D(x_c) + s_2(s_2 + s_3 e^{i\delta}) D(x_t) \right|^2$$

before QCD corrections to the short-distance amplitude, where $x_j = m_j^2/m_W^2$, $j = u, c, t$, and

$$D(x) = \frac{1}{8} \left[1 + \frac{3}{(1-x)^2} - \frac{(4-x)^2}{(1-x)^3} \right] x \log x + \frac{x}{4} - \frac{3}{4} \frac{x}{(1-x)} \tag{1.2}$$

The branching ratio is most sensitive to the values of the KM matrix elements and the top quark mass. Uncertainties in the hadronic matrix element have been largely eliminated by relating this rate to that of $K^+ \to \pi^0 e^+ \nu_e$ and long distance effects have been shown to be negligible[9] for $K^+ \to \pi^+ \nu\bar{\nu}$.

Fig. 1.2 shows the allowed range for the branching ratio versus top quark mass, per neutrino family, with and without QCD corrections, from Ref. 7. Assuming three generations of neutrinos and a top quark mass less than 200 GeV/c^2, the expectation is that

$$BR(K^+ \to \pi^+ \nu\bar{\nu}) = (1 - 7) \times 10^{-10}. \tag{1.3}$$

A recent calculation,[8] which takes into account new measurements of the branching ratio for $K^0_L \to \mu^+\mu^-$, restricts the branching ratio to an even smaller range, $(1 - 2.5) \times 10^{-10}$. The best previous experimental upper limit on this process[10] is 1.4×10^{-7} at the 90% confidence level.

This leaves a large window in which to search for new physics. An experiment sensitive to $K^+ \to \pi^+\nu\bar{\nu}$ will detect the charged pion coming from the K^+ decay recoiling against a weakly interacting system of particles. Hence it will be sensitive to certain types of new particles which include the familon,[11] which has been proposed as being responsible for the breakdown of family symmetry, and supersymmetric particles.[12-13] In addition, the standard model rate may can also be altered with the addition of new virtual particles. For example in models with non-minimal Higgs sectors a virtual charged Higgs can change the flavor of the s quark. Diagrams like these can add to or subtract from the rate for $K^+ \to \pi^+\nu\bar{\nu}$.[14]

Experiment 787 at Brookhaven is expected to address some of these issues. The status and future of the experiment will be described here. The main goal of E787 is to measure the rate of $K^+ \to \pi^+\nu\bar{\nu}$ at the standard model level. In addition, the experiment has sensitivity to other FCNC decays of the charged kaon, in particular the decays $K^+ \to \pi^+\mu^+\mu^-$ and $K^+ \to \pi^+\gamma\gamma$.

2. STRATEGY

The main backgrounds to a measurement of $K^+ \to \pi^+\nu\bar{\nu}$ come from the decays $K^+ \to \mu^+\nu$ $(K_{\mu2})$ with a branching ratio of 64% and $K^+ \to \pi^+\pi^0$ $(K_{\pi2})$ with a branching ratio of 21%. These are both topologically similar to the decay of interest, with one charged track in the final state. In order to deal with the background there are three broad areas of attack. First, one must be able to distinguish pions from muons. One way to do this is to observe the unique decay sequence $\pi^+ \to \mu^+\nu$, $\mu^+ \to e^+\nu\bar{\nu}$. Second, the detector must efficiently veto the photons from π^0 decay. This requires a 4π photon detector. And third, one must exploit the differences between the kinematics of the decay and those of the background.

Fig. 2.1 shows the range spectrum of the daughter charged particles from K^+ decay measured in the kaon rest frame. The two-body modes have a unique range. The three-body decay $K^+ \to \pi^+\nu\bar{\nu}$ peaks at its end point and has a significant fraction of its phase space between the $K_{\pi2}$ and $K_{\mu2}$ peaks. It can be seen that simple kinematic cuts can be used to eliminate background.

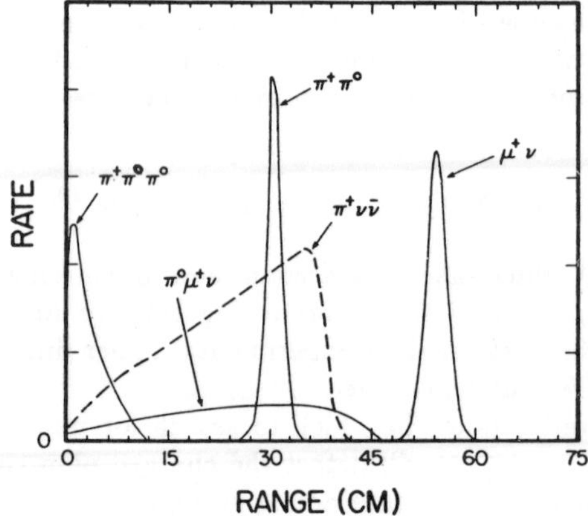

Figure 2.1: Range spectrum in scintillator for charged particles from K^+ decay. The effect of experimental resolution has been included.

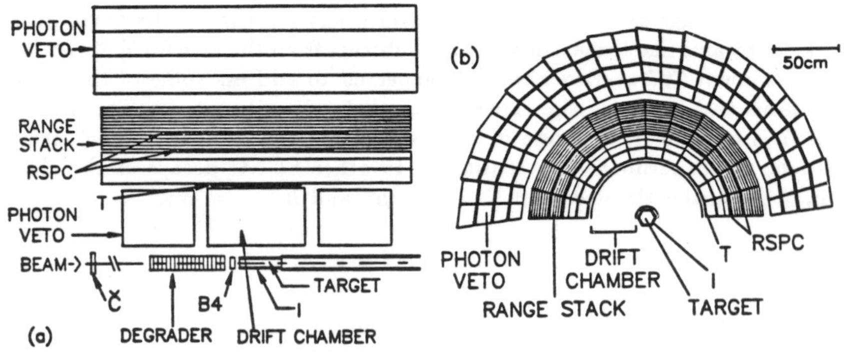

Figure 3.1: Brookhaven Experiment 787. Only the top half of the detector is shown. (a) Side view. (b) End view.

3. DETECTOR

The detector is shown in Fig. 3.1. It is located at the end of a 775 MeV/c separated beam at the AGS. The π:K ratio in the beam is about 2:1. The beam is instrumented with several scintillator hodoscopes, a small MWPC, and a Čerenkov counter. The Čerenkov counter consists of two concentric rings of photomultiplier tubes, constructed to give hits in the outer ring for beam kaons and in the inner ring for beam pions. The beam passes through a beryllium oxide degrader and kaons are ranged out in a scintillating fiber target. The fibers are 2 mm in diameter and are ganged six to a phototube. The fibers are oriented parallel to the beam direction. This orientation ensures that kaons, which are at the end of their range and heavily ionizing, deposit most of their energy in a few fiber clusters giving large pulse height. The charged particle from the decay is minimum ionizing and is travelling roughly transverse to the fibers, leaving very little energy. This contrast aids in the pattern recognition. The sides of the target, where the decay track exits the fiber array, are surround by a set of scintillators (I-counters) which serve as a convenient pick-off point for measuring the kaon decay time.

Surrounding the target and the I-counters is a cylindrical drift chamber. The wires are arranged in five super-layers of cells, each cell containing six sense wires. Super-layers 2 and 4 have stereo angles of approximately $\pm 4°$. Charged particles from K^+ decay are brought to rest in the plastic scintillator range stack, the long dimension of the counters being parallel to the beam direction. The first layer is relatively thin and short and along with the I-counters serves to define the geometric acceptance at the zeroth level in the trigger, roughly 2π steradians. It is followed by 20 layers, 3/4" thick each, ganged into 14 logical layers. Each logical layer is viewed on both ends with photomultiplier tubes. The range stack has 24 azimuthal sectors. It is also instrumented at two radial positions with multiwire proportional chambers with serpentine cathode strips which allow a z (beam direction) position measurement by means of end-to-end timing, with the anode wires providing an x-y measurement.

Outside of the range stack is the lead-scintillator barrel photon veto. It is in total 14 radiation lengths thick and is arranged in logs along the beam direction, with 48 in the azimuthal direction, and 4 in the radial direction. The logs are read-out on both ends with photomultiplier tubes. Completing the solid angle coverage of the photon veto are lead-scintillator endcaps, arranged in 24 azimuthal "petals", which have waveshifter read-out to phototubes.

Surrounding the entire detector is a conventional copper-coil magnet which provides a 1 T field for the momentum measurement.

4. TACTICS

The tagging of the $\pi^+ \rightarrow \mu^+\nu$ decay in the range stack presents a severe experimental challenge. The problem is to resolve two pulses from the range stack counter where the charged particle stops, one pulse from the stop, one from the decay. The difficulty comes from the fact that the lifetime of the charged pion (26 ns) is comparable to the width of the phototube pulse. In addition, the size of the pulse from the decay muon ranging out (4 MeV of kinetic energy) is small compared to the size of the stopping pulse (typically 20-30 MeV).

To accomplish this the range stack is presently instrumented with 180 channels of transient digitizers, which give 8 bits of ADC information every 2 ns. Fig. 4.1 shows a fit to the transient digitizer data for a pion candidate using (a) a single standard pulse shape, and (b) two such pulse shapes. Pions can be separated from muons by comparing the goodness-of-fit between these two hypotheses, along with the requirement that the second pulse have a time and amplitude consistent with $\pi^+ \rightarrow \mu^+\nu$ decay.

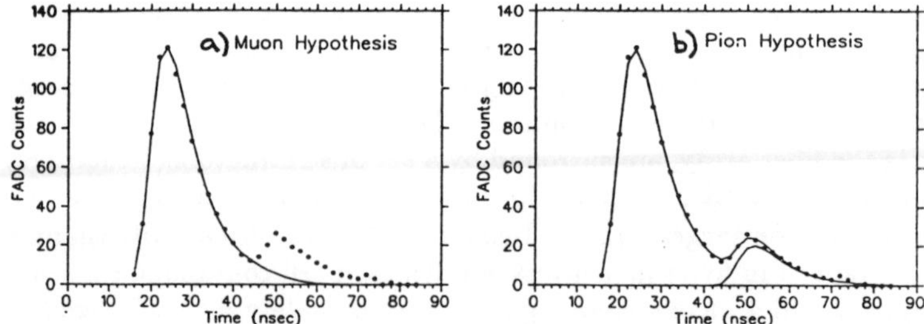

Figure 4.1: Transient digitizer data from the stopping counter. (a) Fit to muon (one pulse) hypothesis. (b) Fit to pion (two pulse) hypothesis.

Although the muon lifetime of 2.2 μs is two orders of magnitude greater than that of the pion, and the end-point of the decay positron spectrum is 53 MeV, it is possible that a muon can decay early with a low energy positron and thus fake the $\pi^+ \to \mu^+\nu$ signature. To reject this source of background, the positron from the decay chain $\pi^+ \to \mu^+\nu$, $\mu^+ \to e^+\nu\bar{\nu}$ is required since an early $\mu^+ \to e^+\nu\bar{\nu}$ decay uses up its positron to fake $\pi^+ \to \mu^+\nu$. The decay positron is also detected with the transient digitizers, which have a long time range, allowing us to record data for several muon lifetimes after the $\pi^+ \to \mu^+\nu$ decay.

Photon detection is accomplished using not only the barrel and endcap photon vetoes, but also the range stack and target. We have measured our π^0 inefficiency to be 2×10^{-6} which agrees well with Monte Carlo calculations. The inefficiency is dominated by photonuclear interactions.

We have three independent measurements of the decay kinematics. First there is the momentum measurement in the cylindrical drift chamber. Second, since nearly all of the material that the decay particle traverses before ranging-out is active—principally the target and range stack—summing the pulse-height information from these detectors gives a measurement of the kinetic energy of the track. Third, by tracking the particle in the target, drift chamber, and range stack, the geometric range is measured. The rms resolutions we obtain in range, kinetic energy, and momentum are 4.2%, 2.9%, and 2.6% respectively, for $K_{\pi 2}$ events. We use the redundancy to guard against non-Gaussian fluctuations in any one of the measured quantities. Also since the relation between energy and momentum and the relation between range and momentum depend on the mass of the particle, we can use the measured quantities to distinguish pions from muons.

5. TRIGGER

The trigger for the experiment was arranged in three levels. What follows are typical rates for our 1988 run. For this period, a typical spill from the AGS yielded 1.5×10^5 stopping kaons.

Level 0 was based on relatively simple coincidence electronics. It required that a pulse in the I-counters, where the decay track left the target, be seen delayed by at least 2 ns with respect to an incoming beam kaon detected with the Čerenkov counter. This delayed coincidence rejected kaon decays in flight. The three innermost logical layers of the range stack were required, with the outer three layers in veto to eliminate the penetrating $K_{\mu 2}$ decays. There was also a veto on analog sums from the photon vetoes, rejecting events with greater than 5 MeV of visible energy in the barrel or with greater than 10 MeV visible in the endcaps. These requirements gave us a rejection factor of 22. This left 6800 events per spill passing level 0, mostly $K_{\mu 2}$ events.

Level 1 further reduced the rate from $K_{\mu 2}$'s by examining a more refined measurement of the range. It took as input z information from the range stack MWPC's, and the number of target elements associated with the charged track, along with the stopping layer information found at level 0. A memory look-up scheme was used to make the decision. Also, level 1 applied a veto on range stack energy greater than 10 MeV outside the azimuthal region that contained the charged particle, to eliminate early photon conversion. The rejection of this level of the trigger was a factor of 15, leaving 450 events per spill.

The level 2 trigger was executed in microprocessors resident in each FASTBUS crate (SLAC Scanner Processors or SSP's).[15] Two cuts were applied. One required that the sum of energy associated with the charged track from the target and range stack be less than that from $K_{\mu 2}$ events. The other read the transient digitizer data and did a crude calculation to identify $\pi^+ \to \mu^+ \nu$ decay candidates. The total area (charge) of the photomultiplier pulse was predicted based on its peak ADC value. The area of the pulse was also calculated directly. The $\pi^+ \to \mu^+ \nu$ decay was seen as pulse area in excess of the prediction. The combined rejection of these cuts was a factor of 30 leaving 15 events to be written to tape per spill.

The live time of all levels of the trigger combined was 70%.

6. OFFLINE ANALYSIS

The offline analysis enforced many of the criteria required by the trigger, only with greater precision, as well as applying additional cuts not easily implemented in hardware. One charged track was required to have been reconstructed in the target, drift chamber, and range stack. The delayed coincidence mentioned above was re-asserted using the time of the decay track measured in the target rather than that in the I-counters. Events were vetoed if they showed a significant signal in the pion Čerenkov counters. The beam counter directly upstream of the target was required to show large energy loss, consistent with a kaon near the end of its range. Photon vetoing was re-applied using more detailed information from ADC data, and with better timing information from TDC's. By requiring that energy be detected at the same time as the charged particle, we were able to reduce the level of vetoes by accidental counts in the detector enough to allow lower energy thresholds than that used in the trigger. Transient digitizer information was subjected to the complete least-squares fit, comparing a muon (one pulse) hypothesis

Table 7.1: Acceptance for $K^+ \to \pi^+\nu\bar{\nu}$.

Delayed coincidence	0.79
Solid angle	0.47
Trigger inefficiency	0.81
Reconstruction	0.74
$K^+ \to \pi^+\nu\bar{\nu}$ spectrum acceptance	0.17
π^+ nuclear int.	
and decay in flight	0.51
$\pi^+ \to \mu^+\nu$ decay tagging	0.52
$\mu^+ \to e^+\nu\bar{\nu}$ decay tagging	0.79
accidental veto	0.70
Overall acceptance	0.0055

to a pion (two pulse) hypothesis. Positrons required to complete the $\pi \to \mu \to e$ decay sequence were demanded. A cut was applied on the relation between range and momentum, requiring consistency with a pion hypothesis. Finally, range, energy, and momentum were required to be in the kinematic region above the $K_{\pi 2}$ peak and below the $K_{\mu 2}$ peak.

7. RESULTS

In the 1988 run we had an exposure of 1.24×10^{10} stopped K^+'s taken over a two week period.

Table 7.1 displays the acceptance of the various requirements. The overall acceptance was 0.55%. Fig. 7.1 shows the candidates before the final cuts on range and kinetic energy have been applied. All of the events shown are consistent with $K_{\pi 2}$ decays where both photons from the π^0 have been missed. No events are seen in the signal region. This gives us a limit:[16]

$$\mathrm{BR}(K^+ \to \pi^+\nu\bar{\nu}) < 3.4 \times 10^{-8} \text{ at the 90\% C.L.} \tag{7.1}$$

The kinematic acceptance for the two body decay $K^+ \to \pi^+ X^0$ is greater than that for $K^+ \to \pi^+\nu\bar{\nu}$ if X_0 is a light particle. If we assume that it is massless we obtain

$$\mathrm{BR}(K^+ \to \pi^+ X^0) < 6.4 \times 10^{-9} \text{ at the 90\% C.L.} \tag{7.2}$$

8. STATUS AND FUTURE

In addition to our effort at measuring $K^+ \to \pi^+\nu\bar{\nu}$, we are able to study other rare processes. We have set a new upper limit on the branching ratio of the decay[17] $K^+ \to \pi^+\mu^+\mu^-$ of 2.3×10^{-7}. We have also obtained a preliminary upper limit on the branching ratio[18] for $K^+ \to \pi^+\gamma\gamma$ of 1.0×10^{-6}. As a by-product of our analysis of $K^+ \to \pi^+\nu\bar{\nu}$, the rate of events that evade photon veto cuts and

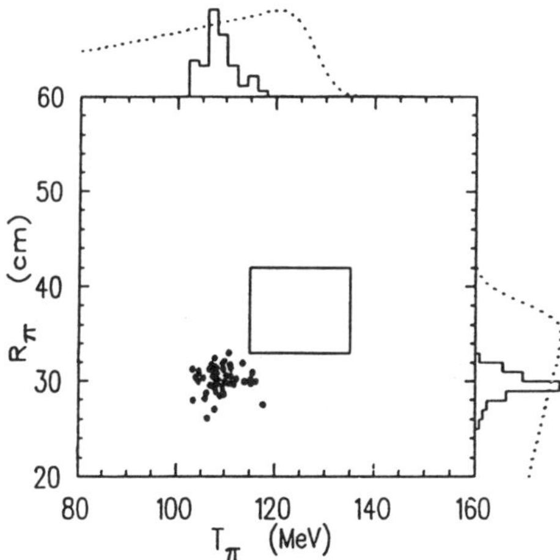

Figure 7.1: Final sample of candidates from the 1988 data before range and kinetic energy cuts have been applied. The box shows the acceptance of these cuts. The dotted lines show the expected spectra of pions from $K^+ \to \pi^+ \nu \bar{\nu}$.

are consistent with a $K_{\pi 2}$ hypothesis gives us a preliminary upper limit on the branching ratio of the process $\pi^0 \to \nu \bar{\nu}$ of 8×10^{-7}. All of these limits are at the 90% confidence level.

The experiment ran in 1989 with many improvements over the conditions in 1988. The charged track energy sum formerly performed in the level 2 microprocessor trigger was done in dedicated trigger logic with an attendant reduction in dead time. The coverage of the range stack with transient digitizer channels was improved, which helped mainly with the efficiency for detecting positrons. An ACP farm[19] was added to the data acquisition. This was used to reduce the volume of transient digitizer data. Most importantly, a combination of longer running time and higher intensity resulted in an increase of almost an order of magnitude in stopped kaons. The analysis of $K^+ \to \pi^+ \nu \bar{\nu}$ from this data is proceeding.

We have just completed another data run (1990). We have installed improvements over our 1989 configuration, most notably the addition of a RISC processor board that handles the trigger function and raw data formatting of the transient digitizers, replacing the SSP's in this role. The gain realized is in both trigger and read-out dead time. We expect that our 1990 run will double our total data sample.

We have recently had a proposal approved for additional running, which includes a new beam line. The new line is designed to increase the kaon flux, through larger acceptance at the entrance to the line, and reduce the pion flux through two stages of electrostatic separation versus the one stage in the present line. We are also preparing efforts to improve the detector to be able to handle the higher rates that will come when the AGS Booster is turned on. With these improvements we hope not only to close the window on exotic physics but also to make a measurement of the branching ratio for $K^+ \to \pi^+ \nu \bar{\nu}$.

REFERENCES

1. S. L. Glashow, J. Iliopoulos, and L. Maiani, Phys. Rev. D **2**, 1285 (1970).

2. T. Inami and C. S. Lim, Prog. Theor. Phys. **65**, 297 (1981).

3. J. Ellis and J. S. Hagelin, Nucl. Phys. **B217**, 189 (1983).

4. J. Ellis, J. S. Hagelin, and S. Rudaz, Phys. Lett. B **192**, 201 (1987).

5. Y. Nir, Nucl. Phys. **B306**, 14 (1988).

6. J. Ellis, J. S. Hagelin, S. Rudaz, and D.-D. Wu, Nucl. Phys. **B304**, 205 (1988).

7. C. O. Dib, I. Dunietz, and F. J. Gilman, SLAC-PUB-4840, March 1989.

8. C. Q. Geng and J. N. Ng, Phys. Rev. D **41**, 2351 (1990).

9. D. Rein and L. M. Sehgal, Phys. Rev. D **39**, 3325 (1989).

10. Y. Asano *et al.*, Phys. Lett. **107B**, 159 (1981).

11. F. Wilczek, Phys. Rev. Lett. **49**, 1549 (1982).

12. M. K. Gaillard, Y.-C. Kao, I-H. Lee, and M. Suzuki, Phys. Lett. **123B**, 241 (1983).

13. S. Bertolini and A. Masiero, Phys. Lett. B **174**, 343 (1986).

14. A. J. Buras, P. Krawczyk, M. E. Lautenbacher, and C. Salazar, MPI-PAE/PTh 52/89, August 1989.

15. H. Brafman, *et al.*, IEEE Trans. Nucl. Sci. **NS-32**, 336 (1985).

16. M. S. Atiya, *et al.*, Phys. Rev. Lett. **64**, 21 (1990).

17. M. S. Atiya, *et al.*, Phys. Rev. Lett. **63**, 2177 (1989).

18. This limit on BR($K^+ \to \pi^+ \gamma \gamma$) assumes a phase space spectrum for the pion momentum.

19. I. Gaines, *et al.*, Compt. Phys. Commun. **45**, 323 (1987).

New Results on the Rare Decays
$K_L \to \pi^0 e^+ e^-$ and $K_L \to e^+ e^- \gamma$

M.P. SCHMIDT, R.K. ADAIR, H.B. GREENLEE, H. KASHA,
E.B. MANNELLI,[a] K.E. OHL, and M.R. VAGINS

Department of Physics
Yale University, New Haven, CT 06511, U.S.A.

E. JASTRZEMBSKI,[b] R.C. LARSEN, L.B. LEIPUNER, and W.M. MORSE

Department of Physics
Brookhaven National Laboratory, Upton, NY 11973, U.S.A.

C.B. SCHWARZ

Department of Physics and Astronomy
Vassar College, Poughkeepsie, NY 12601, U.S.A.

ABSTRACT

We report on new results from Brookhaven AGS experiment 845. A dedicated search for the CP violating decay $K_L \to \pi^0 e^+ e^-$ has revealed no candidates and obtained a preliminary limit for the branching ratio: $B(K_L \to \pi^0 e^+ e^-) < 6 \times 10^{-9}$ (90% C.L.). We also discuss the progress of a determination of the branching ratio and form factor for the Dalitz decay $K_L \to e^+ e^- \gamma$.

1 Introduction

In a continuation of our program of searching for rare or forbidden K_L decays we have carried out a dedicated search for the CP violating decay $K_L \to \pi^0 e^+ e^-$. The design of the detector has in addition made it possible to study the Dalitz decay $K_L \to e^+ e^- \gamma$ and to search for the double Dalitz decay $K_L \to e^+ e^- e^+ e^-$.

The decay $K_L \to \pi^0 e^+ e^-$ has received considerable attention because of its potential for elucidating the mechanisms responsible for the violation of CP invariance. In the Kobayashi-Maskawa (KM) six quark model of CP violation, $K_L \to \pi^0 e^+ e^-$ is expected to receive comparable CP violating contributions from a *direct K_2* decay amplitude and from an *indirect* amplitude due to the CP impurity (ϵK_1) of the K_L state. Calculations within the Standard (KM) Model obtain very small values for the $K_L \to \pi^0 e^+ e^-$ branching ratio, typically[1] in the range $10^{-11} - 10^{-12}$. A CP conserving contribution to $K_L \to \pi^0 e^+ e^-$ through a $\pi^0 \gamma \gamma$ intermediate state is also expected to be small, although there is theoretical uncertainty over its importance relative to the CP violating contributions.[2] Recent measurements[3] provide a limit of $B(K_L \to \pi^0 e^+ e^-) < 4 \times 10^{-8}$. As the search continues down towards the level predicted by the Standard Model, the decay $K_L \to \pi^0 e^+ e^-$ provides a means to search for light scalar particles that couple to $e^+ e^-$ and to detect effects expected in nonstandard models[4] of CP violation.

The Dalitz decay $K_L \to e^+ e^- \gamma$ as well as the double Dalitz decay $K_L \to e^+ e^- e^+ e^-$ provide a means to study the dynamics of the underlying K_L-γ-γ vertex. Assuming a constant form factor, the Kroll-Wada[5] prediction for the branching ratios is $B(K_L \to e^+ e^- \gamma) = 9 \times 10^{-6}$ and $B(K_L \to e^+ e^- e^+ e^-) = 3.6 \times 10^{-8}$. The structure of the form factor is determined by contributions from weak $\Delta S = 1$ nonleptonic transitions between pseudoscalars states ($K_L \to \pi, \eta, \eta' \to \gamma\gamma^*$) and between vector states ($K_L \to K^*\gamma$ with $K^* \to \rho, \omega, \phi \to \gamma^*$). The vector-vector transition, which vanishes for photons on mass shell and therefore does not contribute to $K_L \to \gamma\gamma$, could in principle be important for the K_L Dalitz decays and $K_L \to \mu^+\mu^-$.[6]

2 The Detector

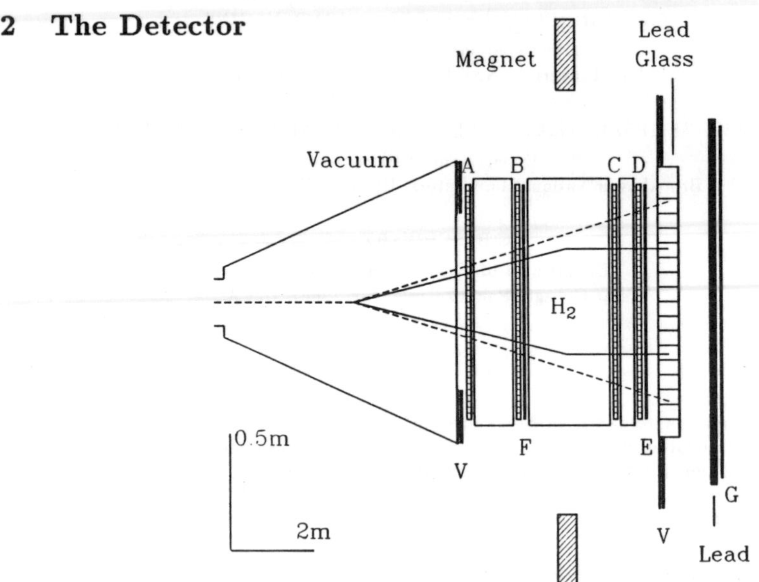

Figure 1: Schematic plan view of the E845 detector. The neutral beam enters from the left. Note the different horizontal and vertical scales.

The measurements presented below were performed at the Brookhaven AGS on K_L decays occurring in a neutral beam produced by the interactions of 24 GeV/c protons with a one interaction length Cu target. The neutral beam had a production angle of 2° and a solid angle of 35μsr. A 2.5 cm lead plug placed in the neutral beam removed high energy gamma rays from the target. About 10 m of shielding and 6 Tesla-meters of sweeping field separated the target from the beginning of a 6 m decay region. The evacuated decay region was terminated by a window consisting of 0.43 mm of Kevlar and 0.13 mm of mylar. Typically about 10^{12} protons per ($\sim$1 sec) pulse were directed onto the target, resulting in a beam flux of 3 $\times 10^8$ neutrons per pulse.

The detector for the experiment (AGS E845), shown in Figure 1, was largely constructed from the apparatus used in our previous search[7] (AGS E780) for $K_L \to \mu e$, $K_L \to ee$ and $K_L \to \pi^0 e^+ e^-$. Charged particle trajectories were determined with four sets of chambers, two upstream (A and B) and two downstream (C and D) of a momentum analysis magnet. The magnet had a 91 cm vertical gap and mean field integral corresponding to a momentum transfer of 114 MeV/c. The chambers had small (3.2 mm) drift cells and provided good resolution (250 μm) and high efficiency (99%) over a 1 m^2 region. In order to maximize the acceptance, the neutral beam passed through the active regions of the chambers. Each chamber set presented less than 10^{-3} of an interaction length of material to the neutral beam.

Electrons and positrons were identified by a hydrogen gas threshold Čerenkov counter operating at atmospheric pressure, as well as by the pattern of energy deposited in a segmented lead glass calorimeter. The Čerenkov counter had a nominal 2 m long radiator volume and was divided into four cells. Each quadrant was viewed by a 12.7 cm diameter EMI-9823B quartz window photomultiplier tube by reflection from an aluminized front-surface mirror with spherical section. The efficiency of the Čerenkov counter was determined to be 92% using electrons (from K_{e3} decays) in the data sample.

Gamma rays were detected in the lead glass array which consisted of 244 blocks of Schott F2 glass, each of dimension 6.4 cm $\times$ 6.4 cm $\times$ 46 cm (length). The blocks were arranged in a square (1 m^2) array with a 12.7 cm $\times$ 38 cm hole in the center to pass the neutral beam. For 1 GeV electrons the spatial resolution was 13 mm, and the energy resolution was $\sigma_E/E = 7\%/E^{1/2} + 1.6\%$ (E in GeV).

Accepted events were required to have two charged tracks consistent with electron identification, and at least 4 GeV of energy deposited in the lead glass array. An electron was identified at the trigger level by the activation of one Čerenkov quadrant, and a downstream scintillation counter (E) in the same quadrant. In addition the energy deposited in the corresponding quadrant of lead glass was required to be greater than 800 MeV, as determined by a fast analog summing network. One of the electrons was required to activate an upstream scintillation counter (F). Discrimination at the trigger level against penetrating pions and muons was obtained with a scintillation counter veto array (G) placed downstream of a 15 cm thick lead wall behind the lead glass array. Scintillation counters (V) covered with 3 radiation lengths of material were used to veto events with gamma rays outside the fiducial detector acceptance.

Unbiased triggers, without particle identification, energy or veto requirements, were collected concurrently, prescaled by a factor of 10,000. From this sample $K_L \rightarrow \pi^+\pi^-\pi^0$ decays were identified and used to determine the resolution of the detector and the sensitivity of the experiment. The unbiased triggers also provided data which were used to monitor the performance of specific elements of the detector (*e.g.* the Čerenkov counter efficiency).

Time and pulse height information was recorded for the trigger counters and calorimeter elements. Times for the Čerenkov counter were determined with a resolution of 680 psec with respect to the event time, as defined by the F and E scintillation counters. Times for lead glass pulses crossing an effective threshold of 300 MeV were determined with a resolution of 1.5 nsec. The timing resolution of the detector was essential for the rejection of backgrounds arising from the overlap of two K decays — for example $K_L \rightarrow \pi e\nu$, with the pion misidentified as an electron, in coincidence with two gamma rays from another K decay such as $K_L \rightarrow 3\pi^0$. The pulse heights of the F and $\check{C}$ counters were recorded to provide additional rejection if necessary of background to $K_L \rightarrow \pi^0 e^+ e^-$ from $K_L \rightarrow 2\pi^0$ and $3\pi^0$ decays with two π^0s suffering highly asymmetric Dalitz decays. In these events, even if only two complete tracks through the analysis magnet were found, the the Dalitz pairs might be revealed by pulse heights observed in the upstream detectors.

The data was collected with a FastBus data acquisition system that rejected uninteresting events in several levels of on-line processing. About 500 triggers per pulse were recorded and about 80 events were written to tape per pulse after passing on-line event selection criteria. The data acquisition live time was typically 90%. In all about 100 million events were collected during the AGS running period in spring of 1989. In what follows we discuss analyses and results from the final 80% of the data taken.

3 Analyses

The off-line event analysis required that two oppositely charged tracks meet, with a distance of closest approach <1.25 cm. Each electron or positron track had to be uniquely associated with a Čerenkov counter pulse occurring within ±2.5 nsec of the event time. The electron momentum (p)

508

was required to be less than the Čerenkov counter threshold for charged pions (8 GeV/c). The energy (E) of an electron as measured in the lead glass had to be consistent with its momentum: $0.75 < E/p < 1.25$. Backgrounds resulting from K_{e3} decays with the pion misidentified were suppressed by requiring the that momentum asymmetry of the charged tracks, $A = | (p_+ - p_-)/(p_+ + p_-) |$, be less than 0.8.

Clusters of energy in the lead glass array not associated with charged particles were identified as gamma rays. Gamma rays with energies above 500 MeV occurring within ± 5 nsec of the event time were included in the analysis. In order to suppress backgrounds from neutral beam interactions, the minimum energy was 700 MeV for γ rays striking within 6 cm of the neutral beam hole in the lead glass array. Each cluster was also required to have a narrow transverse shower profile. The reconstruction of a π^0 from two γ clusters assumed the decay vertex established by the charged particle trajectories.

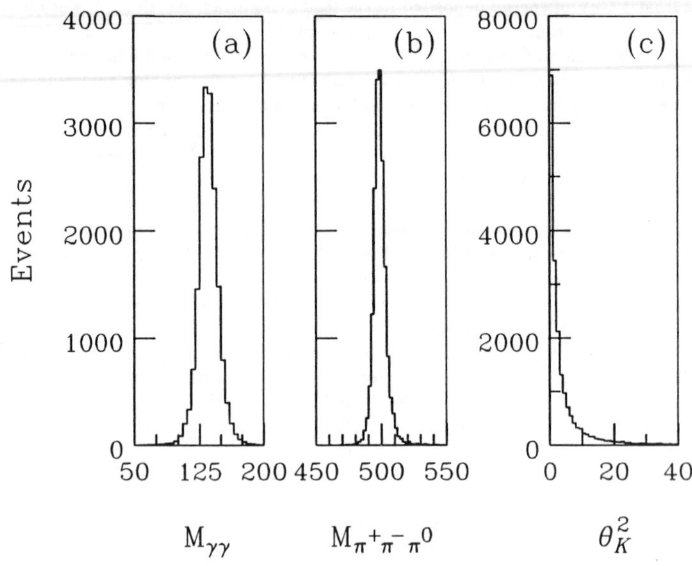

Figure 2: For a sample of $K_L \rightarrow \pi^+\pi^-\pi^0$ decays the distributions of (a) the $\gamma\gamma$ effective mass (MeV/c^2), (b) the $\pi^+\pi^-\pi^0$ effective mass (MeV/c^2), and (c) θ_K^2, the square of the target reconstruction angle (mrad^2).

Additional constraints were imposed in order to reject events arising from coincident K decays. The analysis required that the reconstructed kaon momentum be less than 20 GeV/c. Events were rejected if they exhibited track segments in the A and B chambers originating at the vertex, or if charged particle tracks were found passing through the neutral beam hole in the lead glass array. Finally events were required to have no additional clusters in the lead glass with $E > 500$ MeV and within ± 5 nsec of the event time.

For the two decay modes of interest, specific selection criteria were imposed. The electron-positron pair mass was restricted to be greater than π^0 mass for the $K_L \rightarrow \pi^0 e^+ e^-$ search in order to suppress backgrounds arising from π^0 Dalitz decays. Candidates for $K_L \rightarrow e^+ e^- \gamma$ were accepted only if the γ ray struck the lead glass array more than 6 cm from the beam hole, thereby suppressing backgrounds due to accidentals and neutron interactions.

4   Results

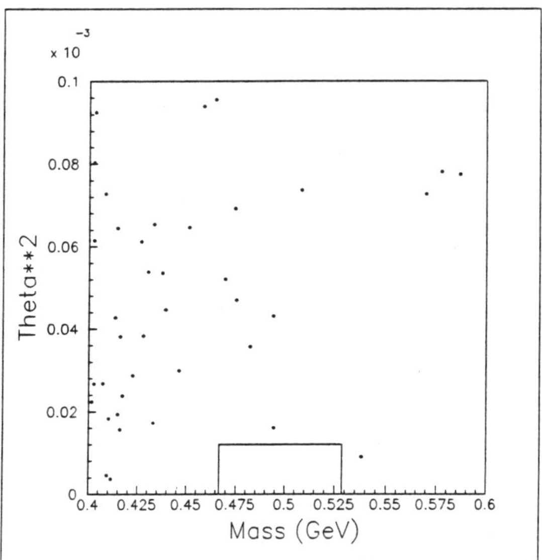

Figure 3: Event scatter plot of the square of the target reconstruction angle vs. the $\pi^0 e^+ e^-$ effective mass.

The resolution of the detector was determined from the study of $K_L \rightarrow \pi^+ \pi^- \pi^0$ decays. Figure 2 shows the relevant kinematic distributions: (a) the $\gamma\gamma$ effective mass, (b) the $\pi^+ \pi^- \pi^0$ effective mass, and (c) θ_K^2, the square of the angle between the line from the target to the decay vertex and the reconstructed kaon momentum vector. Events were required to have a $\gamma\gamma$ pair with an invariant mass within 34 MeV/c^2 of the π^0 mass. With the $\gamma\gamma$ mass constrained to the π^0 mass, events were selected if the $\pi^+ \pi^- \pi^0$ effective mass was within 13 MeV/c^2 of the K^0 mass, and if $\theta_K^2 < 12$ mrad^2. The kinematic constraints constitute cuts for values three standard deviations away from the mean.

Using the $\pi^+ \pi^- \pi^0$ mass resolution of 4.3 MeV/c^2 as a normalization, the uncertainty in mass for the $\pi^0 e^+ e^-$ final state was calculated by Monte Carlo methods to be 10.7 MeV/c^2. The results of the analysis for the $\pi^0 e^+ e^-$ final state are presented as a scatter plot in Figure 3 of the $\pi^0 e^+ e^-$ effective mass and θ_K^2 for events passing the criteria described above. Events from $K_L \rightarrow \pi^0 e^+ e^-$ decays are expected to be concentrated in the (3σ) bounded region shown, corresponding to a $\pi^0 e^+ e^-$ effective mass within 32 MeV/c^2 of the K^0 mass and $\theta_K^2 < 12$ mrad^2. There were no event candidates consistent with the decay $K_L \rightarrow \pi^0 e^+ e^-$.

The limit on the branching ratio is obtained from the following expression:

$$B(K_L \to \pi^0 e^+ e^-) = B(K_L \to \pi\pi\pi)\frac{N(\pi ee)}{N(\pi\pi\pi)}\frac{A(\pi\pi\pi)}{A(\pi ee)},$$

The number of $K_L \to \pi^+\pi^-\pi^0$ events observed including the prescale factor was $N(\pi\pi\pi) = 14997 \times 10,000$. The branching ratio for $K_L \to \pi^+\pi^-\pi^0$ with $\pi^0 \to \gamma\gamma$ is $B(K_L \to \pi\pi\pi) = 0.122$. The (efficiency corrected) acceptances, $A(\pi ee) = 0.41\%$ and $A(\pi\pi\pi) = 1.2\%$, were determined by Monte Carlo methods for kaon momenta greater than 4 GeV/c. A uniform Dalitz plot population is assumed for $K_L \to \pi^0 e^+ e^-$. The single event sensitivity for the $K_L \to \pi^0 e^+ e^-$ branching ratio is then 2.4×10^{-9}.

We have also determined the sensitivity by using $K_L \to 3\pi^0$ decays with a single Dalitz decay $(\pi^0 \to \gamma\gamma)$, which has an effective branching ratio of $B(K_L \to \pi\pi\pi) = 7.6 \times 10^{-3}$. We have observed $N(\pi\pi\pi) = 3904$ of these decays with an acceptance $A(\pi\pi\pi) = 5.7 \times 10^{-6}$, thus obtaining a single event sensitivity of 2.7×10^{-9}.

Based on these two relatively independent measures of the sensitivity and the absence of any candidates, we claim a preliminary upper limit of $B(K_L \to \pi^0 e^+ e^-) < 6 \times 10^{-9}$ (90% C.L.). This represents an increase in sensitivity by 6.5 over previous searches.

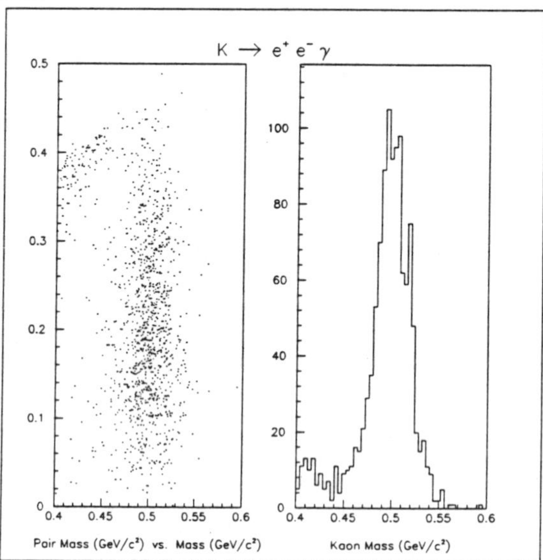

Figure 4: On the left, a scatter plot of the $K_L \to e^+ e^- \gamma$ data as a function of the invariant $e^+ e^-$ pair mass and the $e^+ e^- \gamma$ invariant mass. On the right a projection which emphasizes the signal at the kaon mass.

In addition to the $K_L \to \pi^0 e^+ e^-$ search we have also collected more than one thousand examples of the Dalitz decay $K_L \to e^+ e^- \gamma$. In Figures 4 & 5 we show the invariant mass distributions for events with $\theta_K^2 < 16$ mrad^2. The background events at high pair mass, seen in the upper left hand corner of Figure 4, are understood to arise from K_{e3} decays. Monte Carlo calculations

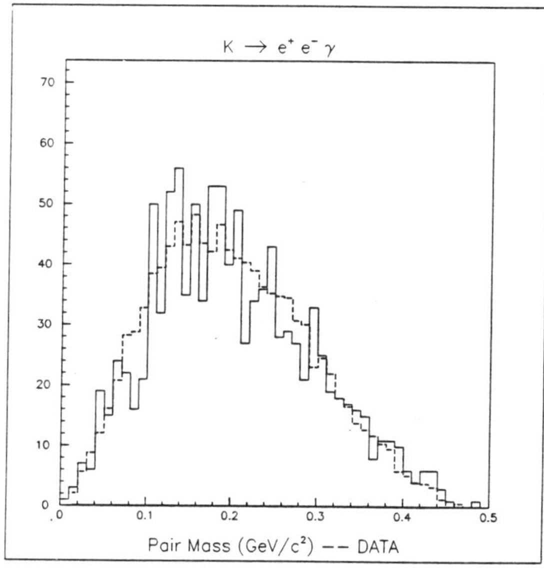

Figure 5: The pair mass distribution (solid histogram) for the $K_L \rightarrow e^+e^-\gamma$ events, after the selection on kaon mass described in the text. The dashed histogram is the Monte Carlo expectation with the form factor parameter $\alpha = 0$. Note that radiative corrections have not been included.

suggest a cut on the $e^+e^-\gamma$ invariant mass depend on the e^+e^- pair mass, that is we require $\mid m_{ee\gamma} - m_K \mid < (860 - m_{ee})/13.7$ (m in MeV). The remaining background is small, about 1%.

In Figure 5 we show the e^+e^- pair mass distribution. The distribution extends well into the high pair masses where the effects of the decay form factor are most important. We have superimposed the Monte Carlo expectation with the form factor dominated by the pseudoscalar-pseudoscalar transition, that is $\alpha = 0$ in the parametrization of Bergström et al.[6] While we can claim that the data disfavor large contributions from vector-vector transitions (corresponding to $\mid \alpha \mid \sim 1$), the determination of the precise value for α requires an accounting of QED radiative corrections in the context of the experiment. These calculations will be finished in the near future.

Acknowledgements

We thank the management and staff at Brookhaven for the strong support that allowed for the expeditious changeover from E780 to E845. We also acknowledge the important contributions to E845 by M. Mannelli and S.F. Schaffner resulting from their efforts on E780. This research is supported by the U.S. Department of Energy under contracts No. DE-AC02-76ER03075 and DE-AC02-76CH00016. One of us (M.P.S.) received additional support from the Alfred P. Sloan Foundation.

References

a. Present address: Dipartimento di Fisica dell'Università, INFN Sezione di Pisa, e Scuola Normale Superiore, 56010 Pisa, Italy.

b. Present address: CEBAF, Newport News, VA 23606.

1. J.F. Donoghue, R. Holstein, and G. Valencia, Phys. Rev. D **35**, 2769 (1987); G. Ecker, A. Pich, and E. deRafael, Nucl. Phys. **B291**, 692 (1987), and **B303**, 665 (1988); C.O. Dib, I. Dunietz, and F.J. Gilman, Phys. Rev. D **39**, 2639 (1989); J. Flynn and L.J. Randall, Nucl. Phys. **B326**, 31 (1989).

2. L.M. Seghal, Phys. Rev. D **38**, 808 (1988); J. Flynn and L.J. Randall, Phys. Letts. **216B**, 221 (1989); T. Morozumi and H. Iwasaki, KEK preprint TH-206 (1988, unpublished).

3. G.D. Barr *et al.*, Phys. Letts. **214B**, 303 (1988); L.K. Gibbons *et al.*, Phys. Rev. Letts. **61**, 2661 (1988).

4. L.J. Hall and L.J. Randall, Nucl. Phys. **B274**, 157 (1986); X.-G. He, B.H.J. McKellar, and N.E. Tupper, University of Melbourne preprint UM-P-88/40 (1988, to be published); J. Flynn and L.J. Randall, Nucl. Phys. **B326**, 31 (1989).

5. N.M. Kroll and W. Wada, Phys. Rev. **98**, 1355 (1955).

6. L. Bergström, E. Massó and P. Singer, Phys. Lett. B **131**, 229 (1983), L. Bergström, E. Massó, P. Singer, and D. Wyler, Phys. Lett. B **134**, 373 (1984), and references therein.

7. H.B. Greenlee *et al.*, Phys. Rev. Lett. **60**, 893 (1988); E. Jastrzembski *et al.*, Phys. Rev. Lett. **61**, 2300 (1988); S.F. Schaffner *et al.*, Phys. Rev. D **39**, 990 (1989).